TECHNIQUES *in* GLYCOBIOLOGY

TECHNIQUES *in* GLYCOBIOLOGY

edited by

R. REID TOWNSEND

University of California—San Francisco
San Francisco, California

ARLAND T. HOTCHKISS, JR.

Agricultural Research Service
U.S. Department of Agriculture
Wyndmoor, Pennsylvania

MARCEL DEKKER, INC.   NEW YORK · BASEL · HONG KONG

Library of Congress Cataloging-in-Publication Data

Techniques in glycobiology / edited by R. Reid Townsend, Arland T.
  Hotchkiss, Jr.
    p. cm.
  Includes index.
  ISBN 0-8247-9822-8 (alk. paper)
    1. Glycoconjugates—Analysis—Laboratory manuals. I. Townsend,
  R. Reid. II. Hotchkiss, Arland T.
  QP702.G577T43 1997
  572'.567—dc21

                                                    97-1913
                                                       CIP

The publisher offers discounts on this book when ordered in bulk quantities.
For more information, write to Special Sales/Professional Marketing at the
address below.

This book is printed on acid-free paper.

MARCEL DEKKER, INC.
270 Madison Avenue, New York, New York 10016
http://www.dekker.com

Current printing (last digit):
10 9 8 7 6 5 4 3 2 1

PRINTED IN THE UNITED STATES OF AMERICA

Preface

Complex carbohydrates are ubiquitous throughout the plant and animal kingdoms. Some of their recognized roles include conferring unique properties on proteins and lipids, functioning as biological semaphores, and, in plants, serving as a nutritional energy reserve and as a means of skeletal support. The structural elucidation of complex carbohydrates remains a challenging analytical problem. Unlike linear biopolymers (e.g., proteins, nucleic acids, cellulose, and chitin), oligosaccharides derived from glycoconjugates and complex carbohydrates are branched structures with different types of covalent linkages between residues. Like most biologically active compounds, many complex carbohydrates are present in trace quantities. However, there is no way to increase the quantity using methods analogous to cloning, or to the polymerase chain reaction. The large number of possible structures requires sensitive methods which can determine monosaccharide composition (including the distinction between D and L sugars), sequence, anomericity, linkage positions, and branching.

Although many structural features can be determined using high-resolution proton nuclear magnetic resonance (^{1}H NMR), in many cases insufficient amounts of a sample cannot be obtained. Chapter 1 describes a new method for analyzing smaller quantities of oligosaccharides using nanoprobe NMR. The greater sensitivity of mass spectrometric methods remains an impetus to develop strategies described in Chapters 2 to 5. The utility of mass spectrometry (MS) for analyzing carbohydrate–protein interactions is detailed in Chapter 6. Chapter 7 incorporates both NMR and MS methods for the structural elucidation of bacterial oligosaccharides.

Microscale analysis and quantification for glycolipids using novel blotting methods, high-performance liquid chromatography (HPLC), and enzyme-linked immunoassays are detailed in Chapters 8 through 10, respectively. The modification of lipids with oligosaccharide chains gives unique characteristics to plasma membranes as inferred from studies described in Chapter 11. The physical properties of large carbohydrate polymers using noncontact atomic force microscopy are described in Chapter 12. Two novel biosensor methods for measuring carbohydrate–protein interactions are described in Chapters 13 and 14.

The glycosylation of proteins in specific Ser or Thr residues significantly alters the conformation of the polypeptide backbones as reviewed in Chapter 15. Computer algorithms that predict the sites of O-glycosylation are presented in Chapter 16 and a novel in vivo method for confirming these predictions is described in Chapter 17. Chapters 18 to 20 review two sensitive approaches, MS and modified Edman sequencing, to determine the sites of O-glycosylation and for characterizing the attached oligosaccharide chains.

Complex carbohydrates invariably occur as an array of different structures, even at a single glycosylation site on a protein. Methods to separate these naturally derived mixtures have evolved considerably in recent years. Complex carbohydrates neither are fluorescent nor absorb significant amounts of UV light for detection during chromatography. Novel labeling approaches are presented in Chapters 21 to 23. New exoglycosidases for use with high-resolution separation techniques are described in Chapter 25. Modern high-performance separations have developed such that a single peak will likely contain one oligosaccharide structure, which is the basis of oligosaccharide mapping. Chapters 26 and 27 outline mapping methods using gel electrophoresis, HPLC, and capillary electrophoresis. New sample preparation protocols for HPLC mapping are detailed in Chapter 28. The use of high-resolution separations for assessing the fidelity of glycosylation of recombinant glycoprotein therapeutics is addressed in Chapters 29 and 30. Chapters 31 and 32 discuss strategies incorporating HPLC, reagent array exoglycosidase analysis, NMR, and MS.

New methods are described that utilize a DNA sequencer (Chapter 33) or a post-column enzyme reactor for high-pH anion exchange chromatography with pulsed amperometric detection (Chapter 34) for chain length analysis of depolymerized amylopectin or other plant polysaccharides. Acidic matrix polysaccharides such as pectin and galactan contribute structural support to plant cell walls. In Chapter 35, a new derivatization method is described for the preparation of 1-amino-1-deoxyalditol acetates, which are used for GC-MS analysis of enantiomeric sugars in the red algal galactan, corallinan. A novel class of pectin esters that may be involved in crosslinking the pectin matrix to other cell wall polysaccharides is characterized in Chapter 36 by using gas chromatography-MS (GC-MS), FTIR microspectroscopy, and immunocytochemical electron microscopy. In Chapter 37, the cellulose and carrageenan structure of red algal cell walls is characterized by using various methods including light and electron microscopy (enzyme-gold and antibody-gold labeling), GC-MS, NMR, x-ray diffraction and a novel reductive hydrolysis procedure that protects acid-labile 3,6-anhydrogalactose residues in carrageenan during composition analysis.

The chapters were selected from presentations at the Third International Glycobiology Symposium on Current Methods, held in San Diego, California. We thank the California Separation Science Society for their support of this meeting series. A special thanks to Dr. Bill Hancock for his enthusiastic support of methods in glycobiology.

R. Reid Townsend
Arland T. Hotchkiss, Jr.

Contents

Contributors

Renaud Albigot Institute of Pharmacology and Structural Biology, National Center for Scientific Research, Toulouse, France

Amir A. Amiri Department of Glycolipid Immunotherapy, John Wayne Cancer Institute at Saint John's Hospital and Health Center, Santa Monica, California

Kalyan Rao Anumula Department of Analytical Sciences, SmithKline Beecham Pharmaceuticals, King of Prussia, Pennsylvania

Louisette J. Basa Department of Pharmacokinetics and Metabolism, Genentech, Inc., South San Francisco, California

Michael Batley School of Chemistry, Macquarie University, Sydney, New South Wales, Australia

Philip M. Bauer Department of Glycolipid Immunotherapy, John Wayne Cancer Institute at Saint John's Hospital and Health Center, Santa Monica, California

Richard Bernasconi Department of Structural Protein Chemistry, Genzyme Corporation, Framingham, Massachusetts

Collette Bigge Oxford GlycoSystems Limited, Abingdon, Oxon, England

Therese Brando Institute of Pharmacology and Structural Biology, National Center for Scientific Research, Toulouse, France

David A. Brant Department of Chemistry, University of California–Irvine, Irvine, California

James Bruce Oxford GlycoSystems Limited, Abingdon, Oxon, England

Laura Cantù Department of Medical Chemistry and Biochemistry, University of Milan, Milan, Italy

Nicholas C. Carpita Department of Botany and Plant Pathology, Purdue University, West Lafayette, Indiana

Wulf Carson Beckman Instruments, Inc., Fullerton, California

Paul K. Cartier III Biotechnology Development Center, Beckman Instruments, Inc., Fullerton, California

Marcelo R. Cases Department of Organic Chemistry, Faculty of Exact and Natural Sciences, University of Buenos Aires, Buenos Aires, Argentina

Alberto Cerezo Department of Organic Chemistry, Faculty of Exact and Natural Sciences, University of Buenos Aires, Buenos Aires, Argentina

Vanna Chigorno Department of Medical Chemistry and Biochemistry, University of Milan, Milan, Italy

Yong K. Cho Department of Quality Control, Test Technology, Baxter Biotech/Hyland Division, Duarte, California

Mario Corti Department of Medical Chemistry and Biochemistry, University of Milan, Milan, Italy

George A. M. Cross Laboratory of Molecular Parasitology, The Rockefeller University, New York, New York

Elena Del Favero Department of Medical Chemistry and Biochemistry, University of Milan, Milan, Italy

Carole Delmas Institute of Pharmacology and Structural Biology, National Center for Scientific Research, Toulouse, France

Kilian Dill Department of Research and Development, Molecular Devices Corporation, Sunnyvale, California

Åke P. Elhammer Department of Biochemistry, Pharmacia & Upjohn Inc., Kalamazoo, Michigan

Jeffrey J. Engstrom Department of Pharmaceutical Chemistry, University of California–San Francisco, San Francisco, California

Arnold M. Falick PerSeptive Biosystems, Inc., South San Francisco, California

Ruth Falshaw Department of Polysaccharide Chemistry, Industrial Research Ltd., Lower Hutt, New Zealand

Vince Farnsworth Advanced Technology Center, Beckman Instruments, Inc., Fullerton, California

Konrad Feichtinger Max-Planck-Institut für Biochemie, Martinsried, Germany

Michael A. J. Ferguson Department of Biochemistry, University of Dundee, Dundee, Scotland

Thomas A. Gerken Department of Pediatrics and Biochemistry, Case Western Reserve University, Cleveland, Ohio

Bradford W. Gibson Department of Pharmaceutical Chemistry, University of California–San Francisco, San Francisco, California

Martine Gilleron Institute of Pharmacology and Structural Biology, National Center for Scientific Research, Toulouse, France

Martin Gohlke Department of Molecular Biology and Biochemistry, Free University of Berlin, Berlin-Dahlem, Germany

Andrew A. Gooley Macquarie University Centre for Analytical Biotechnology, Macquarie University, Sydney, New South Wales, Australia

Michael R. Gretz Department of Biological Sciences, Michigan Technological University, Houghton, Michigan

Ellen P. Guthrie Department of Research, New England Biolabs, Beverly, Massachusetts

András Guttman Beckman Instruments, Inc., Fullerton, California

Fred K. Hagen Department of Dental Research, School of Medicine and Dentistry, University of Rochester, Rochester, New York

Gunnar C. Hansson Department of Medical Biochemistry, Göteborg University, Gothenburg, Sweden

Mark R. Hardy* Oxford GlycoSystems, Inc., Bedford, Massachusetts

Yukio Hasegawa Department of Research and Development, Pharmacia Bioteck K.K., Tokyo, Japan

Paul A. Haynes Laboratory of Molecular Parasitology, The Rockefeller University, New York, New York

Elizabeth Higgins Department of Carbohydrate Chemistry, Genzyme Corporation, Framingham, Massachusetts

*Current affiliation: Genetics Institute, Andover, Massachusetts

Wade M. Hines   Department of Product Research and Development, Per-Septive Biosystems, Inc., Framingham, Massachusetts

Arland T. Hotchkiss, Jr.   U.S. Department of Agriculture, Agricultural Research Service, Eastern Regional Research Center, Wyndmoor, Pennsylvania

Jay-lin Jane   Department of Food Science and Human Nutrition, Iowa State University, Ames, Iowa

Daniel R. Jardine   Macquarie University Centre for Analytical Biotechnology, Macquarie University, Sydney, New South Wales, Australia

Constance M. John   Department of Pharmaceutical Chemistry, University of California–San Francisco, San Francisco, California

Andrew J. S. Jones   Department of Analytical Chemistry, Genentech, Inc., South San Francisco, California

Hasse Karlsson   Department of Medical Biochemistry, Göteborg University, Gothenburg, Sweden

Niclas G. Karlsson   Department of Medical Biochemistry, Göteborg University, Gothenburg, Sweden

Paul A. Keifer   Varian NMR Instruments, Palo Alto, California

Warren C. Kett   School of Chemistry, Macquarie University, Sydney, New South Wales, Australia

Ferenc J. Kézdy   Department of Biochemistry, Pharmacia & Upjohn Inc., Kalamazoo, Michigan

Kristina Kopp   Department of Biotechnical Production, Dr. Karl Thomae GmbH, a Company of Boehringer Ingelheim Pharma Germany, Biberach, Germany

Sadamu Kurono   Glycobiology Research Group, Frontier Research Program, The Institute of Physical and Chemical Research (RIKEN), Wako-shi, Saitama, and Mitsubishi Kasei Institute of Life Sciences, Machida, Tokyo, Japan

Roderic P. Kwok   Department of Quality Control, Test Technology, Baxter Biotech/Hyland Division, Duarte, California

David Landry   Department of Research, New England Biolabs, Beverly, Massachusetts

Peter H. Lipniunas* Department of Pharmaceutical Chemistry, University of California–San Francisco, San Francisco, California

Adriana E. Manzi Cancer Center, School of Medicine, University of California–San Diego, La Jolla, California

Stephen A. Martin Department of Product Development, PerSeptive Biosystems, Inc., Framingham, Massachusetts

Franz J. Mayer-Posner Bruker-Franzen Analytik GmbH, Bremen, Germany

Maureen C. McCann Department of Cell Biology, John Innes Centre, Norwich, England

Theresa M. McIntire Department of Chemistry, University of California–Irvine, Irvine, California

Katalin F. Medzihradszky Department of Pharmaceutical Chemistry, University of California–San Francisco, San Francisco, California

Jean-Claude Mollet Department of Biological Sciences, Michigan Technological University, Houghton, Michigan

Matthew K. Morrell Co-operative Research Centre for Plant Science, Canberra, A.C.T., Australia

Yoshitaka Nagai Glycobiology Research Group, Frontier Research Program, The Institute of Physical and Chemical Research (RIKEN), Wako-shi, Saitama, and Mitsubishi Kasei Institute of Life Sciences, Machida, Tokyo, Japan

Keith Nehrke Department of Dental Research, School of Medicine and Dentistry, University of Rochester, Rochester, New York

Marco Nicolini Department of Medical Chemistry and Biochemistry, University of Milan, Milan, Italy

Milos Novotny Department of Chemistry, Indiana University, Bloomington, Indiana

Rolf Nuck Department of Molecular Biology and Biochemistry, Free University of Berlin, Berlin-Dahlem, Germany

Yoko Ohashi Glycobiology Research Group, Frontier Research Program, The Institute of Physical and Chemical Research (RIKEN), Wako-Shi, Saitama, and Mitsubishi Kasei Institute of Life Sciences, Machida, Tokyo, Japan

---

*Current affiliation: Astra Draco AB, Lund, Sweden

Michael G. O'Shea   Co-operative Research Centre for Plant Science, Canberra, A.C.T., Australia

Cheryl L. Owens   Department of Biochemistry and Molecular and Microbiology, Case Western Reserve University, Cleveland, Ohio

Nicolle H. Packer   Macquarie University Centre for Analytical Biotechnology, Macquarie University, Sydney, New South Wales, Australia

Paola Palestini   Department of Medical Chemistry and Biochemistry, University of Milan, Milan, Italy

Damon I. Papac   Department of Analytical Chemistry, Genentech, Inc., South San Francisco, California

Raj Parekh   Oxford GlycoSystems Limited, Abingdon, Oxon, England

Murali Pasumarthy   Department of Pediatrics, Case Western Reserve University, Cleveland, Ohio

Anthony Pisano   Macquarie University Centre for Analytical Biotechnology, Macquarie University, Sydney, New South Wales, Australia

Roger A. Poorman   Department of Biochemistry, Pharmacia & Upjohn Inc., Kalamazoo, Michigan

Germain Puzo   Institute of Pharmacology and Structural Biology, National Center for Scientific Research, Toulouse, France

M. Janardhan Rao   Department of Test Technology, Baxter Biotech/Hyland Division, Duarte, California

Uwe Rapp   Bruker-Franzen Analytik GmbH, Bremen, Germany

Mepur H. Ravindranath   Laboratory of Glycolipid Immunotherapy, John Wayne Cancer Institute at Saint John's Hospital and Health Center, Santa Monica, California

John W. Redmond*   Macquarie University Centre for Analytical Biotechnology, Macquarie University, Sydney, New South Wales, Australia

Lorri Reinders   Department of Pharmaceutical Chemistry, University of California–San Francisco, San Francisco, California

Anja Resemann   Bruker-Franzen Analytik GmbH, Bremen, Germany

Michel Rivière   Institute of Pharmacology and Structural Biology, National Centre for Scientific Research, Toulouse, France

---

*Current affiliation: Research School of Biological Sciences, Australian National University, Canberra, A.C.T., Australia

Wolfram Schäfer   Max-Planck-Institut für Biochemie, Martinsried, Germany

Michael Schlüter   Department of Biotechnical Production, Dr. Karl Thomae GmbH, a Company of Boehringer Ingelheim Pharma Germany, Biberach, Germany

Yasuro Shinohara   Department of Research and Development, Pharmacia Bioteck K.K., Tokyo, Japan

Sandro Sonnino   Department of Medical Chemistry and Biochemistry, University of Milan, Milan, Italy

Hiroyuki Sota   Department of Research and Development, Pharmacia Bioteck K.K., Tokyo, Japan

Morgan Stefansson   Department of Analytical Chemistry, Uppsala University, Uppsala, Sweden

Carlos A. Stortz   Department of Organic Chemistry, Faculty of Exact Sciences, University of Buenos Aires, Buenos Aires, Argentina

Lawrence A. Tabak   Department of Dental Research and Biochemistry, School of Medicine and Dentistry, University of Rochester, Rochester, New York

Takao Taki   Department of Biochemistry, School of Medicine, Tokyo Medical and Dental University, Tokyo, Japan

Kristina A. Thomsson   Department of Medical Biochemistry, Göteborg University, Gothenburg, Sweden

R. Reid Townsend   Department of Pharmaceutical Chemistry, University of California–San Francisco, San Francisco, California

Robert B. Trimble   Wadsworth Center, New York State Department of Health, and State University of New York at Albany School of Public Health, Albany, New York

Manuela Valsecchi   Department of Medical Chemistry and Biochemistry, University of Milan, Milan, Italy

Anne Venisse   Institute of Pharmacology and Structural Biology, National Centre for Scientific Research, Toulouse, France

Alain Vercellone   Institute of Pharmacology and Structural Biology, National Centre for Scientific Research, Toulouse, France

Rolf G. Werner   Department of Biotechnical Production, Dr. Karl Thomae GmbH, a Company of Boehringer Ingelheim Pharma Germany, Biberach, Germany

Keith L. Williams   Macquarie University Centre for Analytical Science, Macquarie University, Sydney, New South Wales, Australia

Kit-Sum Wong   Department of Food Science and Human Nutrition, Iowa State University, Ames, Iowa

Sharon T. Wong-Madden   Department of Research, New England Biolabs, Beverly, Massachusetts

TECHNIQUES *in* GLYCOBIOLOGY

1

New Frontiers in Nuclear Magnetic Resonance Spectroscopy
Use of a Nano•NMR Probe for the Analysis of Microgram Quantities of Complex Carbohydrates

Adriana E. Manzi
University of California—San Diego, La Jolla, California

Paul A. Keifer
Varian NMR Instruments, Palo Alto, California

I. INTRODUCTION

We describe here the use of a Nano•NMR probe for obtaining one- and two-dimensional ^{1}H nuclear magnetic resonance (NMR) spectra of complex oligosaccharides available in only very small quantities. These results were obtained on glycosaminoglycan (GAG)-type chains isolated from human melanoma cells in culture; however, because there are no intrinsic limitations to the approach, it is applicable to any situation where an NMR analysis of a limited amount of a carbohydrate is required. The GAG-type oligosaccharides used in this study were obtained by adding β-xylosides to the cell culture to prime GAG biosynthesis [1,2]. These acceptors diffuse into the Golgi apparatus, where they compete with endogenous core proteins to make protein-free GAG chains. Freeze and coworkers [3] had characterized several of the GAG chains produced in this manner by human melanoma cells; however, about 10–15% of the anionic oligosaccharides proved to be resistant to the enzymatic digestions used to successfully analyze the majority of oligosaccharides. The use of NMR spectroscopy allowed us to explain the resistance of the minor oligosaccharides to enzymatic digestions and demonstrated the existence of a previously unidentified core-related structure.

A.   Options for Analyzing Small Sample Quantities

Nuclear magnetic resonance is recognized as a powerful analytical tool for determining molecular structures. In many studies of samples of biological interest, however, it is difficult (if not impossible) to obtain sufficient amounts of sample to allow even simple one-dimensional (1D) ^{1}H NMR data to be obtained using conventional NMR hardware. For complex carbohydrates of biological interest, NMR analysis was previously limited to those cases where sufficient material could be isolated from large numbers of cells in culture or many grams of tissues, followed by tedious large-scale purification steps. Even in such cases, the limited quantities tempted the operator to minimize sample dilution by using the minimum volume of solvent to obtain the highest NMR sensitivity. However, when this standard 5-mm NMR tube was inserted into a magnet, the smaller volume of the sample required the tube to be positioned higher to maintain the center of the sample within the primary detection region (the coil) of the probe. Because magnetic field homogeneity is required for high quality NMR data, this setup created the need for "shimming" of the B_0 field (accomplished by adjusting the electric current running through various room temperature shim coils located outside the probe) to minimize the resulting minor field inhomogeneities.

Every material on earth can be characterized by magnetic susceptibility, a measurable physical property that measures how much the material distorts a magnetic field. If a homogeneous sample (e.g., a solution) is placed in a large, mostly homogeneous magnetic field (e.g., an NMR magnet), the resulting field within the sample is still mostly homogeneous and can be perfected easily by routine shimming. This is true for any perfect sphere or infinitely long cylinder; however, as perfect spheres are hard to manufacture, conventional NMR tubes are instead designed to approximate an infinite cylinder. For example, if the receiver coil in the probe is 15 mm long, the sample should be approximately 3 × 15 mm long (45 mm; about 600 μL in a 5-mm tube) to function like an infinite cylinder. Unfortunately, while the sample solution has a certain magnetic susceptibility value, the air and glass around the sample have different ones; at the interfaces of these differing materials, the homogeneous magnetic field is distorted in a complex way, and the distortions are not easily corrected using available shim corrections. While most of the glass sample tube approximates an infinite cylinder, if the bottom of the tube (having a liquid-to-glass-to-air interface) or the meniscus at the top of the sample (having a liquid-to-air interface) gets too close to the receiver coil, the resulting field distortions are not easily correctable and can cause broadened NMR resonances.

When faced with trying to obtain NMR data on a very small quantity of sample, one must maximize sensitivity while maintaining good NMR lineshapes (since broader linewidths themselves will dramatically lessen sensitivity). The following six options can be used to accomplish this goal. (1) The standard volume of solution can be used (600 μL in a 5-mm probe), which will provide the narrowest lineshapes, but if this causes a three-fold dilution of the sample, a nine-fold increase in the total experiment time will result. (2) The solution can be concentrated (to 400 μL) and a shorter column of liquid used; the sample must be centered on the receiver coil, but this increases shimming time and may even make it impossible to shim correctly (resulting in inferior lineshapes, which then lower sensitivity). (3) A shorter column of concentrated solution can be placed between two inert plugs or plungers that have been designed to be susceptibility matched to the solution (this can allow better lineshapes to be obtained, but the susceptibility matching is never perfect for all samples, and any residual trapped air bubbles, no matter how small, will cause their own extensive line broadenings. (4) The smaller, more concentrated sample can be placed in a spherical microcell, although, despite decades of effort, sufficiently perfect spherical microcells are difficult (if not impossible) to make and use well (their imperfections, though small, still make them difficult to shim). (5) The smaller volume of sample can be placed in a smaller-diameter tube to regenerate a longer column of liquid. This typically creates unfavorable filling factors and poor sensitivity in a 5-mm probe, but it does allow very narrow lineshapes to be obtained. This approach originated the scaled-down microprobes holding 1 to 3 mm sample tubes. (6) The sample can be spun at the magic angle (54.7°), in which case all the terms in the spin Hamiltonian, which involve magnetic susceptibility contributions to the line broadening, disappear. This means that sample geometries, menisci, air bubbles, and sample inhomogeneities no longer influence lineshapes. This now allows the entire sample to be placed within the active region of the receiver coil, to generate the highest sensitivity possible, without inducing any undesired line broadening.

Of course, for options 2 through 5, as the sample volume becomes smaller, the difficulties encountered become larger. Changing the vertical position of the NMR tube (option 2) becomes insufficient after a critical minimum solvent volume is reached. The quality of the spectra obtained with options 2 through 4 is often very poor, with broadening near the base of the largest peaks, frequently causing difficulties in observing other small signals in close proximity to these major resonances. For options 2 through 4, any reduction of sample volume also results in an unacceptable loss of spectral resolution; this is especially true for higher-field spectrometers (e.g., 500 MHz and above). On the other hand, if the sample volume is maintained high enough to obtain good resolution, the lessened sensitivity requires many more hours of spectrometer time, and in some cases, the larger volume of solvent allows larger resonances from the solvent or absorbed impurities to complicate spectral assignments.

While the use of microprobes for NMR has been under development for more than 20 years, all commercially available microprobes still have a geometry similar to that of a standard 5-mm probe, in which the tube's axis is oriented along the Z axis of a superconducting magnet. (With modern magnet installations, the decision to spin or not to spin is slowly becoming almost inconsequential.) While microprobes can greatly extend the capabilities of a spectrometer by facilitating studies of ever-smaller sample volumes, the conventional nature of a microprobe causes it to also have a critical minimum volume below which it, too, will not provide high quality data. Typical "micro" NMR tubes (capillary tubes) have volumes in the range 100–150 μL; any attempts to use smaller-volume samples increases the nonideality of the infinite-cylinder approximation, unless a different approach is used to develop a smaller-volume NMR probe (as was outlined in option 6).

B.   Magic-Angle Spinning with Nano•NMR Probes

The use of magic-angle spinning (MAS) in high resolution NMR started around 1982 [4]. Since that early work of Garroway, James Shoolery [5] became interested in applying MAS to the problem of obtaining high resolution spectra from small-volume samples. Conventional design small-volume probes allow high filling factors and enhanced sensitivity, but, as we have outlined, for ever-smaller nonspherical samples the magnetic susceptibility perturbations of the magnetic field become uncorrectable with standard shim systems. In 1990, a small-volume probe having a completely different design was developed at Varian NMR Instruments. This probe, now called a Nano•NMR probe, was designed to constrain all of a small-volume sample to the active region of the receiver coil (to provide the highest possible sensitivity), and then use magic-angle spinning to correct for any unshimmable line broadenings (as caused by air bubbles, sediment, and sample cell imperfections) to provide the narrowest possible linewidths (Fig. 1a). Since the sample is rotated about the magic angle (54.7°), which tips the receiver coil toward the horizontal plane, a more efficient solenoid receiver coil design can also be used to maximize further sensitivity. An analysis of the demagnetization fields for cylindrical samples, as applied to a high resolution small-volume MAS probe, was done by Barbara [6] using the calculations of Mozurkewich et al. [7]. It showed how a very small sample (e.g., 40 μL), when placed in surroundings that were not susceptibility matched, would have a nonspinning linewidth of approximately 1.5

KHz. This suggested that magic-angle spinning of the sample at spin rates > 1.5 KHz should result in narrowed resonances, provided the spin rates are large enough (> 1.5 KHz) to eliminate significant spinning sidebands. It has also been shown how a Nano•NMR probe can provide high quality data using performance that is truly distinct from any other traditional liquids-based or MAS probe [8].

Today, both ^{1}H and ^{13}C Nano•NMR probes, which are most properly thought of as high resolution probes that also do MAS, are available in a variety of field strengths. The uses of these probes have expanded to include not only those small samples requiring increased sensitivity, but also to obtain high resolution NMR spectra on a multitude of heterogeneous samples; of special note are their unique abilities to obtain high resolution NMR data for samples still bound to solid-phase-synthesis resins [8–11]. In this case, the magic-angle spinning also removes those line broadening effects caused by smaller-scale variations in magnetic susceptibility, located *within* phase heterogeneous samples.

The Nano•NMR probes use easy-to-fill Nanocells (Fig. 1b), which provide a maximum sample volume of 40 μL. Any further reduction in the sample volume both reduces the size of the solvent resonances and allows higher effective dynamic range due to smaller solvent artifacts. This has no deleterious impact on sample shimming or linewidths, because the Nano•NMR probe requires minimal reshimming when smaller volumes of sample are used, while maintaining excellent resolution. (A MAS probe is shimmed using z_1, x, and y gradients rather than the more common z_1, z_2, and z_3 gradients.) High quality spectra are routinely obtained on less than 40 μL without any reduction in spectral quality; the only suggested requirement is to have enough ^{2}H in the solvent to obtain a field-frequency lock. The current geometry of the Nano•NMR probe requires the operator to manually place the sample in the probe, and reinsert the probe into the magnet each time the sample is changed. Despite this, when only very small quantities of sample are available for NMR analysis, the time saved by the increased sensitivity far outweighs any time spent manually inserting the sample. These benefits allowed us to acquire an extensive amount of ^{1}H NMR spectral information on several 15 to 25 μg samples, some of which ultimately proved to be mixtures of compounds.

II. MATERIALS AND METHODS

A. Oligosaccharide Isolation and Purification

4-Methylumbelliferyl (4MU)-xylosides were purified from human melanoma cells by Freeze and coworkers (The Burnham Institute) [12]. About 1 × 10^9 melanoma cells grown in 1 mM MU-β-xyloside (XylβMU) were

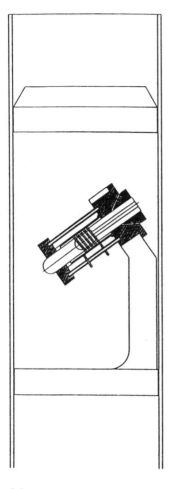

(a)

Figure 1 (a) Cross section of a ^{1}H Nano•NMR probe showing the positioning of the sample tube assembly at the magic angle. (b) Sample tube assembly.

used. The cells were removed by centrifugation, and the medium was lyophilized and then reconstituted in water. Oligosaccharides were isolated from the medium, and those resistant to *Arthrobacter ureafaciens* sialidase and bovine testicular β-glucuronidase were purified (xyloside I) [12]. Monosaccharide compositional analysis and electrospray ionization mass spectrometry were performed [12].

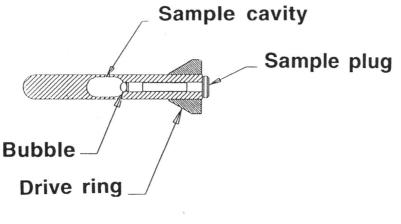

(b)

Figure 1 Continued

B. Preparation of Samples for NMR Spectroscopy

The samples (approximately 25 μg of the anionic xyloside I, or 15 μg of the material neutralized by the combined α-N-acetylgalactosaminidase and β-glucuronidase digestions, xyloside IA) were repeatedly exchanged in D_2O (99.96%, Aldrich Chemical Co., Milwaukee), with intermediate lyophilization, and ultimately dissolved in 40 μL of 99.996% D_2O. Spectra were recorded on a Varian UNITY plus 500-MHz spectrometer using a ^{1}H Nano•NMR probe. The probe temperatures were either 25 or 30°C (as indicated in the figure legends); the spin rates ranged from 1.5 to 2.5 kHz. One-dimensional ^{1}H NMR spectra were obtained in approximately 1000 scans using presaturation of the HOD peak. Two-dimensional totally (scalar) correlated spectroscopy (TOCSY) spectra [13] were obtained using a 10-kHz Malcom Levitt–17 mixing-sequence (MLEV-17) spin lock of either 50 or 100 msec duration, 1.6 sec (sample I) or 1.0 sec (sample IA) of presaturation, 88 (sample I) or 48 (sample IA) scans per t_1 data point, and 300 complex data points in t_1; total measuring time was 30 hr for sample I and 13 hr for sample IA. Two-dimensional double-quantum-filtered correlation spectroscopy (DQFCOSY) spectra [14] were obtained using 300 complex t_1 data points having 48 scans each; measurement times were 18 and 21 hours for samples I and IA, respectively. Two-dimensional rotating frame Overhauser enhancement spectroscopy (ROESY) data [15] for sample I were obtained using a 2-KHz pulsed spinlock of 200 msec duration, and 192 scans for each of the 300 complex t_1 data points (60-hr experiment). All 2D spectra were obtained in a phase-sensitive mode using hypercomplex sampling in t_1, and all included presaturation of the HOD resonance. Line broadening (0.2 Hz) was applied to 1D spectra, whereas Gaussian weightings were applied to 2D spectra. Chemical shifts for xyloside I are given relative to the internal standard DSS (4,4-dimethyl-4-silapentane-1-sulfonate) at 0 ppm. Chemical shifts for xyloside IA are given relative to the residual HOD peak at 4.80 ppm. The standard, 7-hydroxy-4-methylcoumarin β-xyloside (XylβMU), was analyzed using dimethyl sulfoxide DMSO-d_6 as the solvent; in this case, the chemical shifts were referenced to the DMSO multiplet at 2.49 ppm.

III. RESULTS AND DISCUSSION

The 1D ^{1}H NMR spectrum obtained on approximately 25 μg of the anionic xyloside (I) by using a Nano•NMR probe is shown in Fig. 2. This spectrum

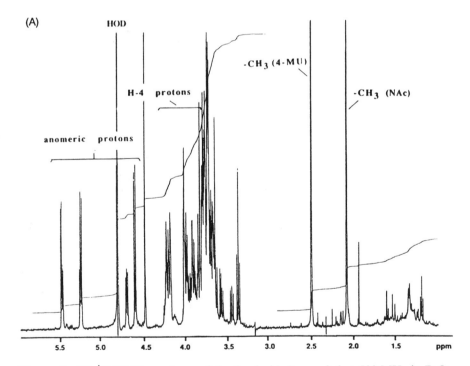

Figure 2 1D ^{1}H-NMR spectrum of the β-xyloside I recorded at 500 MHz in D$_2$O at 25°C: (A) complete spectrum; (B) expansion of the anomeric region; (C) expansion of the aromatic region; (D) expansion of the 3.30 to 4.75 ppm region. (From Ref. 12.)

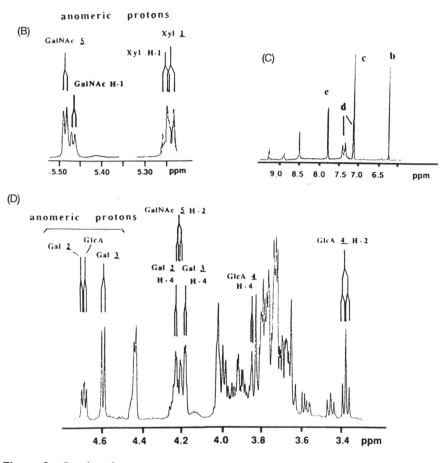

Figure 2 Continued

is consistent with the electrospray ionization mass spectrometry analysis [12], which indicated the presence of a mixture; however, a considerable amount of additional information was obtained from this spectrum. The signals expected for the MU aromatic ring [16], as well as the characteristic regions of the oligosaccharide "reporter groups" [16] are readily recognized. From the integration of the anomeric region, it was possible to conclude that at least two components (in the ratio 2 : 1) are present (Fig. 2A).

The first striking feature of this spectrum was the presence of a pair of α-anomeric resonances at δ 5.480 ($J_{1,2}$ = 3.91 Hz), and δ 5.463 ($J_{1,2}$ = 3.67 Hz) in the ratio 2 : 1 (Fig. 2B). This α-anomeric signal was unexpected because there is no precedent for such a residue in the known sequences of GAG chain cores. The same 2 : 1 ratio was observed for the two pairs of

β-anomeric signals at δ 5.243 ($J_{1,2}$ = 7.57 Hz) and δ 5.254 ($J_{1,2}$ = 7.09 Hz)
(Fig. 2B). Two types of aromatic signals were also observed (Fig. 2C;
resonances *d*). The signal at δ 4.598 was assigned to a β-galactose anomeric
proton (Fig. 2D). The remaining β anomeric signals at δ 4.701 and δ 4.690
were tentatively assigned to another β-galactose residue and a β-glucuronic
acid residue on the basis of the known monosaccharide composition and
the known structure of a GAG chain core region; however, it was not
possible to make respective assignments. The signal at δ 2.496 was assigned
to the MU methyl protons, and that at δ 2.080 to an *N*-acetyl group.

The identity of each residue in the xyloside I sample was determined
using 2D TOCSY NMR data (Fig. 3). The J-connectivity patterns were
initiated at the anomeric region [17,18]. The α-anomeric signal at δ 5.480
exhibited cross peaks with three other signals. This pattern is typical of a
galactose-type ring, where the very small coupling constant between H-4
and H-5 inhibits magnetization transfer to protons beyond H-4, reducing

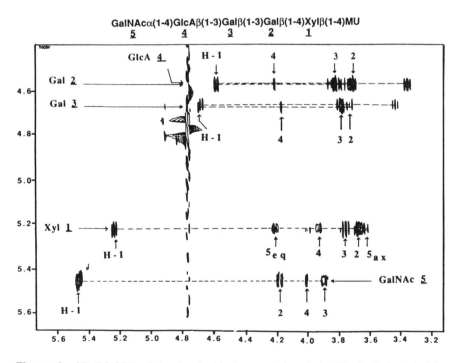

Figure 3 2D TOCSY of the β-xyloside I recorded at 500 MHz in D$_2$O at 25°C.
Expansion of the region between 3.2 and 5.6 ppm showing the connectivities in
each monosaccharide residue. The numbers and letters in the spectra refer to the
corresponding residues in the structure. (From Ref. 12.)

the likelihood of detectable cross-peak signal. Additionally, a comparison of the two TOCSY spectra acquired with 50 and 100 msec mixing times (not shown) made it possible to assign the resonance at δ 4.173 to H-2, the resonance at δ 3.897 to H-3, and the resonance at δ 4.016 to H-4. Very few examples of α-linked N-acetyl galactosamine residues have been reported [19–22], and certainly none of them in GAG chains; therefore, other experiments were carried out to confirm this assignment (to be discussed). The pattern of cross peaks for the anomeric signal at δ 5.243 was similar to that of the xylose ring on the standard MU β-xyloside (not shown); the observed downfield shifts are due to the use of different solvents in both experiments. A detailed analysis of the f_2 axis slices of the 2D TOCSY data obtained using mixing times of both 50 and 100 msec (not shown) allowed assignments to be made for the different resonances [12]. The very complex group of cross-peaks, correlated to the anomeric signals at δ 4.6–4.7, was due to the presence of a mixture of compounds. As shown in Fig. 3, there is a considerable overlap of the signals for the glucuronic acid and galactose residues. By editing the 2D TOCSY subspectra at short intervals across the f_2 axis, two different uronic acid signals could be seen, characterized by the typical H-2 triplets located between 3.3 and 3.5 ppm. These triplets were also observed in the 1D spectrum (Figs. 2A and D). An integration of this 1D spectrum indicated that the glucuronic acid residue having the higher-field H-2 resonance corresponds to the main component of this mixture. With similar editing of the f_2 axis cross sections (or slices) through the anomeric proton resonances of the 2D TOCSY data sets for 50 and 100 msec mixing times (not shown), it was possible to assign all multiplet patterns of Gal-2, Gal-3, and two GlcA residues in both the major and minor components of the mixture. The resonances observed for the two galactose residues and the glucuronic acid residue correspond well with those previously reported for serine-linked GAG-core oligosaccharides [23–26] with the expected influence of the MU aglycone.

While the 1D spectrum contained a signal located at 4.44 ppm (Fig. 2), this peak was not present in the 2D TOCSY spectrum (Fig. 3). This is because the 1D spectrum was acquired with a spin rate of 2240 Hz, which generated a spinning sideband (SSB) "peak" of the large DSS peak near zero, causing a signal 2240 Hz away at 4.44 ppm, whereas the 2D TOCSY experiment was run with a higher spin rate, which moved the SSB to a higher frequency; the typically random amplitude of this SSB during the 2D experiment caused the line of t_1 noise at 4.8 ppm.

To confirm the structure of the major component in sample I, approximately 20 μg of this material was digested with a combination of α-N-acetylgalactosaminidase and β-glucuronidase. The neutralized material was separated from the remaining anionic material on QAE-Sephadex (Pharmacia, Uppsala, Sweden) and desalted on a C-18 cartridge (Rainin, Emery-ville, CA). The neutral component recovered (IA; ~15 μg) was analyzed by 1D and 2D ^{1}H NMR.

The 1D ^{1}H NMR spectrum (not shown) of this neutral product was still a mixture, having anomeric signals for β-xylose at 5.260 and 5.290 ppm in a ratio of 2 : 1; however, the anomeric signal at 5.480 ppm disappeared, as did the methyl signal of the N-acetyl group. This confirmed the assign-ment of the 5.480 ppm resonance to the anomeric proton of an α-GalNAc residue. As expected for this neutral compound, no resonances were ob-served that could be attributed to a glucuronic acid residue. On the con-trary, a clear β-galactose anomeric region showed two resonances at δ 4.660 and 4.715 in a 1 : 1 ratio [12]. Additional resonances were assigned using 2D TOCSY and DQFCOSY data (not shown). The ^{1}H NMR results, there-fore, indicate that one of the components in this mixture has the structure

$$\text{Gal}\beta(1\text{-}3)\text{Gal}\beta(1\text{-}4)\text{Xyl}\beta(1\text{-}4)\text{MU}$$

whereas the other component resembles the starting material (MU-β-xyloside). These results are in agreement with the mass spectrometry data [12].

Therefore, xyloside I contained this MU-linked trisaccharide followed by glucuronic acid and GalNAc residues toward the nonreducing end of the molecule. To complete the primary structural analysis of this oligosaccha-ride, the glycosyl linkage positions, as well as the sequence of the monosac-charide residues in the oligosaccharide, had to be determined. While this could be determined by using heteronuclear multiple-bond correlation (HMBC) data, the ^{1}H Nano•NMR probe is not yet equipped for indirect detection. Another option is to detect the homonuclear through-space con-nectivities (NOEs) between the protons on both sides of the glycosidic linkages. These interresidue NOEs, however, are sometimes smaller than other through-space connectivities within the same molecule and may be difficult to find.

When a 2D rotating frame overhauser enhancement spectroscopy (ROESY) experiment was performed on xyloside I (Fig. 4), interresidue cross-peaks were found between the H-1 of GalNAc and H-4 of GlcA; H-1 of GlcA and H-3 of Gal-3; H-1 of Gal-3 and H-3 of Gal-2; and H-1 of Gal-2 and H-4 of Xyl. In addition, the 3.70-ppm resonance could be assigned to the H-5 signal of the GalNAc residue, because the H-4 and H-5 nuclei are situated close enough in space to generate a cross peak in this spectrum.

The structure of the main component of sample I is therefore

$$\text{GalNAc}\alpha(1\text{-}4)\text{GlcA}\beta(1\text{-}3)\text{Gal}\beta(1\text{-}3)\text{Gal}\beta(1\text{-}4)\text{Xyl}\beta(1\text{-}4)\text{MU}$$

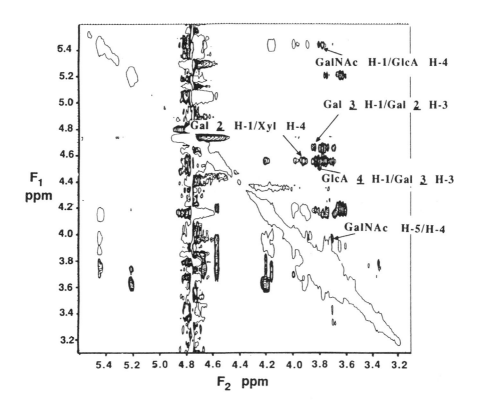

Figure 4 2D ROESY spectrum of the β-xyloside I recorded at 500 MHz in D₂O at 30°C, with a mixing time of 200 msec. Interresidue ROEs are indicated by arrows. ROESY correlations (phase opposite the diagonal) are indicated with 10 contour lines; TOCSY artifacts, and the diagonal, are indicated with a single contour line. (From Ref. 12.)

The novel aspect of this molecule is the presence of a terminal α-GalNAc residue at a position that is normally occupied by β-GalNAc in chondroitin/dermatan sulfate, or by α-GlcNAc in heparin or heparan sulfate chains. An α-GalNAc residue at this critical location may prevent further chain extension or influence the type of chain subsequently added to the common tetrasaccharide core.

IV. CONCLUSIONS

The analysis of the ¹H NMR data obtained using a Nano•NMR probe allowed us to clearly identify the major component of a mixture of oligosaccharides isolated from cells in culture. With about 25 μg of a carbohydrate sample (MW = 1010 amu), and a combination of ^{1}H 1D and homonuclear 2D NMR experiments carried out in aqueous solutions (with presaturation of the HOD peak), we identified a novel GAG-core-like molecule terminating in α-GalNAc. Although definitive assignment of all of the resonances in the ^{1}H NMR spectra of these xylosides was not possible due to the complexity of these mixtures, considerable information was obtained. After the original publication of these findings [12], an enzyme capable of transferring an α-GalNAc residue to a growing GAG chain was discovered in fetal bovine serum [27]. This points to the relevance of structural characterization of complex carbohydrates from biological sources.

The results presented here were obtained using sample quantities that were only one-third to one-tenth of the amount considered to be the minimum quantity required for such analyses when using a standard 5-mm probe. It has since been shown that the Nano•NMR probe can allow some NMR analyses to be performed on solutions having volumes of only 1 to 2 μL, as the only restriction on the minimum volume of solution used is that it provides enough ^{2}H signal for establishing a field lock. Future Nano•NMR probes should allow indirect-detection-type heteronuclear 2D experiments to be performed with the level of sensitivity achieved here for 2D homonuclear experiments. Soon, NMR spectroscopy will be able to provide not only complete primary structural analyses, but also conformational analyses of complex carbohydrates on sample quantities an order of magnitude smaller than those presently used.

ABBREVIATIONS

DMSO	dimethyl sulfoxide
DQFCOSY	Double-Quantum-Filtered COrrelation SpectroscopY
DSS	4,4-dimethyl-4-silapentane-1-sulfonate
GAG	glycosaminoglycan
Gal	galactose
GlcA	glucaromic acid
GalNAc	N-acetyl-2-amino-2-deoxy-galactose
GlcNAc	N-acetyl-2-amino-2-deoxyglucose
HMBC	Heteronuclear Multiple-Bond Correlation spectroscopy
MAS	magic-angle spinning
MLEV 17	Malcom Levitt–17 mixing sequence
MU	4-methylumbelliferyl (7-hydroxy-4-methylcoumarin)
NMR	nuclear magnetic resonance
NOEs	Nuclear Overhauser Enhancements
ROESY	Rotating frame Overhauser Enhancement SpectroscopY
SSB	spinning sideband

TOCSY   TOtally (scalar) Correlated SpectroscopY
Xyl   xylose
XylβMU   MU-β-xyloside

REFERENCES

1. Kjellen, L., and Lindahl, U. (1991). *Ann. Rev. Biochem.* 60, 443–475.
2. Fritz, T. A., Lugemwa, F. N., Sarkar, A. K., and Esko, J. D. (1994). *J. Biol. Chem.* 269, 300–307.
3. Freeze, H. H., Sampath, D., and Varki, A. (1993). *J. Biol. Chem.* 268, 1618–1627.
4. Garroway, A. N. (1982). *J. Magn. Reson.* 49, 168–171.
5. Shoolery, J. N. (1991). personal communication.
6. Barbara, T. M. (1994). *J. Magn. Reson.* Ser. A 109, 265–269.
7. Mozurkewich, G., Ringermacher, H. I., and Bolef, D. I. (1979). *Phys. Rev.* B 20, 33–38.
8. Keifer, P. A., Baltusis, L., Rice, D. M., Tymiak, A. A., and Shoolery, J. N. (1996). *J. Magn. Reson.* A119, 65–75.
9. Fitch, W. L., Detre, G., Holmes, C. P., Shoolery, J. N., and Keifer, P. A. (1994). *J. Org. Chem.* 59, 7955–7956.
10. Keifer, P. A. (1996). *J. Org. Chem.*, 16, 1558–1559.
11. Sarkar, S. K., Garigipati, R. S., Adams, J. L., and Keifer, P. A. (1996). *J. Am. Chem. Soc.*, 118, 2305–2306.
12. Manzi, A. E., Salimath, P. V., Spiro, R. C., Keifer, P. A., and Freeze, H. (1995). *J. Biol. Chem.* 270, 9154–9163.
13. Griesinger, C., Otting, G., Wuthrich, K., and Ernst, R. R. (1988). *J. Am. Chem. Soc.* 110, 7870–7872.
14. Piantini, U., Sorenson, O. W., and Ernst, R. R. (1982). *J. Am. Chem. Soc.* 104, 6800–6801.
15. Kessler, H., Griesinger, R., Kerssebaum, R., Wagner, K., and Ernst, R. R. (1987). *J. Am. Chem. Soc.* 109, 607–609.
16. Sadtler Standard Spectra, Proton NMR Spectra, Spectrum #3220M. Sadtler Research Laboratories, Inc., Philadelphia.
17. van Halbeek, H. (1994). In Lennarz, W. J., and Hart, G. W., eds., *Methods in Enzymology, Vol. 230, Guide to Techniques in Glycobiology.* Academic Press, San Diego, pp. 132–168.
18. Cassels, F. J., and van Halbeek, H. (1995). In Ginsburg, V., ed., *Methods in Enzymology, Vol. 253, Complex Carbohydrates.* Academic Press, San Diego, pp. 69–91.
19. Strecker, G., Wieruszeski, J.-M., Michalski, J.-C., Alonso, C., Boilly, B., and Montreuil, J. (1992). *FEBS Lett.* 298, 39–43.
20. Chai, W., Hounsell, E. F., Cashmore, G. C., Rosankiewicz, J. R., Feeney, J., and Lawson, A. M. (1992). *Eur. J. Biochem.* 207, 973–980.
21. Dabrowski, J., Hanfland, P., and Egge, H. (1987). In Ginsburg, V., ed.,

Methods in Enzymology, Vol. 83, Complex Carbohydrates. Academic Press, San Diego, pp. 69–86.

22. Strecker, G., Wieruszeski, J. M., Michalski, J.-C., and Montreuil, J. (1989). *Glycoconjugate J.* 6, 271–284.

23. Sugahara, K., Yamashina, I., De Waard, P., Van Halbeek, H., and Vliegenthart, J. F. G. (1988). *J. Biol. Chem.* 263, 10168–10174.

24. Sugahara, K., Yamada, S., Yoshida, K., De Waard, P., and Vliegenthart, J. F. G. (1992). *J. Biol. Chem.* 267, 1528–1533.

25. Sugahara K., Ohi, Y., Harada, T., De Waard, P., and Vliegenthart, J. F. G. (1992). *J. Biol. Chem.* 267, 6027–6035.

26. De Waard, P., Vliegenthart, J. F. G., Harada, T., and Sugahara, K. (1992). *J. Biol. Chem.* 267, 6036–6043.

27. Kitagawa, H., Tanaka, Y., Tsuchida, K., Goto, F., Ogawa, T., Lidholt, K., Lindahl, U., and Sugahara, K. (1995). *J. Biol. Chem.* 270, 22190–22195.

2

Analysis of Anionic Glycoconjugates by Delayed Extraction Matrix-Assisted Laser Desorption Time-of-Flight Mass Spectrometry

Bradford W. Gibson, Jeffrey J. Engstrom, and Constance M. John
University of California–San Francisco, San Francisco, California

Arnold M. Falick
PerSeptive Biosystems, Inc., South San Francisco, California

Wade M. Hines and Stephen A. Martin
PerSeptive Biosystems, Inc., Framingham, Massachusetts

I. INTRODUCTION

The characterization and analysis of glycoconjugates by mass spectrometry has undergone a number of improvements in the last 10 years, especially with the development of methods capable of ionizing and analyzing these compounds in their native states [1]. Although a number of protocols have been developed for preparing derivatives of oligosaccharides deemed more suitable for mass spectrometric analysis, such as permethylation and acetylation [2], there has always been a need for methods that could analyze these types of compounds directly. Despite some success, underivatized glycoconjugates and oligosaccharides tend to ionize not as well as their derivatized analogs or other classes of biopolymers of similar mass. This is due, in part, to the high polarity of oligosaccharides, which tends to make them less amenable to mass spectrometric analysis than compounds with significant hydrophobic character. Moreover, chromatographic methods for isolating carbohydrates and glycoconjugates often require buffers containing relatively high levels of salts or other nonvolatile components that tend to interfere with mass spectrometry. Even more troublesome are anionic oligosaccharides and glycoconjugates, the tendency of which to bind cations can greatly increase the difficulties in their final analysis. Even when rigorous desalting is attempted, salt adducts of highly anionic carbohydrates and glycoconjugates remain a serious problem, and even more so as

17

one attempts to isolate and/or purify them at subnanomole levels. In addition, anionic glycoconjugates can contain acidic sugars, such as sialic acid (or N-acetylneuraminic acid, NeuAc) or 2-keto-3-deoxyoctulosonic acid (Kdo), the anionic character of which makes their glycosidic bonds especially labile.

In our laboratory at UCSF, we have been working on methods for the structural and functional characterization of a group of bacterial surface glycolipids, or lipooligosaccharides (LOS), from pathogenic *Haemophilus* and *Neisseria* spp. (see, for example, [3–5]). These LOS are highly complex and anionic in character, and they generally contain several phosphate (*P*) and/or phosphoethanolamine (*PEA*) substituents (Fig. 1). The presence of a highly anionic and hydrophobic Lipid A moiety, containing two or more phosphates, and up to six *N*- and *O*-linked fatty acids, attached to a short but highly branched and polar oligosaccharide, makes these glycoconjugates highly amphipathic and largely insoluble in water and most organic solvents. While our laboratory has developed analytical methods that have demonstrated some success in the analysis of modified or partially degraded LOS and oligosaccharides using liquid secondary ionization mass spectrometry (LSIMS or FAB) [3] and electrospray ionization mass spectrometry (ESI/MS) [4], these techniques may not have the required sensitivity or

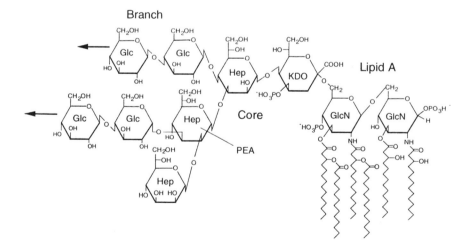

Figure 1 Major lipooligosaccharide obtained from a mutant strain of *Haemophilus influenzae* (strain 276.4) that is also present in the parent wild-type strain, *H. influenzae* A2 [6]. Arrows indicate positions of further oligosaccharide branch extensions in the wild-type strain. The MALDI-TOF spectrum of the *O*-deacylated LOS from this mutant strain is shown in Fig. 7.

versatility needed to analyze the very low level of LOS samples that one can reasonably expect to obtain from human clinical isolates. This latter goal remains one of our most important analytical challenges, as we believe it is essential to identify the precise LOS-glycoforms that are expressed during disease. Without this direct correlation, we are left in the precarious position of trying to determine the structures and functions of biologically relevant LOS from a large set of LOS-glycoforms identified under in vitro growth conditions.

To develop techniques capable of femtomole or lower level analysis of these bacterial glycoconjugates and oligosaccharides derived from human isolates, we have begun to investigate the potential of matrix-assisted laser desorption ionization (MALDI) with a time-of-flight (TOF) mass analyzer. MALDI-TOF has been shown to have exceedingly high sensitivities for peptides and proteins [7,8], as well as some other classes of biological compounds such as oligonucleotides [9–11]. Also, we hoped that a surface-based technique might be better suited for these LOS compounds, especially given their solubility properties.

Despite the potential advantages of MALDI-TOF, we were also aware of some disadvantages of this technique that could possibly compromise its usefulness for this type of glycoconjugate analysis. Some of the adverse effects of conventional MALDI-TOF included a large initial velocity distribution of desorbed analyte ions, which makes for low resolving power, substantial energy losses from collisions in the ionization region, and a high level of background noise in the resulting spectra due to fast fragmentation processes. Furthermore, the desorption and ionization of the analyte depends greatly on matrix and sample preparation protocols. Recently, several groups have reported significant improvements in MALDI-TOF when incorporating a time delay between desorption and ion extraction (Fig. 2) [12,13]. Delayed extraction allows for first order (and possibly higher order) velocity focusing, which yields significantly improved resolution. Additionally, decoupling desorption from ion extraction largely eliminates kinetic energy loss from in-source collisions as well as substantially attenuating background noise from metastable decomposition during ion acceleration. In continuous extraction MALDI, laser power must be kept at or near the ionization threshold to obtain maximum resolving power, but delayed extraction tolerates a much larger range of laser power settings without adversely affecting spectral quality. This latter effect perhaps has its greatest impact with sample/matrix combinations that are more difficult to prepare as homogeneous co-crystals and, in our laboratory, has greatly reduced the time required to perform analyses. In this chapter, we present data acquired using this novel delayed extraction technique for LOS-glycoconjugate analysis.

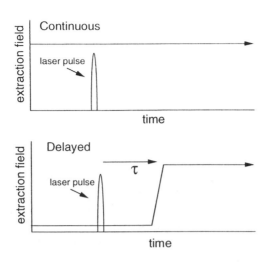

Figure 2   Continuous versus time delayed extraction of ions generated by a single laser pulse. Under delayed extraction conditions, the extraction field is applied after a delay of 100–400 nsec from the time of the laser pulse.

II. MATERIALS AND METHODS

All oligosaccharide and lipooligosaccharide (LOS) samples were prepared according to procedures described in detail elsewhere [3,14] with the exception of LOS from *Salmonella typhimurium*, which was purchased directly from Sigma (St. Louis, Missouri). Lipooligosaccharides were *O*-deacylated with hydrazine at 37°C and precipitated with cold acetone as described previously [3]. An oligosaccharide preparation was prepared from *S. typhimurium* LOS by acetic acid hydrolysis (1% acetic acid, 100°C, 2 hr), treated with aqueous HF to remove phosphate (48% aqueous HF, 3°C, 10 hr), and separated by size-exclusion chromatography.

For all mass spectrometry experiments, a Voyager Elite MALDI-TOF instrument (PerSeptive Biosystems, Framingham, Massachusetts) equipped with a nitrogen laser (337 nM) was used. This instrument has a linear path length of 2 m, and a total reflector length of 3 m (Fig. 3). Except where noted, all experiments were carried out using delayed extraction conditions: 200–350 nsec time delay with a grid voltage of 92–94% of full acceleration voltage (20–30 kV). Spectra were calibrated using an external calibrant or using a one-point internal calibration using software supplied by PerSeptive Biosystems. All intact LOS and *O*-deacylated LOS samples were run in recrystallized 2,5-dihydroxybenzoic acid (DHB) [15]. Oligosaccharides were prepared with a (recrystallized) 100 mM DHB solution containing

33 mM 1-hydroxyisoquinoline, as originally suggested for underivatized oligosaccharides by Mohr et al. [16]. To prepare a soluble form of the intact LOS, we first added 5 mM EDTA followed by several 2- to 3-sec pulses of ultrasonication (microtip with max. 50-W output, Fisher Laboratories). In some cases, small aliquots of O-deacylated LOS and oligosaccharide samples together with the matrix (≈ 5–6 μL) were desalted with cation exchange beads (Dowex 50X, NH_4^+ form) prior to crystallization on the MALDI plates [17].

III. RESULTS AND DISCUSSION

In our initial analyses of LOS by MALDI-TOF, we prepared a partially degraded form of LOS from a commercial preparation of *Salmonella typhimurium* Ra LOS (a mutant rough strain lacking a polysaccharide) that had undergone chemical removal of O-acyl groups from the Lipid A moiety. The rationale for working with an O-deacylated LOS preparation was two-fold. One, removal of O-linked fatty acids on the Lipid A moiety by mild hydrazine treatment of LOS renders them water soluble and, therefore, more likely to co-crystallize with matrices such as DHB. Two, removal of O-linked fatty acids from the Lipid A moiety reduces the heterogeneity of the LOS and enables one to concentrate on the more variable oligosaccharide regions of the LOS molecule. It is our hypothesis that the oligosaccharide regions of LOS encode for specific functions such as cell adhesion,

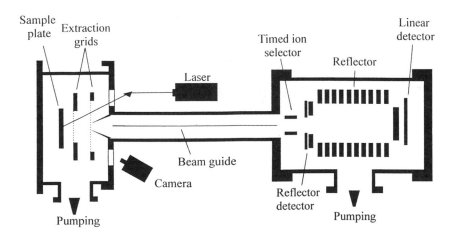

Figure 3 Schematic diagram of a single-stage reflector MALDI-TOF instrument.

invasion, and immune evasion mechanisms that are central to the growth of these organisms in the human host.

The MALDI spectra obtained from these *O*-deacylated LOS preparations, under conventional continuous extraction conditions, were complex with poor resolution of the expected individual glycoforms. For example, Fig. 4 (spectrum A) shows a typical MALDI-TOF spectrum in the positive ion mode obtained from *S. typhimurium* LOS that had first been treated with mild hydrazine to remove *O*-acyl fatty acids on the Lipid A portion. The poor resolution and broad peak shapes of these LOS components in this spectrum resemble to a certain extent those of oligonucleotides, where the combination of metastable decay processes and salt adducts can greatly compromise the quality of the spectra [11]. However, under delayed extraction conditions, the resulting spectra were dramatically improved. As shown in the top Trace of Fig. 4 (spectrum B), multiple LOS-glycoforms and their corresponding salt adducts were readily resolved with significant improvement in peak shapes. In this delayed extraction spectrum, one can

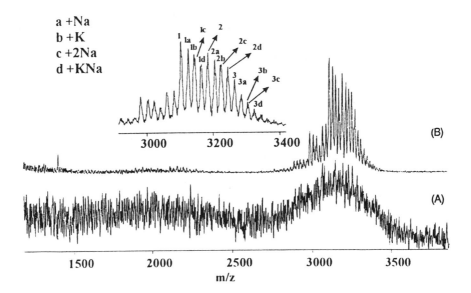

Figure 4 Positive ion MALDI-TOF spectrum of *S. typhimurium* *O*-deacylated LOS run under delayed extraction (top, panel B) or continuous extraction (bottom, panel A) conditions. Inset at top left is an expansion of the *m/z* 3000–3400 region of the delayed extraction spectrum. At least four separate LOS-glycoforms, each with several salt adducts, were observed in this latter spectrum, i.e., M_r = 2982, 3105, 3185, and 3265.

identify glycoforms that can be assigned to the expected major LOS structure (LOS glycoform *1*, shown below with $M_r = 3105$) by the addition of variable levels of phosphate (P, $\Delta M = 80$ Da) and/or phosphoethanolamine (PEA, $\Delta M = 123$ Da):

$$\begin{array}{ccccccccc}
 & \text{Gal} & \text{Hep} & & PPEA & & & & \\
 & \downarrow & \downarrow & & \downarrow & & & & \\
\text{GlcNAc} \rightarrow \text{Glc} \rightarrow \text{Gal} \rightarrow \text{Glc} \rightarrow & \text{Hep} \rightarrow & \text{Hep} \rightarrow & \text{Kdo} \rightarrow & \text{Lipid A*} \\
 & \uparrow & & \uparrow & & & & \\
 & P & & \text{Kdo} & & & & \\
\end{array}$$

However, the loss of phosphate groups through β-elimination of phosphoric acid (-98 Da, H_3PO_4) adds an additional and undesirable level of complexity to the interpretation of the molecular ion region. Follow-up studies on a number of different LOS preparations have clearly shown that this is a very general process in the positive ion mode, leading to the formation of prompt fragments that are 18 Da lower in mass than what a true molecular ion would be for an LOS species containing one less phosphate group, i.e., $(M + H - 98)^+$ for prompt loss of H_3PO_4, compared with the $(M + H - 80)^+$ peak observed for the chemical hydrolysis of a phosphate group, that is, net loss of HPO_3.

In the corresponding negative ion spectrum of the O-deacylated *S. typhimurium* LOS, a much less complicated molecular ion region was observed (Fig. 5). This is due primarily to two factors. First, the degree of β-elimination of phosphoric acid from various deprotonated molecular ion species is considerably less than that observed for protonated molecular ions in the positive ion mode, an effect that has also been observed for phosphopeptides (unpublished observation, [18]). Second, the number of observed salt adducts peaks is also generally reduced. The molecular ion region of the negative ion spectra of these O-deacylated LOS species was further simplified by brief treatment with cation exchange beads, as shown in Fig. 5B, where the sodium adducts to the molecular species are greatly reduced in abundance.

Given the dramatic improvement in the delayed extraction spectra of the O-deacylated LOS, compared with spectra obtained in the continuous extraction mode, we then examined the possibility of analyzing crude LOS preparations that had not undergone O-deacylation. Figure 6 shows the results using the same commercial preparation of *S. typhimurium* LOS from the Ra strain. The resulting spectrum of this intact LOS preparation clearly showed many of the same LOS-glycoforms seen in the MALDI spectra of the O-deacylated preparation, although the peaks due to intact LOS species are 500–1000 Da higher in mass due to the presence of an additional two to four O-linked fatty acids on the Lipid A moiety. Indeed,

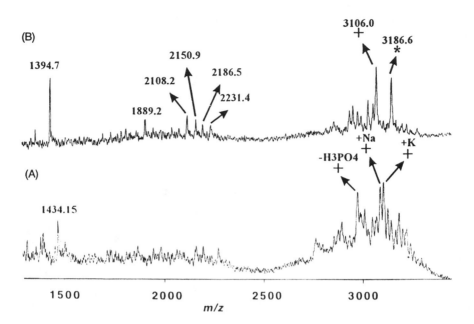

Figure 5 Negative ion MALDI-TOF analysis of *O*-deacylated LOS obtained from *S. typhimurium* Ra strain (A). Spectrum in (B) is same LOS sample after desalting with cation exchange beads (see text). Note the increased abundances of the peaks at *m/z* 3106 and 3186 in B, which are deprotonated molecular ions, $(M-H)^-$, of the major LOS-glycoforms differing by a phosphate group (HPO_3). Also, peaks in the range *m/z* 1800–2300 are prompt fragments originating from the oligosaccharide portions of the LOS precursor ions.

these data show the major LOS-glycoforms to be a series of hexa-acyl diphosphoryl-containing LOS, as well as some less abundant tetra-acyl forms [19,20]. This spectrum provides for the first time the most direct evidence for the fatty acyl substitution state of Lipid A while it is still attached to the oligosaccharide moiety. In conventional studies, Lipid A is first cleaved from the oligosaccharide by mild acid treatment (e.g., 0.1 N HCl, 100°C) prior to its isolation and analysis. Given the potential for this acid treatment to cleave fatty acids from the Lipid A, there has always been some concern over the biological relevance of Lipid A species that are found to be only partially substituted [21].

In an effort to explore the possibility of obtaining additional structural information on LOS preparations using MALDI-TOF under delayed extraction conditions, a series of experiments were carried out to examine the lower-mass fragment ions present in some of these spectra. As shown in

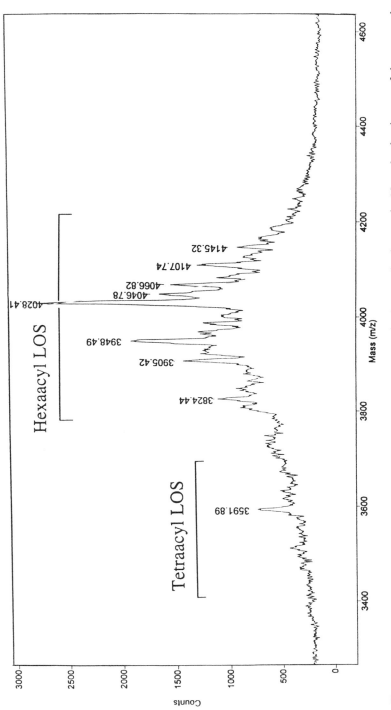

Figure 6 Molecular ion region of *intact* LOS from *S. typhimurium* Ra in the negative ion mode. Note the dominance of the expected hexa-acyl-substituted Lipid A with the small amount of tetra-acyl species at lower masses corresponding to the loss of myristic acid and β-hydroxymyristic acid, e.g., −436 Da, m/z 4028 (hexa-acyl LOS) → 3592 (tetra-acyl LOS).

Fig. 7A, some fragments are present in the initial experiments carried out under continuous extraction conditions. But like the precursor molecular ions, these fragments appear as broad tailing peaks. In contrast, under delayed extraction conditions (and at lower laser power setting), these fragments virtually disappear (Fig. 7B). However, when the laser power is raised, well-resolved prompt fragment ions in the mass range between m/z 900 and 1700 were observed (Fig. 7C). These fragments arise primarily from cleavage at the glycosidic bond between the acidic Kdo sugar and the Lipid A, yielding peaks that are either related to the oligosaccharide or Lipid A portion of the LOS. This fragmentation pathway is followed by additional losses of CO_2 and phosphate from the oligosaccharide fragments, giving rise to the single major diphosphoryl diacyl Lipid A peak at m/z 952, and a series of oligosaccharide fragments in the 1700 to 1300 Da range.

One can also select an individual LOS-glycoform at high mass (for example, m/z 2601) for subsequent post-source decay (PSD) analysis [22]. Under PSD conditions, fragment ions formed in the field-free region of the TOF analyzer from decomposition of a selected precursor ion are focused

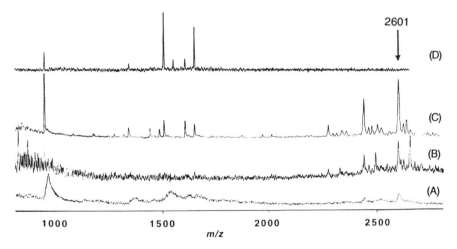

Figure 7 Comparison between prompt and PSD fragmentation of O-deacylated LOS from *H. influenzae* strain 276.4 taken in the negative ion mode. Spectra are as follows: (A) linear with continuous extraction, (B) linear with delayed extraction, (C) linear with delayed extraction but higher laser power to induce prompt fragmentation, and (D) PSD with timed ion selection of m/z 2601 precursor ion. Molecular ions for the O-deacylated LOS are seen in the m/z range of 2200–2700 and differ in mass by 162 Da, the mass increment for a hexose residue. Peaks at lower masses ($m/z \leq 2200$) are either prompt fragments (panels A and C) or PSD fragments (panel D).

by appropriate adjustment of the mirror (reflector) potential. In the PSD spectrum of the precursor ion at m/z 2601, a smaller set of well-defined fragment ions were observed, even though none of the precursor species survived intact (Fig. 7D). Interestingly, these PSD fragments are a subset of those observed in Fig. 7C, where the total LOS-glycoform mixture was subjected to in-source prompt fragmentation through adjustment of the laser power.

Given that we can obtain well-resolved peaks for individual LOS-glycoforms using delayed extraction in the linear mode, it was also apparent that one should gain an equivalent improvement in mass accuracy. Thus, we analyzed a mixture of LOS-glycoforms obtained from *H. ducreyi* strain 35000, where the major glycoform structure and molecular weight had been previously determined [23]. In this experiment, we used the "known" mass of this major glycoform as a single-point calibration for the entire spectrum (see below). The results of this experiment are shown in Fig. 8. In most cases, this single-point calibration allowed for good agreement between experimental and expected masses for the additional six glycoforms, yielding a mass accuracy in this particular LOS mixture of $\leq \pm 0.01\%$ for most

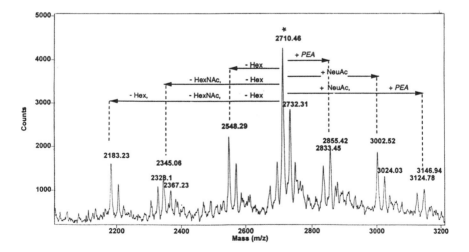

Figure 8 Single-point calibration of negative ion MALDI-TOF spectrum under delayed extraction conditions of *O*-deacylated LOS from *H. ducreyi* strain 35000. The one-point calibrant peak was at m/z 2710.46, corresponding to the deprotonated structure shown in the text with an M_r of 2711.46. The experimental masses of the other LOS species are shown above each peak, and comparisons of expected and calculated masses for the additional LOS-glycoforms are listed in Table 1.

LOS components (Table 1). This increased mass accuracy is important in limiting composition searches*:

$$\text{Gal} \rightarrow \text{GlcNAc} \rightarrow \text{Gal} \rightarrow \text{Hep} \rightarrow \text{Glc} \rightarrow \text{Hep} \rightarrow \text{Kdo(P)} \rightarrow \text{Lipid A*} \qquad M_{r,\text{avg.}} = 2711.5$$
$$\uparrow$$
$$\text{Hep}$$
$$\uparrow$$
$$\text{Hep}$$

Finally, we evaluated delayed extraction MALDI-TOF for the analysis of acidic Kdo-containing oligosaccharides. These oligosaccharides were isolated by mild acid treatment of the intact LOS, which cleaves at the acid labile Kdo bond(s), leaving a single acidic Kdo saccharide attached at the reducing terminus of a branched oligosaccharide(s). For example, when the *S. typhimurium* LOS was hydrolyzed in 1% acetic acid for 1–2 hr at 100°C, a mixture of differentially phosphorylated oligosaccharides was obtained. After dephosphorylation with aqueous HF, a major oligosaccharide species was obtained:

$$\begin{array}{ccc} \text{Gal} & \text{Hep} \\ \downarrow & \downarrow \end{array}$$
$$\text{GlcNAc} \rightarrow \text{Glc} \rightarrow \text{Gal} \rightarrow \text{Glc} \rightarrow \text{Hep} \rightarrow \text{Hep} \rightarrow \text{Kdo} \qquad M_{r,\text{exact}} = 1665.54$$

Under LSIMS conditions, one typically needs up to a nanomole (1–2 μg) for reasonably good detection. Although conventional MALDI-TOF spectra could be obtained using considerably less material (≈ 0.1 μg), the

Table 1 *O*-Deacylated LOS from *H. ducreyi* Strain 35000

Expt. (M − H)⁻	Calc. (M − H)⁻	Mass difference	Sugar differences[a]
2183.23	2182.98	$\Delta m = 0.25$	−Gal, −GlcNAc, −Gal
2345.06	2345.12	$\Delta m = 0.06$	−Gal, −GlcNAc
2548.29	2548.32	$\Delta m = 0.03$	−Gal
2710.46[b]	*2710.46*	$(\Delta m = 0.0)$	—
2833.45	2833.51	$\Delta m = 0.06$	+PEA
3002.52	3001.72	$\Delta M = 0.8$	+NeuAc
3124.78	3124.77	$\Delta m = 0.01$	+PEA, +NeuAc

[a]Sugar differences in LOS species are shown relative to the $M_r = 2711.46$ LOS-glycoform.
[b]Single-point calibration used m/z 2710.46, a chemically defined LOS component [23].

---

*In searching possible compositions for LOS and other glycoconjugates, we use a computer program developed by W. Hines ("Gretta Carbos") that allows for up to eight different structural moieties with associated minimum and maximum numbers and a precision or tolerance of the experimentally determined mass to be considered.

quality of these spectra was low. Even recrystallization of these samples on the probe surface, a technique reported to enhance signal abundance, did not significantly improve the spectral quality (data not shown). However, under delayed extraction conditions where ≈ 0.02 μg or less was applied to the sample probe, abundant peaks were obtained in both the negative and positive ion modes with clearly resolved isotopes (Fig. 9). However, the negative ion spectrum was considerably simpler, containing only two major peaks: one at the expected m/z 1664.5, and a smaller peak 18 Da lower in mass at m/z 1646.6, presumably arising from the loss of water. In contrast, the positive ion spectrum contained no protonated molecular ion signals, but rather a complex series of sodium and potassium salt adducts, where the base peak was the singly sodiated species, MNa^+ at m/z 1688.5. In addition, lower mass peaks were observed that appeared to result from the

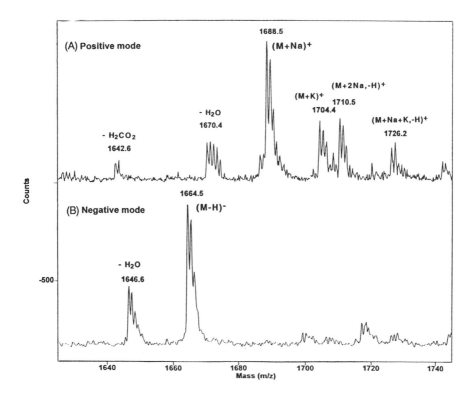

Figure 9 Positive ion (A) and negative ion (B) MALDI-TOF spectrum of dephosphorylated oligosaccharide from *S. typhimurium* Ra strain LOS obtained in the linear mode using delayed extraction.

loss of water $(-H_2O)$ and the elements of formic acid $(-H_2CO_2$ or $-H_2O,$ $-CO)$.

IV. CONCLUSIONS

Delayed extraction represents a major advance over conventional MALDI-TOF mass spectrometry for the analysis of LOS glycoconjugates. In the linear mode, it provides a remarkable increase in mass resolution combined with improved mass accuracy and reduction in metastable decay processes, resulting in far more easily interpretable spectra. The latter feature is particularly significant in the case of glycoconjugates, because of the presence of many labile groups such as sulfate, phosphate, or acidic sugars such as sialic acid or Kdo. Structural information needed for identification of individual glycoforms without prior separation can be obtained through the use of higher laser power to generate prompt fragments, and by using a reflector to analyze PSD fragments. The ease of fragmentation of the labile groups and linkages enhances these processes. The full range of applications of the MALDI-TOF under delayed extraction conditions in the analysis of carbohydrates and glycoconjugates has yet to be explored, but the data presented in the preliminary study in this chapter demonstrates that it is likely to make a major contribution.

ACKNOWLEDGMENTS

This work was supported by grants from National Institutes of Health (AI 31254 and AI 24016; B. Gibson) and from instrumentation kindly provided to B. Gibson by PerSeptive Biosystems, Framingham, Massachusetts.

ABBREVIATIONS

DHB	2,5-dihydroxybenzoic acid
ESI/MS	electrospray ionization mass spectrometry
Kdo	2-keto-3-deoxyoctulosonic acid
LOS	lipooligosaccharides
LSIMS	liquid secondary ionization mass spectrometry
MALDI	matrix-assisted laser desorption ionization
NeuAc	*N*-acetylneuraminic acid
P	phosphate
PEA	phosphoethanolamine
PSD	post-source decay
TOF	time-of-flight

REFERENCES

1. McCloskey, J. A., ed. (1990). *Mass Spectrometry, Vol. 193, Section III. Glycoconjugates.* Academic Press, pp. 539–770.
2. Dell, A. (1990). *Meth. Enzymol.* 193, 647–660.
3. Phillips, N. J., John, C. M., Reinders, L. G., Gibson, B. W., Apicella, M. A., and Griffiss, J. M. (1990). *Biomed. Environ. Mass Spectrom.* 19, 731–745.
4. Gibson, B. W., Phillips, N. J., Apicella, M. A., Campagnari, A. A., and Griffiss, J. M. (1993). *J. Bacteriol.* 175, 2702–2712.
5. Gibson, B. W., Phillips, N. J., John, C. M., and Melaugh, W. (1994). In Fenselau, C., ed., *Mass Spectrometry for the Characterization of Microorganisms.* ACS, Washington, D.C., pp. 185–202.
6. Phillips, N. J., McLaughlin, R., Miller, T. J., Apiecella, M. A., and Gibson, B. W. (1996). *Biochemistry,* 35, 5937–5947.
7. Karas, M., and Hillenkamp, F. (1988). *Anal. Chem.* 60, 2299–2301.
8. Beavis, R. C., and Chait, B. T. (1990). *Proc. Natl. Acad. Sci. USA* 87, 6873–6877.
9. Shaler, T. A., Tan, Y., Wickham, J. N., Wu, K. J., and Becker, C. H. (1995). *Rapid Commun. Mass Spectrom.* 9, 942–947.
10. Fitzgerald, M. C., and Smith, L. M. (1995). *Ann. Rev. Biophys. Biomol. Struct.* 24, 114–140.
11. Juhasz, P., Roskey, M. T., Smirnov, I. P., Haff, L. A., Vestal, M. L., and Martin, S. A. (1996). *Anal. Chem.* 68, 941–946.
12. Vestal, M. L., Juhasz, P., and Martin, S. A. (1995). *Rapid Commun. Mass Spectrom.* 9, 1044–1050.
13. Brown, R. S., and Lennon, J. J. (1995). *Anal. Chem.* 67, 1998–2003.
14. Apicella, M. A., Griffiss, J. M., and Schneider, H. (1994). *Meth. Enzymol.* 235, 242–252.
15. Strupat, K., Karas, M., and Hillenkamp, F. (1991). *Int. J. Mass Spectrom. Ion Proc.* 111, 89–93.
16. Mohr, M. D., Bornsen, K. O., and Widmer, H. M. (1995). *Rapid Commun. Mass Spectrom.* 9, 809–814.
17. Nordhoff, E., Ingendoh, A., Cramer, R., Overberg, A., Stahl, B., Karas, M., Hillenkamp, F., and Crain, P. F. (1992). *Rapid Commun. Mass Spectrom.* 6, 771–776.
18. Gibson, B. W., and Cohen, P. (1990). *Meth. Enzymol.* 193, 480–501.
19. Takayama, K., Qureshi, N., and Mascagni, P. (1983). *J. Biol. Chem.* 258, 12801–12803.
20. Qureshi, N., Takayama, K., Heller, D., and Fenselau, C. (1983). *J. Biol. Chem.* 258, 12947–12951.
21. Chan, S., and Reinhold, V. N. (1994). *Anal. Chem.* 218, 63–73.
22. Kaufmann, R., Spengler, B., and Lutzenkirchen, F. (1993). *Rapid Commun. Mass Spectrom.* 7, 902–910.
23. Melaugh, W., Phillips, N. J., Campagnari, A. A., Tullius, M. V., and Gibson, B. W. (1994). *Biochemistry* 33, 13070–13078.

3

Matrix-Assisted Laser Desorption/Ionization Time-of-Flight Mass Spectrometry of Oligosaccharides Separated by High pH Anion-Exchange Chromatography

Damon I. Papac, Andrew J. S. Jones, and Louisette J. Basa
Genentech, Inc., South San Francisco, California

I. INTRODUCTION

High pH anion-exchange chromatography (HPAEC) with pulsed amperometric detection has proven to be an invaluable tool for the characterization of carbohydrates [1,2]. Unfortunately, the primary information gained from this technique is retention time, and unless authentic standards are run, the identification of the separated oligosaccharides is tenuous. By combining mass spectrometry with this technique, the separated oligosaccharides can be identified with greater certainty. Mass spectrometry can be used to obtain an exact mass of the oligosaccharide. Because most oligosaccharides are composed of five unique monosaccharide units with different incremental masses (fucose, 146 Da; hexose, 162 Da; N-acetylhexosamine, 203 Da; N-acetylneuraminic acid, 291 Da; N-glycolylneuraminic acid, 307 Da), knowledge of the molecular weight can be used to determine the potential composition of the oligosaccharide. Fast-atom bombardment (FAB) [3,4] and electrospray-ionization (ESI) [5,6] mass spectrometry have both been utilized successfully to this end. Matrix-assisted laser desorption/ionization (MALDI) is the most suitable ionization method for the analysis of carbohydrates collected after HPAEC, because MALDI is 10–100 times more sensitive than FAB for detection of underivatized oligosaccharides [7] and is more tolerant of salts than either FAB or ESI [8]. This chapter will describe the implementation of MALDI time-of-flight (TOF) mass spectrometry for the analysis of HPAEC-separated fractions of N-linked oligosaccharides released from recombinant tissue plasminogen activator (tPA).

II. MATERIALS AND METHODS

A. Oligosaccharide Isolation from tPA

Reduction and carboxymethylation of 2 mg of tPA (Genentech, Inc., South San Francisco, California) was performed as previously described [3]. The reduced and carboxymethylated tPA was dissolved in PNGase F digestion buffer (75 mM sodium phosphate [pH 8] containing 5 mM EDTA and 0.02% sodium azide) to a final protein concentration of 5 mg/mL. The N-linked oligosaccharides were released by incubation with (6.4 units/mg protein) glycerol-free PNGaseF (Boehringer Mannheim, Indianapolis, Indiana) overnight at 37°C. The released oligosaccharides were recovered in the supernatant following protein precipitation with 75% ice-cold ethanol. The recovered oligosaccharides, some existing as glycosylamines, were hydrolyzed to the free reducing oligosaccharides by treatment with 13 mM acetic acid at room temperature for 2 hr [9]. Following hydrolysis, the oligosaccharides were dried in a Savant Speed Vac (Farmingdale, New York) and were reconstituted with deionized water to a final concentration equivalent to 2 mg/mL (based on the original protein concentration). Typically, 100 μL of material was loaded onto the HPAEC column.

B. High pH Anion-Exchange Chromatography

The HPAEC was performed on a BioLC system (Dionex, Sunnyvale, California) equipped with a CarboPac PA-1 anion exchange column (4 × 250 mm, Dionex) and a PAD-2 pulsed amperometric detector with a solvent-compatible cell (Dionex). On-line desalting was accomplished by passing the column effluent through an AMMS-II anion micromembrane suppressor (Dionex). The regenerant solution for the AMMS-II was 75 mN H_2SO_4 flowing at 4 mL/min. When needed, the conductivity of the effluent was measured with a CDM-3 conductivity detector (Dionex).

Separation of the oligosaccharides was achieved with a gradient program at a flow rate of 1 mL/min. Solvent A consisted of 0.1 N NaOH, and solvent B consisted of 0.1 N NaOH containing 0.5 M sodium acetate. The column was equilibrated with 5% solvent B for 2 min, and with a two-step linear gradient, the oligosaccharides were eluted by first increasing solvent B to 15% in 5 min, and then to 40% in 25 min.

C. Desialylation

Desialylation of the acidic oligosaccharides was accomplished either chemically or enzymatically. Chemical desialylation entailed acidifying the collected fractions with acetic acid to a final concentration of 10% (v/v). The samples were heated for 3 hr at 80°C and then dried in a Speed Vac.

Enzymatic desialylation was achieved by adding 0.2 U of recombinant *Clostridium perfringens* sialidase (Oxford GlycoSystems, Rosedale, New York) directly to the collected fractions and incubating at 37°C for 12 hr. The final enzyme concentration was 1 U/mL

D. Sample Preparation for MALDI/TOF Analysis

Small columns packed with cation-exchange resin were required to remove the sodium acetate not eliminated by the AMMS-II. The columns were prepared by packing approximately 400 μL of either AG50W-X8 or -X2 resin (hydrogen form, Bio-Rad Laboratories, Hercules, California) into 1.0-mL compact reaction columns fitted with 90-μm polypropylene frits (Amersham Life Sciences, Arlington Heights, Illinois). Typically, 300–500 μL of the collected HPAEC fractions were loaded onto the cation-exchange resin, and the column washed with deionized water (3 × 300 μL). The sample effluent was collected and pooled with the effluent obtained from the three water washes. The samples were subsequently dried by vacuum centrifugation and reconstituted in 20–50 μL of deionized water.

The two matrices used were prepared as follows. The 2,5-dihydroxybenzoic acid matrix (sDHB) was prepared by dissolving 5 mg of 2,5-dihydroxybenzoic acid plus 0.25 mg of 5-methoxysalicylic acid in 1 mL of ethanol/10 mM aqueous sodium chloride 1 : 1 (v/v). The 2′,4′,6′-trihydroxyacetophenone matrix (THAP) was prepared by dissolving 1 mg of 2′,4′,6′-trihydroxyacetophenone in acetonitrile/20 mM aqueous ammonium citrate 1 : 1 (v/v).

Typically, 0.5 μL of analyte was applied to a polished stainless steel target, and 0.5 μL of matrix was then added to the analyte. All samples were dried under vacuum (50 × 10^{-3} Torr).

E. MALDI/TOF Instrument Operating Parameters

The MALDI/TOF mass spectrometer used to acquire the mass spectra was a Voyager Elite (PerSeptive Biosystems, Framingham, Massachusetts). All samples were irradiated with UV light (337 nm) from an N$_2$ laser. Neutral oligosaccharides were analyzed at 25 kV with the reflectron (3-m flight path) in the positive ion mode using sDHB as the matrix. Acidic oligosaccharides were analyzed at 30 kV without the reflectron (2-m flight path) in the negative ion mode using THAP as the matrix.

A two-point external calibration was used for mass assignment of the ions. When neutral oligosaccharides were analyzed, a high mannose oligosaccharide [$(M + Na)^+_{mono} = 1257.42$ m/z] and a fully galactosylated tetraantennary oligosaccharide without core fucose [$(M + Na)^+_{mono} = 2393.95$ m/z] were used to calibrate the instrument. When the acidic oligosaccharides were analyzed, a monosialylated diantennary oligosaccharide without core fucose [$(M - H)^-_{avg} = 1931.8$ m/z] and a trisialylated triantennary oligosaccharide without core fucose [$(M - H)^-_{avg} = 2879.6$ m/z] were used to calibrate the instrument. The oligosaccharide standards used for instrument calibration were obtained from Oxford GlycoSystems.

Typically, spectra from 64 laser shots were summed to obtain the final spectrum. This instrument utilizes a multichannel plate detector. Therefore, to prevent matrix ions from saturating the detector, the detector voltage was held below threshold until ions with a mass-to-change ratio of greater than 800 m/z could strike the detector.

III. RESULTS AND DISCUSSION

A. Desalting of Fractions with an In-Line Anion Micromembrane Suppressor

Although MALDI-TOF/MS tolerates salts and biological buffers, the >100 mM sodium in the HPAEC effluent hampers the acquisition of useful spectra. This problem is diminished by adding an anion micromembrane suppressor (AMMS) in-line after the detector. The AMMS eliminates sodium ions by exchanging the sodium ions for hydrogen ions supplied by the regenerant, H_2SO_4, flowing countercurrent over an anionic membrane. Figure 1 shows how the pH and conductivity of the HPAEC effluent are influenced by one in-line AMMS. Without the AMMS, the pH of the effluent is approximately 13 and the conductivity meter is saturated (>3.8 mS) during the entire gradient.

Ten of the major fractions were collected and reinjected onto the HPAEC to determine the effect that one AMMS had on recovery. The recoveries ranged from 72 to 89% with an average value of approximately 80% as determined by comparing peak heights. No trends in recoveries were observed to suggest that the recovery differed between the charged and neutral oligosaccharides. Although a second AMMS would provide additional sodium ion removal, the loss of oligosaccharides approaches 60%. Even with a second AMMS in-line, additional desalting of the fractions is required for MALDI-TOF analysis of the more highly charged, later eluting, fractions (see Section B). Therefore, a second AMMS is not recommended.

B. Desalting of Fractions with Cation-Exchange Resin

The HPAEC fractions containing the tri- and tetrasialylated oligosaccharides elute later in the gradient when the sodium concentration is approximately 300 mM. One AMMS can only remove up to 200–250 mM of sodium

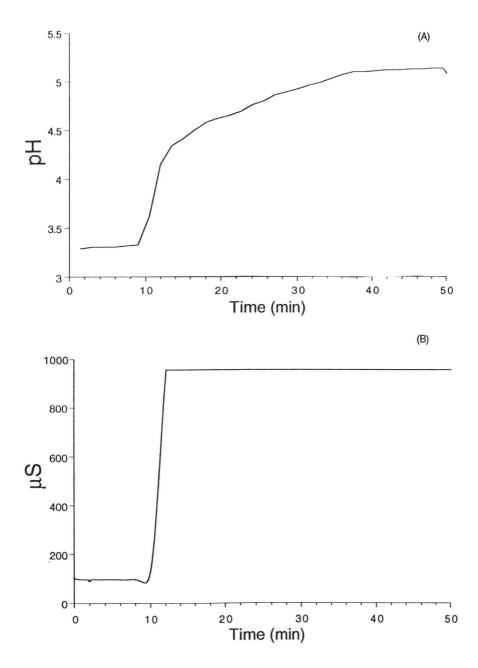

Figure 1 The effect of one in-line anion micromembrane suppressor (AMMS-II) on (A) pH and (B) conductivity. The pH was determined on 1.5-mL fractions that were individually collected, and the conductivity was measured with an in-line conductivity meter.

at a flow rate of 1 mL/min [4]. Unfortunately, the tri- and tetrasialylated oligosaccharides have a greater propensity to form adducts with cations. Therefore, an additional desalting step is necessary to obtain useful spectra for these later eluting fractions.

Desalting the collected fractions with cation-exchange resin packed into 1-mL polypropylene columns eliminated the residual sodium. The percent recovery of sialylated oligosaccharides following desalting with two different cation exchange resins was determined. Figure 2 shows the HPAEC profile obtained for approximately 125 pmol of a disialylated diantennary oligosaccharide first desalted with either X2 (molecular weight exclusion limit, 2700 Da) or X8 (molecular weight exclusion limit, 1000 Da) resin. Figure 2C shows the HPAEC profile of the control sample, which was not desalted on cation-exchange resin. Comparison of the peak areas (16.5 min) revealed a 89% recovery for the sample desalted with X8 resin compared with a 84% recovery when the X2 resin was used.

For all subsequent experiments, 0.5 mL of X8 resin was used to desalt the HPAEC fractions in addition to the on-line desalting achieved with one AMMS.

C. MALDI Analysis of Neutral Fractions

Figure 3 shows the HPAEC profile obtained for the PNGase F released *N*-linked oligosaccharides from 200 μg of tPA. The first three fractions contain the neutral oligosaccharides and are labeled "0-charge."

In positive ion MALDI-TOF, neutral oligosaccharides are observed as the cationized form [10]. Typically, both the $(M + Na)^+$ and $(M + K)^+$ species are observed. The simultaneous presence of both species, which differ by 16 Da, unnecessarily complicates the spectra and reduces sensitivity. Spectra of the neutral oligosaccharides collected following HPAEC separation could be obtained without using cation-exchange resin. However, the cation-exchange resin eliminated potassium ions, which formed potassium adducts with the oligosaccharides. Even after the fractions were desalted with cation-exchange resin, potassium adducts continued to form unless the matrix was fortified with sodium. Figure 4 shows the positive ion spectra obtained for fractions 1, 2, and 3 following desalting on cation-exchange resin. The spectra were acquired using sDHB as the matrix with the reflector on.

The majority of the oligosaccharides observed in the neutral fractions were the high-mannose type (Fig. 4). Chinese hamster ovary–derived recombinant tPA has three *N*-linked glycosylation sites, with one site, Asn-117, containing exclusively high-mannose oligosaccharides [11]. As predicted, the retention time of the high-mannose oligosaccharides increased

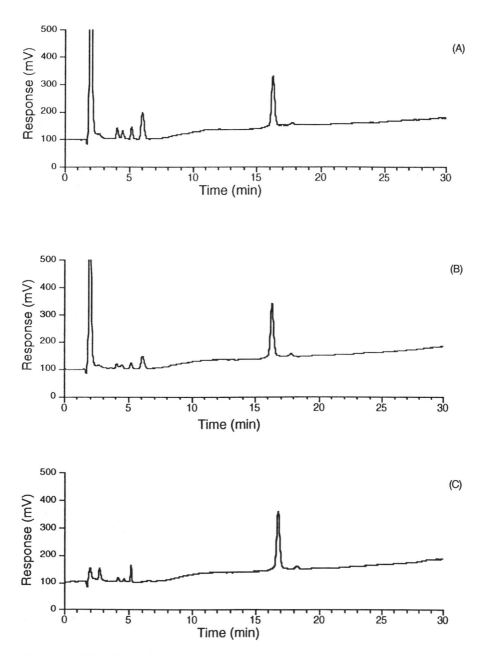

Figure 2 HPAEC profiles of approximately 125-pmol aliquots of fraction 4 reinjected after desalting with either (A) X2 or (B) X8 resin. (C) The same amount of fraction 4 reinjected without prior desalting.

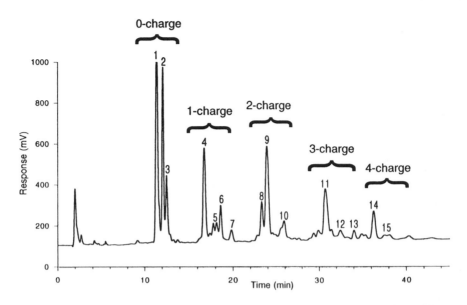

Figure 3   HPAEC profile of a 100-μL injection of PNGase F released oligosaccharides from tPA. The numbers above the peaks correspond to the collected fractions. The charge, which corresponds to the number of sialic acid residues, is also indicated with brackets above the peaks.

as the number of mannose residues increased [12]. Some complex- and hybrid-type oligosaccharides were also observed and were released from the other two sites (Table 1). The molecular weight observed for the neutral oligosaccharides were all within 1 Da of their calculated molecular weight. This is consistent with the anticipated 0.05% mass accuracy anticipated for an externally calibrated instrument operated in the reflector mode. The appearance of the mannose-5 and mannose-6 oligosaccharides in more than one fraction is attributed to carryover during fraction collection and peak broadening that occurs during transit through the AMMS.

D.   Sensitivity of Acidic and Neutral Oligosaccharides

The sensitivity of MALDI-TOF for the detection of acidic oligosaccharides, using sDHB as the matrix, is limited to 1–10 pmol of material applied to the target. Moreover, sDHB can promote loss of sialic acid, which can lead to incorrect assignment of the oligosaccharides found in the HPAEC fractions. In the presence of sodium, sialylated oligosaccharides acquired one sodium for each sialic acid present plus an additional sodium to produce

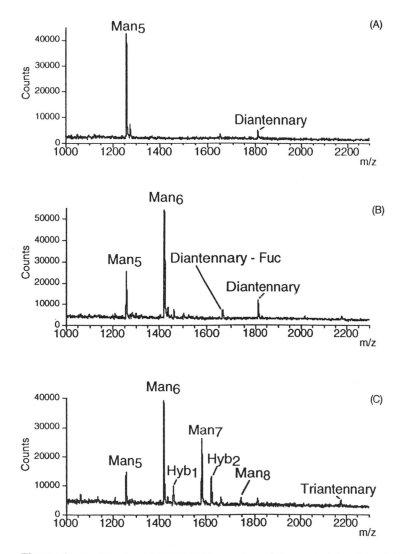

Figure 4 Positive ion MALDI-TOF spectra of the neutral fractions 1 (A), 2 (B), and 3 (C) collected from HPAEC (Fig. 3). The spectra were acquired in the reflector mode using sDHB as the matrix. The m/z values for the peaks are listed in Table 1 with the corresponding structures. The peaks are labeled with the following abbreviations: $Man_{5,6,7,8}$, high-mannose oligosaccharides containing the indicated number of mannose residues; Diantennary-Fuc, fully galactosylated, diantennary, complex oligosaccharide without core fucose; Diantennary, fully galactosylated, diantennary, complex oligosaccharide with core fucose; Hyb_1, agalacto, hybrid oligosaccharide containing five mannose residues; Hyb_2, galactosylated, hybrid oligosaccharide containing five mannose residues; Triantennary, fully galactosylated, triantennary, complex oligosaccharide with core fucose.

Table 1 MALDI-TOF Identification of the Neutral Fractions Collected from HPAEC

Fraction	Observed m/z^a	Calculated m/z^a	Composition[b]
1	1258.1	1258.1	$GlcNAc_2Man_5$
1	1649.5	1648.5	$Gal_1GlcNAc_4Man_3Fuc_1$
1	1810.3	1810.6	$Gal_2GlcNAc_4Man_3Fuc_1$
2	1258.5	1258.1	$GlcNAc_2Man_5$
2	1420.5	1420.2	$GlcNAc_2Man_6$
2	1461.8	1461.3	$GlcNAc_3Man_5$
2	1503.0	1502.4	$Gal_1GlcNAc_4Man_3$
2	1665.3	1664.5	$Gal_2GlcNAc_4Man_3$
2	1811.3	1810.6	$Gal_2GlcNAc_4Man_3Fuc_1$
2	2014.2	2013.8	$Gal_2GlcNAc_5Man_3Fuc_1$
2	2176.5	2176.0	$Gal_3GlcNAc_5Man_3Fuc_1$
3	1258.3	1258.1	$GlcNAc_2Man_5$
3	1420.5	1420.2	$GlcNAc_2Man_6$
3	1461.6	1461.3	$GlcNAc_3Man_5$
3	1582.8	1582.4	$GlcNAc_2Man_7$
3	1624.0	1623.4	$Gal_1GlcNAc_3Man_5$
3	1664.9	1664.5	$Gal_2GlcNAc_4Man_3$
3	1745.6	1744.5	$GlcNAc_2Man_8$
3	1811.6	1810.6	$Gal_2GlcNAc_4Man_3Fuc_1$
3	2176.5	2176.0	$Gal_3GlcNAc_5Man_3Fuc_1$

[a] The m/z values are for the $(M + Na)^+$ ions.
[b] The abbreviations for the monosaccharides are Fuc, fucose; Gal, galactose; GlcNAc, N-acetylglucosamine; Man, mannose.

the positive charge (Fig. 5A). Ions containing only one or two sodium ions were also formed and were observed at lower mass-to-charge ratios (Fig. 5A).

Figure 5 demonstrates that an asialo-diantennary oligosaccharide that contains core fucose has increased sensitivity compared to the corresponding disialylated oligosaccharide. Both positive ion spectra were acquired in the linear mode using sDHB as the matrix. Although the signal-to-noise ratio is comparable in the two spectra, 2.5 pmol of the disialylated oligosaccharide (Fig. 5A) was required to obtain the same signal-to-noise ratio as 80 fmol of the asialo species (Fig. 5B). Omitting the sodium from the matrix and operating the instrument in the negative ion mode offered no significant improvement in sensitivity for the sialylated oligosaccharides (data not shown).

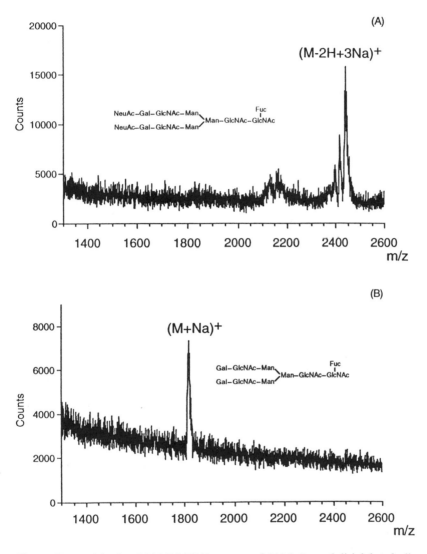

Figure 5 Positive ion MALDI-TOF spectra of (A) 2.5-pmol disialylated, diantennary oligosaccharide with core fucose and (B) 80-fmol asialo, diantennary oligosaccharide with core fucose. Both spectra were acquired in the linear mode using sDHB as the matrix.

E. Desialylation Prior to MALDI Analysis

As illustrated in Fig. 5, neutral oligosaccharides were better detected in sDHB than the sialylated oligosaccharides. One way to overcome this lack of sensitivity was to desialylate the oligosaccharides prior to MALDI-TOF analysis, thereby producing the corresponding neutral species. A caveat to this approach is that one must assume the number of sialic residues present in each fraction. Increased negative charge introduced from sialic acid is known to increase retention time [12]. Therefore, the number of sialic acid residues present in a given fraction can be predicted with reasonable confidence (Fig. 3).

Desialylation of the collected fractions at 80°C for 3 hr in 10% acetic acid effectively removes sialic acid. Figure 6 demonstrates a problem encountered during acid-catalyzed desialylation. Desialylation with acid prior to desalting on cation-exchange resin promoted acetylation of the oligosaccharides. Figure 6 demonstrates two examples of acid-catalyzed acetylation of oligosaccharides from fraction 9 (Fig. 6A) and fraction 11 (Fig. 6B). Acetylation was demonstrated by the appearance of peaks increased in mass by 42 Da. Acetylation can lead to ambiguity in structural assignment, and it diminishes the signal-to-noise ratio by spreading the signal out over multiple peaks. Acetylation was also found to occur if either dilute trifluoroacetic acid or dilute hydrochloric acid were used (data not shown). Apparently, the residual sodium acetate from the AMMS was responsible for the acetylation.

Two methods were employed to eliminate acetylation. The first method simply required that the sample be desalted on cation-exchange resin prior to acid-catalyzed desialyation. Using this approach, the positive ion MALDI-TOF spectrum of fraction 9 yielded a $(M + Na)^+$ with the expected mass-to-charge ratio of 1810.6 (Fig. 7A). Based on the elution time, two sialic acid residues were present on the oligosaccharide in fraction 9 (Fig. 3). Therefore, the structure of the oligosaccharide was a disialylated oligosaccharide with the base structure shown (Fig. 7A).

The sialylated oligosaccharides are recovered in an effluent that contains 50–100 mM sodium acetate at pH ~5 (Fig. 1). The manufacturer's recommended buffer for *C. perfringens* sialidase is 50 mM sodium acetate pH5. Therefore, enzymatic desialylation with sialidase can be performed without addition of buffers or adjustment of pH. A MALDI-TOF spectrum of fraction 4 following sialidase treatment is shown in Fig. 7B. The expected $(M + Na)^+$ ion was observed at 1810.6 m/z. Based on the elution time, fraction 4 contained a monosialylated oligosaccharide with the base structure indicated (Fig. 7B). Both methods of desialylation eliminated the problem of acetylation; however, the use of 1 U/mL sialidase required an overnight incubation.

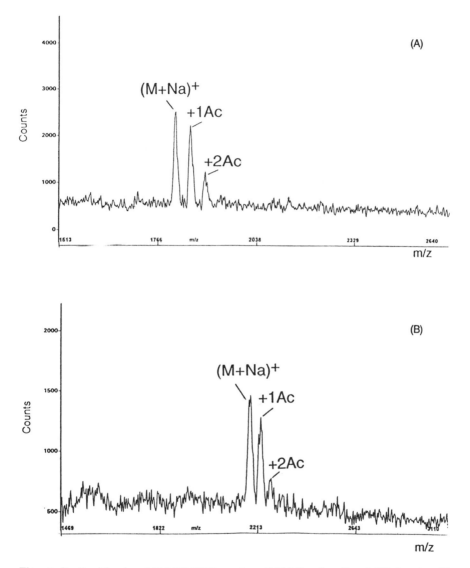

Figure 6 Positive ion MALDI-TOF spectra of (A) fraction 9 and (B) fraction 11 that have been desialylated with 10% acetic acid. The spectra were acquired in the linear mode using sDHB as the matrix. The "Ac" indicates peaks that have incorporated the indicated number of acetyl groups.

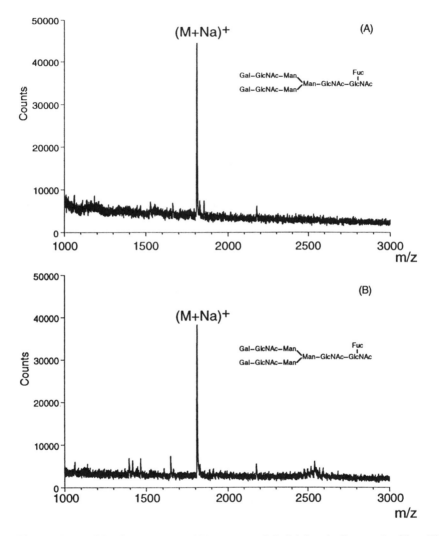

Figure 7 Positive ion MALDI-TOF spectra of desialylated oligosaccharides. The spectra were acquired in the reflector mode using sDHB as the matrix. (A) Fraction 9 was desialylated chemically with 10% acetic acid after first desalting with cation-exchange resin. (B) Fraction 4 was desialylated enzymatically with *C. perfringens* recombinant sialidase.

F. MALDI Analysis of Acidic Fractions

Desialylation prior to MALDI analysis is not an ideal way to analyze HPAEC fractions. The major drawback to this approach is that the number of sialic acid residues must be assumed for oligosaccharides eluting in a particular fraction and may be erroneous. Sulfation and phosphorylation can each introduce a negative charge and thus affect the retention time of the oligosaccharide [12]. The most straightforward approach is to analyze directly the collected fractions without modifying the oligosaccharides.

We have found a matrix that, in contrast to sDHB, provides low femtomole sensitivity for acidic oligosaccharides [13]. $2',4',6'$-Trihydroxyacetophenone (THAP) containing ammonium citrate allows acquisition of negative ion MALDI spectra for 10–25 fmol of sialylated oligosaccharides with negligible loss of sialic acid [13]. The negative ion MALDI-TOF spectra obtained for the three largest acidic fractions separated by HPAEC are shown in Fig. 8. The observed ions are the $(M - H)^-$ species. Clearly, little evidence indicating the loss of sialic acid is observed.

A standard disialylated N-linked oligosaccharide was used to determine the current limit of detection for the analysis of HPAEC fractions using the approach outlined in the next paragraph. Figure 9A shows the HPAEC profile for 50 pmol of a disialylated, diantennary oligosaccharide, with elutes at 24 min. Figure 9B shows the negative ion MALDI-TOF spectrum for the collected fraction using THAP as the matrix. Approximately 500 fmol of sample was applied to the target assuming no losses occurred during sample manipulation. The amount of material applied is less because the recovery following one AMMS is 80%. Although the matrix offers sensitivity in the low femtomole range, this method as described is currently limited to injecting 20 pmol or more of an oligosaccharide onto the HPAEC.

The method currently used for analyzing fractions collected from HPAEC separation is outlined in the following:

1. Desalt HPAEC fractions with one in-line AMMS-II.
2. Desalt 300–500 μL of the collected fraction with 0.5 mL of AG50W-X8 cation-exchange resin.
3. Vacuum dry samples and reconstitute in deionized water to an approximate oligosaccharide concentration of 50–500 fmol/μL.
4. For neutral oligosaccharides, analyze 0.5–1.0 μL of sample with 0.5 μL of sDHB by MALDI-TOF in the positive ion mode.
5. For acidic oligosaccharides, analyze 0.5–1.0 μL of sample mixed with 0.75 μL of THAP by MALDI-TOF in the negative ion mode.

Using the foregoing approach, all the fractions indicated in Fig. 3 were identified except for fraction 15 (Table 2). The mass accuracy for

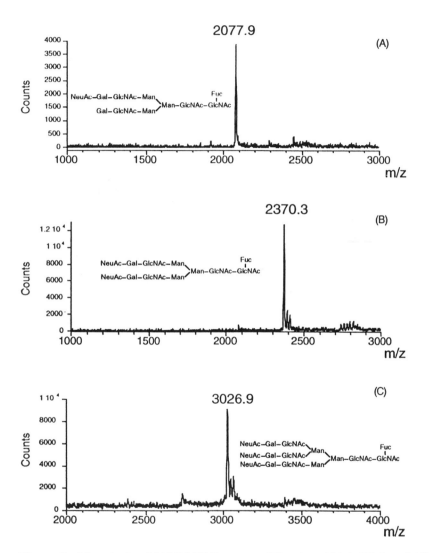

Figure 8 Negative ion MALDI-TOF spectra of fractions (A) 4, (B) 9, and (C) 11. The spectra were acquired in the linear mode using THAP as the matrix.

samples acquired in the linear mode using an external calibration is typically 0.1–0.05%. Except for the weakest signals, this type of mass accuracy is consistent with the m/z values observed. Although better mass accuracy (0.05–0.01%) can be obtained for the sialylated oligosaccharides by operating the instrument in the reflector mode, significant loss (>40%) of sialic

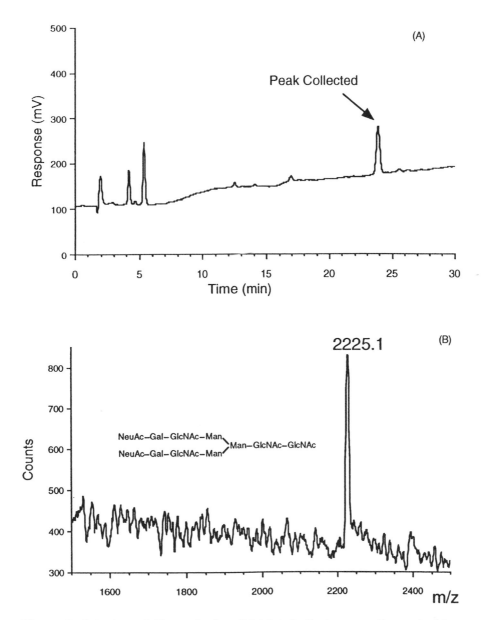

Figure 9 Injection of 50 pmol of a disialylated, diantennary oligosaccharide yielded the corresponding signal in HPAEC (A). Following desalting with both the AMMS-II and X8 resin, the oligosaccharide was detected by MALDI-TOF using THAP as the matrix. The spectrum (B) was acquired in the negative ion mode using the linear detector.

Table 2 MALDI-TOF Identification of the Charged Fractions Collected from HPAEC

Fraction	Observed m/z^a	Calculated m/z^a	Composition[b]
4	2077.9	2077.9	$NeuAc_1Gal_2GlcNAc_4Man_3Fuc_1$
4	2444.2	2443.7	$NeuAc_1Gal_3GlcNAc_5Man_3Fuc_1$
5	1729.1	1728.6	$NeuAc_1Gal_1GlcNAc_3Man_4$
5	1932.5	1931.8	$NeuAc_1Gal_2GlcNAc_4Man_3$
5	2078.6	2077.9	$NeuAc_1Gal_2GlcNAc_4Man_3Fuc_1$
6	1892.2	1890.7	$NeuAc_1Gal_1GlcNAc_3Man_5$
6	2079.7	2077.9	$NeuAc_1Gal_2GlcNAc_4Man_3Fuc_1$
7	1892.7	1890.7	$NeuAc_1Gal_1GlcNAc_3Man_5$
7	2518.1	2515.3	$NeuAc_2Gal_2GlcNAc_4Man_3Fuc_2$
8	2370.2	2369.2	$NeuAc_2Gal_2GlcNAc_4Man_3Fuc_1$
8	2737.4	2734.5	$NeuAc_2Gal_3GlcNAc_5Man_3Fuc_1$
8	3104.4	3099.8	$NeuAc_2Gal_4GlcNAc_6Man_3Fuc_1$
9	2370.3	2369.2	$NeuAc_2Gal_2GlcNAc_4Man_3Fuc_1$
10	2225.5	2223.0	$NeuAc_2Gal_2GlcNAc_4Man_3$
10	2372.2	2369.2	$NeuAc_2Gal_2GlcNAc_4Man_3Fuc_1$
11	3026.9	3025.8	$NeuAc_3Gal_3GlcNAc_5Man_3Fuc_1$
11	3392.0	3391.1	$NeuAc_3Gal_4GlcNAc_6Man_3Fuc_1$
12	3028.4	3025.8	$NeuAc_3Gal_3GlcNAc_5Man_3Fuc_1$
13	3028.4	3025.8	$NeuAc_3Gal_3GlcNAc_5Man_3Fuc_1$
13	4419.4	4414.0	$NeuAc_4Gal_6GlcNAc_8Man_3Fuc_1$
14	3680.2	3682.3	$NeuAc_4Gal_4GlcNAc_6Man_3Fuc_1$
15	N.O.[c]	–	–

[a] The m/z values listed are for $(M - H)^-$ ions.
[b] The abbreviations for the monosaccharides are Fuc, fucose; Gal, galactose; GlcNAc, N-acetylglucosamine; Man, mannose; NeuAc, N-acetylneuraminic acid.
[c] N.O. indicates that no peaks were detected.

acid occurs. This improved mass accuracy can be valuable for discerning compositions that differ by only 1 Da at the 4–5 kDa mass range.

The results from Table 2 are consistent with the anticipated elution order for the identified oligosaccharides. The oligosaccharides containing one sialic acid group elute before those containing two sialic acid residues, which elute before oligosaccharides containing three sialic acid groups. Many ions with the same m/z are observed in different acidic fractions. The separation time between these fractions suggests that these ions represent positional isomers with the same mass and do not result from sample carryover. The 2,4-branched isomer of the trisialylated, triantennary oligosaccharide is known to elute before the 2,6-branched isomer [3]. Fraction 11 is probably the 2,4-branched isomer, based on retention time and m/z. Moreover, because the 2,4-branched isomer is the major form of trisialylated, triantennary oligosaccharide found in tPA [11], the relative proportion of the material is consistent with this assignment.

If we had relied on desialylation to identify the fractions, the major oligosaccharide in fraction 7 would have been incorrectly identified. Peak 7 elutes close to other monosialylated species and would be presumed to contain only one sialic acid. Based on molecular weight and acid-catalyzed desialylation, fraction 7 contains a disialylated oligosaccharide with two fucose residues. Apparently, the additional fucose group causes this oligosaccharide to elute much earlier than expected based on the number of sialic acids present. Indeed, fucose attached to the c-6 position of the reducing end GlcNAc causes oligosaccharides to elute approximately 2 min earlier than the nonfucosylated counterpart under similar HPAEC conditions [3]. Fraction 7 eluted 4 min earlier than the corresponding monofucosylated species, indicating the position of this second fucose has a more dramatic effect on retention time.

IV. CONCLUSIONS

Using the approach we have described, MALDI-TOF can be used to obtain the molecular weight of oligosaccharides separated by HPAEC. Unfortunately, molecular weight information alone is insufficient to discern the positional isomers and anomeric forms of oligosaccharides. However, using a knowledge-based approach, reasonable structures can be proposed. The structures listed in Tables 1 and 2 are based on the molecular weights obtained from MALDI-TOF and the elution order from HPAEC, and they are therefore implied from the known structures of oligosaccharides found on Chinese hamster ovary cell-derived recombinant tPA [11]. Additional structural information can be obtained for oligosaccharides by using this technique in conjunction with specific exoglycosidases. MALDI-TOF in combination with specific exoglycosidases has been used successfully to obtain structural details regarding the linkage position and anomery of the monosaccharides in oligosaccharides [14].

ABBREVIATIONS

AMMS	anion micromembrane suppressor
ESI	electrospray ionization
FAB	fast-atom bombardment
HPAEC	high pH anion-exchange chromatography
MALDI	matrix-assisted laser desorption ionization

sDHB	2,5-dihydrobenzoic acid matrix
THAP	2′,4′,6′-trihydroxyacetophenone matrix
TOF	time-of-flight
tPA	tissue plasminogen activator

REFERENCES

1. Hardy, M. R., and Townsend, R. R. (1988). *Proc. Natl. Acad. Sci. USA* 85, 3289–3293.
2. Townsend, R. R. (1995). In Rassi, E. L., ed., *Carbohydrate Analysis: High Performance Liquid Chromatography and Capillary Electrophoresis.* Elsevier, Amsterdam, pp. 181–210.
3. Basa, L. J., and Spellman, M. W. (1990). *J. Chromatogr.* 499, 205–220.
4. Barr, J. R., Anumula, K. R., Vettese, M. B., Taylor, P. B., and Carr, S. A. (1991). *Anal. Biochem.* 192, 181–192.
5. Conboy, J. J., and Henion, J. (1992). *Biol. Mass Spectrom.* 21, 397–407.
6. Tetaert, D., Soudan, B., LoGuidice, J.-M., Richet, C., Degand, P., Boussard, G., Mariller, C., and Spik, G. (1994). *J. Chromatogr.* 658, 31–38.
7. Huberty, M. C., Vath, J. E., Yu, W., and Martin, S. A. (1993). *Anal. Chem.* 65, 2791–2800.
8. Mock, K. K., Davey, M., and Cottrell, J. S. (1991). *Biochem. Biophys. Res. Commun.* 177, 644–651.
9. Hardy, M. R., and Townsend, R. R. (1994). In Lennarz, W. J., and Hart, G. W., eds., *Methods in Enzymology.* Academic Press, San Diego, vol. 230, pp. 208–225.
10. Stahl, B., Steup, M., Karas, M., and Hillenkamp, F. (1991). *Anal. Chem.* 63, 1463–1466.
11. Spellman, M. W., Basa, L. J., Leonard, C. K., Chakel, J. A., O'Connor, J. V., Wilson, S., and van Halbeek, H. (1989). *J. Biol. Chem.* 264, 14100–14111.
12. Rohrer, J. S. (1995). *Glycobiology* 5, 359–363.
13. Papac, D. I., Wong, A., and Jones, A. J. S. (1996). *Anal. Chem.*, 68, 3215–3223.
14. Sutton, C. W., O'Neill, J. A., and Cottrell, J. S. (1994). *Anal. Biochem.* 218, 34–46.

4

The Fragmentation Behavior of Glycopeptides Using the Post-Source Decay Technique and Matrix-Assisted Laser Desorption Ionization Mass Spectrometry

Uwe Rapp, Anja Resemann, and Franz J. Mayer-Posner
Bruker-Franzen Analytik GmbH, Bremen, Germany

Wolfram Schäfer and Konrad Feichtinger
Max-Planck-Institut für Biochemie, Martinsried, Germany

I. INTRODUCTION

Time-of-flight (TOF) mass spectrometers, coupled to matrix-assisted laser desorption ion sources (MALDI), have in the last few years been used mainly for the determination of molecular weights [1,2]. Hereby, many aspects of biomolecular analysis have been revolutionized, expecially speed, accuracy, and sensitivity of molecular weight determinations. Previously, the different mass spectrometric techniques have been most useful for relating fragmentations, occurring in the mass spectrometer, to molecular structure. Because a significant portion of molecules are fragmented by laser light, investigations by Kaufmann [3,4] have shown that post-source decay (PSD) processes can be detected in a reflectron instrument. The TOF instrument must be equipped with a reflector, preferably of the gridless double-stage design. With this instrument configuration, structure-related fragment ions can be measured. The first experiments have demonstrated the usefulness of PSD for sequencing peptides, one of the main application areas [5,6]. PSD spectra have also been published on linear and branched glycans [7,8] and on *N*-linked glycopeptides [9].

II. MATERIALS AND METHODS

A. Instrument Parameters

All MALDI/TOF measurements were performed on a Bruker REFLEX mass spectrometer (Bremen, Germany). The acceleration voltage was set to 28.5 kV with a reflector voltage of 30.0 kV. For external calibration angiotensin II, ACTH clip 18–39, or bovine insulin were used depending on the mass range. The individual segments of the PSD measurement were calibrated using ACTH clip 18–39. A nitrogen laser (337 nm) with a repetition rate of 3 Hz was used; attenuation was adjusted by a computer-controlled continuously variable density filter. The beam flux was usually near the threshold of ion production. For PSD data, a slightly higher flux was required. The detector signal was digitized by a transient recorder at a 500 MHz sampling rate, and instrument control and data processing were accomplished with the Bruker XMASS software package using a SUN Sparc-Station 5 under UNIX as the operating system.

PSD measurements were performed using the Bruker FAST hardware/software package. Under computer control, the reflector voltages were stepped down according to preselected values, and spectra of individual voltage segments were recorded and stored. Typically, data of 10 to 14 segments were acquired, the segments calibrated and combined into one PSD spectrum. For sequencing of individual components in mixtures a precision precursor ion selector was used with resolution > 100 (MS/MS technique).

B.   Sample Preparation Techniques

In this work 2,5-dihydroxybenzoic acid (DHB) and α-cyano-4-hydroxy-cinnamic acid (4-HCCA) were used as matrices. Both were recrystallized. DHB was dissolved to ~ 10 mg/mL in 30% methanol/water, and 4-HCCA was dissolved in 30% acetonitrile/0.1% trifluoroacetic acid (TFA)/water. For the thin-layer preparation, saturated solutions of 4-HCCA in acetone were applied.

The samples were prepared on a multisample carrying target as follows:

1. A standard preparation was made by mixing a saturated matrix solution and the sample in an Eppendorf vial and transferring an aliqout (0.5–1 μL) onto the target. The sample amounts for routine operation were about 1–10 pmol.
2. The thin-layer preparation as introduced by Mann [10] was used for obtaining quality spectra from small amounts (below 1 pmol level of material on the target). If a sample contained salts or other contaminants, the target was rinsed using water or 0.1% TFA/water. The removal of the contaminants was in most cases successful. In our experience, however, this technique is not optimal for PSD measurements, because the small sample amounts result in depletion of a spot after a few laser shots. So, to record complete PSD spectra, the sample spot

had to be changed. PSD spectra with acceptable signal-to-noise ratios were obtained from sample amounts below 100 fmol.

3. To overcome the obstacle we have described, a double-layer technique was developed that essentially combines the two previously described methods (M. Mann, personal communication). A thin-layer surface of matrix was prepared, and then a small amount of the standard preparation mix was added onto the surface. The percentage of acetonitrile in the solute was important. If it was higher than 30%, in our experience, the matrix surface was completely dissolved and a hole was evident. We, therefore, applied a solution containing 20–25% acetonitrile to the thin layer. A visual inspection of the target using a microscope or in the ion source by visualization optics is extremely helpful, and the PSD analysis is simplified.

C. Glycopeptide Samples

The glycopeptide samples were dried high performance liquid chromatography (HPLC) fractions that were redissolved in water or 0.1% TFA/water. The solutions applied to the target contained 1–10 pmol. Samples using DHB as a matrix were recrystallized on the target with 0.1 μL of ethanol, resulting in a larger spot and a smoother distribution.

III. RESULTS AND DISCUSSION

The glycopeptides are of biological relevance because the glycosylation of proteins is a major post-translational modification and is site dependent. Dipeptidylpeptidase IV (DPP IV 3.4.14.5) [11,12] is a serine protease of 230–280 kDa with two identical subunits, each containing eight N-linked glycosylation sites. Because it was found that DPP IV is identical to CD 26, which plays an important role in the activation of T cells [13], characterization of its individual glycosylation sites and determining the structure of the carbohydrate side chains has become important. In addition, it was postulated that DPP IV is a co-factor necessary for the infection of T cells by HIV I virus [14]. The characterization of the glycoprotein prior to the MALDI mass spectrometry (MS) studies is as follows. Using the isolated protein, monosaccharide analysis by high pH anion-exchange chromatography with pulsed amperometric detection (HPAEC-PAD) was performed. The results were those expected for N-linked glycans of the complex type. The absence of GalN excluded O-glycosylation. The oligosaccharides were released after denaturing the protein with 1 N HCl, neutralizing using 2 N NaOH, and treating with chymotrypsin. The peptides were removed by reversed phase HPLC and the oligosaccharides were desalted using size exclusion chromatography with Sephadex G_{10}. The oligosaccharides were analyzed using HPAEC-PAD. Compared with other proteins such as fetuin, transferrin, and asialofetuin, which were treated according to the same scheme, the chromatogram of DPP IV contained primarily heterogeneous neutral glycans and a small portion of monosialylated oligosaccharides. The chymotryptic digest was problematic because the protein aggregated and precipitated if denaturation conditions were not optimal. However, sufficient material could be isolated from porcine kidney. The characterization of the glycopeptides was achieved after chymotryptic digestion and size exclusion chromatography using a Superose column, equilibrated in 50 mM ammonium acetate and 30% acetonitrile/water pH 5.0. Glycopeptide isolation was performed by reversed phase HPLC (Nucleosil C_{18}) using a gradient of 0.06% TFA in water. Aliquots of peak fractions were treated as follows: (1) deglycosylated using PNGase F and the resulting peptide sequenced to determine the glycosylated site; (2) analyzed for monosaccharide composition; (3) methylated and analyzed by gas chromatography (GC) /MS; and (4) analyzed by fast-atom bombardment MS, if sufficient material was available. MALDI/MS analysis was also performed, and the branching of the glycans was determined prior to MALDI/MS.

Figure 1 shows the glycosylation pattern of Asn-321, where the peptide is a pentapeptide, RAQNY, with an average peptide molecular weight (PMW) of 650.7 Da. The matrix was 4-HCAA. The compositions of the glycans are indicated. The observed masses were consistent with oligomannosidic- and biantennary-type structures. Two species from this mixture were selected for PSD measurements: m/z 1707 (Fig. 2) using 4-HCCA as a matrix, and m/z 2234 (Fig. 3) using DHB as a matrix. The masses were chosen to demonstrate the dynamic range of PSD data. The signal-to-noise ratio of the m/z 1707 spectrum was near the detection limits for some of the glycan fragments. However, it demonstrated that even small amounts of a mixture component can be analyzed by PSD. The PSD spectrum of m/z 2234 had a better signal-to-noise ratio because it represented the main component.

Both PSD spectra displayed abundant signal at m/z PMW + 83 Da. Furthermore, PMW + HexNAc and signals corresponding to PMW + 2 HexNAc residues were apparent. Scheme 1 summarizes the structures of fragments in the glycan core region. The PMW + 83 Da was attributed to a ring cleavage of the inner core HexNAc residue between the ring oxygen and C1 and C2–C3, an $^{0,2}X$ fragment. (The nomenclature of Domon and Costello [15] was used.) All other fragmentations were single bond cleavages, for example, between the pyranose ring and the glycosidic oxygen (Y fragments). In all the glycopeptides we have investigated, no B-type ions

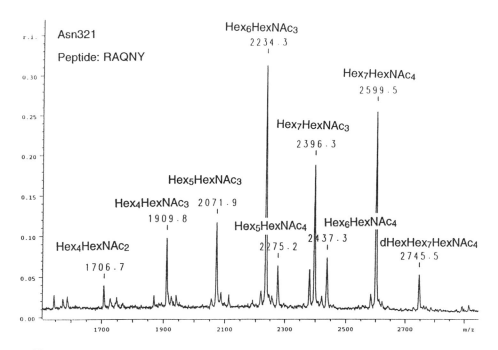

Figure 1 MALDI/mass spectrometry spectrum of glycopeptides containing Asn-321. The carbohydrate composition is annotated at the individual masses. The calculated average PMW was 650.7. The matrix was 4-HCCA.

were detected. The $[M + H]^+$ fragmentations always occurred with charge retention on the peptide portion.

The peptide fragments were accompanied by an elimination of 17 Da (Figs. 2 and 3), which is known for Arg-containing peptides, and was attributed to a loss of ammonia. In the two PSD spectra, insignificant peptide chain fragmentation occurred (Figs. 2 and 3). For example, only the m/z 634.2 and 470.7 signals are due to peptide chain fragmentation, B_5 and B_4, according to Roepstorff nomenclature [16]. The difference was 163.5 Da, which is the mass of a Tyr residue.

Figure 4 displays the MALDI/MS spectrum of the Asn-685-containing glycopeptide using 4-HCCA as a matrix. The molecular weight distribution, representing oligosaccharide heterogeneity, is not as complex as was observed for the Asn-321 glycopeptide. The PSD spectrum of the main component (m/z 2801) in Fig. 5 shows another interesting aspect of fragmentation analysis. The glycopeptide consisted of a hexapeptide,

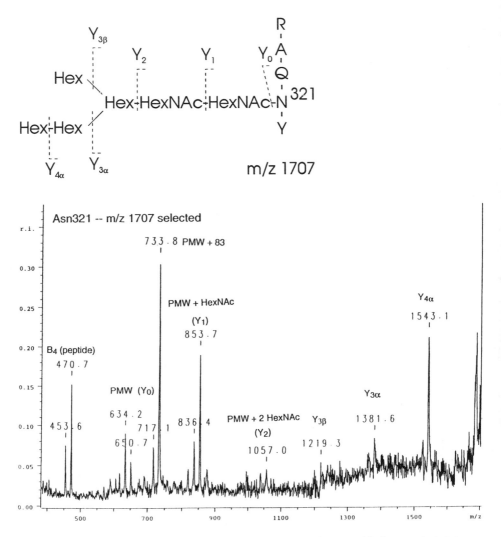

Figure 2 MALDI/PSD mass spectrum of Asn-321 glycopeptide ion (*m/z* 1707).
The fragments are labeled according to the scheme above the spectrum. The matrix
was 4-HCCA.

RNSTVM (average PMW = 706.8 Da), and a fucosylated glycan (Fig. 5).
As expected, the fragments corresponding to a PMW + 203 Da and PMW
+ 406 Da of the chitobiose core were shifted by 146 Da to PMW + 349
Da and PMW + 552 Da. There was also a loss of the Fuc residue from the
signal assigned to PMW + deoxyHex–HexNAc. Interestingly, in the high

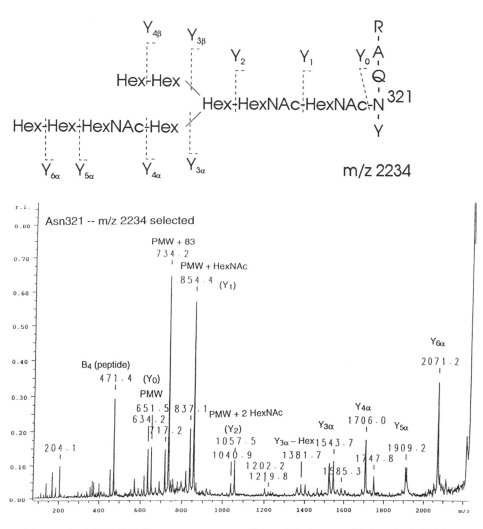

Figure 3 MALDI/PSD mass spectrum of Asn-321 glycopeptide ion (m/z 2234). The fragments are labeled according to the scheme above the spectrum. The matrix was DHB.

mass range of the PSD spectrum of fucosylated glycopeptides and oligosaccharides, groups of doublets with mass differences of 16 Da were observed, indicating a concomitant loss of Hex and Fuc residues.

The third glycosylation site investigated was Asn-150. In this case a dodecapeptide, ITEERIPNNTQW (average PMW = 1500.6 Da), consti-

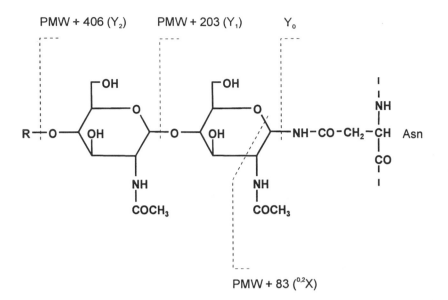

Scheme 1 Fragmentation pattern of the core region in the glycopeptides investigated.

tuted a greater portion of the total molecular weight than the two previously discussed glycopeptides. The matrix was 4-HCCA. The heterogeneity was marked, as Fig. 6 demonstrates. However, by gating it was possible to select the signal at m/z 3960 as the precursor ion for the PSD spectrum (Fig. 7). Because the resolution of the precursor ion selector (gate) is limited to ~ 100, the $[M + H]^+$ peak could not be separated and isolated from the $[M + Na]^+$ species [17]. Therefore, it was assumed that the included $[M + Na]^+$ peak did not influence the fragmentation pattern of the $[M + H]^+$ parent ion. Overall, fragmentation followed the rules described in the previous examples. The expected ions at m/z PMW + 83 Da, + the mass of 1 and 2 HexNAc residues were observed (Fig. 7). Additionally, Y-type fragments were also observed. There is, however, a lack of signals in the range m/z 2100–2900. This may be due to the lower fragmentation efficiencies of larger molecular weight molecules. Further investigations are presently being undertaken.

IV. SUMMARY AND CONCLUSIONS

The fragmentation data of different sized glycopeptides showed the following: (1) The peptide molecular weight can be determined from signals corresponding to the PMW, the PMW + 83 Da, PMW + 83 Da + 203 Da

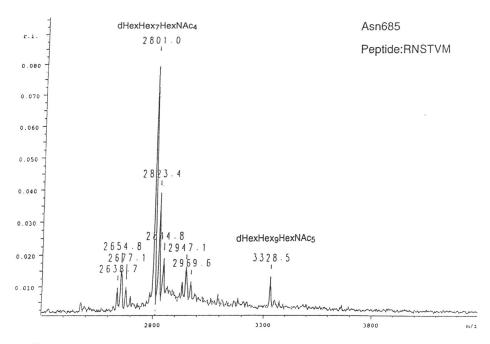

Figure 4 MALDI/MS spectrum of glycopeptides containing Asn-685. The carbohydrate composition is annotated at the individual masses. The calculated average PMW was 706.8. The matrix was 4-HCCA.

(mass of a HexNAc residue), and PMW + 83 Da and 406 Da. (2) In many cases, fragments were accompanied by an elimination of 17 Da, previously described for Arg-containing peptides. The glycopeptides in this study all contained Arg residues. (3) Fucosylation of the chitobiose core gave mass shifts of 146 Da and additional fragments (PMW + 349 Da, deoxyHex–HexNAc and PMW + 552 Da , deoxyHex–HexNAc$_2$. (4) Fucosylated glycopeptides showed a series of paired signals separated by 16 Da from the concurrent loss of a Hex and a Fuc residue. (5) The glycan composition, when available from other analyses, was confirmed by Y-type cleavages. (6) Significant peptide sequence information could not be obtained. (7) Fragments with charge retention on the oligosaccharide chains were not detected. Detailed studies to determine different branching structure were not performed because additional sample would be required. PSD analysis of glycopeptides with masses >4000 Da was not successful. In conclusion, the PSD spectra of glycopeptides showed characteristic fragmentations, which were used to deduce structural features. MALDI/MS with PSD anal-

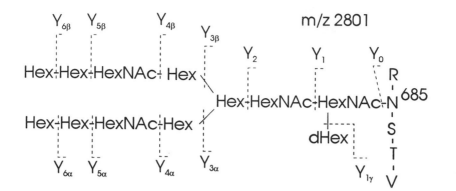

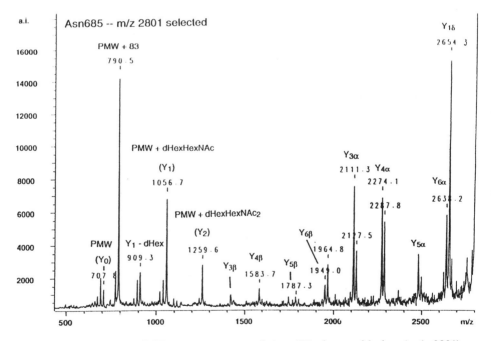

Figure 5 MALDI/PSD mass spectrum of Asn-685 glyopeptide ion (*m/z* 2801). The fragments are labeled according to the scheme above the spectrum. The matrix was 4-HCCA.

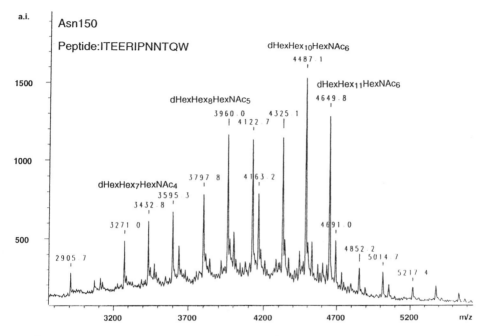

Figure 6 MALDI/MS spectrum of glycopeptides containing Asn-150. The carbohydrate composition is annotated at the individual masses. The calculated average PMW was 1500.6. The matrix was 4-HCCA.

ysis is an extremely valuable tool, particularly when only small amounts of sample are available.

ABBREVIATIONS

DHB	2,5-dihydroxybenzoic acid
DPP IV	dipeptidylpeptidase IV
GC	gas chromatography
4-HCCA	α-cyano-4-hydroxycinnamic acid
HPAEC	high pH anion-exchange chromatography
HPLC	high performance liquid chromatography
MALDI	matrix-assisted laser desorption ionization
MS	mass spectrometry
PAD	pulsed amperometric detection
PMW	peptide molecular weight
PSD	post-source decay
TFA	trifluoroacetic acid
TOF	time-of-flight

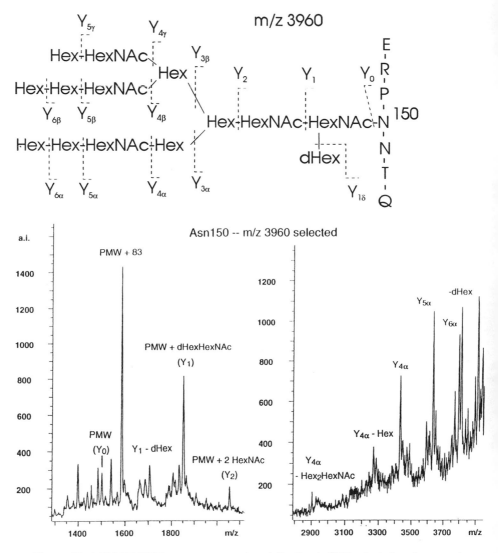

Figure 7 MALDI/PSD mass spectrum (partial) of *m/z* 3960 selected as the parent ion of Asn150. The fragments are labeled according to the scheme above the spectrum. The matrix was 4-HCCA.

REFERENCES

1. Chait, B. T., and Kent, S. B. H. (1992). *Science* 257, 1885–1894.
2. Wang, R., and Chait, B. T. (1994). *Curr. Opin. Biotech.* 5, 77–84.
3. Kaufmann, R. (1995). *J. Biotechnol.* 41, 155–175.
4. Kaufmann, R., Kirsch, D., and Spengler, B. (1994). *Int. J. Mass Spectrom. Ion. Proc.* 131, 355–385.
5. Rapp, U., Mayer-Posner, F. J., Talbo, G., and Mann, M. (1995). *Peptides 1994.* ESCOM, Leiden, pp. 399–400.
6. Kellner, R., Talbo, G., Houthaeve, T., and Mann, M. (1995). Hugli, T., ed., *Techniques in Protein Chemistry VI.* Academic Press, New York, pp. 47–54.
7. Harvey, D. J., Naven, T. J. P., Küster, B., Bateman, R. H., Green, M. R., and Critchley, G. (1995). *Rap. Commun. Mass Spectrom.* 9, 1556–1561.
8. Rouse, J. C., Vath, J. E., and Scoble, H. A. (1995). *Proceedings of the 43rd Annual Conference on Mass Spectrometry and Allied Topics.* Atlanta, p. 359.
9. Huberty, M. C., Vath, J. E., Yu, W., and Martin, S. A. (1993). *Anal. Chem.* 65, 2791–2800.
10. Vorm, E., Roepstorff, P., and Mann, M. (1994). *Anal. Chem.* 66, 3281–3287.
11. Seidl, R., Mann, K., and Schäfer, W. (1991). *Biol. Chem. Hoppe Seyler* 372, 213–214.
12. Seidl, R., and Schäfer, W. (1991). *Prep. Biochem.* 21, 141–150.
13. Bristol, L. A. (1992). *J. Immunol.* 149, 367–372.
14. Callebaut, C., Krust, B., Jacotot, E., and Hovanessian, A. G. (1993). *Science* 262, 24428–24432.
15. Domon, B., and Costello, C. E. (1988). *Glycoconj.* 5, 397–409.
16. Roepstorff, P. (1984). *Biomed. Mass Spectrom.* 11, 601–603.
17. Mayer-Posner, F. J., and Rapp, U., unpublished results.

5

Application of Electrospray Ionization Mass Spectrometry to Unusual Phosphorylated Oligosaccharides of *Trypanosoma cruzi*

Paul A. Haynes and George A. M. Cross
The Rockefeller University, New York, New York

Michael A. J. Ferguson
University of Dundee, Dundee, Scotland

I. INTRODUCTION

Trypanosoma cruzi is the etiological agent of Chagas disease, which is endemic in much of Central and South America. The parasite undergoes distinct biochemical and morphological changes during a complex life cycle involving reduviid bugs and mammals. The biochemical changes that occur during differentiation between stages are important factors in establishing infection of insect vectors, invading mammalian cells, and evading the host immune system. A glycoprotein of molecular weight 72 kDa, GP 72, was first identified on the surface of insect stage epimastigote cells using a carbohydrate-specific monoclonal antibody, WIC 29.26 [1]. This was subsequently found to be present on the surface of all *T. cruzi* strains examined, despite variations in posttranslational modification that caused some masking of glycan epitopes [2,3].

The exact function of GP 72 is still unclear, despite considerable study and a number of recent advances. It has been demonstrated that metacyclogenesis can be inhibited by WIC 29.26 in a model system, which indicates GP 72 may play a significant role in differentiation, possibly by interaction with lectins of the insect gut [4]. The determination of the structure of the carbohydrates carried by GP 72 and recognized by WIC 29.26 may provide valuable insights into the nature of this host–parasite interaction.

Studies on the carbohydrate structure of GP 72 have been limited by its very low abundance, which has been estimated at 0.04% of total cell protein [5]. There is also considerable difficulty in purifying a single glyco-

67

protein using an antibody that reacts with several glycoproteins carrying the same carbohydrate epitope [6]. In our current work [7], we have concentrated on the spectrum of carbohydrates that react with WIC 29.26 antibody, rather than from a single glycoprotein.

Early studies indicated that the glycosylation recognized by WIC 29.26 is unusual, as it contains high levels of phosphate and is rich in xylose, fucose, and rhamnose [5,8]. We have confirmed earlier compositional data and obtained detailed structural information about several major component glycans purified from WIC 29.26 affinity–purified glycopeptides. In particular, we have employed electrospray ionization mass spectrometry (ESI/MS) and tandem electrospray ionization mass spectrometry (ESI/MS/MS), in both positive and negative ion modes, to elucidate the structure of a highly unusual cyclic phosphate substituent at the nonreducing terminus of several glycans.

II. METHODS

A. WIC 29.26 Affinity Purification of Glycopeptides

Trypanosoma cruzi epimastigote cells of strain Y-NIH were grown to late log phase in Liver infusion tryptose (LIT) media supplemented with 10% fetal calf serum [9], harvested by centrifugation, and lysed at 10^9 cells/mL in 1% deoxycholate with heating and sonication. Lysates were digested with 0.5 mg/mL pronase for 48 hr. Glycopeptides were immunoprecipitated batchwise with WIC 29.26–sepharose beads. After extensive washing, glycopeptides were eluted with 50 mM diethylamine, pH 11.5. Eluates were immediately frozen and lyophilized, and redissolved in water. The glycopeptides were then passed through a BIO-Gel P2 (BIO-RAD, Hercules, CA) column (50 × 1 cm), eluted with water, collected in the void volume, and concentrated under vacuum. A second desalting step was performed by passing the glycopeptides over an Oxford Glycosystems (Abingdon, United Kingdom) GlycoMap 1000 carbohydrate analysis system.

B. Monosaccharide Compositional Analysis

Carbohydrate compositional analysis was performed by gas chromatography mass-spectrometry, GC/MS of trimethylsilyl glycosides. Samples were mixed with *scyllo*-inositol as an internal standard, dried and subjected to methanolysis in 50 μL of 0.5 M HCl in dry methanol for 4 hr at 85°C. Methanolysates were re-*N*-acetylated prior to derivatizing with trimethylsilane reagent. Aliquots of 1 μL were analyzed by GC/MS using an Econocap SE-54 column (Alltech, Deerfield, IL) (30 m × 0.25 mm) with He carrier gas at 0.5 mL/min. The temperature program used was 140°C (2 min)–

250°C at 5°C/min to 300°C at 15°C/min, held for 20 min. Electron impact mass spectra were collected from linear scanning over m/z 40–800, and quantification was based on empirically determined response factors and integration of total ion current chromatograms.

C. Highly pH Anion-Exchange Chromatography

Oligosaccharides were identified and quantified using a Dionex (Sunnyvale, CA) high pH anion-exchange chromatography (HPAEC) system fitted with a Dionex CarboPac PA-1 column and a pulsed amperometric detector (PAD). Oligosaccharides were separated using a gradient of NaOAc in 150 mM NaOH. The gradient program was 10 mM initially, then linear gradients to 30 mM at 15 min, 200 mM at 40 min, 500 mM at 55 min, and held at 500 mM to 60 min. This program separates carbohydrates over a wide range of both size and charge, with monosaccharides eluting at approximately 7.5 min.

D. Chemical and Enzymatic Treatments of Glycopeptides and Oligosaccharides

Mild acid hydrolysis was performed in 40 mM trifluoroacetic acid at 100°C for 8 min. Samples were cooled on ice and dried under vacuum. Alkaline-catalyzed β-elimination with concurrent reduction was performed in 100 mM NaOH and 800 mM $NaBH_4$ overnight at 37°C. Alkaline-catalyzed β-elimination was also performed by incubation in 100 mM NaOH overnight at 37°C, and subsequent alkaline phosphatase digestion was performed after neutralization with boric acid. Aqueous hydrofluoric acid (HF) dephosphorylation was performed in 48% (v/v) aqueous HF at 0°C for 48 hr.

E. Electrospray Ionization Mass Spectrometry

Data were collected on a VG Quattro Mass Lynx Spectrometer (VG Analytical, Manchester, United Kingdom). Samples were dissolved in 50% acetonitrile, 5 mM ammonium acetate and introduced via an injection loop. Cone voltage was set at 30 V for molecular ion determinations. For positive ion tandem mass spectrometry, cone voltage was set at 26 V and collision energy was applied as a gradient from 94 to 49 V over the mass range m/z 40 to 800. For negative ion tandem mass spectrometry, cone voltage was set at 40 V and collision energy at 150 V.

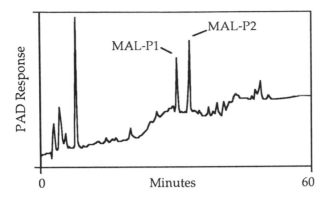

Figure 1 HPAEC separation of MAL oligosaccharides released from WIC 29.26 affinity–purified glycopeptides by mild acid hydrolysis. The two oligosaccharides that were further studied, designated MAL-P1 and MAL-P2, are indicated.

III. RESULTS

A. Compositional Analysis of Glycopeptides

Glycopeptides were prepared from *T. cruzi* epimastigotes and purified by immunoprecipitation with the monoclonal antibody WIC 29.26. Monosaccharide compositional analysis showed that the WIC 29.26 affinity–purified glycopeptides contained fucose, rhamnose, xylose, and galactose in the approximate ratio 1 : 1 : 2 : 3. This confirmed the highly unusual nature of the carbohydrates, as described previously [5]. The possibility of the galactose being present as galactofuranose residues was investigated by compositional analysis of aliquots of the sample, following periodate oxidation and reduction. This treatment converts galactofuranose to arabinose [10]. This resulted in the loss of greater than 80% of the galactose, and in the appearance of arabinose. Thus, the majority of the galactose is present in the highly acid labile furanose configuration.

B. Release of Oligosaccharides from Glycopeptides

Aliquots of the purified glycopeptides were subjected to (1) digestion with peptide *N*-glycosidase F; (2) base-catalyzed β-elimination with concurrent reduction; and (3) base-catalyzed β-elimination followed by alkaline phosphatase treatment and reduction. None of these treatments liberated significant amounts of oligosaccharides. In contrast, however, a series of glycans, designated as MAL (mild acid liberated) oligosaccharides, were released by mild acid hydrolysis (Fig. 1). These consisted of a single major component, which eluted very early on HPAEC and presumably contained monosaccharides, and several later eluting peaks, which were subsequently shown to be negatively charged. These data suggest the presence of phosphodiester linkages between peptide and oligosaccharides, as these hydrolysis conditions have been shown to quantitatively cleave hexose-1-PO$_4$ linkages while leaving glycosidic linkages intact [11]. The two major negatively charged peaks, designated MAL-P1 and MAL-P2, were analyzed in further detail.

C. Alkaline Phosphatase and Aqueous Hydrofluoric Acid Treatment of MAL Oligosaccharides

An aliquot of the MAL oligosaccharides was treated with alkaline phosphatase to screen for the presence of phosphomonoesters. No change was observed in the size or retention time of the HPAEC peaks, which showed that the glycans did not contain any exposed phosphomonester groups, which are sensitive to alkaline phosphatase. To confirm that the negative charge on the oligosaccharides was due to the presence of phosphate groups, an aliquot was treated with 48% aqueous hydrofluoric acid (HF). Greater than 90% of the negatively charged oligosaccharides were converted to small neutral species. This proved that the majority of oligosaccharides contain phosphorylated substituents that can be removed by HF. As the major products eluted in positions corresponding to monosaccharides, it appears that HF treatment also causes significant degradation of the dephosphorylated oligosaccharides.

D. Monosaccharide Compositional Analysis and ESI/MS Molecular Ion Analysis of MAL-P1 and MAL-P2

Aliquots of MAL-P1 and MAL-P2 were analyzed for monosaccharide composition by methanolysis and GC/MS. The results (Table 1) showed that the two oligosaccharides are very similar in composition, with MAL-P2 containing an additional galactose. This was supported by the observation that MAL-P2 could be rehydrolyzed under the same acid conditions to produce MAL-P1 and monosaccharides. Moreover, the composition of these oligosaccharides is very similar to that of the starting glycopeptides, which indicates that the MAL oligosaccharides probably represent the majority of carbohydrate originally attached to the glycopeptides.

MAL-P2 was subjected to ESI/MS in negative ion mode and an (M − H)$^-$ ion of m/z 1120.8 was observed (Fig. 2A), and in positive ion mode a prominent ammonium adduct of m/z 1139.8 was produced (Fig. 2B). These results were consistent with a monoisotopic mass of $M = 1122$, as isotopic abundances indicated the ions were singly charged. MAL-P1 was analyzed under the same conditions, and it produced (M − H)$^-$ ions at

Table 1 Electrospray Ionization Mass Spectrometry and Compositional Analysis
Data of MAL-P1 and MAL-P2

Component	Formula mass	MAL-P1 composition	MAL-P1 theoretical[a]	MAL-P2 composition	MAL-P2 theoretical[a]
Rhamnose	146	0.7	1 = 146	1.0	1 = 146
Fucose	146	1.3	1 = 146	1.2	1 = 146
Xylose	132	1.7	2 = 264	1.7	2 = 264
Galactose	162	2.3	2 = 324	3.2	3 = 486
Cyclic-PO_4	62		1 = 62		1 = 62
Terminus	18		1 = 18		1 = 18
Total			$M = 960$		$M = 1122$

[a]Theoretical values are based on oligosaccharide structures deduced from mass spectrometry
data.

m/z 959, and an ammonium adduct of m/z 978 along with a sodium adduct
of m/z 982 in positive ion mode (data not shown). These results are consistent with a monoisotopic mass of $M = 960$, one hexose residue less than
MAL-P2.

E. Tandem Mass Spectrometry of MAL-P1 and MAL-P2

The m/z 1120.8 ion of MAL-P2 was subjected to negative ion mode ESI/
MS/MS collision analysis, which produced ions of m/z 79 and 97, corresponding to $(PO_3)^-$ and $(H_2PO_4)^-$ (Fig. 2C). The m/z 959 ion of MAL-P1
produced identical fragments under the same conditions (data not shown).
These fragment ions could only be produced using collision at a much
higher energy than was necessary for producing the same ions from a man-
nose-6-phosphate standard.

The m/z 1139.8 ammonium adduct ion of MAL-P2 was subjected to
positive ion mode ESI/MS/MS collision analysis, which produced two major fragment ions at m/z 225.2 and 357.2 (Fig. 2D). These differ by 132,
which corresponds to one pentose residue. The m/z 978 ion of MAL-P1
produced identical fragment ions under the same conditions (data not
shown). The ion at m/z 225.2 was the smallest monosaccharide fragment
produced, and was attributed to a hexose residue substituted by a cyclic
phosphate group. Not only does this substituent have the correct mass, it is
consistent with the results in previous experiments: (1) it is a negatively
charged group that is resistant to alkaline phosphatase; (2) it is sensitive to
aqueous HF treatment; and (3) it requires very high energy negative ion
ESI/MS/MS collision to produce phosphoanion fragments.

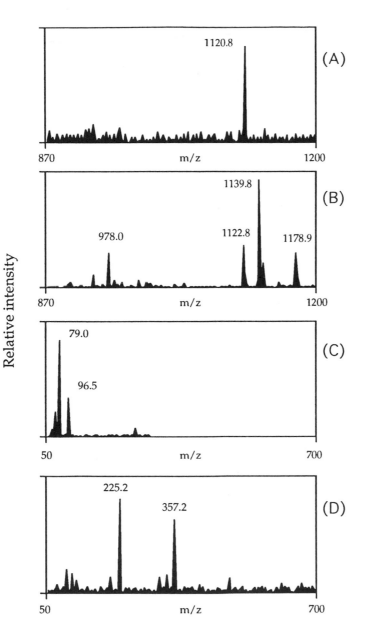

Figure 2 Electrospray ionization mass spectrometry of oligosaccharide MAL-P2:
(A) pseudomolecular ion of $(M - H)^- = m/z$ 1120.8 in negative ion mode; (B)
pseudomolecular ions in positive ion mode, where m/z 1139.8 is $(M + NH_4)^+$,
m/z 1122.8 is $(M + H)^+$, and m/z 1178.9 appears to be $(M + NH_4 + K)^+$; (C)
daughter ions of m/z 1120.8 ion in (A), where m/z 79.0 is $(PO_3)^-$ and m/z 96.5 is
$(H_2PO_4)^-$; (D) daughter ions of m/z 1139.8 in (B), where m/z 225.2 is a protonated
hexose-cyclic-phosphate residue and m/z 357.2 is a pentose linked to a protonated
hexose-cyclic-phosphate residue.

The identification of these fragment ions allowed the first tentative assignment of structures to the observed molecular ions. The mass of 960 Da (MAL-P1) can be accounted for by a structure consisting of two hexoses, two pentoses, two deoxyhexoses, and a cyclic phosphate. As mentioned previously, the mass of 1122 Da (MAL-P2) is 162 mass units larger, which corresponds to the addition of one hexose residue. As the fragmentation patterns of the two species are identical, it is reasonable to assume that they have a common core structure differing only by the addition of a single hexose residue. The structures of MAL-P1 and MAL-P2 inferred from the mass spectrometry data are also in close agreement with the compositional data (Table 1).

IV. DISCUSSION

The highly unusual nature of the oligosaccharide structures described in this chapter make them refractory to analysis by many conventional techniques. The putative phosphodiester linkage between oligosaccharide and peptide has been reported previously in both Dictyostelium [12,13] and Leishmania [14,15]. In the case of Leishmania-secreted acid phosphatase, for example, glycans are present in phosphodiester-linked repeating units, which can be depolymerized using identical mild acid hydrolysis conditions to those we have used for liberating MAL oligosaccharides. The depolymerized glycan subunits contain a 6-phosphate substituent at the nonreducing terminus, which readily can be identified by digestion with alkaline phosphatase. In contrast, the MAL glycans described in this chapter appear to undergo a rearrangement of the 6-phosphate substituent following, or concurrent with, the mild acid hydrolysis. This is likely to be a consequence of the 6-phosphate attachment to a galactofuranose residue, which contains two exocyclic hydroxyl groups in close proximity to the phosphate, allowing the formation of a highly stable five-membered ring structure. The unusually stable nature of the cyclic phosphate group was confirmed by the fact that it is unchanged by treatments that open the inositol-1,2-cyclic phosphate ring. It is also possible, although highly unlikely, that the cyclic phosphate group could be part of the intrinsic oligosaccharide structure, as described in a *Vibrio cholerae* capsular polysaccharide [16].

It was only by the use of ESI/MS and tandem ESI/MS/MS in both positive and negative ion mode that we were able to elucidate the proposed structure of the nonreducing terminus of the MAL oligosaccharides. The use of ESI/MS/MS induced fragmentation is certain to become increasingly important in the analysis of unusual oligosaccharide modifications such as those we have described.

ABBREVIATIONS

ESI/MS	electrospray ionization mass spectrometry
ESI/MS/MS	tandem electrospray ionization mass spectrometry
GC/MS	gas chromatography-mass spectrometry
HPAEC	high pH anion-exchange chromatography
LIT	Liver Infusion Tryptose
MAL	mild acid liberated
PAD	pulsed amperometric detector

REFERENCES

1. Snary, D., Ferguson, M. A. J., Scott, M. T., and Allen, A. K. (1981). _Mol. Biochem. Parasitol._ 3, 343–356.
2. Kirchhoff, L. V., Hieny, S., Shiver, G., Snary, D., and Sher, A. (1984). _J. Immunol._ 133, 2731–2735.
3. Schechter, M., Stevens, A. F., Luquetti, A. O., Snary, D., Allen, A. K., and Miles, M. A. (1986). _Infect. Immun._ 53, 547–556.
4. Sher, A., and Snary, D. (1982). _Nature_ 300, 639–640.
5. Ferguson, M. A. J., Allen, A. K., and Snary, D. (1983). _Biochem. J._ 213, 313–319.
6. Cooper, R., Inverso, J. A., Espinosa, M., Nogueira, N., and Cross, G. A. M. (1991). _Mol. Biochem. Parasitol._ 49, 45–60.
7. Haynes, P. A., Ferguson, M. A. J., and Cross, G. A. M. (1996). _Glycobiology_, in press.
8. Snary, D., Ferguson, M. A. J., Allen, A. K., Miles, M. A., and Sher, A. (1987). _Cell surface glycoproteins of Trypanosoma cruzi._ Presented at NATO ASI Series H11.
9. Bone, G. J., and Steinert, M. (1956). _Nature_ 178, 308–309.
10. De Lederkremer, R. M., Casal, O. L., Alves, M. J. M., and Colli, W. (1980). _FEBS Lett._ 116, 25–29.
11. McConville, M. J., Homans, S. W., Thomas-Oates, J. E., Dell, A., and Bacic, T. (1990). _J. Biol. Chem._ 265, 7385–7394.
12. Gustafson, G. L., and Milner, L. A. (1980). _J. Biol. Chem._ 255, 7208–7210.
13. Haynes, P. A., Gooley, A. A., Ferguson, M. A. J., Redmond, J. W., and Williams, K. L. (1993). _Eur. J. Biochem._ 216, 729–737.
14. Ilg, T., Overath, P., Ferguson, M. A. J., Rutherford, T., Campbell, D. G., and McConville, M. J. (1994). _J. Biol. Chem._ 269, 24073–24081.
15. Ilg, T., Stierhof, Y. D., McConville, M. J., and Overath, P. (1995). _Eur. J. Cell Biol._ 66, 205–215.
16. Knirel, Y. A., Paredes, L., Jansson, P.-E., Weintraub, A., Widmalm, G., and Albert, M. J. (1995). _Eur. J. Biochem._ 232, 391–396.

6

Analysis of Carbohydrate–Protein Interactions by Mass Spectrometry

Yoko Ohashi, Sadamu Kurono, and Yoshitaka Nagai
The Institute of Physical and Chemical Research (RIKEN),
Wako-shi, Saitama, and Mitsubishi Kasei Institute of Life Sciences,
Machida, Tokyo, Japan

I. INTRODUCTION

It has been recognized recently that interactions between proteins and carbohydrate chains play important cell biological roles, particularly in cell-cell interactions and the signal transduction mechanisms. For example, noncovalent binding of cholera toxin B-subunit (CTB) with the ligand GM1a is well recognized [1–4]. Although there are now several methods to analyze such complexes, we investigated the use of electrospray ionization mass spectrometry (ESI/MS) [5] to determine the specificity and stoichiometry of CTB binding. Being an MS method that does not produce fragmentation unless skimmer voltage is applied, ESI/MS has been used to measure the mass of intact proteins and common glycolipids, such as the tetrasialoganglioside GQ1b [6] and polysulfated Le-type glycoconjugates [7,8]. ESI/MS is also one of the two methods currently available for analyzing proteins that may play important roles when complexed with sugar chains.

In 1992, we reported the molecular weight of a CTB-GM1a noncovalent complex, recorded as a deconvoluted ion peak from an ESI spectrum using a quadrapole mass spectrometer equipped with an electrospray ion source [9] (Fig. 1). However, the question remained whether the observed complex was specifically bound as occurs in the cell membrane.

In this chapter, we report investigations on the effects of (1) desalting, (2) ionization with different organic solvents, and (3) heat applied to spray and evaporate solvents in the ion source. We also performed control experiments using glycoconjugates analogous to GM1a.

II. MATERIALS AND METHODS

Typically, 50 μL of CTB buffer solution (40 pmol/μL) was incubated with 100 μL of a GM1a aqueous solution (200 pmol/μL) at 37°C for 3 h. The

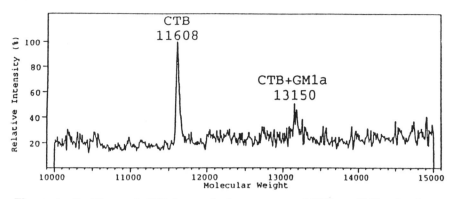

Figure 1 Positive mode ESI deconvolution spectrum of CTB + GM1a after incubated under physiological conditions.

solution was then desalted by centrifugation through a membrane (molecular cut: 10,000) and washing the retentate twice with 50 μL water. Thus desalted sample was dissolved in various solvents such as water and methanol (1 : 1) and water and acetonitrile (1 : 1). Positive mode ESI specta were recorded by injecting the sample solution into the ESI ion source of a TSQ 700 quadrapole mass spectrometer at a flow rate of 1 μL/min. The spray potential was (−)3.5 kV, and the drying gas was nitrogen, which was heated typically to 240°C at the indicator, but probably around 100°C near the sample droplets. Deconvolution on ESI/MS [10] was performed with a computer program, to obtain the molecular weight of a sample from multiply charged ions.

III. RESULTS AND DISCUSSION

Table 1 shows control glycoconjugate samples we investigated as possible ligands, which were classified as follows: (A) glycoconjugates having the same sugar chain as GM1a, but a different aglycon moiety; (B) gangliosides with an additional sialic acid; (C) monosialogangliosides with a shorter sugar chain than GM1a; (D) neutral glycosphingolipids; and (E) sialic acid and related derivatives. Table 1 also shows the qualitative binding of each sample to CTB. Many of the control samples were weakly and nonspecifically bound to CTB, especially when the control sample had the same sugar chain as GM1a (oligo-GM1a) as seen in category (1). However, a deconvolution program installed in the data system was effective only for the natural ligand GM1a, the complex of which alone had the ion intensity adequate for the deconvolution algorithm; all control analogs showed ion

Table 1   Binding of Cholera Toxin to Related Glycoconjugates

Sugar analog	Apparent binding	Sugar analog	Apparent binding
CTB + GM1a	+ + + +		
A) oligo-GM1a[a]	+	D) GA1	−
A) oligo-GM1a-n-octyl	+ + +	D) LacCer	−
A) oligo-GM1a-DIG[b]	+ +	D) GalCer(d 18:1/C 16:0)	−
B) GD1a	+ + +	D) GalCer(d 18:1/C2:0)	−
B) GD1b	+ + +	E) βNeuAc	−
C) GM2	+ +	E) αNeuAc(OMe, CO₂Me)	−
C) GM3	+ +	E) αNeuAc(OMe)	−
C) GM4	−	E) αNeuAc(TDG[c])	−

A) Glycoconjugates having the same sugar chain as GM1a, but different aglycon moieties.
B) Gangliosides with an additional sialic acid.
C) Monosialo-gangliosides with a shorter sugar chain than GM1a.
D) Neutral glycosphingolipids.
E) Sialic acid and derivatives.

[a]Sugar chain of GM1a.
[b]1,2:3,4-di-*O*-isopropylidene-α-D-galactopyranose.
[c]1,2-di-*O*-*n*-tetradecyl-3-*O*-(α-*N*-acetylneuraminyl)-*sn*-glycerol.

intensities of the complexes too low to allow use of the deconvolution program. In other words, there were certain differences in the binding strength between the specific ligand GM1a's and control glycoconjugate analogs to CTB when examined by mass spectrometry. The binding tendency thus observed was similar to that shown by other assay methods [11,12]. Control experiments were also performed using heat- or acid-denatured CTB and unrelated proteins. Unrelated proteins such as lysozyme, cytochrome C, or trypsinogen showed no evidence of the binding to GM1a. Moreover, heat- or trifluoroacetic acid (TFA)–denatured CTB failed to display the bound complex signal. The noncovalent complex with GM1a was detected only for native CTB. This suggests that conformation(s) required for specific binding survives the ESI process.

    We next investigated whether ESI/MS could be used to measure the affinity of complexes. However, the intensity for the molecular ion of the CTB–GM1a complex was not consistent with previously reported analytical data [11,12]. Apparently, the specific noncovalent binding of CTB with GM1a, which is initially formed by incubation in solution, dissociates either during or before the electrospray process. The cholera toxin B-subunit is thought to assume a pentamer structure [13]. In the foregoing studies, the pentamer structure for the specific binding with GM1a was not detected by ESI/MS. Experimental conditions adverse to weak intermolecular interactions between adjacent monomers as well as between protein and ligand molecule were reexamined. The low salt concentrations compatible with ESI/MS may change the native tertiary structure of CTB needed for the specific, noncovalent binding. Moreover, the organic solvent that is added to the aqueous sample solution prior to electrospray ionization to control the surface tension of the droplets may be a factor. The elevated temperature to evaporate charged droplets may also influence complex stability.

We next investigated using milder conditions for ESI/MS. A 5 pmol/ μL aqueous solution of myoglobin (MW 17,571) is a frequently used reference samples for ESI/MS, but it is always observed as apomyoglobin (MW 16,955) (Fig. 2A), obviously losing the prosthetic heme group under standard positive mode ESI conditions. However, the molecular weight of the holomyoglobin was observed after deconvolution, by switching the solvent from aqueous 0.5% acetic acid/methanol (1 : 1) to an aqueous ammonium acetate solution (Fig. 2B). In addition, the temperature of the drying gas was reduced from 240°C (estimated temperature around the sample was about 60°C) to 100°C (sheath gas and sheath liquid needed) as described in Section II. These results show that methanol and possibly hot gas (even though it may take only microseconds for the sample species to pass through the heated area) significantly affected the conformation of apomyoglobin, which holds heme in the pocket-like structure under the physiological environment.

We next applied these conditions to the binding of CTB to GM1a. The modified ESI conditions not only failed to give the molecular weight of the CTB pentamer, but also neither the complex (CTB + GM1a) nor the monomeric CTB were observed.

We also investigated the effect of matrix-assisted laser desorption ionization (MALDI) mass spectrometry on complex stability. This method is used to ionize and detect polymer molecules as large as several hundred kilodaltons. The sample is dissolved in an aqueous solution, but the matrix is typically dissolved in an aqueous organic solution, and aliquots of both are mixed and dried together on the laser target. The laser energy is absorbed into the matrix, which has an absorption maximum at the laser wavelength, and is then transferred to the sample molecule so as to ionize the sample molecules with protonation. Our idea was to determine if MALDI was capable of ionizing the CTB pentamer structure, as MALDI tolerates greater amounts of salt. Figure 3 shows the positive mode MALDI spectrum of CTB, using 4-hydroxyazobenzene-2'-carboxylic acid as the matrix. The spectrum exhibited a series of CTB multimers, which displayed

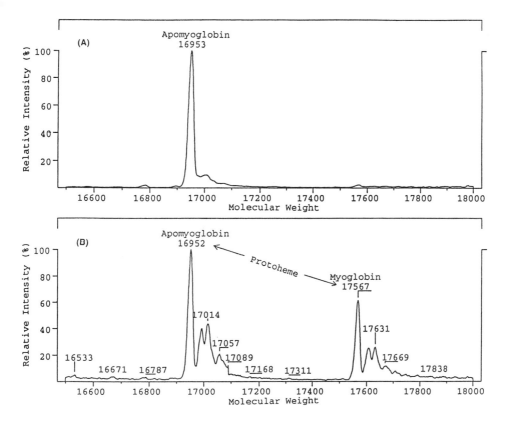

Figure 2 Positive mode ESI spectrum of myoglobin (calculated molecular weight: 17,571). (A) Solvent: equivolume mixture of 0.5% acetic acid and methanol; N_2-drying gas: nominally 240°C. (B) Solvent: ammonium acetate aqueous solution; N_2 drying gas: nominally 100°C. Isopropyl alcohol was used as the sheath liquid in addition to the sheath gas of nitrogen in the latter case.

lower ion intensities for successively larger complexes, rather than showing the pentamer at a higher intensity. The experiment to obtain spectra of the bound complex of CTB with GM1a also failed, with CTB and GM1a detected at their respectively expected masses. Doubly charged ions of CTB and desialylated fragments of GM1a were also observed. In addition to the higher energy for laser desorption, the sample protein is likely denatured by the matrix or by the solvent for matrix. An efficient water-soluble matrix may enable the use of MALDI to detect weakly bonded protein–ligand complexes.

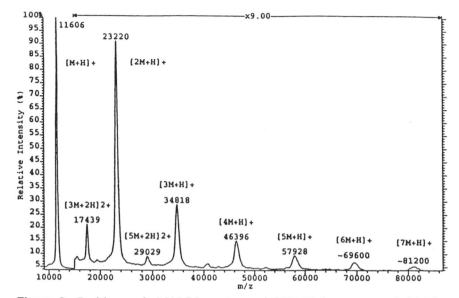

Figure 3 Positive mode MALDI spectrum of CTB. N$_2$ laser was used. Matrix: 4-hydroxyazobenzene-2′-carboxylic acid.

IV. SUMMARY AND FUTURE DIRECTIONS

Although we have had some success in the structure elucidation of complex glycoconjugates [6–8] as well as covalently bound glycoconjugates with peptides [14], we have not determined the optimal conditions to detect specific, noncovalent binding between a protein and its liganded ganglioside by mass spectrometry. GM1a, the natural ligand for CTB, showed the highest deconvoluted peak intensity of the complex in ESI spectra, but other similar glycoconjugates also exhibited complexation signals to some extent, depending on the sugar-chain structures. Glycosides of oligo-GM1a and extended sugar-chain analogs like GD1a and GD1b also formed complexes with CTB, whereas neutral glycoconjugates, sialic acid, and its simple glycosides failed to show detectable complex ions. These observations were in agreement with trends from more time-consuming conventional assays.

It is worth noting that recently reported nanospray ESI requires an extremely low flow rate for sample introduction into the ESI ion source. In our case, the ESI flow rates were typically 1 μL/min, whereas flow rates of nanoliters per minute are used for nanospray ESI/MS. Consequently, the smaller diameter of droplets requires less heat for evaporation. This factor may prove to be more significant than high sensitivity for binding studies.

ACKNOWLEDGMENTS

Precious samples of oligo-GM1 analogs were generous gifts from Y.-T. Li of Tulane University (New Orleans). The authors are grateful to T. Ogawa of RIKEN for sialic acid and its derivatives. We are also most thankful to H. Komura and T. Williams of JASCO International as well as to J. Letcher of FISONS, who kindly made available the TofSpec for our MALDI/ mass spectrometry studies. This work was partially supported by a Grant-in-Aid for Scientific Research on Priority Area No. 06240247 from the Ministry of Education, Science and Culture of Japan.

ABBREVIATIONS

CTB	cholera toxin B-subunit
ESI/MS	electrospray ionization mass spectrometry
MALDI	matrix-assisted laser desorption ionization
MW	molecular weight
TFA	trifluoroacetic acid

REFERENCES

1. Holmgren, J., Lonnroth, I., and Svennerholm, I. (1973). *Infect. Immun.* 8, 208–214.
2. Cuatrecasas, P. (1973). *Biochemistry* 12, 3547–3558.
3. Cuatrecasas, P. (1973). *Biochemistry* 12, 3558–3581.
4. Fishman, P. H., Moss, J., and Osborn, J. (1978). *Biochemistry* 17, 711–716.
5. Yamashita, M., and Fenn, B. (1984). *J. Chem. Phys.* 88, 4451–4459.
6. Ii, T., Ohashi, Y., and Nagai, Y. (1995). *Carbohydr. Res.* 273, 27–40.
7. Ii, T., Ohashi, Y., Ogawa, T., and Nagai, Y. (1996). *J. Mass Spectrom. Soc. Jpn.* 44, 183–195.
8. Ii, T., Ohashi, Y., Ogawa, T., and Nagai, Y. (1996). *Glycoconj. J.* 13, 273–283.
9. Ohashi, Y., Hirabayashi, Y., Kanai, M., and Nagai, Y. (1992). *Biological Mass Spectrometry Kyoto '92, Abstracts,* 207.
10. Mann, M. (1990). *Org. Mass Spectrom.* 25, 575–587.
11. Wuand, G. and Ledeen, R. (1988). *Anal. Biochem.* 173, 368–375.
12. Karlsson, K.-A. (1989). *Ann. Rev. Biochem.* 58, 309–350.
13. Sixma, T. K., Pronk, S. E., Kalk, K. H., Wartna, E. S., van Zanten, B. A. M., Witholt, B., and Hol, W. G. J. (1991). *Nature* 351, 371–377.
14. Zhao, M., Yoneda, M., Ohashi, Y., Kurono, S., Iwata, H., Ohnuki, Y., and Kimata, K. (1995). *J. Biol. Chem.* 270, 26657–26663.

7

Mycobacterial Lipoarabinomannans
Structural and Tagging Studies

Carole Delmas, Anne Venisse, Alain Vercellone, Martine Gilleron, Renaud Albigot, Therese Brando, Michel Rivière, and Germain Puzo
National Center for Scientific Research, Toulouse, France

I. INTRODUCTION

Lipoarabinomannans (LAM) are ubiquitous components of the mycobacterial cell walls [1]. According to their structures, LAM are classified in two types, namely, AraLAM and ManLAM. AraLAM were first isolated from an unidentified fast growing mycobacterial species [2], whereas ManLAM were found in slow growing strains such as *Mycobacterium tuberculosis*, Erdman strain [3,4], and in the *Mycobacterium bovis* BCG (Bacille Calmette–Guérin) [5,6] vaccine. The structural models of AraLAM and ManLAM reveal that they are composed of two homopolysaccharides, namely, the D-mannan and the D-arabinan (Fig. 1). The mannan is a linear oligosaccharide composed by 6-*O*-linked α-D-Man*p*s (mannopyranose) with side chains containing one single unit of α-D-Man*p* attached at the position C2. At its reducing end, this mannan core is glycosylated by the phosphatidyl *myo*-Ins (inositol) anchor. Three major structural motifs composed the D-arabinan: (1) a linear oligosaccharide of 5-*O*-linked α-D-Ara*f* (arabinofuranose), (2) side chains attached via the C3, and finally, (3) terminal disaccharidic unit of β-D-Ara*f*-(1 → 5)-α-D-Ara*f* → . The major structural difference between AraLAM and ManLAM is the presence of small mannooligosaccharide units capping the arabinan side chains of the ManLAM (Fig. 1). More recently, though, it was established that AraLAM from an unidentified fast growing mycobacterial species [2,7] and from *Mycobacterium smegmatis* (unpublished results) contain phosphoinositide caps. The AraLAM and ManLAM present a large spectrum of immunological activities. Some of these activities are specific. For example, ManLAM bind murine and human macrophages via the mannose receptor [8,9] and activate human lymphocyte Tαβ CD4⁻ CD8⁻ [10], whereas AraLAM stimulate the precoce genes of the macrophage activation [11,12] as well as the

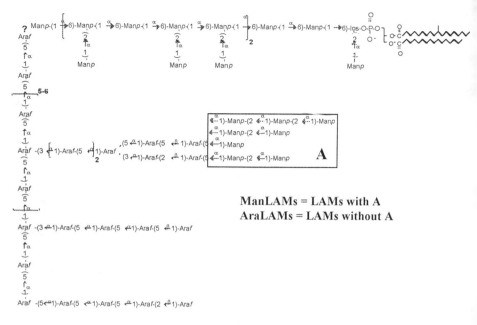

Figure 1 Structural model of ManLAM and AraLAM.

tumor necrosis factor (TNF)-α secretion [13]. In this chapter, we focus on the analytical approaches developed for the structural determination of LAM.

II. MATERIALS AND METHODS

A. LAM Purification

Mycobacterium bovis BCG cells (85 g dry weight) were delipidated by six successive extractions with chloroform/methanol (1 : 1, v/v) (Fig. 2). Bacterial residue was then refluxed five times in 50% (v/v) ethanol. Ethanol extracts were concentrated and lyophilized to give a complex mixture (2 g), containing proteins, polysaccharides as glucans, arabinomannans and lipoglycans as LAM, lipomannans (LM) and phosphatidyl-inositol mannosides (PIM). Glucans and proteins were degraded by enzymatic digestion. The extract was then dissolved in 50 mM sodium phosphate buffer (pH 7) containing 50 mM, NaCl and treated with α-amylase from porcine pancreas (Boehringer Mannheim, Indianapolis) for 2 hr at 37°C. Trypsin (2%, w/w of protein) was then added in two aliquots and first incubated 5 hr at 37°C and then overnight. The sample was then brought up to 2% triton X-100

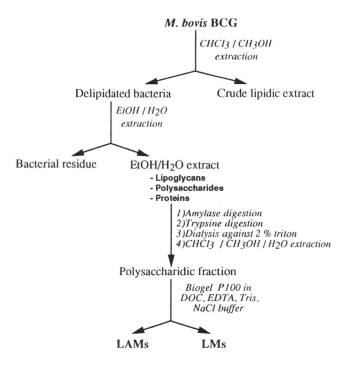

Figure 2 *M. bovis* BCG LAM purification.

and dialyzed against 2% triton X-100. This step eliminates maltooligosaccharides and peptides generated by enzymatic digestions and also PIM by dissociating aggregates. Triton X-100 was then extracted with chloroform/methanol/water (2 : 1 : 1, v/v/v). The water-soluble fraction (400 mg) was freeze-dried and incubated in 10 mM Tris-HCl (pH 8) buffer containing 200 mM NaCl, 0.25% deoxycholate, and 1 mM EDTA buffer and loaded onto a Bio-Gel P-100 column equilibrated with the same buffer. Fractions from the eluant were analyzed by sodium dodecyl sulphate (SDS) polyacrylamide gel electrophoresis. LAM-containing fractions (20 mg) were pooled, dialyzed three times against 10 mM Tris-HCl (pH 8) and finally against mQ water before lyophilization.

B. Hydrolysis

Dried polysaccharides (50 to 100 μg) in Teflon-lined, screw-capped vials were hydrolyzed with 100 to 150 μL of TFA 2N at 80°C in a heating block or a sand bath for 2 hr. After cooling at room temperature, the reaction mixture was dried under a nitrogen stream; in some cases, drying was achieved either under vacuum or with a second evaporation after the addition of 100 μL of methanol. A similar procedure, replacing TFA with 1 N anhydrous methanolic HCl was used to produce the methyl glycosides (methanolysis), which are in some cases more stable. To improve the Ins recovery, the same protocol can be used changing only the reaction conditions: HCl 6 N (constant boiling grade), 110°C during 18 hr.

C. Gas Chromatographic Analysis

After methanolysis, the dried methyl glycosides, in Teflon-lined, screw-capped vials, were resuspended in a 100-μL solution of pyridine/hexamethyldisilazane/trimethylchlorosilane (8 : 4 : 2, v/v/v/), and the vial was immediately tightly capped. After 30 min at room temperature, the reagents were evaporated under a gentle stream of nitrogen until the sample was dried. This step must be carried out with attention (not too long), because the silylated derivatives are volatile and moisture-sensitive. Thus, after drying, the products were immediately resuspended in 100 μL of hexane. The sample (1 μL) was then deposited on the needle of a gas chromatographic (GC) Ross injector and dried for 1 min under a helium stream before injection. The gas chromatograph was a Girdel series 30 equipped with an OV-1 (0.3 μm film thickness, Spiral, Dijon, France) fused silica capillary column (25 m length × 0.22 mm i.d.) and flame ionization detector. An injector temperature of 260°C and an oven temperature program from 100 to 280°C (3°C/min) were used.

D. Carbohydrate High pH Anion-Exchange Chromatography Analysis

After hydrolysis, the sample was dried and resuspended in (18 MΩ) deionized water (Millipore) and injected into the high pH anion-exchange chromatography (HPAEC)-DX-300 system (Dionex Corp., Sunnyvale, California) consisting of a Dionex Gradient Pump, an Eluent Degas Module, and a pneumatic Rheodyne injector. The pulsed amperometric detector (PAD) was set to integrate amperometry mode with a waveform for carbohydrate detection [gold electrode without pH reference, E1 = 0.1 V (0.5 sec), E2 = 0.6 V (0.1 sec), E3 = -0.6 V (0.05 sec)]. This system was equipped with a CarboPac PA-1 column and guard column (250 × 4 mm; 25 × 4 mm) and controlled via the Dionex Advanced Computer Interface with AI-450 software. For monosaccharide composition analysis, the mobile phase was a 20 mM NaOH solution (prepared from a 50% solution of sodium hydroxide from Baker).

A major drawback of Pulse Amperometric Detection is the high sodium hydroxide concentration needed for electrochemical detection. Thus, for preparative purpose, the presence of such NaOH concentrations may result in the degradation of the sample, and the alkali must be removed for further analysis. To overcome these problems during the purification of the mannan core by anion-exchange chromatography, only part (15 to 20%) of the eluant was monitored by the detector after post column addition of sodium hydroxide (see schematic diagram, Fig. 3). The sample (10 mg/mL) was repeatedly injected (50 μL) onto the CarboPac PA-1 column eluted (1mL/min) in isocratic conditions (0.04 M sodium acetate [ACS grade, Merck] in (18 MΩ) water) for 10 min after which a linear gradient from 0.04 to 0.2 M sodium acetate over 10 min was applied. Regeneration of the

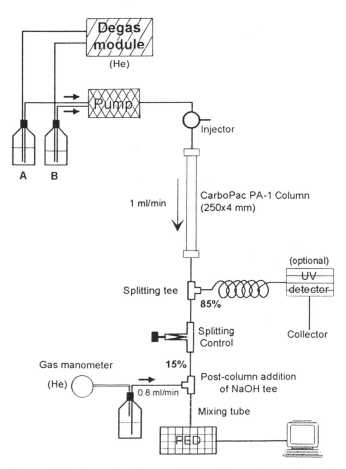

Figure 3 Schematic diagram of the anion-exchange chromatography system used for the purification of Ma-I and Ma-II.

column was achieved by elution with pure water for 15 min followed by an equilibration step of 10 min with the initial eluant. The PAD parameters were set as described for the monosaccharide analysis. After the column, a first split followed by a variable flow valve was used to send the major part ($\approx 80\%$) of the eluent to a fraction collector, while the remaining ($\approx 20\%$) was directed to the detector after post-column addition of 0.8 mL/min of a O,SM sodium hydroxide solution (final flow rate : 1mL/mm).

E. Nuclear Magnetic Resonance Spectroscopy

Prior to [1]H Nuclear Magnetic Resonance (NMR) spectroscopy, samples were repeatedly lyophilized in deuterium oxide (D_2O). Then, they were dissolved in 0.5 mL of D_2O (99.96% purity) at a concentration of 2 mg/mL in a 200 × 5 mm NMR tube. All the spectra were recorded at 303 K on a Bruker AMX-500 spectrometer equipped with an Aspect X32 computer and a 5-mm ^{1}H broadband tunable probe with reversal geometry.

F. Fluorescent Activated Cell Sorting Analysis

All experiments were performed using a fluorescent activated cell sorting (FACS) scan (Beckton Dickinson Immunocytometry Systems, San Jose, California). Preliminary experiments with propidium iodide staining permitted a "morphological window" to define viable cells: 10^4 of these forward scatter/side scatter–gated cells were acquired per experiment. Cells positive for cell markers (fluorescein isothiocyanate [FITC] labeling) were selected using channel FL1 and analyzed on channel FL2 for (phycoerythrin–streptavidin)-biotinylated ManLAM, both with logarithmic amplifiers. Fluorescence intensity parameters were kept constant for all leukocyte analyses. Fluorescence intensity histograms were recorded as relative fluorescence on a logarithmic scale and analyzed using Lysis II software and an HP9000 Series 340 computer station (Hewlett Packard, Fort Collins, Colorado). A difference between means of fluorescence intensities, obtained in experiments with or without biotinylated ManLAM, was calculated and used to represent quantitatively the histograms.

III. RESULTS AND DISCUSSION

A. LAM Purification and Characterization

LAM was characterized by its behavior in SDS-PAGE (polyacrylamide gel electrophoresis) (Fig. 4), showing a broad band around 30 kDa. To determine more precisely the LAM molecular weights, the *M. bovis* BCG ManLAM was analyzed with a matrix-assisted laser desorption ionization time-of-flight (MALDI-TOF) mass spectrometer, equipped with a nitrogen

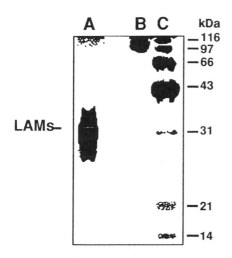

Figure 4 SDS-PAGE of LAM (A) and deacylated LAM (dLAM) (B) from *M. bovis* BCG, and of protein markers (C). The sample amount was 5 μg. The gel was stained with a silver stain containing periodic acid.

pulsed laser [5]. To a 2,5-dihydroxybenzoic acid solution (solubilized in 30 μL of water/ethanol, 9 : 1 at a concentration of 10 μg/mL) was added 3 μL of an aqueous solution of ManLAM (10 μg/mL). This mixture (5 μL) was applied to the target and dried at room temperature before analysis. In positive mode, the mass spectrum (Fig. 5) in the mass range 5 to 50 kDa is dominated by two peaks centered at m/z 8080 and 17,400. The latter peak is broad with a width at half height of 4 to 5 kDa, whereas the MALDI spectra of proteins of similar molecular mass exhibited peak width of 10 Da. The broad signals likely reflect, in agreement with the SDS-PAGE analysis, the molecular heterogeneity of the ManLAM fraction. The peak at m/z 17.4 kDa was tentatively assigned to single charged monomene pseudomolecular ions; however, its attribution to either double charged molecular ions M^{2+} or dimeric molecular ions $(2M^+)$ could not be excluded. However, this last assignment was not supported by a peak localized at half mass (8.7 kDa). A similar spectrum was obtained in the negative mode using sinnapinic acid as a matrix. Moreover, in the MALDI mass spectrum of the deacylated ManLAM, only one peak, centered at m/z 16,700, was observed. Taken together, these data suggest that the average ManLAM molecular mass is 17 kDa with a molecular heterogeneity of 4 kDa, primarily from the number of monosaccharide units (estimated between 90 and 130). Thus, the low mass resolution of MALDI-TOF and the molecular heterogeneity of the fraction precluded a precise determination of the molecular mass of each species.

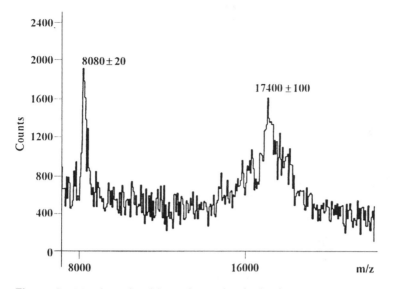

Figure 5 Matrix-assisted laser desorption ionization mass spectrum of a mixture of LAM and LM from *M. bovis* BCG in 2,5-dihydroxybenzoic acid as matrix.

B. Two-Dimensional ¹H–¹³C Heteronuclear Multiple Quantum Correlation Spectroscopy of ManLAM and AraLAM

The mannooligosaccharide caps that typify the ManLAM can be differentiated from the mannan core by the 2-*O*-linked α-D-Man*p* units [3–5]. So, their identification, by routine methylation analysis, proves that the analyzed LAM belong to the ManLAM class, whereas their absence defines them as AraLAM. We used a nondegradative approach two-dimensional, (2D) ¹H–¹³C heteronuclear multiple quantum correlation (HMQC) NMR spectroscopy, to differentiate ManLAM from AraLAM.

Figure 6A shows the anomeric region of the one-dimensional (1D) ¹³C NMR spectrum of *M. bovis* BCG ManLAM. The resonances between 108 and 110 ppm are from the α-Ara*f* units. The intense downfield resonance, I, was assigned to the 5-*O*-linked α-Ara*f*, whereas resonances II and III were attributed, based on their chemical shifts, to the 3,5-di-*O*-linked and 2-*O*-linked α-Ara*f*s, respectively. Indeed, a substitution at the C2 causes a C1 upfield of 1 to 2 ppm. Likewise, the intense signals, IV and VII, were tentatively assigned to the C1 of t(terminal)-α-D-Man*p*s and 2,6-di-*O*-linked α-D-Man*p* residues [5]. As shown by the ¹H–¹³C HMQC experiments (Fig. 7), the latter C1 correlates with two anomeric proton signals, revealing the presence of two populations of 2-*O*-linked α-Man*p* units. One corresponds to the 2,6-di-*O*-linked α-D-Man*p* units from the mannan core, whereas the other one can be assigned to the 2-*O*-linked α-Man*p* units of the terminal Man residues linked to the arabinose chain. This assignment was supported by the t-Man*p* C1 correlation with two anomeric proton signals, which also demonstrated the presence of two types of t-Man*p* units, one arising from the mannan core and the other one from the mannooligosaccharide termini. Finally, resonances, V and VII, were assigned to the t-β-Ara*f*s and 6-*O*-linked α-Man*p* residues, respectively. From the ¹H–¹³C HMQC experiments, it was proposed that the LAM from *M. bovis* BCG belong to the ManLAM class.

This strategy was also applied to the LAMs from *M. smegmatis*, a fast growing mycobacterial species. Figure 6B shows the anomeric region of the 1D ¹³C spectrum. Comparative analysis showed anomeric resonances at the same chemical shifts, but with major differences in the relative intensity of signals V and VII. Resonance V, assigned to the C1 of the β-Ara*f*s, was more intense in the LAM from *M. smegmatis*, whereas resonance VII, which typified the 2-*O*-linked α-Man*p*s was less intense. From the ¹H–¹³C HMQC spectrum, it was established that resonance VII correlated with one anomeric proton signal attributed to the 2,6-di-*O*-linked α-Man*p* residues. Likewise, only one correlation was observed for resonance IV, assigned

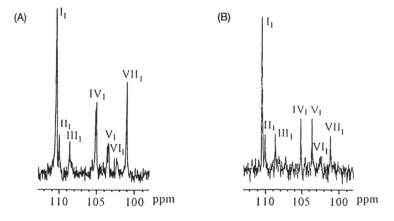

Figure 6 One-dimensional ¹³C anomeric region (δ 98–113) of ManLAM from *M. bovis* BCG (A) and AraLAM from *M. smegmatis* (B) at 500 MHz. Glycosyl residues are labeled in roman numerals and their carbons in arabic numerals. I: 5-*O*-linked α-Ara*f*s; II: 3,5-di-*O*-linked α-Ara*f*s; III: 2-*O*-linked α-Ara*f*s; IV: t-α-Man*p*s; V: t-β-Ara*f*s; VI: 6-*O*-linked α-Man*p*s; VII: 2-*O*-linked and 2,6-di-*O*-linked α-Man*p*s.

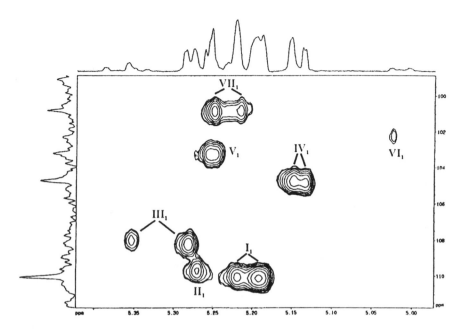

Figure 7 Anomeric region (δ_{f2} 5.00–5.40 and δ_{f1} 100–112) of the 2D ^{1}H–^{13}C HMQC from *M. bovis* BCG ManLAM at 500 MHz. Glycosyl residues are labeled in roman numerals and their carbons in arabic numerals. I: 5-*O*-linked α-Ara*f*s; II: 3,5-di-*O*-linked α-Ara*f*s, III: 2-*O*-linked α-Ara*f*s; IV: t-α-Man*p*s; V: t-β-Ara*f*s; VI: 6-*O*-linked α-Man*p*s; VII: 2-*O*-linked and 2,6-di-*O*-linked α-Man*p*s.

to the t-α-Man*p* residues. These results demonstrate that Man*p* units are restricted to the mannan core, revealing that the LAM from *M. smegmatis* belong to the AraLAM class.

C. Structural Characterization by Fast-Atom Bombardment Mass Spectrometry of the Mannooligosaccharide Caps from *M. bovis* BCG ManLAM

The LAM was submitted to a five-step analytical approach [5] involving (1) mild hydrolysis (0.1 N HCl, 90°C, 15 min), (2) gel filtration chromatography on Bio-Gel P4, (3) derivatization of the included oligosaccharides with the chromophore *p*-aminobenzoic ethyl ester (ABEE), (4) fractionation of the ABEE derivatives by reverse phase high performance liquid chromatography (HPLC) monitored by UV, and (5) analysis by fast-atom bombardment mass spectrometry (FAB/MS) and tandem fast-atom bombardment mass spectrometry (FAB/MS/MS). The products (1 mg) of

ManLAM hydrolysis were applied to a Bio-Gel P4 in 0.1 N acetic acid. Molecules in the void volume were found to contain, by routine monosaccharide analysis, Ara and Man in the ratio 1/20, indicating that this fraction is mainly composed of the mannan core. The included fraction was dried, dissolved in 10 µL of water followed by the addition of 40 µL of the ABEE reagent solution (165 mg of ABEE and 35 mg of sodium cyanoborohydride in 350 µL of methanol and 40 µL of glacial acetic acid). The glass vials were capped, vortexed, and heated at 80°C during 90 min. The crude reaction mixture was partitioned in 2 mL of CHCl₃ and water. The aqueous phase containing the ABEE oligosaccharide derivatives (Fig. 8) was lyophilized and analyzed by reverse phase HPLC. The C_{18} column (5 µm Spherisorb) was eluted by a gradient of 5 to 80% of acetonitrile in water with a flow rate of 2 mL/min. The HPLC profile (Fig. 9) showed a major peak, f, with a retention time of 20 min, characteristic of an ABEE monosaccharide derivative. This sample was then analyzed by FAB/MS in positive mode using as a mixture of glycerol and thioglycerol as the matrix. The FAB mass spectrum showed two intense peaks at m/z 300 and 322, assigned to protonated $(M + H)^+$ and cationized $(M + Na)^+$ molecular ions of Ara-ABEE. With this method, we were able to assay each component, and, for example, f represented 80% of the P4 included fraction. In addition to peak f, less intense well-resolved peaks were observed (Fig. 9). Among them, peak b predominated (15.5 min) and represented 12% of the total fraction. The FAB/MS spectrum of b showed again two peaks $(M + H)^+$ and $(M + Na)^+$ at m/z 624 and 646, respectively, establishing that b is a trisaccharide containing one Ara$f$ and two Man$p$ units. The fragment ions were missing, but the mass analysis ion kinetic energy (MIKE) spectrum (Fig. 10) of the precursor ions m/z 624 showed two inter-residue fragment ions Y1 (m/z 300) and Y2 (m/z 462) arising from the loss of anhydro Man$p$1 → 2Man$p$ and anhydro-Man$p$, respectively. The following structure was proposed for the mannooligosaccharide b: Man$p$-(1 → 2)-Man$p$-(1 → 5)-Ara$f$. Similarly, peak a was assigned to a linear tetrasaccharide composed of three Man$p$s and one Ara$f$, whereas peaks c and d were

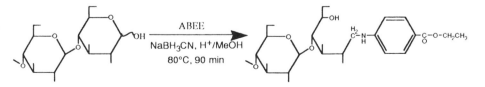

Figure 8 Derivatization of oligosaccharides with ABEE (*p*-aminobenzoic ethyl ester).

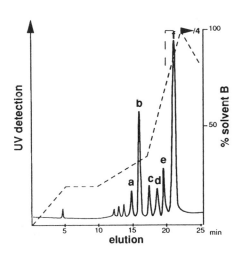

Figure 9 Reversed phase HPLC profile (semipreparative column C_{18}, 9.5 × 250 mm, 5-μm Spherisorb ODS2) of ABEE derivatives obtained from LAM of *M. bovis* BCG after mild hydrolysis (0.1 N HCl, 15 min, 90°C), gel filtration, and derivatization with aminobenzoic ethyl ester. Solvent A: 5% acetonitrile in water; solvent B: 80% acetonitrile in water.

attributed to a heterodisaccharide and to an arabinose homodisaccharide, respectively. Thus, these data unambiguously demonstrate that the arabinan side chains are capped by mannose (16%), and di- (70%) and tri- (14%) mannosides.

D. Arabinan Domain Structure

The arabinan backbone was established by methylation analysis. In this method, the ManLAM permethylation step was critical. The Ciucanu and Kerek [14] protocol using solid sodium hydroxide was performed. To 200 μg of dried ManLAMs in a Teflon-lined, screw-capped tube were added 100 μL of anhydrous dimethyl sulfoxide and one pellet of finely powdered NaOH. The mixture was then sonicated for 15 min followed by the addition of 300 μL of methyl iodide, and sonication was repeated. After 2 hr, 300 μL of water and sodium thiosulfate were added to neutralize the iodine. The permethylated ManLAM was extracted directly from the mixture by $CHCl_3$. The Ciucanu reaction was repeated twice in order to obtain complete ManLAM permethylation. These derivatives were hydrolyzed with 2 N TFA at 100°C for 2 hr and then reduced to alditols, peracetylated, and analyzed by GC and gas chromatography mass spectrometry (GC/MS).

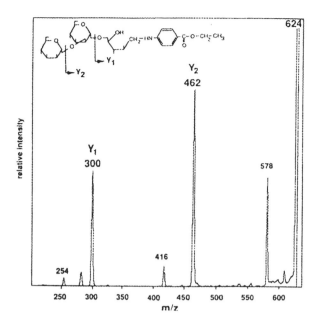

Figure 10 Positive ion MIKE spectrum of the protonated molecular ion (m/z 624) of fraction b from the HPLC profile (Fig. 8) of oligosaccharide-ABEE obtained from LAM of *M. bovis* BCG. The matrix was glycerol/thioglycerol.

Except for the 5-*O*-linked Ara, it was proposed, from the data summarized in the Table 1, that all the Ara and Man are in the furanose and pyranose forms, respectively. The arabinan domain is composed of a main chain formed of 5-*O*-linked α-D-Ara*f*s and side chains linked at position C3. These side chains contain 2-, 5- and 3,5-di-*O*-linked α-D-Ara*f* residues and are capped either by β-D-Ara*f* or mannooligosaccharides. The degree of the mannose caps was estimated from the ratio t-β-D-Ara*f* / 3,5-di-*O*-linked α-Ara*f* + 1. In the case of the ManLAM from *M. bovis* BCG, this ratio indicates that 80% of the arabinan chains are capped by mannooligosaccharides (Fig. 1).

E.   Structural Analysis of the *M. bovis* BCG Mannan Core

1.   *Partial Hydrolysis*

To examine the structure of the D-mannan domain and to locate the presence of the phosphatidyl inositol anchor at its reducing end, the mannan core was prepared from the deacylated ManLAM (dManLAM) by mild acidic hydrolysis. This procedure takes advantage of the higher lability to

Table 1   Methylation Analysis of LAM from *M. bovis* BCG

Full name of the partially methylated alditol acetate	Abbreviated name of the glycosyl residue	Mol %
2,3,5-tri-*O*-Me-1,4-di-*O*-Ac-arabinitol	t-Ara*f*	2
3,5-di-*O*-Me-1,2,4-tri-*O*-Ac-arabinitol	2-*O*-linked Ara*f*	7
2,3-di-*O*-Me-1,4,5-tri-*O*-Ac-arabinitol	5-*O*-linked Ara*f*	37
2-*O*-Me-1,3,4,5-tetra-*O*-Ac-arabinitol	3,5-di-*O*-linked Ara*f*	10
2,3,4,6-tetra-*O*-Me-1,5-di-*O*-Ac-mannitol	t-Man*p*	24
3,4,6-tri-*O*-Me-1,2,5-tri-*O*-Ac-mannitol	2-*O*-linked Man*p*	10
2,3,4-tri-*O*-Me-1,5,6-tri-*O*-Ac-mannitol	6-*O*-linked Man*p*	3
3,4-di-*O*-Me-1,2,5,6-tetra-*O*-Ac-mannitol	2,6-di-*O*-linked Man*p*	8

mild acidic conditions of the arabinofuranosyl residues. To 10 mg of dea-cylated ManLAM (prepared as previously described [5]), which had been lyophilized in 100 × 8 mm Teflon-lined, screw-capped tube, 1.5 mL of a solution of 0.1 N HCl (constant boiling grade) was added. The tube was tightly capped immediately and placed in a sand bath at 100°C for 15 to 20 min. The reaction was then stopped by cooling in an ice bath and neutral-ized by adding an appropriate volume of NaOH solution from a 0.5 N solution to minimize the volume.

2.   Purification

Isolation of the mannan core from the hydrolyzed arabinose residues was then performed by gel filtration. The solution (≈ 2 mL) was deposited gently onto a Bio-Gel P-4 (Bio-Rad medium bead size) column (50 cm × 2.5 cm i.d.), equilibrated with water. Elution was performed with water at 6 mL/hr with a peristaltic pump, and the eluant was monitored with a refractive index detector. Four-milliliter fractions were collected with a fraction collector. Under these conditions, the mannan core eluted as a single peak in the void volume (V_0), whereas the arabinose co-eluted with NaCl (arising from neutralization) in an included peak $>2V_0$. The fraction corresponding to the excluded material was then pooled, dried by lyophili-zation, and after hydrolysis, analyzed for monosaccharides.

3.   Monosaccharidic Composition

Monosaccharidic composition analysis has been reviewed by R. K. Merkle and I. Poppe [15], who discussed several current methods. We routinely conduct these analyses either by GC (after methanolysis and trimethylsilyla-tion) or by HPAEC using a CarboPac PA-1 column. The latest technology, coupled with PAD (DX 300 Dionex System) appears to be more sensitive for carbohydrates and does not require prederivatization. However, GC analysis is a better approach for analyzing Gro (glycerol) and Ins. More acidic hydrolysis conditions (6 N HCl, 110°C, 18 hr) are necessary to liberate the Ins from the P-Ins (inositol-1-phosphate) [16]. Thus, compositional analysis of the *M. bovis* BCG mannan core showed that it was mainly composed of Man, with low amounts of Ara (Ara/Man < 2/20 determined by GC or HPEAC) and Ins (Ins/Man < 1/50 determined by GC).

4. Liquid Secondary Ion Mass Spectrometry Analysis

To define further the structure of the mannan core, liquid secondary ion mass spectrometry (LSIMS) on ZAB 2E mass spectrometer was performed. The sample ($\approx 10 \mu g$) was mixed directly on the target with a drop of the appropriate matrix. For positive ion detection mode, the matrix consisted of a mixture of thioglycerol with 10% acetic acid. Only thioglycerol was used for negative ion detection. Spectra were generated by a 35-kV ccsium ion beam with an acceleration voltage of 8 kV. In the high mass range, the negative ion detection mode mass spectrum [17] revealed the presence of an intense peak at m/z 3037.9, preceded by two lower intensity peaks at m/z 3015.9 and 2999.8. The mass differences of 22 and 38 amu, respectively, between these two ions and the major one suggested the presence of sodium and potassium adducts. This assumption was confirmed either by doping the matrix with KI, which induces the extinction of all sodium containing ions, or by recording the spectrum in the positive ion detection mode. In this case, the LSIMS spectrum of the mannan core was dominated by a major ion at m/z 3039.6, in agreement with the following adduct: $(M + Na + K - H)^+$. The average molecular mass of the major compound is 2978.4 Da agreeing with the following composition: 16 (Man + Ins), 1 Ara, 1 Gro, and 2 phosphates. However, the positive ion detection mode LSIMS spectrum also revealed a signal at m/z 3539.3 (Fig. 11). This ion was assigned to a second cationized molecular species. The mass difference of 499.8 amu observed between the ions at m/z 3539.3 and 3039.6 cannot be explained only by a difference in the number of monosaccharide units. From the molecular weight of 3517 Da, the following composition was proposed: 11 (Man + Ins), 2 Ara, 1 Gro, and 1 phosphate. The mannan core mixture was next purified by anion-exchange chromatography (AEC) monitored by PAD.

5. Anion-Exchange Chromatography of the Mannan Core

Using preparative AEC-PAD as described in Section II, the homogeneous mannan core purified by Bio-Gel P-4 gel filtration was fractionated into two peaks eluting near the void volume, Ma-II (3 min) and Ma-I (4.5 min) (data not shown) [17]. Only small quantities of material were detected using the sodium acetate gradient (between 15 and 20 min). Because the

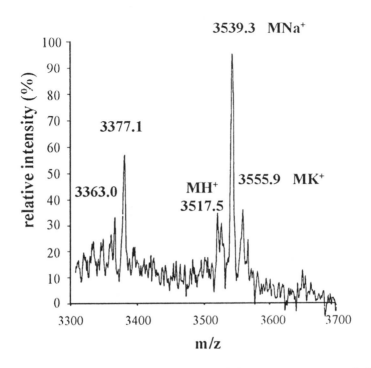

Figure 11 Partial LSIMS spectrum of the mannan core recorded in positive ion detection mode using thioglycerol–acetic acid 10% as a matrix.

separation on the quaternary ammonium CarboPac PA-1 column at neutral pH depends on the oligosaccharide charge-to-mass ratio, this difference in the retention time is in agreement with a difference in the charge state of the mannan core. Quantitative GC analysis of Ma-I gave a molar ratio Ara/ Man and Ins/Man of 1/20, whereas in the case of Ma-II the ratios were found to be 2/20 and 1/80, respectively. Moreover, determination of the phosphate content [18] revealed nearly five times as much phosphorus in Ma-I (0.7 mol per mol of Ma-I) as in Ma-II (0.15 mol per mol of Ma-II). The LSIMS spectrum of Ma-I, in positive ion detection mode, showed a major ion at m/z 3539.3, previously observed in the spectrum of the mixture of mannan core (Fig. 11). However, LSIMS analysis of the Ma-II was unsuccessful in either positive or negative ion modes and may be associated with the low amount of phosphate in this fraction. Thus, to assess those differences and to define more precisely the structure of Ma-I and Ma-II, these were analyzed by high resolution NMR spectroscopy.

6. Nuclear Magnetic Resonance Spectroscopy

This section is not intended to review NMR spectroscopy for carbohydrate structural analysis (for this purpose, see [19] and [20]), but only to present the application of this technique to the analysis of these oligosaccharides.

The overall strategy of analysis of the mannan cores by NMR spectroscopy involved different experiments using one- (1D) and two-dimensional (2D) homonuclear and heteronuclear correlation spectroscopy. The ^{1}H 1D spectra of both Ma-I and Ma-II were very similar and differed only in the relative intensity of some reporter signals resonating in the anomeric proton region (between 4.6 and 5.3 ppm). To assign these signals accurately, an exhaustive examination of Ma-I and Ma-II was performed by ^{1}H–^{1}H correlation (COSY and TOCSY) spectroscopy. Correlation spectroscopy (COSY) allows the localization of the carbohydrate H-2 resonance from the anomeric proton signal. The total correlation spectroscopy (TOCSY, also termed HOHAHA) experiment is probably the most interesting experiment for carbohydrate analysis, as it reveals the correlations between all the protons coupled by scalar interactions, which in the case of carbohydrate means all the protons of a monosaccharide unit. Such a network of correlations enables the determination of the ring proton spin system, and analysis of the coupling constants of the corresponding proton signals identifies the type of monosaccharide residues. However, in the case of complex oligosaccharides, where the proton spectra are difficult to analyze, 2D ^{1}H–^{13}C heteronuclear correlation experiments (HMQC, heteronuclear multiple bound connectivity spectroscopy [HMBC], and HMQC-TOCSY) are often required. These experiments, designed to observe scalar magnetization transfers between the low natural abundance ^{13}C and ^{1}H atoms, provide complementary information, sometimes not obtainable by other techniques. In the present case, due to strong overlapping of correlation signals in the 2D ^{1}H–^{1}H homonuclear correlations spectra, ^{1}H–^{13}C heteronuclear correlation experiments were necessary to assign unambiguously the major proton signals observed in the anomeric proton resonance region. By this strategy, the most shielded signal at δ 4.915 ppm was attributed to the 6-O-linked α-D-mannopyranoside units. In the same way, the signals at δ 5.06, 5.11, and 5.14 ppm were assigned to the t-Man$p$, Ara$f$, and 2,6-Man$p$ residues, respectively. Moreover, the TOCSY spectrum of Ma-I allowed the identification of the complete Ins proton spin system, demonstrating that the signal at 3.34 ppm observed in the Ma-I 1D spectrum corresponded to the Ins H-2 (Table 2). Analysis of the chemical shifts of these Ins proton signals revealed also that the H-1, H-2, and H-6 resonate at lower fields than those of the Ins-1-P standard [21]. These deshieldings are in agreement with glycosylation on both C-2 and C-6. The presence of a mannose residue on the C-2 of the Ins was confirmed by ^{1}H–^{1}H dipolar

Table 2　Chemical Shifts[a] of Some Protons and Carbons of Ma-I

	H-1	C-1	H-2	C-2	H-3	C-3	H-4	C-4	H-5	C-5	H-6	C-6
6-Manp	4.915 4.93	102.16	4.035 4.005	72.73	3.85	73.44	3.75	69.5		73.75		68.35
t-Manp	5.06	104.9	4.095	72.75	3.831	73.29	3.683	69.5	3.775	76.12	3.74–3.83 3.96–3.9	63.8
Araf	5.11		4.16		3.967		4.07					
2,6-Manp	5.14 5.15	100.89	4.055 4.045	81.45	3.945 3.97	73.35	3.83	69.3	3.825	74.05		68.35
Ins	4.16		3.34		3.62		3.67		3.40		3.87	

[a]Chemical shifts are given in ppm and were measured in D_2O at 303 K.

correlation spectroscopy (rotating frame nuclear Overhauser spectroscopy, ROESY), which showed a spatial proximity between the Ins H-2 and Man$p$ H-1. This analysis confirmed the lower amount of Ins in Ma-II (Table 3), as was previously shown by the quantitative GC analysis. Moreover, on basis of the relative integration values measured for the 6-Man$p$ H-1 signal versus t-Man$p$ H-1, it appeared that the Ma-II was more branched than the Ma-I. Phosphorus 1D NMR spectroscopy also confirmed the absence of phosphorus in Ma-II, in agreement with the phosphorus assay. The ^{31}P 1D NMR spectrum of Ma-I revealed a single resonance at δ −0.1. This signal was shown not to be affected by the pH in the pD range from 6 to 10 [22], suggesting a phosphodiester group. The phosphate substituents were further investigated by ^{1}H–^{31}P heteronuclear correlation experiments (HMQC, HMBC, and HMQC-TOCSY). Figure 12A shows the ^{1}H–^{31}P HMQC-TOCSY spectrum of Ma-I recorded at 500 MHz, and Fig. 12B shows a region of the Ma-I 2D TOCSY spectrum at the frequency (δ_{f1} 4.34) of the Ins H-2, showing the complete myo-Ins spin system. The magnetization transfers observed in the ^{1}H–^{31}P HMQC-TOCSY experiment are schematically represented in Fig. 12C. This experiment consisted of irradiating the phosphorus atom by a pulse at its frequency and then observing the transfer of energy (magnetization) to the first neighboring protons (^{3}J coupled proton) and from these to the other protons of the spin system. From these data, we concluded that the phosphate is linked to the Ins. Moreover, the ^{1}H–^{31}P HMQC spectrum, which only shows the ^{3}J coupled protons, confirmed the position of the linkage on the C-1 of the Ins (data not shown). In the same way, the second substituent of the phosphate was identified as a glycerol [17] from the characteristic chemical shifts of the protons of this spin system [22].

Thus, from detailed LSIMS and NMR analyses of the mannan core, we unambiguously confirmed the presence of the phosphatidyl-inositol an-

Table 3   Relative Intensity of the Reporter Proton Signals of Ma-I and Ma-II

Reporter signals	Chemical shifts[a]	Ma-I	Ma-II
2,6-di-O-linked Man$p$ H-1	5.14	11.1	10.9
Ara$f$ H-1[b]	5.11		
t-Man$p$ H-1	5.06	9.3	9.5
6-O-linked Man$p$ H-1	4.92	4.7	3.1
Ins H-2	4.34	1	0.25

[a]Measured at 303 K in D$_2$O.
[b]These two signals were too close to be integrated separately.

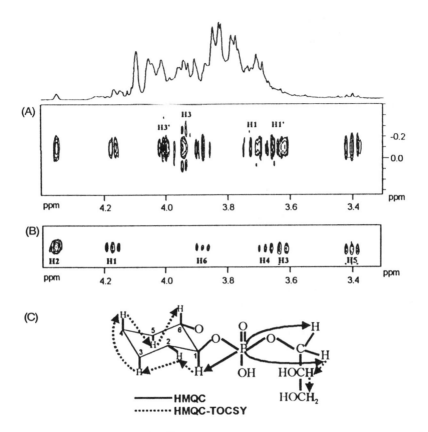

Figure 12 Spectrum (^{1}H–^{31}P HMQC-TOCSY) of Ma-I revealing the two substituent proton spin system of the phosphate moiety (A). The ^{1}H–^{31}P HMQC-TOCSY spectrum was recorded with spectral windows of 4500 Hz in the F2 dimension (^{1}H) and 1000 Hz in the F1 dimension (^{31}P) and 4096 × 64 (TPPI) point data matrix with 120 scans per t_1 value. Relaxation delay was 1.5 sec, and the mixing time 120 msec. The original data matrix point was zero-filled to 4K × 1K points, and multiplied by a $\pi/2$-shifted squared sine-bell (SSB = 2) function in both dimensions before Fourier transformation. (B) Partial ^{1}H–^{1}H TOCSY spectrum of Ma-I showing the correlations of the Ins H-2 with all the protons of the Ins spin system. Magnetization transfer in the 2D TOCSY experiment was performed by a 120-msec MLEV-17 sequence. The spectrum was recorded with a spectral width of 4500 Hz in both dimensions, a 4096 × 512 data matrix, and 16 scans for each t_1 increment. The data matrix was processed as in panel A. (C) Diagram of the magnetization transfers observed in the ^{1}H–^{31}P HMQC (line) and ^{1}H–^{31}P HMQC-TOCSY (dashed line) experiments.

chor at the reducing end of Ma-I. More interesting was the observation of a mannan core (Ma-II) devoid of such an anchor, as well as a component bearing two phosphates (observed in the negative mode LSIMS spectrum of the crude mannan core) and which was not found after anion-exchange chromatography. One possible explanation for the lack of recovery of this compound could be its low relative abundance. This assumption is not in contradiction with the high intensity of its pseudomolecular ion observed in LSIMS, as it is well known that similar products can have different ionization efficiencies. To date, the biological significance of this heterogeneity in the phosphorylation state of the mannan core is unknown. However, our understanding of the *M. bovis* BCG ManLAM properties will have to take into account the possible variation of the phosphorylation and, thus, the charge state of this molecule.

F. ManLAM Tagging

To explore ManLAM interactions with immune cells, we reported a procedure to label polysaccharides devoid of reducing monosaccharides with a fluorescent marker [9]. The biotin was covalently linked to the ManLAM allowing the formation of stable complexes with streptavidin coupled to a fluorophore. Then, the interaction of the biotinylated ManLAM with the immune cells was characterized by flow cytofluorometry. The ManLAM oxidation step to generate aldehydic groups was optimized in order to preserve the structural and functional integrity of the native ManLAM. The structural integrity was measured by GC determination of the Man$p$s and Ara$f$s, whereas function was tested by ELISA and immunoblot using a monoclonal antibody (mAb) (L9) known to bind the ManLAM. The following optimal conditions were determined: 0.5 mg/mL of ManLAM (30 nmol) in 100 mM sodium acetate (pH 5.5) was oxidized with 1 mM sodium periodate at 0°C in the dark. This amount of periodate would cleave a maximum of 50% of the ManLAM monosaccharides, as 30 nmol of LAM contained approximately 3 μmol of monosaccharides. The 10 min reaction time was selected because GC analysis indicated no significant amount of oxidized monosaccharide, whereas ELISA showed a 15% decrease in activity. For up to 2 hr, the mAb reactivity drastically decreased. The periodate excess was reacted with sodium sulfite (1 mL of an 80 mM aqueous solution). The biotin derivative was then covalently coupled to the oxidized ManLAM via the aldehyde groups, generated by the oxidation reaction. To the oxidized ManLAM was added 1 mL of a 15 mM solution of biotinamidohexanoyl hydrazide in 100 mM sodium acetate (pH 5.5) at room temperature. After 1 hr, the reaction mixture was dialyzed against phosphate-buffered saline (10 mM sodium phosphate, 150 mM NaCl, pH 7.4) at 4°C.

The biotinylated ManLAM showed the same SDS-PAGE behavior as the native one. After electrotransfer to nitrocellulose, this band was shown to be coupled to avidin using an alkaline phosphatase coupled reaction. One biotinyl residue per five molecules of ManLAM was determined by the 4′-hydroxyazobenzene-2-carboxylic acid dye-displacement method (ImmunoPure HABA kit; Pierce) [23].

G. Interaction of the Biotinylated ManLAM with Immune Cells

Murine splenocytes were obtained from 8- to 12-week-old female C57/6 mice [9]. The mice were killed by cervical dislocation, their spleens were removed, and single cell suspensions were prepared in Iscove's modified Dulbecco medium (IMDM; Gibco BRL). The erythrocytes were lysed by incubation in a hypotonic buffer (150 mM NH_4Cl, 10 mM $KHCO_3$, 0.1 mM EDTA) for 5 min at 37°C. The cells were washed twice with IMDM and suspended to a density of 10^6 cells/mL in IMDM supplemented with 10% heat-inactivated fetal calf serum (30 min at 56°C), penicillin, streptomycin, glutamine, and 50 μM 2-mercaptoethanol (Gibco). Splenic cells (10^6) were incubated in plastic tubes with or without 10 μg of labeled ManLAM in a final volume of 1 mL of culture medium for 90 min at 37°C. After incubation, the cells were washed twice with cold PBS buffer and resuspended in 100 μL of PBS containing 1% BSA, 0.02% NaN_3 with appropriate dilutions of both mAb coupled to fluorescein isothiocynate (FITC: anti-phagocytes, anti-B-lymphocytes, or anti-T-lymphocytes) and phycoerythrine–streptavidin conjugate to detect biotinylated-ManLAM. The cells were incubated for 30 min at 4°C and washed twice in PBS, 1% BSA, 0.02% NaN_3, and finally suspended in PBS, 1% paraformaldehyde, 0.02% NaN_3 for cytofluorimetry.

Figure 13 shows the phycoerythrine fluorescence of $B220^+$ (B-lymphocytes), $H57^+$ ($\alpha\beta$-T-lymphocytes), and $Mac1^+$ (macrophages and granulocytes) cells, incubated with or without biotinylated ManLAM. Both T- and B-lymphocytes incubated with labeled LAM showed an increase in fluorescence of +100 to +150 FU (arbitrary fluorescence units) relative to those without ManLAM, whereas phagocytic cells presented a greater increase in fluorescence, ~ +800 FU. The LAM-dose/effect titration was down with various splenic cells, and ManLAM-binding to phagocytes was still detectable at 1 μg/mL, whereas on T- and B-cells at these concentrations, no significant binding was observed. This dose/effect analysis corroborates the aforementioned data, which showed a higher interaction of ManLAM with granulocytes and macrophages, that is, phagocytes, than with T- and B-lymphocytes. The binding to phagocytes was totally inhibited

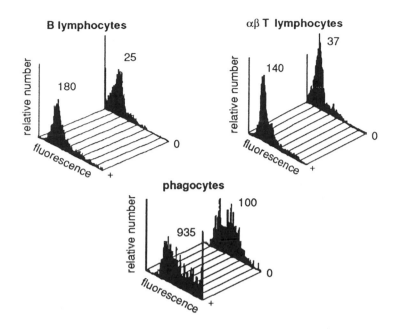

Figure 13 Flow cytometry analysis of representative binding experiments with biotinylated ManLAM to the surface of splenic cells. Cells were incubated for 90 min at 37°C with (+) or without (0) biotinylated ManLAM (10 μg/mL), stained for 30 min with the appropriate FITC-mAb and streptavidin–phycoerythrin, and analyzed by FACS. The numbers indicate the mean fluorescence intensity in phycoerythrin of the selected cell population.

by a 100-fold molar excess of native unlabeled ManLAM, indicating the specificity of the binding assay. This binding was partly inhibited by temperature, trypsin, EDTA, and mannan, suggesting that the ManLAM from *M. bovis* BCG was bound to the macrophage mannose receptor.

ABBREVIATIONS

ABEE	p-aminobenzoate ethyl ester
Ara*f*	arabinofuranose
AraLAM	LAM with arabinofuranosyl termini
BCG	Bacille Calmette–Guérin
COSY	correlation spectroscopy
dLAM	deacylated LAM
dManLAM	deacylated ManLAM

FAB/MS	fast-atom bombardment mass spectrometry
FAB/MS/	tandem fast-atom bombardment spectroscopy
FACS	fluorescent activated cell sorting
FITC	fluorescein isothiocyanate
GC	gas chromatography
GC/MS	gas chromatography mass spectrometry
Gro	glycerol
HMBC	heteronuclear multiple bound connectivity spectroscopy
HMQC	heteronuclear multiple quantum correlation spectroscopy
HPAEC	high pH anion-exchange chromatography
HPLC	high performance liquid chromatography
IMDM	Iscove's modified Dulbecco medium
Ins	inositol
LAM	lipoarabinomannans
LM	lipomannans
LSIMS	liquid secondary ion mass spectrometry
mAb	monoclonal antibody
MALDI-TOF	matrix-assisted laser desorption ionization time-of-flight
ManLAM	LAM with mannosyl units capping the arabinan ends
Man*p*	mannopyranose
MIKE	mass analysis ion kinetic energy
NMR	nuclear magnetic resonance
PAD	pulsed amperometric detector
PAGE	polyacrylamide gel electrophoresis
PIM	phosphatidyl-inositol mannosides
P-Ins	inositol-1-phosphate
ROESY	rotating frame nuclear Overhauser spectroscopy
t	terminal
TMS	trimethylsilyl ester
TNF	tumor necrosis factor
TOCSY	total correlation spectroscopy

REFERENCES

1. Daniel, T. M. (1984). In Kubica, G. P., and Wayne, L. G., eds., *The Myco-bacteria. A Sourcebook*. Marcel Dekker, New York, pp. 417–465.
2. Chatterjee, D., Bozic, C. M., McNeil, M., and Brennan, P. J. (1991). *J. Biol. Chem.* 266, 9652–9660.
3. Chatterjee, D., Lowell, K., Rivoire, B., McNeil, M., and Brennan, P. J. (1992). *J. Biol. Chem.* 267, 6234–6239.

4. Chatterjee, D., Khoo, K.-H., McNeil, M., Dell, A., Morris, H. R., and Brennan, P. J. (1993). *Glycobiology* 3, 497–506.
5. Venisse, A., Berjeaud, J. M., Chaurand, P., Gilleron, M., and Puzo, G. (1993). *J. Biol. Chem.* 268, 12401–12411.
6. Prinzis, S., Chatterjee, D., and Brennan, P. J. (1993). *J. Gen. Microbiol.* 139, 2649–2658.
7. Khoo, K.-H., Dell, A., Morris, H. R., Brennan, P. J., and Chatterjee, D. (1995). *J. Biol. Chem.* 270, 12380–12389.
8. Schlesinger, L. S. (1993). *J. Immunol.* 150, 2920–2930.
9. Venisse, A., Fournié, J. J., and Puzo, G. (1995). *Eur. J. Biochem.* 231, 440–447.
10. Sieling, P. A., Chatterjee, D., Porcelli, S. A., Prigozy, T. I., Mazzaccaro, R. J., Soriano, T., Bloom, B. R., Brenner, M. B., Kronenberg, M., Brennan, P. J., and Modlin, R. L. (1995). *Science* 269, 227–230.
11. Roach, T. I., Barton, C. H., Chatterjee, D., and Blackwell, J. M. (1993). *J. Immunol.* 150, 1886–1896.
12. Roach, T. I., Chatterjee, D., and Blackwell, J. M. (1994). *Infect. Immun.* 62, 1176–1184.
13. Chatterjee, D., Roberts, A. S., Lowell, K., Brennan, P. J., and Orme, I. M. (1992). *Infect. Immun.* 60, 1249–1253.
14. Ciucanu, I., and Kerek, F. (1984). *Carbohydr. Res.* 131, 209–217.
15. Merkle, R. K., and Poppe, I. (1994). In Lennarz, W. J., and Hart, G. W., eds., *Methods in Enzymology, Vol. 230, Guide to Techniques in Glycobiology.* Academic Press, San Diego, pp. 1–15.
16. Roberts, W. L., Kim, B. H., and Rosenberry, T. L. (1987). *Proc. Natl. Acad. Sci. USA* 84, 7817–7821.
17. Venisse, A., Rivière, M., Vercauteren, J., and Puzo, G. (1995). *J. Biol. Chem.* 270, 15012–15021.
18. Bartlett, G. R. (1959). *J. Biol. Chem.* 234, 466–468.
19. Van Halbeek, H. (1994). In Lennarz, W. J., and Hart, G. W., eds., *Methods in Enzymology, Vol. 230, Guide to Techniques in Glycobiology.* Academic Press, San Diego, pp. 132–167.
20. Hård, K., and Vliegenthart, J. F. G. (1993). In *Glycobiology. A Practical Approach.* IRL Press, Oxford, pp. 223–242.
21. Johansson, C., Kördel, J., and Drakenberg, T. (1990). *Carbohydr. Res.* 207, 177–183.
22. Glushka, J., Cassels, F. J., Carlson, R. W., and van Halbeek, H. (1992). *Biochemistry* 31, 10741–10746.
23. Green, N. M. (1965). *Biochem. J.* 94, 23c–24c.

8

Microscale Analysis of Glycosphingolipids by Thin-Layer Chromatography Blotting

Takao Taki
Tokyo Medical and Dental University, Tokyo, Japan

I. INTRODUCTION

Glycosphingolipids, as membrane components, have been shown to play important roles in the modulation of cell differentiation and proliferation, cell-to-cell recognition in inflammation and organogenesis, and other physiological functions [1–4]. Structural studies of glycosphingolipids are a prerequisite for understanding the relationships between their chemical structures and functions in cellular activities. To characterize glycosphingolipid structures, they must be purified to homogeneity by DEAE and multiple silica bead column chromatographies. These purification steps are time consuming, and the yields of the purified glycosphingolipids are always low.

We have established a new method, called thin-layer chromatography (TLC) blotting, in which various glycosphingolipids and phospholipids can be transferred from a high performance TLC (HPTLC) plate to a polyvinylidene difluoride (PVDF) membrane [5–7]. TLC blotting has the following advantages for lipid analysis: (1) the transfer of lipids from an HPTLC plate to a PVDF membrane is rapid (within 1 min); (2) the procedure is simple; (3) the transfer is quantitative; (4) chemical degradation does not occur; (5) microscale purification of glycosphingolipids can be done; and (6) structural analysis by mass spectrometry is possible without further sample handling. In this chapter, the TLC blotting method and its application to microscale analysis of glycosphingolipids are described.

II. MATERIALS AND METHODS

A. TLC Blotting

Standard glycosphingolipids consisted of GlcCer, LacCer, Gb_3Cer, Gb_4Cer, nLc_4Cer, GM3, $I^3SO_3GalCer$ (HSO_3-$3Gal\beta(1$-$1)Cer$), and a ganglioside mixture (GM1a, GD1a, GD1b, and GT1b) (abbreviations are given in legends to Figs. 2 and 7). Each glycosphingolipid (3 μg) was separated on an HPTLC plate (E. Merck, Darmstadt, Germany) using the following solvent system: chloroform/methanol/0.2% $CaCl_2$ (60/35/8 by volume). After developing, the HPTLC plate was dried thoroughly with a hair dryer and then dipped into blotting solvent (isopropanol/0.2% $CaCl_2$/methanol, 40/20/7 by volume) for 20 sec. The HPTLC plate was placed on a glass plate and covered first with a PVDF membrane sheet (Clear Blot Membrane P, ATTO Co., Ltd., Japan) and then a glass microfiber filter sheet (GF/A, Whatman International Ltd., Maidstone, England). All the layers were pressed evenly for 30 sec with an iron at 180°C. The PVDF membrane was dried after separation from the plate, and the transferred glycosphingolipids were made visible by dipping the membrane into orcinol–sulfuric acid reagent containing 50% (v/v) methanol and placing on a heating plate at 100°C. The membrane was washed with distilled water to remove all traces of sulfuric acid.

B. Immunostaining of Glycosphingolipids on a PVDF Membrane After TLC Blotting

Immunological detection of the glycosphingolipids on PVDF membranes was performed after drying and overnight incubation with an anti-glycosphingolipid monoclonal antibody (4°C in a nylon bag). The membrane was rinsed in phosphate-buffered saline (PBS) and then incubated with a horseradish peroxidase-conjugated anti-mouse IgM antibody (Cappel Laboratories, Westchester, Pennsylvania). After incubation at room temperature for 1.5 h, the membrane was rinsed with PBS. The glycosphingolipids on the membrane were detected with a Konica Immunostaining HRP Kit (IS-50B, Konica Co., Tokyo).

C. Purification of Glycosphingolipids and Phospholipids by TLC Blotting

1. Two-Dimensional Thin-Layer Chromatography

For purification using TLC blotting, glycosphingolipids were separated by two-dimensional TLC [6]. The solvent system (A) used for neutral glycosphingolipids (both dimensions) was chloroform/methanol/0.2% $CaCl_2$ (60/35/8 by volume). Solvent B (chloroform/methanol/0.2% $CaCl_2$ (55/45/10 by volume) and solvent C (propanol/H_2O/NH_4OH (75/25/5 by volume) were used in the first and second directions, respectively, for the separation of acidic glycosphingolipids. The glycosphingolipids were made visible with primuline reagent (0.0001% in acetone/water, 4/1 by volume). The bands were then marked with a colored pencil under UV light (365 nm). The glycosphingolipids and color marks on the HPTLC plate were then transferred by TLC blotting to a PVDF membrane. The glycosphingolipids in the marked areas could be extracted with small amounts of methanol and assessed for purity by TLC.

2. Extraction of Glycosphingolipids from the PVDF Membrane

The marked glycosphingolipids on the HPTLC plate were dipped in blotting solvent for 20 sec, and the glycosphingolipids were transferred as previously described. Any glycosphingolipids remaining on the HPTLC plate (detected by UV light) can be blotted onto another PVDF membrane. The PVDF membrane was then washed with water to remove the primuline reagent and dried. The marked areas were excised, and each piece was placed in a test tube. The neutral glycosphingolipids were extracted with 500 μL of chloroform/methanol (2/1, by volume) and sonication for 30 sec. For the acidic glycosphingolipids, sequential extractions with 500 μL of methanol and then chloroform/methanol (2/1 by volume) were performed. It was necessary to remove PVDF membrane contaminants. The purified, extracted lipid was dried and redissolved in chloroform/methanol (9/1 by volume) and then applied to a Pasteur pipette column (1 cm in height) of Iatrobeads (Diayatoron Co., Ltd., Tokyo). The contaminants were eluted with chloroform/methanol (9/1 by volume) and then the glycosphingolipids were eluted with chloroform/methanol/water (30/60/8 by volume). An aliquot of the extracted glycosphingolipids was analyzed by TLC.

D. Mass Spectrometric Analysis of Glycosphingolipids After Transfer to PVDF Membranes

Glycosphingolipids were developed on an HPTLC plate and visualized with primuline reagent and UV light, as described in Section II. C. The PVDF membranes, containing the transferred bands and marker pigment, were washed with water to remove the primuline reagent. A circle (2 mm in diameter) was excised from the marked bands and placed on the probe tip. One microliter of triethanolamine was added to the membrane on the probe tip and negative secondary ion mass spectra were obtained on a Finnigan TSQ70 spectrometer. The Cs$^+$ beam was at 20 kV, and the ion multiplier and conversion dynode were at 1.5 and 20 kV, respectively.

III. RESULTS AND DISCUSSION

A. TLC Blotting

Figure 1 shows a scheme of the TLC blotting method. The profile of the glycosphingolipids separated by HPTLC and then transferred onto the PVDF membrane is shown in Fig. 2. All glycosphingolipids were made

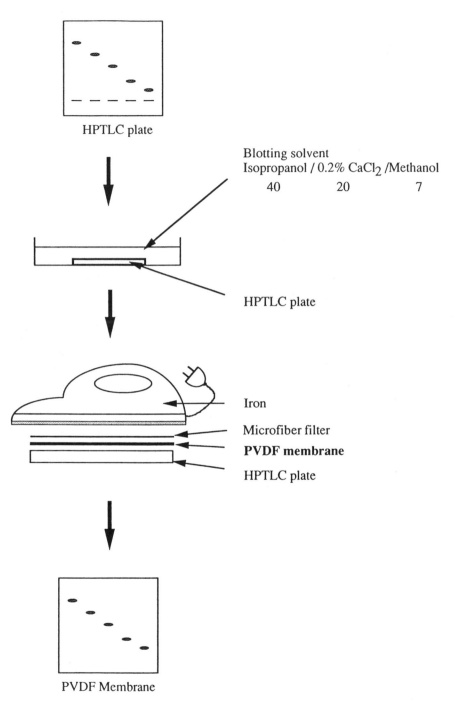

HPTLC plate

Blotting solvent
Isopropanol / 0.2% $CaCl_2$ /Methanol
40   20   7

HPTLC plate

Iron
Microfiber filter
PVDF membrane
HPTLC plate

PVDF Membrane

Figure 1 The TLC blotting method.

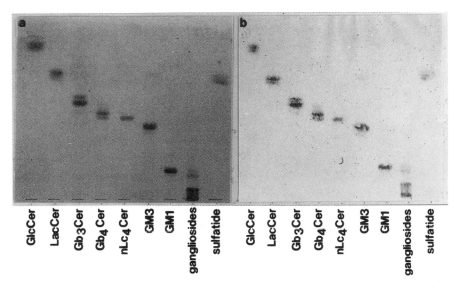

Figure 2 Glycosphingolipids blotted onto PVDF membranes. (a) A 3-μg sample of each glycosphingolipid was separated by TLC. (b) The glycosphingolipids, developed on the HPTLC plate, were blotted onto a PVDF membrane. The glycosphingolipids were visualized by dipping the membrane into a modified orcinol–H_2SO_4 reagent and then heated at 100°C. GlcCer = Glcβ(1-1)Cer; LacCer = Galβ(1-4)Glcβ(1-1)Cer; Gb$_3$Cer = Galα(1-4)Galβ(1-4)Glcβ(1-1)Cer; Gb$_4$Cer = GalNAcβ(1-3)Galα(1-4)Galβ(1-4)Glcβ(1-4)Cer; nLc$_4$Cer = Galβ(1-4)GlcNAcβ(1-3)Galβ(1-1)Cer; GM3 = NecAcα(2-3)Galβ(1-4)Glcβ(1-1)Cer; GM1a = Galβ(1-3)GalNAcβ(1-4)(NeuAcα(2-3))Galβ(1-4)Glcβ(1-1)Cer.

visible, as on the HPTLC plate, with orcinol–H_2SO_4 reagent. As no glycosphingolipids were detected on the HPTLC plate after blotting, most of the glycosphingolipids were transferred from the plate to the membrane. Because glycosphingolipids were present on the side of the membrane opposite to that in contact with the HPTLC plate, it was assumed that they passed through the membrane by capillary action during heating. Resorcinol–HCl and α-naphthol reagent are also useful for staining glycosphingolipids present on membranes, but these spray reagents should be diluted twofold with methanol. The color of the glycosphingolipids on the HPTLC plate stained with orcinol–H_2SO_4 faded with time, whereas the color on the PVDF membrane was stable for more than a year. This is attributed to the fact that further degradation of the visible compounds was stopped by removing the sulfuric acid from the membrane. Figure 3 shows the dose-dependent transfer of various glycosphingolipids by this method.

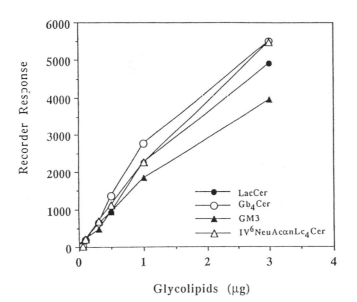

Figure 3   Dose-dependent blotting of glyosphingolipids to PVDF membranes. IV^6-NeuAcαnLc$_4$Cer = NeuAcα(2-6)Galβ(1-4)GlcNAcβ(1-3)Galβ(1-4)Glcβ(1-1)Cer; for other abbreviations, see legend of Fig. 2.

B. Immunostaining of Glycosphingolipids Transferred to PVDF Membranes

Figure 4 shows the dose-dependent immunological detection of transferred glycosphingolipids. The lower limit of detection was 0.3 ng of glycosphingolipid on the PVDF membrane, much lower than the amount detectable on HPTLC plates (~ 1.6 ng). This result suggests that glycosphingolipids developed on the HPTLC plate were then concentrated on one side of the PVDF membrane by TLC blotting and that no degradation of the epitopes occurred during the blotting.

C. Preparation of Glycosphingolipids and Phospholipids by TLC Blotting

TLC blotting can be used for purification of glycosphingolipids [6]. Figure 5 shows a scheme of preparative TLC blotting. Glycosphingolipids separated by two-dimensional TLC were made visible with primuline reagent; then bands were marked with a colored pencil under UV light. The glycosphingolipids on the HPTLC plate were then transferred by TLC blotting

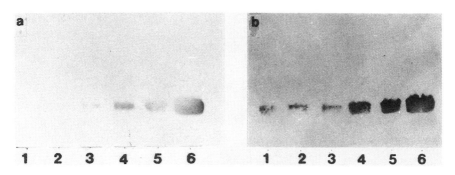

Figure 4 Immunological detection of IV^{6}NeuAcαnLc$_4$Cer blotted onto a PVDF membrane. Various amounts of IV^{6}NeuAcαLc$_4$Cer were separated by TLC (a) and then blotted onto a PVDF membrane (b). Lanes 1 to 6 correspond, respectively, to 0.32, 1.6, 8, 40, 200, and 1000 ng of IV^{6}NeuAcαnLc$_4$Cer. Immunostaining was done as described in Section II.

together with the color marks. The labeled areas were extracted with small amounts of methanol and monitored for purity by TLC.

By combining two-dimensional TLC and TLC blotting, we separated 20 glycosphingolipids that migrated as homogeneous bands on an HPTLC plate (Fig. 6). The same technique was used to purify the acidic glycosphingolipids from bovine brain and phospholipids (Fig. 7). During the purification process, there was no detectable decomposition of the glycosphingolipids.

The yields of 13 different glycosphingolipids were in the range 68.2–92.6% with a mean yield of 82.3%. For the preparation of glycosphingolipids, two sheets of PVDF membrane were placed over the HPTLC plate, because when large amounts of glycosphingolipids were used for TLC blotting, some lipids passed through the single membrane and were observed on the glass filter paper.

The purification method for glycosphingolipids by TLC blotting is simple and quantitative. The primuline reagent and drawing pencil were used to mark the bands on the HPTLC plate. This made it possible to detect lipids blotted to the PVDF membrane. The markings transferred to the membrane, locating the glycosphingolipids. The primuline reagent was removed by washing the membrane with water.

Purification of glycosphingolipids extracted from limited amounts of clinical samples or cultured cells is difficult, primarily because silica bead chromatography has been the sole preparative method. It is necessary, however, to repeat the chromatography many times to purify glycosphingolipids to a homogeneous band on an HPTLC plate, and in each purification

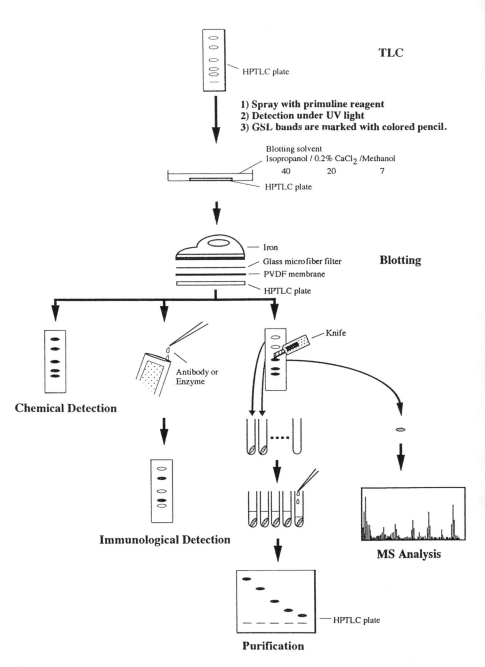

Figure 5   Glycosphingolipid analyses using TLC blotting.

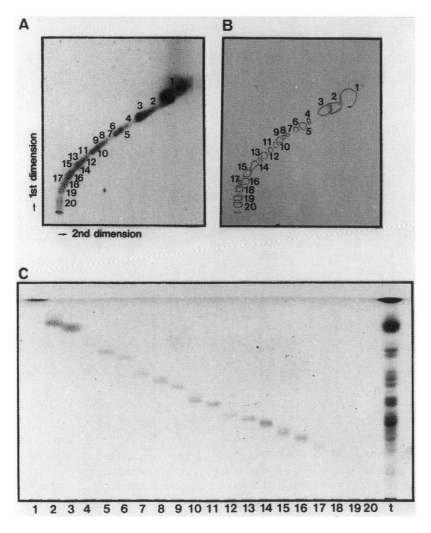

Figure 6 Purification of neutral glycosphingolipids from human meconium. The neutral glycosphingolipid fraction (125 μg) from human meconium was applied to two HPTLC plates and separated by two-dimensional TLC. One plate was sprayed with orcinol–H_2SO_4 reagent (A), the other with primuline reagent. The glycosphingolipid bands, visualized with UV light, were marked with a drawing pencil. In all, 20 bands were marked and subjected to TLC blotting. The locations of the transferred glycosphingolipids were identified by the markings that were transferred from the HPTLC plate (B). Each area on the PVDF membrane was excised, and the lipid was extracted and analyzed by TLC (C).

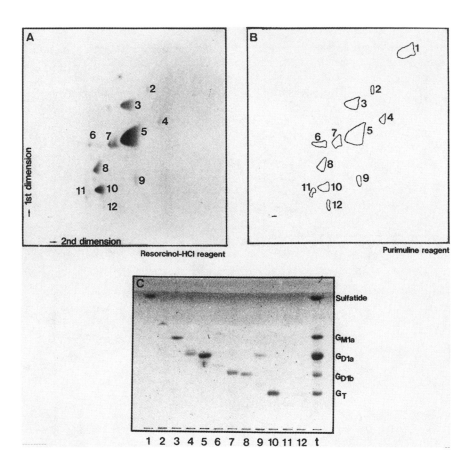

Figure 7 Purification of bovine brain gangliosides. The total acidic fraction from bovine brain (100 μg) was applied to two HPTLC plates and separated by two-dimensional TLC. One plate was sprayed with resorcinol reagent (A), the other was used for TLC blotting (B). The purity of the glycosphingolipids, extracted from the PVDF membrane, was determined by TLC (C). Twelve bands were observed. Nine different gangliosides (bands 2–10) together with a sulfatide (band 1) were separated by the TLC blotting technique. Bands 11 and 12 were in insufficient quantities to be detected with resorcinol reagent. G_{Mla} = Galβ(1-3)GalNAcβ(1-4)(NeuAcα(2-3))Galβ(1-4)Glcβ(1-1)Cer; G_{Dla} = NeuAcα(2-3)Galβ(1-3)GalNAcβ(1-4)(NeuAcα(2-3))Galβ(1-4)Glcβ(1-1)Cer; G_{Dlb} = Galβ(1-3)GalNAcβ(1-4)(NeuAcα(2-8)NeuAcα(2-3))Galβ(1-4)Galβ(1-1)Cer; G_{Tlb} = NeuAcα(2-3)Galβ(1-3)GalNAcβ(1-4)(NeuAcα(2-8)NeuAcα(2-3))Galβ(1-4)Glcβ(1-1)Cer.

step, significant amounts of sample are lost. Moreover, the yield of purified glycosphingolipid is further reduced because glycosphingolipid fractions eluted from the column are usually contaminated with only a few fractions containing the purified glycosphingolipid. The TLC blotting method overcomes these difficulties and is useful for purifying small amounts of glycosphingolipids.

D. Direct Mass Spectrometric Analysis of Glycosphingolipids Transferred by TLC Blotting

Microscale analysis of glycosphingolipids that combines TLC blotting and mass spectrometry has been developed [7]. Glycosphingolipids were transferred to PVDF membrane from an HPTLC plate by TLC blotting, after which a small piece of the glycosphingolipid band was excised and placed on the mass spectrometer probe tip covered with triethanolamine matrix. The sample was then analyzed by secondary ion mass spectrometry (SIMS). The structural analysis of all glycosphingolipids separated on the HPTLC plate was performed without further purification.

To optimize the conditions for TLC blotting/SIMS, experiments were performed with GM1a. First, the time required to dissolve GM1a into the triethanolamine matrix from the PVDF membrane was determined. A piece of the GM1a-blotted PVDF membrane was dipped in triethanolamine on the target and bombarded, after which the time-dependent change in the intensity of the protonated molecular ion was measured. The signal intensity was most intense within 3 min, indicating that GM1a partitioned into the matrix within a short time. The effect of temperature on the dissolution of GM1a into matrix from the PVDF membrane was next examined. There was no difference in signal intensity from room temperature to 100°C. Next, the lower limit of detection was examined. With GM1a, 0.1 μg could be analyzed. As only a part of the transferred band ($\sim 1/3$) on the PVDF membrane can be attached to the probe tip, repeat SIMS is possible.

Figure 8 shows one example of TLC blotting/SIMS. The spectra show deprotonated molecular ions and fragmentation patterns with low noise. TLC blotting/SIMS is an improvement to the TLC method reported by Kushi et al. [8]. In summary, TLC blotting/SIMS has the following advantages: (1) analysis can be done without column chromatography purification; (2) the fragmentation patterns of the spectra are apparent with low noise; (3) the method is simple, with no special devices required; (4) it is easy to specifiy which sample is to be analyzed, as the separated glycosphingolipid bands are visible; and (5) sample loss is negligible.

The TLC blotting/SIMS method has been applied to the analysis of incubation products. We characterized individual products that were

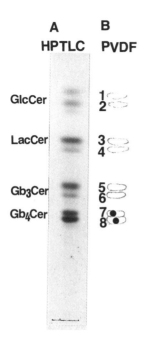

Figure 8 Secondary ion mass spectrometry of glycosphingolipids on a PVDF membrane. Two sets (1 μg each) of glycosphingolipids were separated on the same plate. The plate was cut in half; then one half was sprayed with orcinol–H$_2$SO$_4$ reagent to confirm the locations of the glycosphingolipids (A), and the other was subjected to TLC blotting/SIMS (B). Mass spectra of the glycosphingolipid bands (Nos. 7 and 8) are shown in the next page (C). The residue losses from the parent ion are indicated in the spectra. See legend of Fig. 2 for abbreviations.

generated from GalNAc–GD1a by incubation with clostridial sialidase in the presence of saposin B or GM2 activator protein [9]. Specific recognition of N-acetylneuraminic acid in the GM2 epitope by the GM2 activator protein was elucidated in this study. Recently, we have studied the glycosphingolipid composition of three kinds of tumor cell lines with different metastatic potentials. All the glycosphingolipids detected primuline reagent were characterized by TLC blotting/SIMS. In these experiments, the amounts of glycosphingolipids extracted from 10^7 cells were sufficient for analysis. Thus, TLC blotting/SIMS enabled the characterization of glycosphingolipid structures from very limited amounts of samples without significant loss of lipid.

Furthermore, we have developed a bacterial binding assay using TLC blotting [10]. A suspension of bacteria, radiolabeled with ^{35}S-methionine,

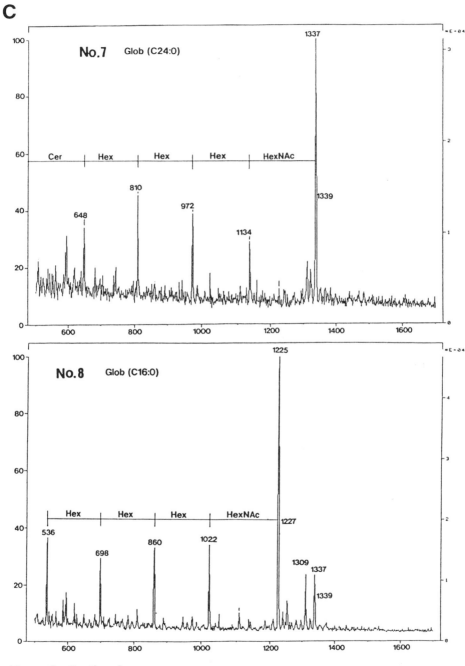

Figure 8 Continued

was layered and incubated over glycosphingolipids that had been transferred to a PVDF membrane. The radiolabeled bands, made visible with an image analyzer, were analyzed directly with SIMS. The carbohydrate structure and ceramide species were then elucidated.

IV. SUMMARY AND FUTURE DIRECTIONS

We have established a simple, microscale analytical method for glycosphingolipids using TLC blotting, as summarized in Fig. 5. The transferred glycosphingolipids can also be used for binding studies with virions, toxins, and lectins. Incubation systems with enzymes on the membrane seem to be a new approach for searching for activators and inhibitors as well as for new enzymes. Taking advantage of the stable binding of glycosphingolipids to the PVDF membrane, we have developed new methods to detect enzymes involved in glycosphingolipid metabolism [11]. The described procedures are useful not only in glycosphingolipid research, but also for any naturally occurring materials and drug metabolites that can be analyzed by TLC.

ABBREVIATIONS

DEAE	diethylamino ethanol
HPTLC	high performance thin-layer chromatography
PBS	phosphate-buffered saline
PVDF	polyvinylidene difluoride
SIMS	secondary ion mass spectrometry
TLC	thin-layer chromatography

REFERENCES

1. Hakomori, S. (1981). *Ann. Rev. Biochem.* 50, 733–764.
2. Spiegel, S., and Fishman, P. H. (1985). *Science* 230, 1285.
3. Hakomori, S. (1990). *J. Biol. Chem.* 265, 18713–18716.
4. Zeller, C. B., and Marchase, R. B. (1992). *Am. J. Physiol.* 262, C1341–C1355.
5. Taki, T., Handa, S., and Ishikawa, D. (1994). *Anal. Biochem.* 221, 312–316.
6. Taki, T., Kasama, T., Handa, S., and Ishikawa, D. (1994). *Anal. Biochem.* 223, 232–238.
7. Taki, T., Ishikawa, D., Handa, S., and Kasama, T. (1995). *Anal. Biochem.* 225, 24–27.
8. Kushi, Y., and Handa, S. (1985). *J. Biochem.* 98, 265–268.

9. Li, S.-C., Wu, Y.-Y., Sugiyama, E., Taki, T., Kasama, T., Casellato, R., Sonnino, S., and Li, Y.-T. (1995). *J. Biol. Chem.* 270, 24246–24251.
10. Kasama, T., Hisano, Y., Nakajima, M., Handa, S., and Taki, T. (1996). *Glycoconj. J.*, 13, 461–469.
11. Isobe, T., Naiki, M., Handa, S., and Taki, T. (1996). *Anal. Biochem.*, 236, 35–40.

9
High-Performance Liquid
Chromatography of Gangliosides

**Sandro Sonnino, Manuela Valsecchi, Paola Palestini,
Marco Nicolini, and Vanna Chigorno**
University of Milan, Milan, Italy

I. INTRODUCTION

Gangliosides are normal components of the plasma membrane of vertebrate cells [1] and are particularly abundant in the nervous system. Their concentration at the level of the cortex gray matter is about one-tenth that of total phospholipids [2]. They are asymmetrically located in the outer lipid layer of the membrane [3] and exhibit strong amphiphilic properties [4]. Gangliosides are glycosphingolipids constituted by a hydrophylic sialic acid–containing oligosaccharide and a hydrophobic ceramide portion, connected by a glycosidic linkage. The oligosaccharide portion protrudes from the outer membrane surface, and the ceramide moiety is inserted into the lipid core of the membrane. Gangliosides occur in nature in a large number of different species. The number and chemical features of sialic acid, and the number, type, and sequence of the individual sugars of the neutral oligosaccharide core are the basis for a wide variety of chemical structures. Moreover, a ganglioside with a homogeneous oligosaccharide portion can be a mixture of several molecular species differing in their ceramide composition (fatty acid of different hydrocarbon chain length, with or without double bonds or hydroxyl groups; long chain bases of different chain length, with or without double bonds).

Glycosphingolipids are assumed to serve as recognition markers at the cell surface [5] and can be involved in the various expressions of cell social behavior. Increasing evidence implies that gangliosides serve as membrane receptors and in the membrane-mediated transfer of information [6]. Their highly differentiated oligosaccharide chains provide a variety of sites for specific interactions with extracellular ligands. These interactions are followed by intramembrane events resulting in the activation of adenylate-cyclase and other systems controlling protein phosphorylation and dephosphorylation. A peculiar surface behavior of gangliosides, promoted by their binding to external ligands, is their self-association to form permanent "clusters" or "patches" [2]. The formation of ganglioside-rich phases on the membrane can lead to dramatic changes in the local organization of membranes (determination of fusogenic conditions; formation of micellar structures and lipid channels) [7]. These structural changes can induce conformational changes of membrane-bound proteins, and they probably constitute the molecular basis for the cascade of reactions that transduce signals through the membrane. Both the oligosaccharide and the lipid portions of the ganglioside molecule are involved in the aforementioned events. In fact, the formation of ganglioside clusters relies mainly on mutual interactions (hydrogen bonds; dipole–dipole interactions) within the polar head groups of gangliosides, such interactions being facilitated by the presence of cross-linking agents. However, although the possible influence of the ceramide portion of gangliosides on cluster formation cannot be excluded, it is expected that the ceramide portion plays a crucial role in changes of the membrane architecture that directly affect the inner hydrophobic core of the membrane and the interactions with membrane-bound proteins.

The maintenance of a constant composition in the sugar and lipid portions of glycosphingolipids appears to be essential for membrane function. This regularity is guaranteed by the genetic control of biosynthesis of the enzymes involved in glycosphingolipid metabolism, and by the fine regulation of their activities. A dramatic demonstration of the importance of this phenomenon is given by malignant cells where gene irregularities lead to different chemical expressions of cell surface glycolipids, such glycolipids being intimately related to the abnormal social behavior of cells [8,9].

II. MATERIALS AND METHODS

A. Preparation of Ganglioside Mixtures and Standard Gangliosides

The ganglioside mixtures were extracted from the total brain by the tetrahydrofuran-phosphate buffer procedure and purified by partitioning with diethyl ether [10]. Standard gangliosides homogeneous in both the oligosaccharide and ceramide moieties were prepared by liquid chromatography and high performance liquid chromatography (HPLC) [11].

Gangliosides were assayed as bound Neu5Ac by the resorcinol-HCl method [12,13], pure Neu5Ac being used as the reference standard.

B. Structure of Gangliosides

Table 1 shows the structures of the main gangliosides in the central nervous system (CNS). The oligosaccharides are linked to the ceramide moiety through a glucosidic linkage, and the structure of these gangliosides can be

Table 1 Structure of the Main Gangliosides from the Central Nervous System

GM3, II^{3}Neu5AcLacCer, α-Neu5Ac-(2-3)-β-Gal-(1-4)-β-Glc-(1-1)-Cer

GM2, II^{3}Neu5AcGgOse$_3$LacCer, β-GalNAc-(1-4)-[α-Neu5Ac-(2-3)]-β-Gal-(1-4)-β-Glc-(1-1)-Cer

GM1, II^{3}Neu5AcGgOse$_4$Cer, β-Gal-(1-3)-β-GalNAc-(1-4)-[α-Neu5Ac-(2-3)]-β-Gal-(1-4)-β-Glc-(1-1)-Cer

GD1a, IV^{3}Neu5AcII^{3}Neu5AcGgOse$_4$Cer, α-Neu5Ac-(2-3)-β-Gal-(1-3)-β-GalNAc-(1-4)-[α-Neu5Ac-(2-3)]-β-Gal-(1-4)-β-Glc-(1-1)-Cer

GD1b, II^3(Neu5Ac)$_2$GgOse$_4$Cer, β-Gal-(1-3)-β-GalNAc-(1-4)-[α-Neu5Ac-(2-8)-α-Neu5Ac-(2-3)]-β-Gal-(1-4)-β-Glc-(1-1)-Cer

GT1b, IV^{3}Neu5AcII^3(Neu5Ac)$_2$GgOse$_4$Cer, α-Neu5Ac-(2-3)-β-Gal-(1-3)-β-GalNAc-(1-4)-[α-Neu5Ac-(2-8)-α-Neu5Ac-(2-3)]-β-Gal-(1-4)-β-Glc-(1-1)-Cer

O-Ac-GT1b, IV^{3}Neu5AcII^3(Neu5, 9Ac$_2$-Neu5Ac)GgOse$_4$Cer, α-Neu5Ac-(2-3)-β-Gal-(1-3)-β-GalNAc-(1-4)-[α-Neu5, 9Ac$_2$-(2-8)-α-Neu5Ac-(2-3)]-β-Gal-(1-4)-β-Glc-(1-1)-Cer

GQ1b, IV^3(Neu5Ac)$_2$II^3(Neu5Ac)$_2$GgOse$_4$Cer, α-Neu5Ac-(2-8)-α-Neu5Ac-(2-3)-β-Gal-(1-3)-β-GalNAc-(1-4)-[α-Neu5Ac-(2-8)-α-Neu5Ac-(2-3)]-β-Gal-(1-4)-β-Glc-(1-1)-Cer

O-Ac-GQ1b, IV^3(Neu5Ac)$_2$II^3(Neu5, 9Ac$_2$-Neu5Ac)GgOse$_4$Cer, α-Neu5Ac-(2-8)-α-Neu5Ac-(2-3)-β-Gal-(1-3)-β-GalNAc-(1-4)-[α-Neu5, 9Ac$_2$-(2-8)-α-Neu5Ac-(2-3)]-β-Gal-(1-4)-β-Glc-(1-1)-Cer

Neu5Ac, 5-acetamido-3,5-dideoxy-D-*glicero*-D-*galacto*-nonulosonic acid

Cer, ceramide, N-acyl-sphingosine

C18-sphingosine, (2S, 3R, 4E)-2-amino-1,3-dihydroxy-octadecene; C20-sphingosine, (2S, 3R, 4E)-2-amino-1,3-dihydroxy-eicosene

The ganglioside and sialic acid nomenclatures are those proposed by Svennerholm [14] and Schauer [15], respectively, and are in accordance with the recommendations introduced by the IUPAC-IUB Commission on Biochemical Nomenclature in 1977 [16].

schematically obtained by sequential addition of neutral sugars and sialic acid units to the trisaccharide α-Neu5Ac-(2-3)-β-Gal(1-4)-β-Glc belonging to the GM3 structure. Neutral sugars are present in the β configuration, and sialic acid in the α one.

"Sialic acid" is a code to indicate all the derivatives of neuraminic acid (5-amino-3,5-dideoxy-D-*glycero*-D-*galacto*-nonulosonic acid). The three most represented structures are the N-acetyl-, N-glycolyl-, and N-acetyl/N-glycolyl-9-O-acetyl-derivatives, the N-acetyl-derivative (Neu5Ac) being the main structure in humans.

The lipid moiety of gangliosides is called ceramide and consists of a long chain amino alcohol, generally called sphingosine, connected to a fatty acid by an amide linkage. Ganglioside ceramide shows wide structure variability. The long chain bases are 18-, or 20-carbon-atom structures, mainly containing a *trans* double bond at position 3-4, and with the natural configuration 2S, 3R = 3D(+)*erythro* [(2S, 3R, 4E)-2-amino-1,3-dihy-droxy-octadecene or (2S, 3R, 4E)-2-amino-1,3-dihydroxy-eicosene].

Fatty acids vary from 14 to 26 carbon atoms and may contain unsaturated as well as α-carbonyl hydroxyl groups. In CNS gangliosides, stearic acid covers 80–90% of the total fatty acid content.

C. Normal-Phase HPLC: Separation of Ganglioside Species Homogeneous in the Oligosaccharide Portion

Analytical high performance liquid chromatography was used for the separation of individual gangliosides, the first attempts requiring a prior derivatization of gangliosides with strong UV absorbing probes [17,18]. The adoption of a proper solvent system enabled us to follow the elution pattern by direct assay of UV absorption in the eluate.

A 1–50 nmol portion of calf brain ganglioside mixture, as lipid-bound sialic acid, was dissolved in 5–10 μL of acetonitrile–water, 1 : 1 by volume, in a microtube and introduced into a syringe-loading sample injector equipped with a 50-μL loop. The microtube was washed several times with 5–10 μL of acetonitrile–water, 1 : 1 by volume, the washings being added to the previous sample in order to minimize loss of material. Gangliosides were then chromatographed [19] on a LiChrosphere-NH$_2$ column, 250 $\times$ 4 mm i.d., 5 μm average particle diameter (Merck, Darmstadt, Germany). The separation was carried out at 20°C with a gradient of the following solvent mixtures: solvent A, acetonitrile–5 mM phosphate buffer, pH 5.6 (83 : 17); solvent B, acetonitrile–20 mM phosphate buffer, pH 5.6 (1 : 1). The gradient elution program was as follows: 7 min with solvent A; 53 min with a linear gradient from solvent A to solvent A–solvent B (36 : 64); 20 min with a linear gradient from solvent A–solvent B (66 : 34) to solvent A–solvent B (36 : 64). The flow rate was 1 mL/min, and the elution profile was monitored by flow-through detection of UV absorbance at 200 nm. A complete analysis took 80 min. Before the next analysis cycle, the column was washed with solvent B for 10 min and then equilibrated with solvent A for 15 min.

D. Reversed Phase HPLC: Separation of Ganglioside Species Homogeneous in Both the Oligosaccharide and Lipid Portions

Gangliosides from the CNS belong mainly to the ganglio-series and are characterized by a high content of stearic acid (80–90% of the total fatty acid content) and the presence of both the molecular species containing C18- and C20-sphingosine. We have developed a rapid and highly resolving reversed phase HPLC procedure able to separate, without prior derivatization, ganglioside mixtures into ganglioside species homogeneous in both the oligosaccharide and lipid moieties [11,20].

A 1–50 nmol portion of calf brain ganglioside mixture, as lipid-bound sialic acid, was dissolved in 5–10 μL of acetonitrile–water, 1 : 1 by volume, in a microtube and introduced into a syringe-loading sample injector equipped with a 50-μL loop. The microtube was washed several times with 5-10 μL of acetonitrile–water, 1 : 1 by volume, the washings being added to the previous sample in order to minimize loss of material. The gangliosides were then chromatographed on a reversed phase LiChrosphere RP8 column, 250 $\times$ 4 mm i.d., 5 μm average particle diameter (Merck). The separation was carried out at 20°C with a gradient of the following solvent mixtures: solvent A, acetonitrile–water (85 : 15); solvent B, acetonitrile–5.5 mM phosphate buffer, pH 7, (15 : 85). The gradient elution program was as follows: 10 min with solvent A–solvent B (50 : 50); 15 min with a linear gradient from solvent A–solvent B (50 : 50) to solvent A–solvent B (70 : 30); 15 min with a linear gradient from solvent A–solvent B (70 : 30) to solvent A–solvent B (93 : 7); 20 min with solvent A. The flow rate was 0.5 mL/min, and the elution profile was monitored by flow-through detection of UV absorbance at 200 nm. A complete analysis took 60 min. Before the next analysis cycle, the column was washed with solvent A for 30 min and then equilibrated with solvent A–solvent B (50 : 50) for 30 min.

III. RESULTS AND DISCUSSION

In this chapter we report the analysis of native ganglioside mixtures using HPLC procedures. The ganglioside mixtures were composed of molecular species differing in both the oligosaccharide and lipid moieties.

The use of a rather polar LiChrosphere–NH$_2$ column allowed us to utilize the polarity of the gangliosides to fractionate them. This allowed the separation of gangliosides differing in the number of neutral sugars or sialic acid residues. Isomeric gangliosides such as GD1a and GD1b could also be separated. Thus, GM3 and GQ1b gangliosides were the first and the last to be eluted, by increasing the percentage of water and the ionic strength of the elution solvent system. All the molecular species that differed in the lipid moiety but had the same oligosaccharide chain were eluted with similar retention times, giving rise to a single peak. Figure 1 shows an example of the HPLC procedure applied to the fractionation of the gangliosides from neuronal cells in culture.

The use of the reversed phase LiChrosphere RP-8 column allowed us to separate the ganglioside species, due to polarity differences in both the oligosaccharide and the ceramide portions. The procedure can be applied to ganglioside mixtures from different sources; but it must be noted that

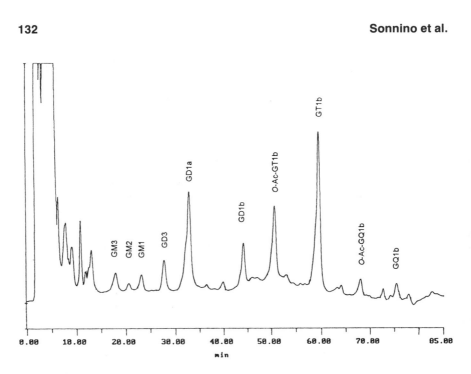

Figure 1 Normal phase HPLC separation of the ganglioside mixture from rat cerebellar granule cells at day 15 in culture.

the more heterogeneous the ganglioside ceramide moiety, the more complex the identification of each chromatographic peak. This is the procedure we used to analyze a ganglioside mixture from nervous cells where the gangliosides show a very simple ceramide composition. In the nervous system, the ganglioside long chain bases (LCBs) are C18- and C20-sphingosine (94–95% of the total LCB content), and the ganglioside fatty acid is stearic acid (>90% of the total fatty acid content). Thus, gangliosides homogeneous in the oligosaccharide chain are mainly split into two peaks containing C18- or C20-sphingosine, respectively. Figure 2 shows an example of the reversed phase HPLC procedure.

Although gangliosides show a maximum UV absorption at 195 nm, column elution was followed at 200 nm, where the ganglioside absorption corresponded to 80% of that recorded at 195 nm. This is necessary to avoid zero-line variation caused by variation in the solvent absorption at 195 nm during the gradient program. The UV absorbance responses were linear with ganglioside content up to 20 nmol. The lowest amount of each ganglioside considered suitable for quantification was 0.08–0.2 nmol as a function of the absorption relative molecular response (Table 2). All the sugars, as well as the lipid moiety, contributed to the UV absorption, the main

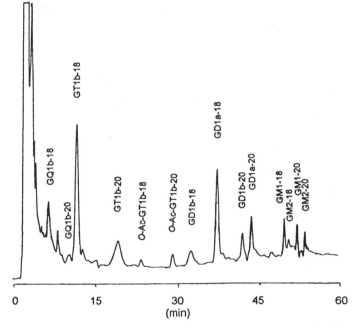

Figure 2 Reversed phase HPLC separation of the ganglioside mixture from rat cerebellar granule cells at day 22 in culture.

Table 2 Relative Molar Responses (rmr) of Standard Gangliosides (± S.D., $n = 6$), as Determined by Absorption at 215 nm

Ganglio-side	Long chain base			
	C18:1	C18:0	C20:1	C20:0
GM3	0.484 ± 0.010	0.397 ± 0.011	0.530 ± 0.013	0.440 ± 0.008
GM2	0.512 ± 0.006	0.423 ± 0.010	0.553 ± 0.011	0.461 ± 0.016
GM1	0.563 ± 0.041	0.482 ± 0.034	0.608 ± 0.023	0.538 ± 0.026
GD1a[a]	1.000 ± 0.014	0.913 ± 0.022	1.044 ± 0.018	0.953 ± 0.018
GD1b	1.050 ± 0.024	0.960 ± 0.019	1.098 ± 0.022	1.005 ± 0.028
GT1b	1.553 ± 0.088	1.461 ± 0.075	1.599 ± 0.055	1.511 ± 0.061
GQ1b	2.001 ± 0.091	1.919 ± 0.078	2.050 ± 0.071	1.961 ± 0.074

[a]The molecular species of GD1a containing C18:1 long chain base = 1.000.

contribution deriving from the sialic acid residues. Table 2 shows the relative molecular responses of several gangliosides containing stearic acid and C18- or C20-sphingosine. Also reported are the values for the molecular species containing sphinganine, C18:0 or C20:0; these are present in the ganglioside mixtures as minor compounds (about 5% of the total).

In conclusion, the described procedures — for the separation of nonderivatized gangliosides homogeneous in the oligosaccharide chain or in both the oligosaccharide and ceramide moieties — fulfill all the method requirements for the separation and quantification of gangliosides. Moreover, in both HPLC procedures, adaptations of the elution gradient allow the analysis of ganglioside mixtures from several sources, showing different composition and complexity.

ACKNOWLEDGMENT

This work was partially supported by grants from the Consiglio Nazionale delle Ricerche (CNR), Italy (Grant 93.02246.PF39 Target Project "ACRO").

REFERENCES

1. Wiegandt, H. (1985). *New. Compr. Biochem.* 10, 199–266.
2. Tettamanti, G., Preti, A., Cestaro, B., Masserini, M., Sonnino, S., and Ghidoni, R. (1980). *ACS Symp. Ser.* 128, 321–343.
3. Steck, T. L., and Dawson, G. (1974). *J. Biol. Chem.* 249, 2135–2142.
4. Sonnino, S., Cantù, L., Corti, M., Acquotti, D., and Venerando, B. (1994). *Chem. Phys. Lipids* 71, 21–45.
5. Hakomori, S.-I. (1981). *Ann. Rev. Biochem.* 50, 733–764.
6. Brady, R. O., and Fishman, P. H. (1979). *Adv. Enzymol.* 50, 303–323.
7. Tettamanti, G., Sonnino, S., Ghidoni, R., Masserini, M., and Venerando, B. (1985). In Corti, M., and Degiorgio, V., eds., *Physics of Amphiphiles: Micelles, Vescicles and Microemulsion.* XC Corso, Società Italiana di Fisica, Bologna, pp. 607–636.
8. Hakomori, S.-I., and Kannagi, R. (1973). *JNCI* 71, 231–251.
9. Nudelman, E., Hakomori, S.-I., Kannagi, R., Levery, S., Yeh, M. Y., Hellstrom, K. E., and Hellstrom, I. (1982). *J. Biol. Chem.* 257, 12752–12756.
10. Tettamanti, G., Bonali, F., Marchesini, S., and Zambotti, V. (1973). *Biochim. Biophys. Acta* 296, 160–170.
11. Gazzotti, G., Sonnino, S., Ghidoni, R., Kirschner, G., and Tettamanti, G. (1984). *J. Neurosc. Res.* 12, 179–197.
12. Svennerholm, L. (1957). *Biochim. Biophys. Acta* 24, 604–611.
13. Miettinen, J., and Takki-Lukkainen, J. T. (1959). *Acta Chem. Scand.* 13, 856–858.

14. Svennerholm, L. (1980). *Adv. Exp. Biol. Med.* 125, 11.
15. Schauer, R., ed. (1982). *Sialic Acid: Chemistry, Metabolism and Functions.* Springer-Verlag, Wien–New York.
16. IUPAC-IUB Commission on Biochemical Nomenclature (1977). *Lipids* 12, 455–468.
17. Kundu, S. K., and Scott, D. D. (1982). *J. Chromatogr.* 232, 19–27.
18. McCluer, R. H., and Evans, J. E. (1983). *J. Lipid Res.* 14, 611–617.
19. Gazzotti, G., Sonnino, S., and Ghidoni, R. (1985). *J. Chromatogr.* 348, 371–378.
20. Valsecchi, M., Palestini, P., Chigorno, V., Sonnino, S., and Tettamanti, G. (1993). *J. Neurochem.* 60, 193–196.

10

A Sensitive Cell Suspension Enzyme-Linked Immunosorbent Assay Measuring the Density of Cell Surface Ganglioside and Carbohydrate Antigens

Mepur H. Ravindranath, Amir A. Amiri, and Philip M. Bauer
John Wayne Cancer Institute at Saint John's Hospital and Health Center, Santa Monica, California

I. INTRODUCTION

Ganglioside and carbohydrate antigens (Ag) exposed on the cell surface are targets for passive immunotherapy using murine [1,2], human [3,4], and genetically engineered monoclonal anti-carbohydrate antibodies (Ab) [5]. They are also the basis for active specific immunotherapy using vaccines made of either purified carbohydrate Ag [6,7] or whole cells that overexpress carbohydrate Ag [8,9]. One of the prerequisites for targeting tumor cell-surface ganglioside and carbohydrate Ag for specific immunotherapies (passive and active) is a high density of expression on the cell surface [1,10].

Biochemical assays measure both intra- and extracellular ganglioside and carbohydrate Ag but are not useful for specific quantitation of cell surface Ag. Although a number of analytical procedures (high performance liquid chromatography, mass spectrometry, nuclear magnetic resonance spectroscopy, infrared spectroscopy) [11] and immunological procedures (double immunodiffusion in agar, hemagglutination, hemolytic plaque-forming assay, complement fixation, quantitative precipitation reactions, immunoadherence, and enzyme-linked immunosorbent assay [ELISA]) are used to characterize ganglioside and carbohydrate Ag, they are not employed to estimate specific cell-surface epitopes. Surface ganglioside and carbohydrate Ag can be measured by tritium-labeling galactose and galactosamine residues in glycolipids and glycoproteins [12,13], but this assay does not quantify the density of a specific cell-surface ganglioside or carbohydrate epitope. Cell-surface localization of surface Ag has been achieved by monoclonal Ab specific for a distinct epitope on the carbohydrate moiety (for a list, see Refs. 11 and 14), and the Ab binding can be monitored by radiolabeled protein A or a second Ab directed to different domains of the antibody (Fc, heavy and light chains, μ chains), but the assay is qualitative. A certain degree of quantitation is achieved using flow cytometry. Flow cytometry detects the monoclonal Ab–carbohydrate complexes on the cell surface and determines the number and percentage of cells expressing the carbohydrate Ag (number of events) and the mean fluorescent intensity (MFI). The fluorescent intensity may vary with the aggregate properties of the fluorescent dye and the Ag. In spite of its remarkable sensitivity, flow cytometry has not been used to quantitate the density of cell-surface ganglioside and carbohydrate Ag, because appropriate standards are not available. Furthermore, the technique is tedious and expensive.

In this chapter, we describe a sensitive cell-suspension ELISA (Cs-ELISA) to quantitate cell-surface ganglioside and carbohydrate Ag on cells expressing multiple carbohydrate epitopes. The density of cell-surface ganglioside and carbohydrate Ag is determined by comparing the concentration of monoclonal Ab bound to cell surface Ag with that bound to varying concentrations of purified Ag on the solid matrix of appropriate microtiter plates [25]. The assay uses *only* those monoclonal Ab that are *monospecific* for carbohydrate epitopes and can detect the highly purified epitope at a concentration lower than 1 pmol of Ag. The following criteria were used to validate this assay. First, when appropriately titrated, the monoclonal Ab showed linear affinity to the epitope (purified Ag) at concentrations ranging from 0.1 to 10 pmol. Second, the sensitivity of the assay was least affected by the presence of carbohydrate structures closely related to the target Ag. Third, more than 80% of the cells used in the assay were viable, as determined by a dye exclusion technique. Fourth, the optimal density of the cells required for each assay was empirically determined for each Ab. We used the Cs-ELISA to monitor routinely the density of ganglioside and carbohydrate Ag in suspensions of freshly resected tumor biopsies and on cultured tumor cell lines that had undergone passages after various treatments (such as cryopreservation, irradiation, and exposure to carcinogens, cytokines, steroids, and other drugs) or gene transfection. The assay was also used to monitor the incorporation of exogenous ganglioside and glycolipid adjuvants (e.g., monophosphoryl lipid A).

II. MATERIALS AND METHODS

A. Murine Monoclonal Ab

Purified murine monoclonal anti-GD3 Ab Mel-1 from clone R24, IgG3 (Signet Laboratories, Dedham, Massachusetts), the hybridoma of 14.G2a (for anti-GD2 Ab, IgG2a) (a gift from Ralph Reisfeld, Scripps Institute,

La Jolla, California), the ascites of anti-GM3 Ab M2590 (IgM) (Cosmo Bio Co., Ltd. Toyo Koto-Ku, Tokyo), and murine anti-MPL Ab 8A1 (IgG1) (a gift from Centocor, Malvern, Pennsylvania) were used. The culture super-natants of 14.G2a were stored in a refrigerator or frozen without any signif-icant difference in activity. All Ab were stored at 4°C until use. All Ab were diluted with phosphate-buffered solution (PBS), pH. 7.4, containing 4% human serum albumin (HSA).

B. Reagents for Cs-ELISA

Most of the treatments for the Cs-ELISA were carried out in polypropylene microfuge tubes, and the final steps were performed in microtiter plates. In addition, standard gangliosides and carbohydrate Ag were coated routinely in microtiter plates. We have empirically screened a variety of microtiter plates for assessing differences in antiganglioside Ab binding [16]. Our first choice is Falcon and Pro-Bind 3915 (Becton Dickson Labware, Lincoln Park, New Jersey); our second choice is Immunolon-1 (Fisher Scientific, Pittsburgh).

Throughout the analyses, the use of glass and polystyrene tubes should be avoided. The following plastiware was used: polypropylene tubes (15 and 50mL); microcentrifuge tubes (1.5 mL) (Fisher Scientific); serolog-ical pipettes (10 mL) (Becton Dickinson Labware); Costar octapipette or multichannel variable pipette; Gilman micropipettes (10 μL; 200 μL); mi-cropipette white tips (Intermountain Scientific, Kaysville, Utah) that were soaked in ethanol before using for plating gangliosides.

The following reagents were used as working solutions: Dulbecco's phosphate-buffered saline (10 × DPBS modified) (without calcium and magnesium) (JRH Biosciences, Lenexa, Kansas); HSA, 25% solution (Bax-ter Healthcare Corporation, Hyland Division, Glendale, California); dou-ble-distilled water; o-phenylenediamine dihydrochloride (Cat. No. 5974SA, Gibco BRL, Life Technologies, Gathersburg, Missouri) (discard if oxi-dized, i.e., color changes to yellowish); citrate–phosphate buffer (pH, 5.0) prepared from stock solutions of citric acid (0.1 M, 19.21 g in 1 L) and dibasic sodium phosphate (0.2 M, 53.65 g $Na_2HPO_4 \cdot 7H_2O$ or 71.7 g $Na_2HPO_4 \cdot 12H_2O$ in 1 L). For a pH 5.0 buffer, 24.3 mL of 0.1 M citric acid was added to 25.7 mL of 0.2 M dibasic sodium phosphate. Thirty percent H_2O_2 (Sigma, St. Louis) was used; it was stored in the refrigerator and discarded 3 months after opening.

The following purified glycolipids were used as Ag for coating the plates: GM3, GM2, GM1, AsialoGM1, GD1a, GD1b, and GD3 (Fluka Biochem., New York, and/or Sigma), GD2 (Advanced Immunochem, Long Beach), and monophosphoryl lipid A (MPL) (Ribi Immunochem.,

Montana). The peroxidase-coupled second Ab, peroxidase-conjugated affinipure F(ab)$_2$ fragment goat anti-mouse IgG (H + L), and goat anti-mouse IgM (μ) (minimum cross reaction to human, bovine, and horse serum proteins) were from Jackson ImmunoResearch (Westgrove, Pennsylvania).

C. Preparation of Tumor Biopsies

Human tumor biopsies were surgically resected, rinsed, and stored in RPMI-1640 containing 4% HSA at 4°C. The specimens were transferred to a plastic petri dish under a laminar flow hood. The tumor was minced using sterile scissors and scalpels until almost homogeneous. The minced tissue was transferred into sterile trypsinizing flasks containing 35 mL of DNAse/collagenase (Collagenase Type I, EC 3.4.24.3) from *Clostridium histolyticum* (1.4g) and DNAse Type II (EC 3.1.21.1) from bovine pancreas (2000 Kunitz units, 100 mg) (Sigma) in 1 L of RPMI-1640, filtered through a 0.4-μm filter). The flask was placed on a stirring plate set at 400–500 rpm in a incubator at 37°C for 2 h, longer if not digested sufficiently. After thorough digestion, the tissue was strained through a piece of sterile gauze into a 50-mL polypropylene centrifuge tube and rinsed with 10mL of RPMI assay medium. The cell suspension was centrifuged at 300 × g for 10 min. The supernatant was decanted, and the cell pellet was resuspended in RPMI-1640. For pellets of 0.2 mL or less, 10 mL of RPMI-1640 was added. The cells were gently vortexed until the cell suspension was uniform. The cell suspension (10 mL) was overlaid in a 50-mL polypropylene tube containing 35 mL of cold Ficoll-Paque (Pharmacia Biotech., Code No. 17-0840-03) and centrifuged at 200 × g for 10 min and at 400 × g for 20 min, so that erythrocytes and dead cells precipitated to the bottom. Tumor cells free of erythrocytes were recovered from the interface. This population also included a low percentage (3–10%) of tumor-Ag negative leukocytes. The recovered cells were washed twice with cold 4% HSA–RPMI-1640. The cells were mildly vortexed between washes. Twenty-five microliters of cell suspension was mixed with 25 μL of sterile 0.1% trypan blue in RPMI-1640 (original dye was 0.4% in 0.81% sodium chloride and 0.06% potassium phosphate, dibasic) (Sigma). The cells were counted in a hemocytometer (Neubauer Ruled).

D. Cryopreservation of Cells

Tumor cell suspensions from biopsy specimens or tissue cultures were washed in RPMI–7.5% HSA. After removing the supernatant, an appropriate amount of freezing solution (RPMI-1640, 7.5% HSA, 10% dimethyl sulfoxide [DMSO]) was added for a cell suspension of 10^7 cells/mL. Cells were in DMSO at room temperature for no longer than 20 min. One milliliter of cell suspension was added to each vial for cryopreservation. The cells were frozen in a programmed liquid nitrogen freezer at a rate of 1 °C per minute. When the temperature dropped below -30°C, the vials were transferred immediately to a -80°C liquid nitrogen freezer. The cells may be gamma irradiated for 43 min (15 kRad) after cryopreservation and in the presence of liquid nitrogen. The irradiation maintained the viability of the cells but prevented division and proliferation.

E. Thawing and Viability of Irradiated and Cryopreserved Cells

For recovery of cryopreserved cells, the vials were taken from the liquid nitrogen freezer and transferred to a 37 °C water bath for 15 to 30 sec. The vials were removed from the water bath when they were partially thawed. They were further thawed at room temperature and then transferred to a 15-mL polypropylene tube with a Pasteur pipette. Nine milliliters of RPMI or RPMI–4% HSA was added dropwise. The cells were allowed to settle for 5 min and then centrifuged at 4 °C for 10 min at 300 × g. After removing the supernatant, the cells were suspended in fresh RPMI, gently tapped, and vortexed. The viability of the cells was monitored by the trypan blue dye exclusion technique as previously described. The cells were counted in a hemocytometer after staining with 0.1% trypan blue in RPMI-1640. The viability of cells used in the Cs-ELISA was 92 ± 4.5 ($n = 19$).

F. Protocol for Assessing the Density of Cell Surface Ganglioside and Carbohydrate Ag

1. Prerequisites for the Assay

The following was determined for every batch of Ag and Ab obtained commercially or prepared in the laboratory: (1) the purity of the ganglioside Ag by thin-layer chromatography (TLC) [15]; (2) the specificity of the monoclonal Ab to a single epitope in solid phase direct immunoassay; (3) linearity less of the Ab response (appropriately diluted) at Ag concentrations of < 10 pmol; and (4) dilutions of the Ab for solid phase immunoassay using box titration, to determine cell-surface epitopes. Identical conditions (buffers, reagents, and substrates) were used for the immunoassay of both purified Ag and cells. The reproducibility of the absorbancy values for the purified antigenic epitope in solid phase immunoassay at concentrations comparable with those found on the cell surface was determined.

2. Assessment of the Specificity of the Anti-Ganglioside Monoclonal Ab

Essentially the protocol was the same as described elsewhere [16]. Microtiter plates used for Ag coating were rinsed in absolute ethanol; uniformity of the uncoated polystyrene wells was assessed from the absorbancy at 650 nm of the plates containing ethanol. Wells were coated with 100 pmol of gangliosides in 100 μL of ethanol. Plates were dried in vacuo for at least 2 days. The Ag-coated plates were blocked with 4% HSA in PBS, pH 7.2, and incubated at 37°C for 90 min. This procedure not only blocks the Ag-free sites, it also removes excess soluble Ag. The Ab diluted with blocking buffer was incubated at 37°C for 1 h to prevent aggregation and minimize the interactions characteristic of IgG3 Ab. After removing the blocking buffer, 100 μL of diluted Ab was added to wells and incubated at 37°C for 90 min. Before washing the plates, the Ab was removed. Wells were washed with PBS containing 0.1% Tween-20 and 0.1% HSA. Four individual washings were performed. Manual washing is preferred over mechanical or spray washing to minimize inconsistencies. Peroxidase-coupled second Ab was diluted 1:5000. If the background values were too high, the Ab was diluted to 1:10,000 or 1:15,000. For purposes of comparison, the same dilution should be used throughout all analyses. The Ab (100 μL) was added to each well and incubated at 37°C for 1 h. The plates were washed four times. Hydrogen peroxide (11.4 μL of 30%) was added to the substrate solution. Hydrogen peroxide was thoroughly mixed with substrate (19.3 mg of o-phenylenediamine dihydrochloride in 25 mL of citrate–phosphate buffer, pH 5.0). Substrate solution (50 μL) was added to each well, and the plates were incubated at room temperature in the dark for 45 min. Sulfuric acid (60 μL of 6 N) was then added to each well and absorbancy was measured at 490 and 650 nm. The following controls were used: *Blank*, well without Ag and first Ab but treated with second Ab and substrate; *Background*, well without Ag treated with first and second Ab and substrate; *Negative Control*, Ag-coated plates treated with class matched nonspecific isotypes of primary antibodies (for Mel-1, IgG3; for 14.G2a, IgG2a; for M2590, IgM; and for 8A1, IgG). All values were adjusted against the background. The experimental values were corrected with the negative control.

3. Sensitivity of the Anti-Carbohydrate Monoclonal Ab

The ganglioside Ag and MPL were serially diluted as follows: 100 μL of absolute ethanol was added to each row of the microtiter plate, omitting row A and column 1; the Ag (100 pmol in 100 μL) was added to rows A and B omitting column 1. The Ag in row B was mixed with the micropipette, and 100 μL of Ag from row B was transferred to row C. Mixing and transfer were continued through row G. Finally, 100 μL of ganglioside in ethanol was removed from row G. Antigens were transferred with ethanol-pretreated white tips to prevent adhesion of micelles. Row H was not coated with Ag and was used as the background control. Plates were dried in vacuo for at least 2 days. The Ag-coated plates were blocked with buffer containing PBS (pH 7.0–7.2) with 4% HSA and incubated at 37°C for 90 min. The Ab (primary or isotype for negative control) was diluted with blocking buffer, and incubated at 37°C for 1 h to prevent aggregation and minimize Ab interactions. A series of dilutions was prepared ranging from 1:100 to 1:6400 or 1:1000 to 1:64,000. Antibody at dilution 1:100 was added to column 2; Ab at 1:200 was added to column 3. Other dilutions were added to the other columns. Antibody diluted to 1:100 (or 1:1000) was used for background subtraction. Affinity-purified murine monoclonal Ab (Mel-1 from R24 clone) (100 μL) was added at a dilution of 1:100. Other monoclonal Ab were used at different dilutions as will be described in the figure legends. For negative control, class-matched isotypes were added at the same dilution as their respective primary Ab. Before washing the plates, the Ab was removed. Wells were washed four times with buffer (PBS, pH 7.2–0.1% Tween-20–0.1% HSA). The rest of the procedure was as described in Section II. F.2.

G. Cell Surface Expression of Ganglioside and Carbohydrate Ag

Three preparations of cells were used: (1) cells recovered from tumor biopsies; (2) cryopreserved and irradiated cells; and (3) freshly harvested (with EDTA–dextrose) cells. Trypsin was not used for harvesting the cells because it affects Ag expression [12,13]. All cells were suspended in RPMI–4% HSA, counted for viability, and centrifuged; the cell pellet was reconstituted in fresh RPMI–HSA. Cells (0.2 or 0.5 × 10^6) in 60 μL of solution in each vial (microcentrifuge tubes) were separated as either *background* (treated only with the second Ab), *experimental* (treated with first and second Ab), or *negative control* (treated with class matched isotypes of the primary Ab). The first Ab (120 μL), for example, Mel-1 at a dilution of 1:100, suspended in PBS–4% HSA, was added to experimental vials and incubated for 1.5 h in an ice bath shaker at < 10°C. The cells were washed gently with PBS–4% HSA. After incubation, cell suspensions were gently vortexed. Approximately 1.25 mL of PBS–HSA was added to each vial and centrifuged. After centrifugation, the supernatant was removed by low vacuum, leaving about 100 to 150 μL of suspension. The procedure was repeated three times. After the third wash, the supernatants were removed without disturbing the pellet. Then, 150 μL of a 1:5000 dilution of peroxidase-coupled goat anti-mouse Ab was added, and cells were then suspended by gentle vortexing, and incubated for 1 h in a ice bath shaker at $<10°C$. The vials were washed two times with PBS–HSA as previously described. The contents were then transferred into fresh microcentrifuge vials containing 1 mL of of PBS–4% HSA, and centrifuged to remove as much supernatant as possible without disturbing the pellet. This step avoids detection of Ab attached to the vials. The substrate solution (50 μL) was added to each vial, vortexed gently, and incubated in the dark at room temperature for exactly 40 min, and then centrifuged for 5 min at 4°C. ELISA plates were prepared by adding 60 μL of 6 N H_2SO_4 to each well. The substrate supernatant of each vial was then transferred to the wells of the microtiter plates containing the acid. The absorbancy was measured at 490 and 650 nm. The following controls were used: *Blank*, cells treated with substrate but not reacted with first or second Ab; *Background*, cells treated only with peroxidase-coupled second Ab; *Negative Control*, cells treated with nonspecific isotypes corresponding to primary Abs. All values were adjusted against the background. The experimental values were corrected for the negative control. It is important that these isotypes be purified and incapable of binding to the purified antigenic epitopes of corresponding primary antibodies. Turbidity can be avoided by centrifuging the cells after incubation with substrate at >3000 rpm and by carefully pipetting the supernatant. During the assay, ELISA plates coated with 1.5 pmol of ganglioside were employed. The procedure described in Section II. F.2. was carried out using *only* the reagents for Cs-ELISA. *Use of Tween-20 or any other detergent in the wash buffer was avoided.* With 1.5 pmol of purified GD3, the Mel-1 (225 μg/mL) at a dilution of 1:100 gave an absorbancy of 0.6. With 0.78 pmol of purified GD2, the 14.G2a culture supernatant (frozen or stored at 4°C for 6 months), at a dilution of 1:4, gave an absorbancy of 1. At least 10 vials coated with 1.5 pmol of each Ag were used as a standard. The coefficient of variation was $<5\%$ when comparing the mean with the mean obtained with the Cs-ELISA. Means obtained with Cs-ELISA were less than 3 × standard deviation.

A summary of recommended guidelines is as follows: The purity of the commercial glycolipids should be assessed with TLC. The specificity and sensitivity of the monoclonal Ab at different dilutions of the ganglioside (with a starting dilution at 100 pmol) should be determined. At least 1 pmol of ganglioside should be prepared accurately. The Ab dilution required for Cs-ELISA should be evaluated by a box titration of different dilutions of the Ab against different concentrations of the purified Ag and cells. The standard (1.5 pmol) should be prepared to assess the reproducibility of the assay. The coefficient of variation should be $<5\%$. The viability of the cells should be not less than 85%. Interassay variability should be assessed for Cs-ELISA with different aliquots of the same preparation.

H. Cryptic and Immunologically Exposed Cell-Surface Carbohydrate Ag

Cell-surface carbohydrates may be cryptic (not exposed to Ab recognition) due to masking by intrinsic and extrinsic cell-surface membrane proteins [17,18] and other complex glycolipids [19] which can be detected after mild trypsinization [17–20]. For this purpose, 0.5×10^6 cells (60 μL) were treated with 0.01% trypsin without EDTA (300 μL) and incubated at 37°C for 1 h. The carbohydrate epitopes were analyzed after inhibiting trypsin activity by adding RPMI–4% HSA (300 μL) and then washing three times with the same buffer. Trypsinized cells were then tested for viability and Ab binding.

I. Cs-ELISA Monitoring of Incorporated Glycoconjugates onto Tumor Cell Surfaces

Glycolipids (GD3 or MPL) (25, 50, 75 μg) in ethanolic solution were added to polypropylene tubes and evaporated to dryness in vacuo. After drying, 200 μL of RPMI was added and the tubes were vortexed (2 min) and sonicated (15 min) in three cycles. To each tube, 50 μL of 1×10^6 cells were added, and the cells were incubated at 37°C for 1 h. After three washings with PBS–HSA, monoclonal Ab was added and cells were incubated at 7.4°C for 1 h. For controls, cells were added to ethanol-treated tubes (75 μL of ethanol added and evaporated to dryness). The controls were divided into three groups: Group I (*blank*) was treated only with the peroxidase-conjugated second Ab; Group II (*background*) was treated in the same way as glycolipid-incorporated cells; and Group III (*negative control*) glycolipid-incorporated cells were treated with nonspecific isotypes of primaries at the same dilutions as the primary Abs. All values were adjusted against the background. The experimental values were corrected for negative control.

III. RESULTS

A. Specificity of Monoclonal Ab

The glycocalyx (cell-surface glycoproteins and glycolipids) of tumor cells may be characterized by aberrant glycosylation of O-linked and N-linked carbohydrate residues of glycoproteins, such as MUC-1 in ductal carcinoma of the breast [19], and overexpression of normally underexpressed or absent Ag such as GD3, *O*-AcGD3, and GD2 in human melanoma [6,11, 15,21]. Immune recognition of these carbohydrate epitopes depends on their cell-surface density [1,14]. For measurement of the density of cell-surface carbohydrate epitopes, we compared the specificity and affinity of three murine monoclonal Ab: (1) affinity-purified IgG3 Ab to GD3 (Mel-1); (2) ascites of IgM Ab directed to GM3 (M2590); and (3) culture supernatant of IgG2a Ab directed to GD2 (14.2Ga). Mel-1 recognizes only disialo-lactosyl (NeuAcα(2,8)NeuAcα(2,3)Galβ(1,4)Glu) residues of the ganglioside GD3; it fails to recognize the monosialo-lactosyl residues of GM3 or the disialo-lactosyl residues of other gangliosides having GalNAc (as in GD2) or Galβ(1,3)GalNAc (as in GD1b). Mel-1 also recognized, to a minor extent, GT1a and GQ1a [22]. The specificity of Mel-1 was maintained at different dilutions (Fig. 1A), and the high affinity of Mel-1 for GD3 was not affected by the presence of GM3 or other gangliosides (Fig. 2A), suggesting that the Ab is capable of specific recognition of GD3 even in the presence of other related gangliosides. Monoclonal Ab 14.G2a recognized a disialo-lactosyl epitope containing GalNAc, as in NeuAcα(2,8)-NeuAc(GalNAc1,4)α(2,3)Galβ(1,4)Glc of GD2), but not in the absence of GalNAc, as in the sialyllactosyl residues of GD3 (Fig. 1B). Monoclonal Ab 14.2Ga also did not bind to Galβ(1,3)GalNAc containing disialyllactosyl units of GD1b or GalNAc containing monosialolactosyl residues of GM2 (Fig. 1B). The afinity of 14.G2a GD2 was not affected by the presence of other gangliosides. The monoclonal Ab M2590 binds specifically to GM3 [23]. However, M2590 failed to show the fine specificity for GM3 in our ELISA method (Fig. 1C). The lack of specificity of M2590 could reflect the fact that it is obtained from ascites rather than from culture supernatant or from purified form. The binding of M2590 to GM3 was also affected by the presence of complex gangliosides (Fig. 2B), suggesting that the Ab may not be able to bind carbohydrate epitopes in the presence of other components of the glycocalyx. Thus, M2590 from ascites may not be suitable for quantitation of the density of GM3 on cells that co-express other complex carbohydrate structures (e.g., human melanoma cells), but it may be useful to measure the density of GM3 in cultured murine melanoma B16 cells, which express only GM3 [23,24]. This antibody must be affinity purified before use in Cs-ELISA. The use of ascites should be avoided in Cs-ELISA.

B. Assessment of Monoclonal Ab Sensitivity

The sensitivity of monoclonal Ab was assessed by titrating against different concentrations of their respective purified gangliosides in microtiter plates.

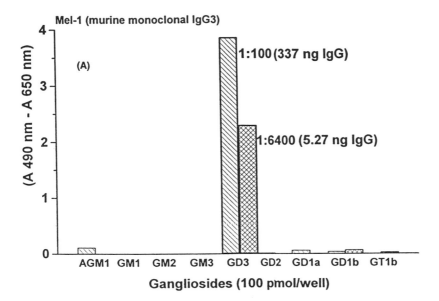

Figure 1 Epitope specificity of anti-carbohydrate murine monoclonal Abs: (A) affinity-purified Mel-1 (IgG3); (B) culture supernatant of 14.G2a (IgG2a); (C) ascites of M2590 (IgM). The glycolipids in ethanol were coated on the wells of microtiter plates (Falcon. Probind: 3915), dried overnight in vacuo and were blocked with PBS–4% HSA (pH 7.2) at 37°C for 1 h. The diluted Ab was incubated in a water bath at 37°C for 1 h before adding to the wells. After incubation, the wells were washed (four times) with PBS–0.4% HSA containing 0.1% Tween-20. Peroxidase-conjugated goat anti-mouse IgG F(ab)$_2$ was diluted at 5000-fold. *O*-phenylene diaminedihydrochloride in citrate–phosphate buffer (pH 5.0) with H$_2$O$_2$ was used as the substrate. The oxidation of the substrate was stopped with 6 N H$_2$SO$_4$. The absorbancy was measured at two wavelengths, 490 and 650 nm. The values were corrected for background (absorbancy of the Ab added to wells without Ag) and negative controls (isotype matched).

Antibodies were tested at different dilutions: Mel-1 at 1:100, 14.2Ga at 1:10 and 1:4, and M2590 at 1:1000. Mel-1 at a dilution of 1:100 bound 1.6 pmol of ganglioside, with an absorbancy (at 490–650nm) of 1 (Fig. 3A), and the culture supernatant of 14.G2a bound the same amount at a dilution of 1:4 (Fig. 3B). M2590 at a dilution of 1:100 bound to 1.6 pmol of GM3 to give an absorbancy of 1 (Fig. 3C). The lowest detection limit of these Ab with an absorbancy of 0.250 at 490–650 nm ranged from 100 to 150 fmol/well.

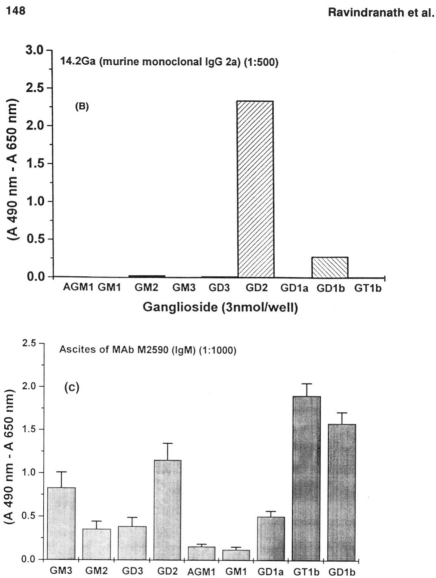

Figure 1   continued

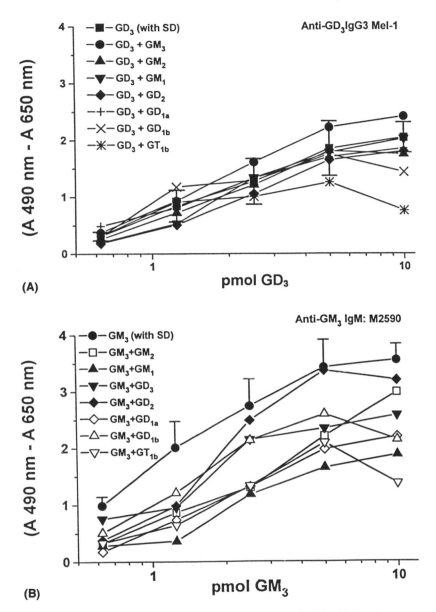

Figure 2 Epitope specificity of Mel-1 (A) and M2590 (B) in the presence of related complex carbohydrates. Plates were coated with either 12.5 pmol of serially diluted GD3, or with equimolar concentrations of GM3, GM2, GM1, GD2, GD1a, or GD1b, or GT1b in ethanol. Plates were evaporated to dryness in vacuo and analyzed for Ab specificity. Other details of the experimental protocols are described in Section II. Error bars show the standard deviation of five determinations.

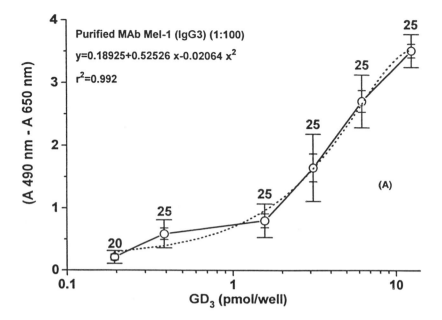

Figure 3 Titration of anti-carbohydrate murine monoclonal Ab against purified Ab. (A) affinity-purified Mel-1 (IgG3); (B) culture supernatant of 14.G2a (IgG2a); (C) ascites of M2590 (IgM). Purified gangliosides, dissolved in ethanol, were serially diluted in the wells of a microtiter plate (Falcon. Probind: 3915). Details of the experimental protocol are described in the legend for Fig. 1. Values obtained by two investigators on different days were pooled to obtain the indicated sample size (n = 25). The error bars represent the standard deviation. The dotted (A & B) or thin (C) line is the fit to the indicated equation. The dilutions of the Ab are given in parenthesis.

C. Titration of Ab Against Cell-Surface Carbohydrate Epitopes

The cell surface of cultured and biopsied human melanoma cells commonly expresses GD3, GD2, and GM3 along with other carbohydrate epitopes of the glycocalyx, whereas cultured murine melanoma cells (B16, F1, or F10) express exclusively GM3 [23,24]. Therefore, we have titrated Ab against a known number of tumor cells. For Mel-1, we have used three human melanoma cell lines (M10-v, M101, and M24), and for M2590 we have used mouse melanoma cells (B16-F10).

To quantitate the density of cell surface carbohydrates, we had to carry out ELISA of purified gangliosides. Usually, in conventional ELISA of purified gangliosides, we use Tween-20 (0.1–0.5%) in the wash buffer.

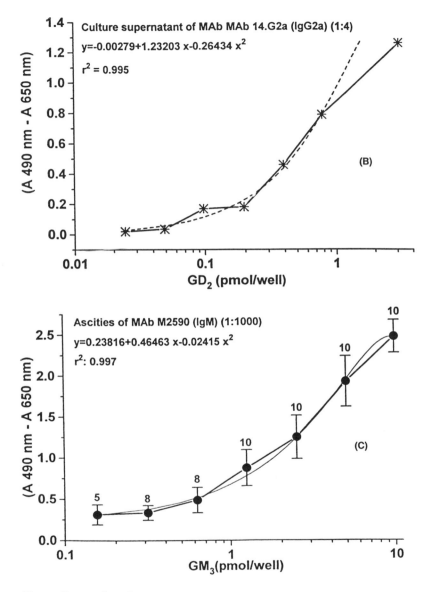

Figure 3 continued

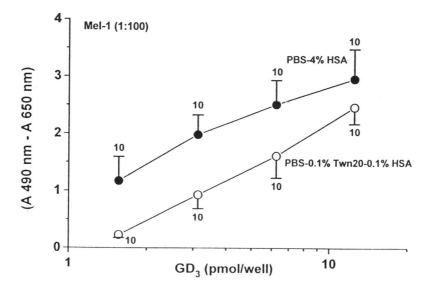

Figure 4   Effect of 0.1% Tween-20 in wash buffer.

In Cs-ELISA studies, the detergent Tween-20 was not used. We, therefore, have tested the ELISA absorbancy of purified gangliosides without Tween-20. Figure 4 shows that removal of Tween-20 affected the standard curve of Mel-1. GD3 (1.34 pmol) gave an absorbancy of 1 when the plates were washed without Tween-20, and an absorbancy of 0.107 when the plates were washed with Tween-20. Encouraged by the linearity of the response at varying concentrations of the ganglioside, we have carried out conventional ELISA for purified gangliosides without Tween-20. The values obtained were used as a reference for Cs-ELISA values. The results of titrating Mel-1 against human melanoma cells and titrating M2590 against B16 mouse melanoma cells are presented in Fig. 5. The results showed that Mel-1, at a dilution of 1:100, and M2590, at a dilution of 1:1000, are suitable for the respective cell lines. When Mel-1 was titrated against cell surface GD3 of the three human melanoma cell lines, we found that M10-V showed the highest affinity followed by M101 and then M24. The relative affinity of melanoma cell lines for 14.G2a was the same as that observed for Mel-1. Based on the titrations of M2590 against B16 mouse melanoma cells, we selected a dilution of 1:1000 as optimal for the Cs-ELISA of B16 cells.

D.   Cell Density Affects Binding of Ab

Figure 6A shows that the binding of Mel-1 (1:100) increased as cell density increased from 0.15×10^6 to 0.50×10^6 for the cryopreserved and irradi-

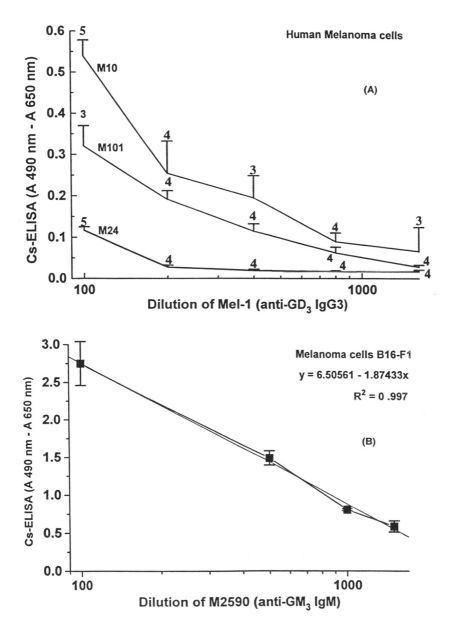

Figure 5 Titration of Mel-1 and M2590 against melanoma cells. One million cells for Mel-1 (A) and 0.2 million cells for M2590 (B) were used for titration. The Ab were serially diluted in PBS–4% HSA. Before serial dilutions, the Ab (diluted 1 : 100) was incubated at 37°C for 1 h. Mean and standard deviations for each of the human melanoma cells used are shown by the error bars. The sample size for each dilution is shown above the error bars for Mel-1. The sample size for M2590 was 5. The linear fit is indicated by the straight line and the parameters.

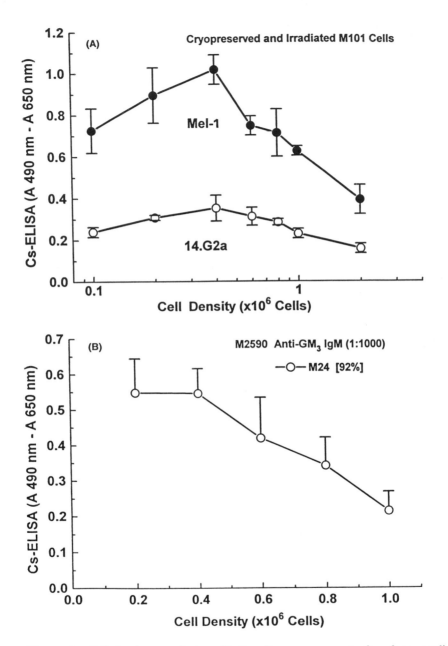

Figure 6 Cell density as a factor affecting the measurement of surface ganglio-sides. Mel-1 (1 : 100), 14.G2a (1 : 4) (A) and M2590 (1 : 1000) (B) were added to different concentrations of cells and incubated as described in Section II. Oxidation of the substrate *ortho*-phenylenediamine dihydrochloride (OPD) was measured in a microtiter plate at 490 nm and corrected against 650 nm. The circle and error bars refer to mean and standard deviation of four determinations. Cryopreserved and irradiated M101 (viability 92%) were used for Mel-1 and 14.G2a (A). Freshly harvested M24 melanoma cells with 89% viability were used for M2590 (B).

ated human melanoma cells (M101), as was found earlier for freshly obtained human melanoma cells [25]. M2590 showed higher values at both 0.2 × 10⁶ and 0.5 × 10⁶ cells (Fig. 6B). At higher cell densities, we expected the absorbancy to continue to increase because of a greater amount of cell-surface gangliosides, or to remain the same if the amount of Ab was less than the available cell-surface gangliosides. However, the results obtained with Mel-1, 14.G2a, and M2590 showed decreases in absorbancy at cell densities above 0.5 × 10⁶. We also observed this pattern with freshly harvested cells, as well as with cells after cryopreservation and irradiation.

These results emphasize the need to assess the optimal cell concentration for maximal binding before assessing cell surface Ag by Cs-ELISA. For a volume of 120 μL of diluted Ab, 0.2 or 0.5 × 10⁶ cells gave maximum binding. The decrease in absorbancy at cell densities higher than 0.5 × 10⁶ could be due to impaired access of the monoclonal Ab to the entire cell surface. This decrease could also be due to the formation of bridges between cells by the Ab. The concentration of the Ab and the temperature of the assay may govern the rate of Ab binding to the cell surface.

E. Viability Affects Quantitation of Cell-Surface Carbohydrates

During the course of our study, we assessed cell surface expression of GD3 and GD2 on cryopreserved and irradiated human melanoma cell lines. We selected M10-v cells for this purpose because the cryopreserved and irradiated M10-v cells differed in viability. M10-v cells are more sensitive to irradiation than M101 or M24 melanoma cells. Furthermore, M10-v cells did not show any significant variation in the quantitative expression of GD3 and GD2 in relation to different passage numbers (Fig. 7). Therefore, we tested whether viability would affect the cell-surface density of gangliosides. Figure 8 shows the Cs-ELISA absorbancy from Ab interaction with different proportions of viable and nonviable cells using Mel-1. An inverted bell-shaped polynomial regression curve with a significant r^2 was observed with both Mel-1 and 14.G2a [25]. In both cases, the absorbancy was highest when cells were mostly viable and lowest when cells were approximately 55% viable. A larger number of dead cells gave high absorbancies due to infiltration of the Ab. The Ab concentration in dead cells was lowest when viability was 55%. An increase in absorbancy with increasing viability indicates that the presence of dead cells in a viable cell population (> 60%) can affect Ab binding to cell surfaces. Based on these observations, we concluded that the viability of cells is a very important factor in quantitation of the density of cell-surface carbohydrate epitopes, and that there is a need to monitor viability during measurement of cell-surface Ag. Other studies

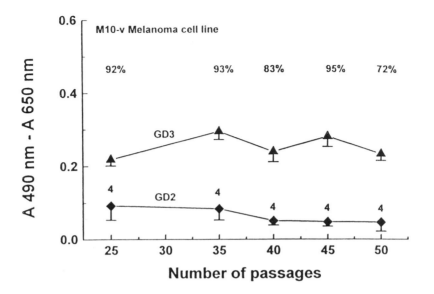

Figure 7 The cell-surface expression of human melanoma cell lines M10-v at different passages. Cells after passages 25 to 50 were cryopreserved. These cells were thawed as described in Section II. A half million cells were used for assay with Mel-1 (1 : 100 dilution) and 14.G2a (1 : 4 dilution). The viability of cells at each passage is given as a percentage. Each point represents the mean of four values. Standard deviations are indicated by the error bars.

demonstrated that cryopreservation alone or even formalinization of cells [25], which is known to induce intramolecular methylene bridges in the cell-surface proteins, did not affect the measurement of cell density.

F. Quantitation of Cell-Surface Carbohydrate Ag in Melanoma Cells

Using monospecific and sensitive monoclonal Ab directed against GD3 and GD2, we quantified the density of cell-surface GD3 and GD2. We used cells with maximum viability (89–92%) at an optimal density (0.5×10^6). We plotted the absorbancy values obtained with the Cs-ELISA against the absorbancy values obtained with purified gangliosides in a routine ELISA. We measured the density of GD3 and GD2 on tumor cells derived from tumor biopsies of melanoma patients. The density of cell-surface GD3 was always higher than the density of GD2 (Fig. 9A). The ratio of cell-surface GD3:GD2 in tumor cells recovered from tumor biopsies ranged from 5:1 to 100:1. Cell-surface GD2 in tumor biopsy–derived cells was much lower than

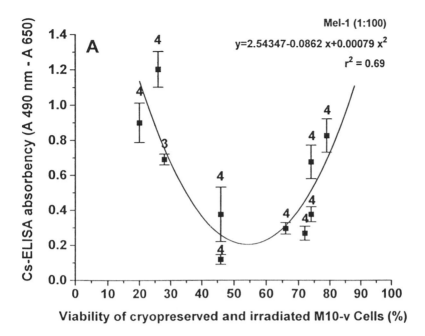

Figure 8 Viability as a factor affecting the measurement of the cell-surface gangliosides. Mel-1 (1 : 100 dilution) (100 μL) was added to 1 million cryopreserved and irradiated M10-V cells reconstituted in RPMI–4%HSA and incubated for 90 min as described in Section II. Sample size is indicated above the error bars. Closed squares refer to the mean, and the error bars indicate the standard deviation. The polynomial regression line is plotted.

that from tissue culture, a finding in agreement with observations made using biochemical methods [25–27]. We observed that the densities of GD3 and GD2 on the melanoma cell M10-v were 5 and 0.5 pmol/million cells, respectively (Fig. 9B). The densities of these two gangliosides on the cell surface of M101 cells were <3 and 0.5 pmol/million cells, respectively. The ratio of GD3:GD2 was 12:1 in M10-v cells and 7:1 in M101 cells. Interestingly, M24 cells did not show cell-surface expression of GD3 and GD2. Our biochemical studies validate these findings [25]. It is already known that M24 cells are deficient in GD3 and GD2 but are rich in GM3 and GM2. We have used M2590 for measuring the density of cell surface GM3 of cultured mouse melanoma B16–F10 cells, which is known to express only GM3 [24]. The density of cell surface GM3 in different batches of B16 cells ranged from 0.7 to 1.2 pmol/million cells.

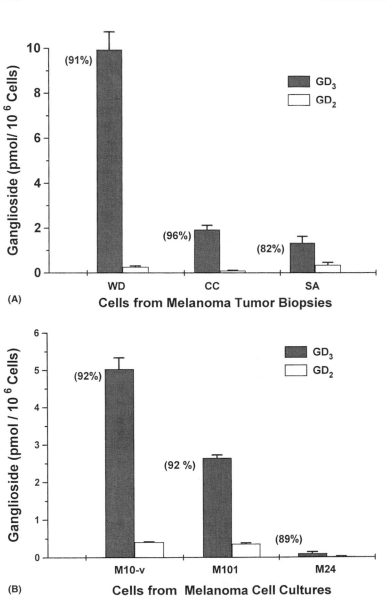

Figure 9 Quantitation of cell-surface GD3 and GD2 in human melanoma cells recovered from human tumor biopsies (A) or harvested from tissue culture (B). The procedure for recovery of tumor cells from tumor biopsies is described in Section II. Cells (0.5×10^6) were suspended in 60 μL of RPMI-4% HSA. The volume of diluted Ab was 120 μL. After

G. Incorporation of GD3 onto the Surface of GD3-Deficient M24 Melanoma Cells

We also used the Cs-ELISA for monitoring cell surface incorporation of glycolipids. We incubated exposed human melanoma cell line M24 (1×10^6 cells) with known concentrations of exogenous GD3 in a shaker bath at 37°C for 1 h. Figure 10 shows that the exogenous GD3 that was incorporated onto cells was at least three times higher than that found on the surface of untreated cells and that incorporation was dose dependent. Encouraged by these results, we assessed the feasibility of incorporating a glycolipid adjuvant onto the tumor cells and monitoring the level of incorporation. We selected MPL because it is a potent immunostimulant that induces an IgM Ab response against cell-surface gangliosides [28,29].

H. Cs-ELISA to Monitor Incorporation of a Potent Bacterial Glycolipid Adjuvant onto Tumor Cells

Monophosphoryl lipid A (MPL) is a nontoxic derivative of lipid A from *Salmonella*. Murine monoclonal Ab 8A1 specifically recognizes MPL. In our ELISA, monoclonal Ab 8A1 did not bind to any of the glycolipids tested (for the list of glycolipids tested, see Fig. 1). The assay was also very sensitive at different dilutions of the Ab. The Ab at a dilution of 1:32,000 gives an absorbancy of 1 for 320 fmol of MPL added to the well (Fig. 11). We incorporated MPL onto tumor cells as was done for exogenous GD3. Using monoclonal Ab 8A1 at a 1:32,000 dilution, we determined the concentration of MPL required to maximize cell association. The Cs-ELISA previously described was used to measure the density of MPL incorporated onto the cell surface of the melanoma cell M101. We found that about 0.350 pmol of MPL was incorporated onto M101 cells after exposure to 44

---

vortexing, cells were incubated in a water bath at 7.4°C for 90 min. Cells were washed three times in RPMI–4% HSA. Then, 100 μL of peroxidase-coupled second Ab (1 : 5000) was added. After vortexing and incubation as described previously, the cells were washed three times with RPMI–4% HSA. Following the final wash, the cells were transferred to fresh tubes, to avoid any interaction with the first and second Abs attached to the tubes. Measurements were done as described in Section II. Values were compared with the absorbancy at known concentrations of purified gangliosides at the same dilution of the Ab. The density of cell-surface gangliosides is calculated by comparing the absorbancy of Ab binding to cells with the binding of the Ab to known quantities (pmol) of highly purified gangliosides. Percentage viability is indiacted in parentheses.

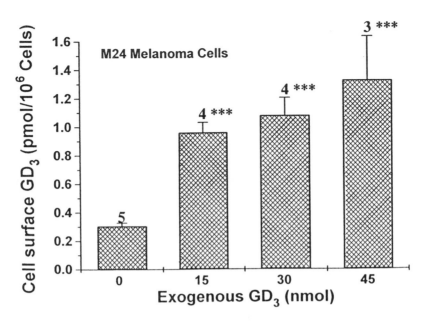

Figure 10 Incorporation of the melanoma-associated ganglioside GD3 onto the surface of GD3-deficient M24 melanoma cell line. One hour after incubating 1×10^6 cells with 250 μL of GD3 at varying concentrations, the cells were washed and incubated with Mel-1 (1 : 100). GD3 is expressed as pmol/10^6 cells. ***Level of significance: $p < 0.001$. Numbers refer to sample size.

nmol of exogenous MPL. Because the mode of cell-surface association may play an important role in immunopotentiation of MPL, we determined whether MPL bound to the lipid membrane or to cell-surface proteins. For this purpose, we trypsinized the M101 cells before and after exposure to exogenous MPL. We found that about 30% of the MPL was associated with trypsinized cells. We also observed that about 35% of MPL remained on the cell surface even after trypsinization, suggesting that MPL is bound to both the lipid membrane and cell-surface proteins.

IV. DISCUSSION

Biochemical estimation of the gangliosides and carbohydrate Ag of tumor cells measures both cytosolic and membrane-bound fractions. Although isolated cell membranes can be used to quantitate the membrane-bound carbohydrates, the isolation procedures may cause shedding of the Ag, particularly glycolipids that are restricted to the outer membrane of the

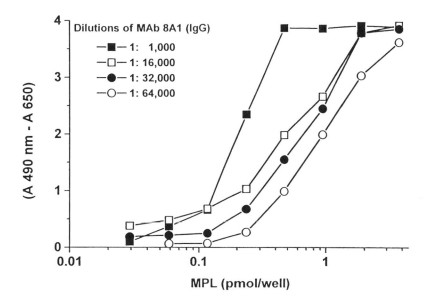

Figure 11   Box titration of murine monoclonal IgG 8A1 against different concentrations of monophosphoryl lipid A (MPL).

lipid bilayer. The immunological reactivity of the cell-surface Ag can be documented by a variety of methods. However, they are not useful to quantitate the density of specific cell-surface carbohydrate Ag. Labeling the cell surface galactose and *N*-acetylated sugars may be useful for measuring the total cell-surface density of carbohydrates, but the assay is not specific for particular carbohydrate epitopes. Flow cytometry employs monoclonal Ab to demonstrate cell-surface expression of the carbohydrate Ag. Flow cytometry measures the percentage of cells expressing a particular Ag in a cell population and the mean fluorescent intensity (MFI), which provides qualitative information on the density and topography of Ag. However, to quantitate the density of the Ag, purified standards of comparable concentration need to be used. Liposomes expressing Ag or microbeads carrying Ag may be used for this purpose. However, flow cytometry is tedious, expensive, and not easily accessible for routine monitoring. In this regard, the Cs-ELISA is a useful method for quantitation of the cell-surface density of carbohydrates.

In Cs-ELISA, the absorbancy obtained after the cell-surface binding of a monoclonal Ab is plotted against the linear standard slope obtained with purified ganglioside in a conventional ELISA under identical conditions (such as dilution of Ab and washing procedures). One may anticipate that the attachment of gangliosides to microtiter plates may not be quantitative. Because gangliosides do not form micelles in ethanol, the antigens may be uniformly suspended (particularly after sonication) at nano-, pico-, and femtomolar concentrations. The linearity of Ab binding in relation to different concentrations of the Ag indicates a predictable close-dependent variation in the Ab binding to the Ag. Loss of linearity at lower concentrations shows loss of sensitivity of the Ab. Loss of linearity at concentrations above 100 nmol indicates an Ag excess state. At the dilution of Ab used in this study, the linear variation in the Ab binding confirms that the concentration of Ag molecules varies strictly in relation to the added Ag and indicates that most of the Ag is attached to the well. The attachment of gangliosides to microtiter plates is indeed quantitative in the zone of linearity [25]. However, at nonlinear concentrations (such as at lower femtomolar ranges and at nanomolar ranges exceeding 100 nmol), the estimated percentage of Ag bound to plates may not be accurate and cannot be assessed by this assay. Therefore, we restrict comparison of the amount of Ab bound to cells only in the zone of linearity of the standards [25].

Our study documents that the efficacy of Cs-ELISA, for cells expressing multiple carbohydrate epitopes, depends on (1) the specificity of the monoclonal Ab for a carbohydrate epitope in the presence of closely related epitopes, (2) the ability of the monoclonal Ab to detect picomole to femtomole amounts of cell-surface carbohydrates, (3) the viability of the cells, and (4) the cell density. In this study, we measured Ab binding at $-10°C$, a temperature at which cellular uptake of Ab is minimal. The reliability of our assay depends on the irreversible nature of Ab binding to the cell surface. It has been well documented that Ab binding to the cell surface is predominantly irreversible, even at 37°C [30].

Using the Cs-ELISA, we measured the density of GD3 and GD2 epitopes in human melanoma cells and GM3 in mouse melanoma cells. Anti-GM3 IgM Ab M2590 could not be used for measuring GM3 in human melanoma cells, because it is not specific for GM3, and its binding to GM3 is affected by the presence of other gangliosides with longer sugar chains. Using the Cs-ELISA, we measured the ratio of GD3 and GD2 in M10-v (12:1) and in M101 cells (7:1). Human melanoma cell line M24 showed a very low level of cell surface GD3 and GD2. Biochemical results indicate that the values obtained by Cs-ELISA are not higher than the total amounts found in cells. More reliable evidence to suggest that the assay measures the absolute quantity of gangliosides is provided by GD3-incorporation experiments done on M24 cells. All assays in this study clearly document the absence or presence of a negligible amount of GD3 in the M24 melanoma cell line. Dosimetric incorporation of GD3 onto the cell surface of M24 suggests that the Cs-ELISA measures the absolute quantity.

Because M24 cells are deficient in GD3, we have incorporated exogenous GD3 onto these cells and monitored the incorporation with the Cs-ELISA. Using the same approach, we also monitored incorporation of MPL onto the tumor cells. We could incorporate 0.350 pmol of MPL onto 1 million M101 cells expressing about 2.5 pmol of GD3. The ratio of adjuvant (MPL) to the Ag (GD3) was 7:1. We also incorporated MPL onto trypsinized cells. Further, we determined the amount of MPL residues on the cells after trypsinization. These studies not only document the incorporation of MPL onto the bilayer lipid membrane, they also suggest various uses of Cs-ELISA in experimental cell research.

V. CONCLUSION

The Cs-ELISA is a valuable tool to assess carbohydrates expressed on the cell surface. The success of the assay depends on using specific monoclonal Ab with sensitive detection limits. The assay should be performed with an optimal cell density and with a cell population of $>80\%$ viability. The presence of a minor fraction (3–10%) of Ag-negative cells among tumor cells recovered from tumor biopsies does not interfere with the assay or evaluation of the density of cell-surface tumor-associated carbohydrate Ag. The Cs-ELISA may be useful for (1) characterizing tumor cells from biopsies and cultures; (2) selecting cell populations with a high expression of cell-surface carbohydrate residues for the development of carbohydrate-based whole cell vaccines; (3) identifying specific carbohydrate epitopes targeted by active and passive immunotherapies; (4) monitoring the effects of cell preservation treatments; (5) monitoring batch-to-batch consistency during whole cell cancer vaccine production; (6) studying interactions between carbohydrate residues of tumor cells and natural killer cells or cytotoxic lymphocytes; and (7) examining the relationship between tumor-associated carbohydrates and prognosis.

ACKNOWLEDGMENTS

The authors are greatly indebted to Donald L. Morton for his munificent support, encouragement, and advice throughout this investigation. He also provided tumor tissue surgically removed from melanoma patients and melanoma cell lines for Cs-ELISA. The authors are grateful to Gwen Berry for invaluable editorial assistance. This chapter is dedicated to Sen-itiroh Hakomori of the Biomembrane Institute, Seattle, Washington, whose work on tumor associated carbohydrates is the source of inspiration for our research. This work is supported by Wrather Family Foundation, Los Angeles, CA.

ABBREVIATIONS

Ab	antibody(ies)
Ag	antigen(s)
Cs-ELISA	cell-suspension ELISA
DMSO	dimethyl sulfoxide
DPBS	Dulbecco's phosphate-buffered saline
ELISA	enzyme-linked immunosorbent assay
HSA	human serum albumin
MFI	mean fluorescent intensity
MPL	monophosphoryl lipid A
PBS	phosphate-buffered solution
RPMI-160	Rosewell Park Memorial Institute series 1640
TLC	thin-layer chromatography

REFERENCES

1. Cheung, N. V., Lazarus, H., Miraldi, F. D., Abromowsky, C. R., Kallick, S., Saarinen, U. M., Spitzer, T., Strandjord, S. E., Coccia, P. F., and Berger, N. A. (1987). *J. Clin. Oncol.* 5, 1430–1440.
2. Vadhan-Raj, S., Cordon-Cardo, C., Carswell, E., Mintzer, D., Dantis, L., Duteau, C., Templeton, M. A., Oettgen, H. F., Old, L. J., and Houghton, A. N. (1988). *J. Clin. Oncol.* 6, 1636–1648.
3. Irie, R. F., and Morton, D. L. (1986). *Proc. Natl. Acad. Sci. USA* 83, 8694–8699.
4. Irie, R. F., Matsuki, T., and Morton, D. L. (1989). *Lancet* I, 786–787.
5. Reisfeld, R. A., Mueller, B. M., Handgretinger, R., Yu, A. L., and Gilles, S. D. (1994). *Prog. Brain Res.* 101, 201–212.
6. Livingston, P. O., Ritter, G., Oettgen, H. F., and Old, L. J. (1989). *Gangliosides and Cancer.* (ed: Oettgen, H. F.) VCH Publishers, New York, pp. 293–300.
7. Livingston, P. O. *1995 Immunol. Rev.* 145, 147–166.
8. Morton, D. L., Foshag, L. J., Hoon, D. S. B., Nizze, A., Famatiga, E., Wanek, L. A., Chang, C., Davtyan, D. G., Gupta, R. G., Elashoff, R., and Irie, R. F. (1992). *Ann. Surg.* 216, 463–482.
9. Ravindranath, M. H., and Morton, D. L. (1991). *Intl. Rev. Immunol.* 7, 303–329.
10. Reisfeld, R. A., and Schrappe, M. (1990). *Therapeutic Monoclonal Abs.* (eds. Borrebaeck, C. A. K., and Larrick, J. W.) Stockton, New York, pp. 57–74.
11. Wiegandt, H. (1985). *Glycolipids.* (ed. Wiegandt, H.) Elsevier, Amsterdam, pp. 199–260.
12. Gahmberg, C. G., and Hakomori, S.-I. (1973). *Proc. Natl. Acad. Sci. USA* 70, 3329–3335.
13. Gahmberg, C. G., and Hakomori, S.-I. (1973). *J. Biol. Chem.* 248, 4311–4317.

14. Irie, R. F. and Ravindranath, M. H. (1990). *Therapeutic Monoclonal Abs.* (eds. Borrebaeck, C. A. K., and Larrick, J. W.) Stockton, New York, pp. 75–94.

15. Ravindranath, M. H., Paulson, J. C., and Irie, R. F. (1988). *J. Biol. Chem.* 263, 2079–2086.

16. Ravindranath, M. H., Ravindranath, R. M. H., Morton, D. L., and Graves, M. C. (1994). *J. Immunol. Meth.* 169, 257–272.

17. Hakomori, S.-I. (1973). *Adv. Cancer Res.* 18, 265–315.

18. Urdal, D. L., and Hakomori, S.-I. (1983). *J. Biol. Chem.* 258, 6869–6874.

19. Matzku, S., Brocker, E. B., Bruggen, J., Dippold, W. G., and Tilgen, W. (1986). *Cancer Res.* 46, 3848–3854.

20. Xing, P. X., Reynolds, K., Tjandra, J. J., Tang, X. L., Purcell, D. F. J., and McKenzie, I. F. C. (1990). *Cancer Res.* 50, 89–96.

21. Ravindranath, M. H. and Irie, R. F. (1988) *Malignant Melanoma: Biology, Diagnosis, and Therapy.* Kluwer Academic, Boston, pp. 17–43.

22. Tai, T., Kwashima, I., Furukawa, K., and Lloyd, K. O. (1988) *Arch. Biochem. Biophys.* 261:51–55.

23. Hirabayashi, Y., Sugimoto, M., Ogawa, T., Matsumoto, M., Tagawa, M., and Taniguichi, M. (1986). *Biochim. Biophys. Acta* 875, 126–128.

24. Yogeeswaran, G., Stein, B. S., and Sebastian, H. (1978). *Cancer Res.* 38, 1336–1344.

25. Ravindranath, M. H., Bauer, P. M., Cornillez-Ty, C., Garcia, J., and Morton, D. L. (1996). *J. Immunol. Meth.*, 197, 51–67.

26. Tsuchida, T., Ravindranath, M. H., Saxton, R. E., and Irie, R. F. (1987). *Cancer Res.* 47, 1278–1281.

27. Morton, D. L., Ravindranath, M. H., and Irie, R. F. (1994). *Prog. Brain Res.* 101, 251–278.

28. Ravindranath, M. H., Morton, D. L., and Irie, R. F. (1994). *J. Autoimmun.* 7, 803–816.

29. Ravindranath, M. H., Brazeau, S. M., and Morton, D. L. (1994). *Experientia.* 50, 648–653.

30. Kyriakos, R., Shih, L. B., Ong, G. L., Patel, K., Goldgenberg, D. M., and Mattes, M. J. (1992). *Cancer Res.* 52, 835–842.

11
Scattering Techniques for the Study of Aggregative Properties of Gangliosides

Laura Cantù, Mario Corti, and Elena Del Favero
University of Milan, Milan, Italy

I. INTRODUCTION

Well-known powerful techniques are nowadays available for the study of the properties of glycomolecules; these include mass spectrometry, electromicroscopy, chromatography, and electrophoresis. Nevertheless, these techniques may be inadequate when the system either cannot be perturbed from its solution environment or has to be studied as a function of different solution parameters. In these cases, radiation scattering techniques can be quite useful because they do not perturb the system. In general radiation is scattered by inhomogeneities in a medium. Therefore, the scattered radiation from a macromolecular solution bears information about the macromolecules themselves.

Different kinds of radiations can be used, including visible light, mostly from laser sources, x-rays from laboratory equipment or from synchrotron radiation sources, and cold neutrons, available at various research nuclear reactor facilities. The purpose of this chapter is to present these different techniques concisely and to give some examples of their application to the study of the aggregation properties of gangliosides in solution.

II. METHODS

Radiation probes matter at a length scale of the order of its wavelength λ. Visible light and x-rays have quite different wavelengths, of order 5000 and 1.5 Å, respectively. Neutrons can also be considered as radiation. For instance 0.0045 eV cold neutrons from a nuclear reactor facility have a de Broglie wavelength of about 10 Å. In a scattering experiment, which consists of measuring the amount of radiation scattered elastically at a given angle by the sample, the important parameter is not the wavelength, but the modulus of the momentum transfer q; this is a combination of the wavelength and the observation angle θ. The mathematical formalism that describes the scattering process is identical for the different radiations in terms of the parameter $q = (4\pi/\lambda) \sin \theta/2$. The probing length of the radiation is more precisely determined by $1/q$ and not by λ alone. Therefore, at very small scattering angles, x-ray or neutron scattering can be equivalent to light scattering at large angles. Indeed, there is a region of dimensions in which all three types of radiations can be used, as shown pictorially in Fig. 1. Interestingly enough, the overlapping region covers dimensions typical of biological macromolecules in solution.

Light [1,2], x-ray [3], and neutron [4] scattering techniques have already been extensively discussed in the literature and the scheme for interpretating the scattering data is well established. The excess scattering over the solvent scattering, the intensity $I(q)$, can be expressed as

$$I(q) = A(\text{contrast})^2 cMP(q)S(q) \tag{1}$$

where A is a constant connected with the experimental apparatus in use, c is the macromolecular concentration, M is the macromolecular mass, if the solution is monodisperse or the weight average one if the solution is polydisperse, $P(q)$ is the form factor of the macromolecule, and $S(q)$ the structure factor of the solution [5–7]. The factorization of $P(q)$ times $S(q)$ in Eq. (1) is strictly valid when macromolecular orientation can be decoupled from intermacromolecular interactions. The contrast, reported in parenthesis, is a term that depends on the radiation used for the scattering experiment. The "optical contrast" for light scattering experiments is given by the refractive index increment, dn/dc, of the solution. It is connected to the difference of electrical polarizability of the macromolecules with respect to the solvent. The refractive index increment depends, therefore, on the overall features of the macromolecules and is not easily calculable a priori. It is nevertheless measurable independently by means of standard refractometers. For x-rays, the contrast is determined by the difference in electron densities of the macromolecules and the solvent. It is calculable from the known chemical composition of the macromolecule and the density of the solution. Similarly, the "neutron contrast" depends on the chemical composition of the macromolecule and the density of the solution, but it arises

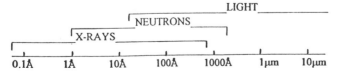

Figure 1   Probing length for the different forms of radiation.

from a completely different kind of interaction. X-rays interact with the electrons of the atoms, whereas neutrons are scattered by their nuclei. The neutron contrast is therefore given by the neutron scattering cross section of each individual nucleus of the atoms making up the macromolecule.

The structure factor takes into account the spatial distribution of the macromolecules in the sample. It arises from the interference of the radiation scattered from different macromolecules. For dilute solutions in which the average distance between macromolecules is much larger than their interaction range and therefore macromolecules can occupy freely any position in the sample, the structure factor $S(q) = 1$. In the opposite limit, in which macromolecules are in fixed position such as in a lattice, the structure factor is a strongly varying function of q and gives rise to the well-known Bragg peaks. In the intermediate cases, $S(q)$ depends on the macromolecular interactions and therefore on their charge if the interactions are Coulombic [5–7].

The form factor $P(q)$ is a measure of the interference of radiation scattered from different points inside the macromolecule. In the limit of macromolecules very small with respect with the radiation wavelength, $P(q) = 1$. This happens, for instance, for small proteins such as lysozyme when studied with visible light. If the same protein is instead probed with the shorter wavelengths of x-rays or neutrons, $P(q)$ is a sensitive function of the protein dimension, shape, and internal inhomogeneities.

The structure factor $S(q)$ depends only on the values of q with which measurements are performed, whereas in general, the form factor depends on the radiation used in the scattering experiments not only for the q values, but also for the contrast, which may be different in the various regions of the macromolecule. In fact, the expression of Eq. (1) is strictly valid for homogeneous macromolecules only, because the contrast term has been factored out of the form factor. For inhomogeneous macromolecules, the contrast has to be explicitly included in the expression that calculates the form factor as an interference process among the radiation scattered by different points inside the macromolecule [8]. For instance, the form factor for an inhomogeneous sphere of radius R made of an internal part of contrast ρ_1 extending up to a radius r and an external shell of contrast ρ_2 is proportional to:

$$\left\{ \rho_2 \left[\frac{3R^3(\sin qR - qR\cos qR)}{q^3R^3} \right. \right.$$
$$\left. \left. - \frac{3r^3(\sin qr - qr\cos qr)}{q^3r^3} \right] + \frac{\rho_1 3r^3(\sin qr - qr\cos qr)}{q^3r^3} \right\}^2 \tag{2}$$

whereas the form factor for an homogeneous sphere of radius R is simply

$$P(q) = \left[\frac{3(\sin qR - qR \cos qR)}{q^3 R^3} \right]^2 \qquad (3)$$

Similar expressions can be written also for ellipsoids of revolution [8]. As it will be described later, the contrast variation effects can be quite useful in the study of self-aggregation of gangliosides.

Equation (1) describes the so-called static scattering measurements, that is, the average scattered intensity independent of the Brownian motion of the macromolecules. With temporally coherent laser light sources now available, it is possible to study also the dynamics of the solution, that is, to measure the diffusion coefficient of the macromolecules and hence their hydrodynamic radius [2].

Diffusion of macromolecules depends not only on their volume but also on their shape and solvation. Therefore, combination of static and dynamic laser light scattering measurements can provide a simple tool to investigate the shape of macromolecules when the latter are small compared with the wavelength of light [9]. Finally, experiments with polarized light can give useful information about the optical anisotropy of the scattering medium [1].

III. RESULTS AND DISCUSSION

A. Self-Aggregation of Gangliosides

Gangliosides [10,11] are double-tailed amphiphilic molecules, like phospholipids, in which a ceramide lipid portion, comprising a sphingosine and a fatty acid with roughly 20 carbons each, carries a rather bulky headgroup made up of several sugar rings, some of which are sialic acid residues. In aqueous solutions, the simultaneous request to preserve and to avoid contact with water leads the amphiphilic ganglioside molecules to organize themselves so as to create distinct hydrophilic and hydrophobic domains, such as the surface and the core of a globular micelle or the outer and inner parts of a bilayer. In general, the packing of amphiphiles in an aggregate obeys some simple rules following the consideration that, except for surface roughness, no water can exist inside the hydrophobic domain, so that one of its dimensions cannot exceed twice the monomer hydrophobic length, whereas contact with water is required by the hydrophilic headgroup. This requirement is very stringent and can be handled theoretically in the frame of the "opposing forces" model [12]. According to this model, surface tension and hydrophobic forces are summarized in an attractive interaction on the hydrocarbon–water interface. The repulsive forces among headgroups, electrostatic, hydration, and steric, are also added up to an equivalent repulsive interaction acting on the same interface. Therefore, two net opposite forces are applied at this surface, one tending to decrease, the other to increase the interfacial area per headgroup exposed to the aqueous phase; the two forces balance in accord with an "optimal surface area" a_o, which is assumed to be the area at the interface required by the monomer in the aggregated structure. Following these observations, it is quite useful to describe the packing of amphiphilic molecules into an aggregate by a simple geometrical model [13].

The hydrophobic part of an amphiphile is schematized as occupying inside the aggregate a place with a shape of a truncated cone, identified by three parameters: volume V, length l, and the area a_o at the interface. They are summarized in a packing parameter or shape factor $P = V/a_o l$, which assumes the limiting values $1/3$ for a true cone and 1 for a cylinder. The corresponding aggregated structures are spheres and bilayers. Vesicles, liposomes, and in general, membrane-like structures are formed when $1/2 < P < 1$. Micelles are formed when $1/3 < P < 1/2$. In between these values, higher P values identify larger and more asymmetrical micelles. An example of an amphiphile that forms rather spherical micelles is the 12-carbon chain sodium dodecyl sulfate surfactant (SDS) in water. Its packing parameter is quite close to $1/3$, $P = 0.37$, with $V = 350 \text{ Å}^3$, $l = 16.7 \text{ Å}$, and $a_o = 57 \text{ Å}^2$. On the other hand, the amphiphile that typically forms vesicles and bilayers is egg lecithin, which has a packing parameter $P = 0.84$, with $V = 1063 \text{ Å}^3$, $l = 17.5 \text{ Å}$, and $a_o = 70 \text{ Å}^2$. These geometrical arguments seem rather simple and convincing, mostly as concerns the limiting values $1/3$ and 1 for the packing parameter corresponding, respectively, to the spherical micelle case with small aggregation number (< 100) and the bilayer case, which includes the well-known liposome-forming amphiphiles. Gangliosides provide a quite interesting example of amphiphilic molecules that give packing parameters around $1/2$, the value at which the transition between globular type of aggregation and bilayer structure occurs [14]. It is clear that, in this region, the final shape of the aggregate is very sensitive to small changes in the characteristics of the individual molecules, so that even minor geometrical effects can be detected. The large hydrophobic volume can be balanced by the steric dimensions of the headgroup, so that some gangliosides can aggregate into micelles and not into bilayers as normally happens for double-tailed lipids.

Light, neutron, and x-ray scattering experiments on aqueous solutions of various gangliosides in the millimolar range give information about their aggregation properties, such as the aggregation number, the hydrodynamic radius of the aggregates, their shape, and polydispersity [14]. Figure 2 shows a schematic representation of the monosialoganglioside series GM1, GM2, and GM3. The ganglioside nomenclature is taken from Svennerholm [15]. The sequence is obtained by progressively reducing the number of

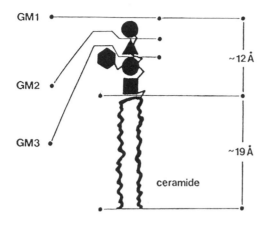

Figure 2 Representation of the monosialoganglioside series GM1, GM2, and GM3.

sugars in the main chain. The reduction of the hydrophilic headgroup dimension has a strong effect on the aggregative properties, as shown in Table 1. An increase in the aggregation number and a drastic change from micellar to vesicular structure are observed in the sequence. The transition from micelle to vesicle takes place when the packing parameter goes through the value 1/2. The ganglioside GM4 is like GM3 without the initial glucose group. Table 1 shows that, as the number of sugar groups decreases, a smaller interfacial area a_0 is required by the molecule in the aggregate, which becomes less curved and bigger with larger aggregation number. Notice that in the vicinity of the micelle to vesicle transition the micellar aggregation number is very sensitive to small changes in the interfacial area. From GM1 to GM2, the aggregation number increases by 50% for only a few Å^2 of variation in interfacial area. This means that the aggregation number of the ganglioside micelle can be a good indicator of small changes in the headgroup, such as the ones that may come either from different ramification points in the sugar chain or from changes in the internal interaction among sugars [14]. For sake of comparison, Table 1 reports data also for the more complex ganglioside GT1b, which has two more sialic acids than GM1. Its micellar dimension and aggregation number are even smaller than those of GM1.

Ganglioside micelles cannot be spherical. In fact, with the usual physical model of a micelle made of a hydrocarbon core with the hydrophilic headgroups projecting out into the water, the shortest semi-dimension of the hydrophobic core cannot exceed the extended length of the hydrophobic

Table 1 Physico-Chemical Properties of Various Gangliosides

Ganglioside	Monomer molecular weight, MW	No. of Sugars in headgroup, S	Aggregation type	Aggregation number, N	Interface area per monomer, $a_0(\text{Å}^2)$	Hydrodynamic radius, $R_H(\text{Å})$	Packing parameter, P	Micellar axial ratio
GT1b	2142	7	Micelle	176	100.8	53.2	0.405	1.7
GM1	1560	5	Micelle	301	95.4	58.7	0.428	2.3
GM2	1398	4	Micelle	451	92	63	0.445	2.9
GM3	1195	3	Vesicle	14,000	80	250	>0.5	
GM4	1015	2	Vesicle	18,000	80	270	>0.5	

chains of the ganglioside molecule. Indeed, for GM1, taking the aggregation number $N = 300$ with a density 0.87 g/cm^3 of the hydrophobic core [14], the core volume is 290,000 Å^3. This would correspond to a sphere of radius 41 Å, much larger than the actual length of the hydrophobic group, which is approximately 20 Å. By assuming a simple shape of ellipsoid of revolution, micelles turn out to be oblate (disklike) with an axial ratio increasing with the aggregation number. The measured hydrodynamic radius discriminates between the prolate and oblate shape [16]. The axial ratios reported in Table 1 are estimated from light scattering data with the foregoing method.

The shorter wavelengths of neutron and x-ray scattering experiments allow a more direct determination of the micellar shape by means of the form factor $P(q)$, which is different from unity at these wavelengths. Figures 3 and 4 present the scattered intensity versus q for neutrons and x-rays, respectively, for a dilute solution of GM1 micelles with added salt. In this case, intermicellar interactions are negligible, $S(q) = 1$, and therefore the scattered intensity is proportional to the form factor. The two experimental curves are quite different, giving direct evidence that micelles are not homogeneous, as expected from micellar models. In fact, the lipid core and the

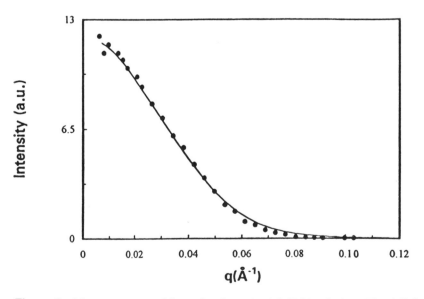

Figure 3 Neutron scattered intensity for a 3 mM GM1 solution. The full line is calculated from the form factor, see text. Experiments were performed with the D17 instrument at ILL (Grenoble, France).

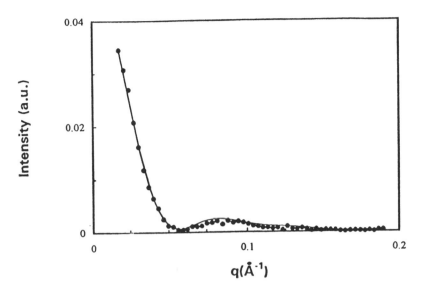

Figure 4 X-ray scattered intensity for the solution described in Fig. 3. The full line is the calculated form factor, see text. Experiments were performed with the D22 instrument at LURE (Orsay, France).

sugar layer, having different contrasts for the two types of radiation, give rise to different interference effects inside the micelle. A radial profile of the contrasts for the GM1 micelles is drawn in Fig. 5. The micelle is seen by x-rays as a hollow ellipsoid, because the lipid electron density is similar to the one of water and different from that of the sugar layer. Its form factor presents, therefore, an oscillation on the tail, which gives sensitivity to the fitting process of a model shape to the experimental data. There is, on the other hand, not too much difference in neutron contrast between the two parts of the micelle with respect to the surrounding solvent, deuterated water for experimental reasons. Therefore, the neutron form factor is a monotonic decreasing function of q that depends mostly on the overall shape of the micelle and less on its internal structure.

The full lines of Figs. 3 and 4 are the form factors calculated for the GM1 micelles, modeled as inhomogeneous oblate ellipsoids. The only free parameters of the fit are the axial ratio of the core and the hydration of the hydrophilic layer. In fact, the other parameters necessary to complete the fit are known, including the hydrophobic and hydrophilic volumes of the ganglioside monomer and the electron densities of the hydrophobic and hydrophilic portions of the micelle fixed by the chemical composition and

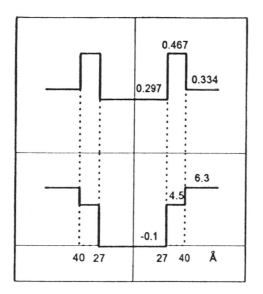

Figure 5 Radial profile of x-ray (top) and neutron (bottom) contrasts for the GM1 micelle. The x-ray and neutron contrasts are expressed in units of $e^-/Å^3$ and 10^{10} cm^{-2}, respectively. The profile is drawn across the minor axis of the ellipsoid.

the measured density values [14]. The micellar aggregation number $N = 300$ is also fixed by the light scattering measurements [14]. The axial ratio of the hydrophobic core comes out to 2.1, which confirms the light scattering estimates rather well (see Table 1). The long and short semi-axes are estimated at 56 and 27 Å, respectively. The hydrophilic layer thickness is 13 Å, with a hydration of 15 water molecules per ganglioside molecule. An important check of the correctness of the fitting parameters for GM1 micelles comes from the excellent agreement between the value of the hydrodynamic radius of the ellipsoid resulting from the fit, 58.2 Å, calculated by means of the Perrin formulas [2], and the value measured experimentally by means of the dynamic light scattering technique, 58.7 Å.

B. Micellar Charge

Gangliosides may hydrolyze in solution, giving rise to charged micelles that interact electrostatically with each other if the ionic strength of the solution is not so high as to screen completely the Coulomb repulsion. If observed with scattering techniques, light, x-rays, or neutrons, the charge of the micelles can be inferred by performing experiments in which the ganglioside concentration is kept constant while the ionic strength of the solution is modified by adding salts, such as NaCl. In fact, the neutron or x-ray scattering pattern at low angles is influenced by the presence of the structure factor $S(q)$, which has a characteristic oscillating behavior in the presence of interactions and levels off progressively as salt is added to the solution. If the light scattering technique is used, the same structure factor has the effect of depressing the scattered intensity to an extent that depends on the range and intensity of repulsive interactions. The q dependence of $S(q)$ is lost, because with light one probes only very low q, where $S(q)$ is practically constant and approximately equal to $S(0)$; but as salt is added to the solution, the scattered intensity increases until it reaches a saturation value when repulsion is completely screened. Figure 6 shows the scattered light intensity at fixed low q as a function of the ionic strength for a 3 mM solution of GM1. Figure 7 shows the neutron scattering data for the same 3 mM GM1 solution with the full q dependence of $I(q)$. Crosses and dots refer to the solution with no added salt and with 100 mM NaCl, respectively. For high salt, micelles behave independently because the Coulomb interaction range is small compared with the micellar average distance. In this case, $S(q) = 1$ and the scattered intensity represents just the micellar form factor. When no salt is added, the ionic strength of the solution is very low; any residual ionic strength is derived from micellar dissociation. In this case, the electrostatic interaction potential is scarcely screened and

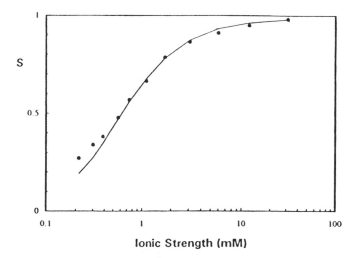

Figure 6 Scattered light intensity (S) as a function of the ionic strength for a 3 mM solution of GM1. The full line is the calculated intensity for a micellar charge of 48 electron units.

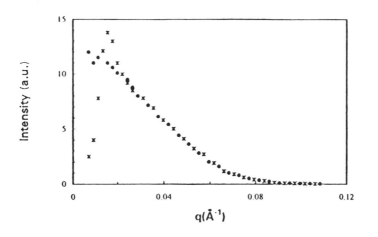

Figure 7 Dependence of the scattered neutron intensity $I(q)$ for a 3 mM GM1 solution. Crosses and dots refer to the solution with no added salt and with 100 mM NaCl, respectively.

extends to intermicellar distances. This causes the depression of the intensity at low q and oscillations. Both light and neutron scattering results as a function of the ionic strength of the solution can be theoretically reproduced by constructing the radial distribution function $g(r)$ of the micelles in the Hypernetted Chain (HNC) approximation [17], using for the pair interaction potential a form consisting of a hard core repulsion plus a screened Coulomb potential. The charge Q of the micelles generating the interaction potential is the only parameter, which is determined by the fitting procedure. The experimental data are reproduced by using a constant charge of 48 electron units per micelle in the full range of ionic strengths [18,19]. The neutron scattering data of Fig. 7 give also the important information that the GM1 ganglioside micelle does not change its size and shape when salt is added. This can be deduced from the perfect superposability of the high q data, indicating that the form factor is the same in the two cases. This independence of the micellar size and the salt concentration may at first seem surprising, as the aggregation number of ionic amphiphiles is known to depend appreciably on the concentration of added salt [12,20], due to variations in the electrostatic repulsion between the headgroups. Indeed, this may not be the case for gangliosides, because their sugar headgroups are bulky with the possibility of hydrogen bonding among them.

C. Mixed Micelles

Mixed micelles may be formed with gangliosides. When a synthetic surfactant, the nonionic $C_{12}E_8$, is added to a GM1 solution, mixed micelles are formed with a molecular weight that depends on their molar ratio. Figure 8 shows the dependence of the mixed micelle molecular mass when different amounts of $C_{12}E_8$ are added to a 0.8 mM solution of GM1. The molecular mass monotonically decreases from the value of the pure GM1 micelle to that of the pure $C_{12}E_8$ micelle [21]. The full line is the prediction of a thermodynamic model that involves regular mixing of the two amphiphiles in the micelle [22]. The agreement is quite good. This means that the non-ionic surfactant mixes randomly with GM1 in the micelle. This may not be the case when two gangliosides are mixed. In fact, mixed micelles of GM2 and GT1b do not follow the predictions of ideal mixing, as can be seen from the full line in Fig. 9. In the figure, X is the mole fraction of GT1b, the total concentration being fixed at 0.9 mM. For $X = 0.2$ the mixed micelle molecular mass is even larger than the ones of both the individual micelles. Nonrandom mixing means obviously that some clustering occurs in the mixed micelle. A careful comparison of light and neutron scattering results [23] suggest that GT1b preferentially locates in regions of larger curvature of the globular nonspherical mixed micelle, namely at the edges, whereas GM2 is more able to fit in the flatter aggregate regions.

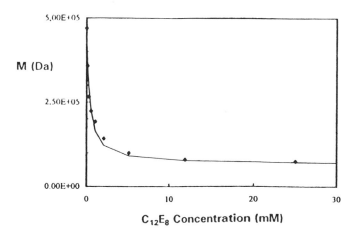

Figure 8 Dependence of the mixed micelle molecular mass when different amounts of $C_{12}E_8$ are added to a 0.8 mM solution of GM1. The full line is the prediction of a thermodynamic model that involves regular mixing of the two amphiphiles in the micelle.

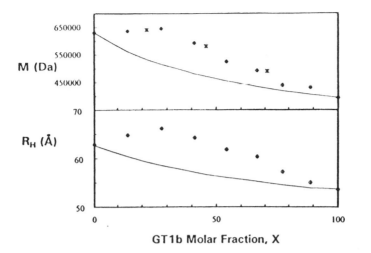

Figure 9 Molecular mass and hydrodynamic radius R_H of mixed micelles of GM2 and GT1b as a function of the mole fraction X of GT1b (total concentration = 0.9 mM). The full line gives the prediction for ideal mixing.

Some geometrical effects can also be observed when a ganglioside is mixed with phospholipids. Figure 10 shows some laser light scattering data regarding the molecular mass and hydrodynamic radius of egg yolk lecithin vesicles for increasing amounts of GM1 [24]. The presence of GM1 allows smaller vesicles because the ganglioside has a larger headgroup area that lecithin. Vesicles no longer form when the percentage of GM1 is higher than about 20%.

The light scattering study of the time of formation of mixed micelles gives strong support to the theoretical prediction that natural ganglioside critical micelle concentrations (cmc's) are low (10^{-9}–10^{-8}M) and within roughly one order of magnitude across the whole series of gangliosides [14]. Figure 11 shows the evolution toward equilibrium of mixed micelles of GM2 and GT1b starting from individual micelles [25]. The final configuration is reached in about 10 hr. The full line represents the theoretical behavior calculated according to a model that attributes the mixing process to monomer transfer from one micelle to another via free molecules in solution. The equilibration time is then dictated by the exchange rate between aggregates and solution, and in particular it is expressed as a bilinear function of the exchange rates of the two molecular species, so that the quicker one will dominate [25]. To obtain long equilibration times, it is then necessary that the two rates be slow and quite similar. Because the

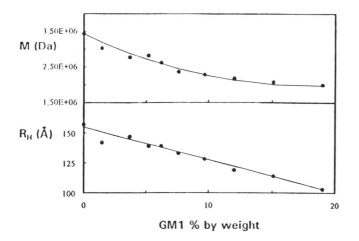

Figure 10 Molecular mass and hydrodynamic radius R_H as a function of GM1 molar fraction for egg yolk lecithine/GM1 mixed monolamellar vesicles. Full lines are drawn to guide the eye through data points.

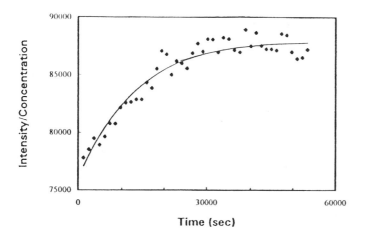

Figure 11 Scattered intensity of mixed micelles of GM2 and GT1b starting from individual micelles. Full line is the theoretical prediction.

exchange rate is closely connected to the cmc of the amphiphile, the observed behavior is an indirect proof that both GM2 and GT1b cmc's are low and, say, within one order of magnitude.

D. Vesicles of Gangliosides

As shown in Table 1, micelle formation is no longer possible when the oligosaccharide chain of gangliosides becomes too small. In fact, GM3 and GM4 form bilayer-type structures. It has been found that the ganglioside GM3 forms vesicles spontaneously in water solution [26]. This is a quite unusual feature for bilayer-forming amphiphiles, such as phospholipids, because they normally require external energy to close up into a vesicle. Indeed, the solution properties of GM3 are quite interesting. Vesicles form and break rapidly and coexist with a small concentration of large bilayer fragments. This dynamic equilibrium of GM3 aggregates in solution can only be studied by nonintrusive techniques, such as light [26] or x-ray [27] scattering techniques. Data show that the vesicle distribution has a mean diameter that changes slightly with temperature, going from 265 Å at 7°C to 245 Å at 50°C. Equilibration times of the aggregate distribution are found to be small, because temperature scans are fully reversible and quite independent of scan rate. This is an indication that monomer exchange is not the leading mechanism responsible for the assessment of the distribution to different equilibrium conditions, as monomer exchange involves much longer equilibration times in ganglioside micelles [25]. The existence of the large lamellar fragments could only be inferred from the light scattering data because their number concentration is so small, of the order of 1 in 10,000 vesicles. Their large dimension, of the order of a few thousand angstroms, gives rise in fact to an angular dependence of the scattered light intensity that discriminates them against vesicles [28]. More precise characterization of such lamellar fragments has been obtained with depolarized laser light scattering measurements [26,29]. In fact, if aggregates are formed by optically anisotropic molecules, such as the ganglioside GM3, the scattered light can have a depolarized component. Depolarized light scattering theory predicts that for thin spherical vesicles, even if made up of anisotropic molecules, no depolarization can occur, due to internal cancellation effects [30]. On the other hand, this does not happen if the aggregate has a lamellar or a disk structure. Depolarization measurements are therefore quite useful to selectively observe fragments of bilayers in the presence of a large population of spherical vesicles, because vesicles do not contribute to the depolarized scattering intensity.

The spontaneous vesicle formation and the coexistence with large lamellar fragments is a peculiar feature of GM3 ganglioside solutions, which suggests interesting physical considerations. Spontaneous vesicle formation cannot be explained with energetic considerations; entropic effects have to be considered. In fact, the two GM3 monolayers of the bilayer, being equal, have the same spontaneous curvature [26,30], and therefore flat lamellar structures are energetically favored. Some energy cost has to be paid to curve the bilayer into a closed vesicle, and the system is said to be frustrated. However, if the bending modulus [26,30], of the bilayer is low with respect to the thermal energy $k_B T$, that is, of the order of 10^{-14}–10^{-13} erg, the gain in entropy of mixing due to the formation of small vesicles can easily compensate for this energy cost. Indeed, the bending rigidity of phospholipid vesicles is of the order of 10^{-12} erg, which is far too high to allow vesicle formation without supply of external energy. Such important characteristics of gangliosides to confer a low bending rigidity of the bilayers can be qualitatively understood again by simple geometrical arguments. The oligosaccharide chain headgroup of gangliosides has a lateral extension that is larger than the corresponding ones of the phospholipid headgroups, phosphatidylcholine, for example. On the other hand, the hydrophobic dimensions are about the same. It is reasonable, therefore, to think that in the bilayer the sugar headgroups exert a greater lateral force than the corresponding choline groups, which do not exceed the hydrocarbon chain lateral dimension. This can reduce the bending rigidity, as it is known that the compression elasticity of the hydrocarbon chains is quite high. In other words, the restoring force due to the sugar headgroups compression upon bending is smaller than the one coming from the hydrocarbon chains. A confirmation of this is given by the comparison of the interfacial area for GM3, which is 80 Å ^{2}, see Table 1 and that for egg lecithin, which is only 70 Å ^{2}.

The GM3 system frustration may also be connected with the fact that vesicles are in thermodynamic equilibrium with the larger aggregates. Indeed, the small population of lamellar fragments is not energetically favored against infinite lamellar sheets due to the unfavorable edge energy; but, here again, it can be stabilized entropically. In other words, energetically unfavored single-component vesicles and lamellar fragments can be found in water solution because of entropic effects, and their coexistence arises from the equivalence of their energy cost.

REFERENCES

1. Fabelinskii, I. L. (1968). *Molecular Scattering of Light*. Plenum, New York.
2. Corti, M. (1985). In Degiorgio, V., and Corti, M., eds., *Physics of Amphiphiles: Micelles, Vesicles and Microemulsions*. North Holland, Amsterdam, pp. 121–151.

3. Glatter, O. (1982). In Glatter, O., and Kratky, O., eds., *Small Angle X-Ray Scattering*. Academic Press, London, pp. 119–196.
4. Squires, G. L. (1978). *Introduction to the Theory of Thermal Neutrons*. Cambridge University Press, Cambridge.
5. Corti, M., and Degiorgio, V. (1981). *J. Phys. Chem.* 85, 711–716.
6. Belloni, L. (1991). In Lindner, P., and Zemb, T., eds., *Neutron, X-Ray and Light Scattering*. North Holland, Amsterdam, pp. 135–155.
7. Pusey, P. N., and Tough, R. J. (1985). In Pecora, R., ed., *Dynamic Light Scattering Applications of Photon Correlation Spectroscopy*. Plenum, New York.
8. Guinier, A., and Fournet, G. (1955). *Small Angle Scattering of X-Rays*. Wiley Interscience, New York.
9. Corti, M., and Degiorgio, V. (1978). *Chem. Phys. Lett.* 53, 237–241.
10. Tettamanti, G., Sonnino, S., Ghidoni, R., Masserini, M., and Venerando, B. (1985). In Degiorgio, V., and Corti, M., eds., *Physics of Amphiphiles: Micelles, Vesicles and Microemulsions*. North Holland, Amsterdam, pp. 607–636.
11. Hakomori, S. I. (1981). *Ann. Rev. Biochem.* 50, 733–764.
12. Tanford, C. (1980). *The Hydrophobic Effect*. Wiley, New York.
13. Israelachvili, J. N. (1990). *Intermolecular and Surface Forces*. Academic Press, New York.
14. Sonnino, S., Cantù, L., Corti, M., Acquotti, D., and Venerando, B. (1994). *Chem. Phys. Lipids* 71, 21–43.
15. Svennerholm, L. (1964). *J. Lipid Res.* 5, 141–145.
16. Cantù, L., Corti, M., Sonnino, S., and Tettamanti, G. (1986). *Chem. Phys. Lipids* 41, 315–328.
17. Goldstein, D. (1975). *States of Matter*. Prentice Hall, New York.
18. Cantù, L., Corti, M., and Degiorgio, V. (1987). *Faraday Discuss. Chem. Soc.* 83, 287–295.
19. Cantù, L., Corti, M., Degiorgio, V., Piazza, R., and Rennie, A. (1988). *Prog. Coll. Pol. Sci.* 76, 216–220.
20. Minero, C., Pramauro, E., Pellizzetti, E., Degiorgio, V., and Corti, M. (1986). *J. Phys. Chem.* 90, 1620–1624.
21. Corti, M., Degiorgio, V., Ghidoni, R., and Sonnino, S. (1982). *J. Phys. Chem.* 86, 2533–2537.
22. Clint, J. (1975). *J. Chem. Soc.* 71, 1327–1333.
23. Cantù, L., Corti, M., and Degiorgio, V. (1990). *J. Phys. Chem.* 94, 793–795.
24. Masserini, M., Sonnino, S., Giuliani, A., Tettamanti, G., Corti, M., Minero, C., and Degiorgio, V. (1985). *Chem. Phys. Lipids* 37, 83–97.
25. Cantù, L., Corti, M., and Salina, P. (1991). *J. Phys. Chem.* 95, 5981–5983.
26. Cantù, L., Corti, M., Del Favero, E., and Raudino, A. (1994). *J. Phys. II France* 4, 1585–1604.
27. Maurer, N., Cantù, L., and Glatter, O. (1995). *Chem. Phys. Lipids*, 78; 47–53.
28. Cantù, L., Corti, M., Lago, P., and Musolino, M. (1991). In Schmitz, ed.,

Photon Correlation Spectroscopy: Multicomponent Systems, Proc. SPIE 1430. Washington, D.C., pp. 144–159.

29. Cantù, L., Mauri, M., Musolino, M., Tomatis, S., and Corti, M. (1993). *Prog. Colloid Polymer Sci. 93*, 30–32.
30. Aragon, S. R., and Elwenspoek, M. (1982). *J. Chem. Phys. 77*, 3406–3412.

12

Imaging Carbohydrate Polymers with Noncontact Mode Atomic Force Microscopy

Theresa M. McIntire and David A. Brant
University of California–Irvine, Irvine, California

I. INTRODUCTION

Electron microscopy (EM) has played an important role in determinations of the structure and function of biopolymers [1,2]. Crystalline polysaccharides have been studied extensively with various EM techniques to determine the features of organized polysaccharide structures. For example, crystalline amylose [3], chitosan [4], chitin [5], and cellulose [6] have recently been investigated. Transmission electron microscopy (TEM) has been applied to isolated carbohydrate polymer molecules to determine polysaccharide shapes, the distributions of molecular weights and end-to-end distances, and the inherent chain stiffness of the polymer chains [7,8]. Typically, it has been necessary in TEM studies of individual carbohydrate polymers to use contrast enhancement procedures that limit the spatial resolution to 1–2 nm.

Scanning probe microscopies involve rastering an atomically sharp probe tip across the surface of a sample and measuring changes in electron tunneling current (scanning tunneling microscopy, STM), or changes in electrostatic, magnetic, capillary, or van der Waals forces sensed by the probe tip (atomic force microscopy, AFM) [9]. Probes sensitive primarily to changes in van der Waals forces have proven to be effective for imaging biological samples, including individual molecules and molecular assemblies. In contrast to TEM, it is usually not necessary to apply heavy metal contrast enhancing agents, and the resolution of the technique may therefore exceed that of TEM, particularly in the dimension normal to the surface of the substrate (freshly cleaved mica or highly oriented pyrolytic graphite) that underlies the sample of interest.

Atomic force microscopy techniques sensitive to the repulsive van der Waals forces operative at very small tip-to-sample distances often result in sample disruption or, indeed, in sweeping the sample of interest out of the field of observation. Contamination of the probe tip is also a common problem in such cases. These techniques are often referred to as *contact mode* AFM. It is also possible to conduct AFM experiments in the regime of tip-to-sample distances where the attractive van der Waals interactions between tip and sample are dominant. These attractive forces are generally smaller than those that bind the sample to the substrate, and the tip can be rastered over the sample repeatedly without sample disturbance or displacement. All attractive regime methods involve use of a vibrating probe tip. The resonant frequency and amplitude of the vibrations are sensitive to the attractive interactions of sample and tip and, hence, to the tip-to-sample distance. Monitoring of the frequency and/or amplitude as a function of the raster scan then results in a topographic map of the substrate and molecules or molecular assemblies that may lie on it. This is termed *noncontact mode* AFM (NCAFM) [10,11], or AC mode to reflect the AC character of the signal detected.

The bacterial polysaccharide xanthan [12,13] and the algal polysaccharide carrageenan [14] have been imaged using STM. Conventional contact mode AFM has been applied successfully to the carbohydrate polymers xanthan [15] and acetan [16]. NCAFM has been used to image linear and circular triple helical forms of the fungal polysaccharide scleroglucan [17,18]. We report in this chapter on use of NCAFM as carried out using the AutoProbe CP atomic force microscope (Park Scientific Instruments, Sunnyvale, California) to image several microbial and fungal polysaccharides. Variants of this experiment exist, including intermittent contact mode or TappingMode [19] as implemented by Digital Instruments (Santa Barbara, California).

II. PRINCIPLES OF NCAFM METHODOLOGY

A schematic diagram of an NCAFM instrument is shown in Fig. 1. A cantilever incorporating the sharpened probe tip (the cantilever/tip or, in short, the tip) is attached to a piezoelectric bimorph that oscillates in response to an applied AC voltage at frequency ω. The bimorph/cantilever assembly is arranged to oscillate in the z dimension normal to the substrate surface (xy plane). The amplitude of the cantilever vibrations is monitored using a position-sensitive photodiode (PSPD) that detects laser light reflected from the upper surface of the cantilever. The substrate surface bearing a sample (the substrate/sample or, in short, the sample) is mounted on a piezoelectric crystal (scanner) that can be driven in the xy plane to raster the tip across the sample surface. The mounting crystal can also be driven in the z dimension to respond through a feedback loop to changes in the amplitude of the cantilever vibrations that occur as the tip-to-sample

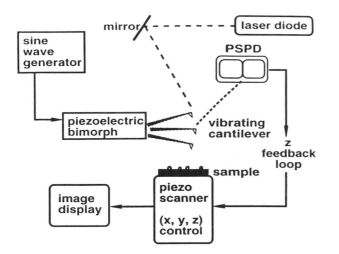

Figure 1   A schematic diagram of an NCAFM instrument with components described in the text.

distance changes during transits of the tip above the xy plane. The voltages that drive excursions of the mounting crystal are converted to the x, y, and z dimensions of the image produced by the instrument.

Figure 2 shows an approximation to the van der Waals potential energy of interaction $V(z)$ between the tip and the sample over some specific point $x'y'$ in the xy plane. A plot of the force $F(z) = -\partial V(z)/\partial z$ associated with the interaction is also shown in Fig. 2. Specifically, z is measured from the upper sample surface ($z = 0$) vertically to the point of the tip ($z > 0$). As the tip is rastered over a sample surface that is not flat, the distance z will vary with position $x'y'$ due to the relief of the surface, and the force $F(z)$ will vary. The magnitude of $F(z)$ will depend on position $x'y'$ not only because of variations in z, but also because the constitution of the surface below the tip may change and, hence, affect the potential energy $V(z)$. The analysis does not account separately for changes in $V(z)$ with position $x'y'$. Attention is focused on changes in $F(z)$ that occur with changes in $x'y'$ regardless of the origin of such changes.

In the absence of any voltage applied to the piezoelectric bimorph, let the rest position of the tip above some point on the sample surface be z_0. The vertical position of the tip is adjusted so as to place z_0 at some distance z in the attractive domain of the potential ($F(z) < 0$) in order to correspond to the operating conditions of the NCAFM experiment. Expanding $F(z)$ to first order about z_0 gives

$$F(z) = F(z_0) + f(z_0)\zeta \tag{1}$$

(a)

(b)

(c)

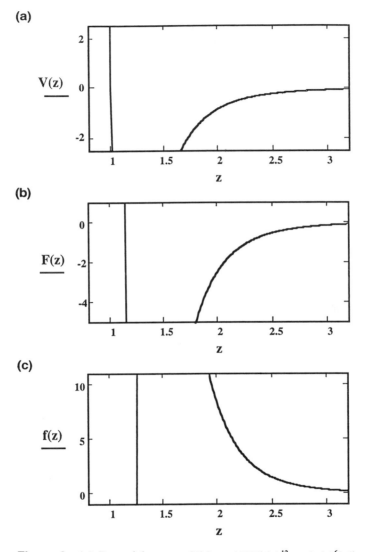

Figure 2 (a) Potential energy $V(z) = 4V_0[(j/z)^{12} - (\sigma/z)^6]$ (in aJ) of van der Waals interaction between the probe tip and some feature of the sample plotted against the vertical tip-to-sample distance z (in nm). (b) Vertical force $F(z) = -\partial V(z)/\partial z$ (in nN) acting between the tip and some feature of the sample. (c) Force gradient $f(z) = \partial F(z)/\partial z$ (in nN/nm, or N/m) along the vertical line connecting the tip and some feature of the sample. Extrema in $V(z)$, $F(z)$, and $f(z)$ are truncated to facilitate display of behavior near $z = 3$ nm. Parameter values ($V_0 = 13.28$ aJ $= 8000$ kJ/mol, $\sigma = 1$ nm) correspond to the interactions of several tens of atoms in tip and sample.

where $f(z_0)$ is $\partial F(z)/\partial z$ evaluated at $z = z_0$ and $\zeta = z - z_0$ is the linear displacement from z_0. The force gradient $f(z)$ is also plotted in Fig. 2. Notice that $f(z_0)$ may be negative, zero, or positive depending on where z_0 is chosen on the attractive branch of $V(z)$.

Still in the absence of any voltage applied to the piezoelectric bimorph, the equation of motion for the cantilever/tip can be approximated by

$$\mu \frac{\partial^2 z}{\partial t^2} = -k(z - z') - \gamma \frac{\partial z}{\partial t} + F(z) \tag{2}$$

where μ is the effective mass of the cantilever/tip, k is the effective Hookean force constant for bending of the cantilever/tip, and γ is the effective frictional coefficient that describes the dissipative elements of the mechanical system. The position z' is the rest of the tip in the absence of any interaction with the substrate/sample.

Application of a voltage to the bimorph can be understood to perturb z'. When the applied voltage is oscillatory with harmonic frequency ω, we can express the time dependent $z'(t)$ by

$$z'(t) = z' + Z \exp(i\omega t) \tag{3}$$

where Z is the amplitude of the piezoelectric driver and z' is the rest position of the tip in the absence of the oscillatory perturbation and the absence of interactions with the sample.

Substitution of Eqs. (1) and (3) into Eq. (2) and elimination of z in favor of the displacement ζ yields

$$\mu \frac{\partial^2 \zeta}{\partial t^2} = -k[z_0 + \zeta - z' - Z \exp(i\omega t)]$$

$$- \gamma \frac{\partial \zeta}{\partial t} + F(z_0) + f(z_0)\zeta \tag{4}$$

At equilibrium in the absence of any voltage applied to the bimorph, the force exerted on the tip by the sample is exactly balanced by the Hookean restoring force of the cantilever/tip, and so

$$F(z_0) = k(z_0 - z') \tag{5}$$

and Eq. (4) reduces to

$$\mu \frac{\partial^2 \zeta}{\partial t^2} = -k[\zeta - Z \exp(i\omega t)] - \gamma \frac{\partial \zeta}{\partial t} + f(z_0)\zeta \tag{6}$$

Steady state solution of Eq. (6) [9] yields (for small γ) a resonant frequency ω_0' for the tip given by

$$\omega_0' = \left(\frac{k'}{\mu}\right)^{1/2} \tag{7}$$

The quantity $k' = k - f(z_0)$ is the effective force constant of the cantilever/tip as modulated by its interactions with the sample. As z_0 becomes large and $f(z_0)$ approaches 0 (see Fig. 2), ω_0' approaches $\omega_0 = (k/\mu)^{1/2}$, the natural resonant frequency of the cantilever/tip far from the substrate/sample. Notice that ω_0' can be greater than, less than, or equal to ω_0 depending on choice of z_0, which dictates the sign and magnitude of $f(z_0)$. For NCAFM one chooses z_0 to render $f(z_0) > 0$ as will be illustrated. This means that the interactions between the tip and sample make it easier to bend the cantilever/tip toward the surface as z diminishes. Care must be exercised not to let k' become too small, or anharmonic behavior may ensue.

The amplitude $A(\omega, z_0)$ of the tip vibration is largest when the driver frequency $\omega = \omega'_0$, as shown by

$$A(\omega, z_0) = \frac{Z\omega_0^2}{[(\omega_0'^2 - \omega^2)^2 + \omega^2\gamma^2/\mu^2]^{1/2}} \tag{8}$$

$$= \frac{Zk}{[(k' - \omega^2\mu)^2 + \omega^2\gamma^2]^{1/2}}$$

The dependence of $A(\omega, z_0)$ on ω is explicit; dependence on z_0 is through the dependence of k' or ω_0' on $f(z_0)$ [see Eq. (7)]. The amplitude $A(\omega, z_0)$ is plotted against ω in Fig. 3 for two values of z_0 in the range where $f(z_0) > 0$. Here $z_{01} > z_{02}$, and z_{02} corresponds to a vertical position sufficiently above the highest points on the surface of the sample to avoid collisions with the oscillating tip. The plots show that as z_0 decreases from z_{01} to z_{02}, ω_0' shifts toward smaller values. The amplitude of the tip oscillations (<1 nm) is seen to be consistent with avoidance of collisions between the tip and the sample surface, provided the relief of the surface is of the order of 1 nm, as it is for a polysaccharide chain.

In the NCAFM experiment a "set point" height z_{01} and corresponding resonant frequency ω'_{01} are established to achieve maximal sensitivity. It is evident from Fig. 3 that sensitivity is enhanced if the driver frequency ω is offset by a few kilohertz to the high frequency side of ω'_{01} [20]. As the tip is rastered across the sample surface, the amplitude of tip oscillations will diminish as the tip passes over the higher points due to the dependence on z_0 of the $A(\omega, z_0)$ versus ω response curve. The instrument detects the amplitude modulations as xy rastering proceeds, and it drives the z component of the piezoelectric crystal upon which the substrate/sample is mounted in the direction that restores the original set point amplitude of tip oscillation. The z driver voltage of the mounting crystal is converted to the z dimension of the image of the substrate/sample surface. Effectively, it is

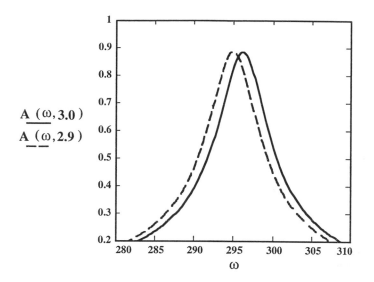

Figure 3 The amplitude $A(\omega, z_0)$ (in nm) plotted against driver frequency ω (in kHz) for two values of z_0 [$z_{01} = 3.0$ nm (solid line) and $z_{02} = 2.9$ nm (dashed line)] from Eq. (8). Van der Waals potential energy parameters from Fig. 2, cantilever characteristics ($k = 13$ N/m, $\mu = 1.44 \times 10^{-10}$ kg, $\gamma = 8.665 \times 10^{-7}$ kg/sec), and bimorph driver amplitude $Z = 0.0175$ nm.

the force gradient $f(z)$ that is mapped. In the example shown, which was calculated using realistic values of the experimental parameters, it is seen that the instrument is sensitive to variations in z of about 1 Å. Contact between tip and sample is avoided by appropriate control of set point height z_{01} and tip oscillation amplitude $A(\omega, z_0)$ through adjustment of the driver amplitude Z.

III.   POLYSACCHARIDE SAMPLE PREPARATION

A.   Choice of Polysaccharides

The polysaccharides selected for study are (1) the (1 → 6)-branched (1 → 3)-β-D-linked fungal glucan scleroglucan [21,22], (2) the bacterial hetero-polysaccharide xanthan, which possesses a cellulosic backbone to which is appended a three-sugar side chain on every second backbone glucose [23,24], and (3) the linear bacterial heteropolysaccharide gellan, which displays a regularly repeating four-sugar repeat unit [25,26]. All of these materials exist in their native states as multiple-stranded helices, and the resulting chain thickness contributes to the ease with which they may be imaged by NCAFM. Scleroglucan (also known as schizophyllan and lentinan, according to biological source) exists as a triple-stranded helix [27]. Xanthan [28] and gellan [29] are double-stranded helices.

B. Preparation of Polysaccharide Solutions

1. Scleroglucan

Scleroglucan powder (Actigum CS 6, Lot No. 26, CECA S.A., Velizy, France) was dissolved at 5 g/L in aqueous sodium azide (0.02%) (bacteriocide) and allowed to stir overnight at room temperature. The solution was centrifuged at 8000 rpm for 1.5 h to remove cell debris and unsolubilized polymer. The supernatant was sonicated (20 kHz, 375 W, 1/2-in. tip) at 0°C for 3 h, and centrifuged and filtered through 0.45-μm filters to remove particulates. The resulting solution was then fractionally precipitated [30] with acetone to produce 10 fractions of differing mean molar mass. Each of the fractions was redissolved in water, dialyzed exhaustively against distilled water, and recovered by freeze-drying. Fractions chosen for study were redissolved in water at 1 mg/mL. Some unfractionated scleroglucan was also investigated. Schizophyllan (kindly provided by Kengo Tabata, Taito Co., Ltd., Tokyo) was dissolved in water at approximately 1 mg/mL and used without further processing.

Some of the redissolved scleroglucan samples were subjected to a denaturation–renaturation procedure: Aliquots in sealed 10-mL microreaction vials were first heated to 160°C for 15 min to disrupt the native triple helix [31]. These vials were then annealed at 70°C for 18 h, and quenched to room temperature in an ice bath. The annealing temperature was chosen to be within the region of stability of the native triple helix but high enough to reduce kinetic hindrances to reassembly and reorganization of the triple helix [31].

2. Xanthan

Xanthan power (Keltrol, Lot No. 76735A, The Nutrasweet Kelco Co., San Diego, California) was dispersed at approximately 6 g/L in 0.1 M sodium chloride containing sodium azide (0.02%) overnight at room temperature. The solution was sonicated (20 kHz, 375 W, 1/2-in. tip) at 0°C for 20 min in two 500-mL aliquots. The combined batches were centrifuged at 8000 rpm for 2 h, filtered through a series of filters (5.0 to 0.45 μm), and dialyzed for several days. Subsequently, sodium chloride was added to the xanthan solution to give a final concentration of 0.05 M, and the solution was fractionally precipitated [30] into fractions of differing mean molar mass using isopropanol. The fractions were redissolved in distilled water, dialyzed exhaustively against distilled water, and stored in solution at 4°C at a concentration of appproximately 2.0 mg/mL until used.

3. Gellan

Gellan (KelcoGel, Lot No. 79119A, The Nutrasweet Kelco Co.) was dissolved at 5 g/L in aqueous sodium azide (0.02%) and allowed to stir overnight at room temperature. The solution was sonicated (20 kHz, 375 W, 1/2-in. tip) for 2.5 h at 0°C, centrifuged at 8000 rpm for 1.5 h, filtered through 0.45-μm filters, and stored in solution at 4°C at a concentration of approximately 2.7 mg/mL until used. Before investigation, small batches were dialyzed against water.

C. Preparation of Samples for Microscopy

Biopolymer solutions, prepared as previously described at approximately 1 mg/mL, are diluted with distilled water to a final polymer concentration of 4–30 μg/mL. In some instances, to be described, the volatile salt ammonium acetate is added to adjust ionic strength. Aliquots of these diluted solutions are deposited by spraying [8,32] onto freshly cleaved mica (Ted Pella, Inc., Redding, California) and air dried. Mica provides an atomically flat substrate free of artifacts found on other commonly used scanning probe microscopy substrates such as highly oriented pyrolytic graphite. After air drying for a few hours, samples are ready to be imaged by NCAFM. The time interval between spraying of the sample and display of images can be as short as 1 h.

D. Imaging Procedures

Specimens were examined using a Park Scientific AutoProbe CP scanning probe microscope equipped with an NCAFM probe head. A piezoelectric scanner with a range up to 10 μm was used for all images. The scanner was calibrated in the xy directions using a 1.0-μm grating, and in the z direction using several height standards. The tips used were V-shaped silicon 2-μm cantilevers (Ultralevers, Model No. APUL-20-AU-25, Park Scientific Instruments), with a force constant of 13 N/m and resonant frequency ω_0 of approximately 300 kHz. The oscillation frequency ω of the cantilever/tip was offset from ω_0 to higher frequencies by a few kilohertz. The oscillation amplitude was kept to a minimum to avoid tip–sample interactions. All measurements were performed in air at ambient pressure and humidity. Images were stored either as 256 × 256 or 512 × 512 point arrays and analyzed using AutoProbe image processing software supplied by Park Scientific Instruments. Figures 4–11 illustrate the capabilities of NCAFM for imaging a variety of polysaccharide molecules. The technique has proven successful for the nondestructive, high resolution imaging of fragile samples in air.

IV. RESULTS AND DISCUSSION

A. Scleroglucan/Schizophyllan

Scleroglucan and schizophyllan are water-soluble, nonionic, high molecular weight extracellular polysaccharides with regular comblike branching of single glucose residues. Scleroglucan is the name applied to capsular polysaccharides excreted by the genus *Sclerotium*, whereas schizophyllan is a cell wall component of the fungus *Schizophyllan commune*. Glucans of this class are well-known immunoadjuvants used for cancer therapy in some countries; immunostimulatory activity is related to the triple-stranded structure [33].

Figure 4 shows an NCAFM image of native schizophyllan applied to mica as a solution in water. The molecules are well separated. The apparent alignment, which exaggerates the stiffness of the molecular structure, is believed to be due to drying artifacts [13] or to effects of interaction of the polymer with features of the underlying mica substrate [17]. Scleroglucan, diluted in water but unprocessed other than for removal of protein material, is shown in Fig. 5. Images obtained for the two polymers are very similar. A tracing across one scleroglucan molecule in Fig. 5 illustrates the capacity of NCAFM to measure the thickness of the macromolecule normal to the surface of the mica substrate. For the trace shown, the thickness is approximately 1 nm. This result is approximately 60% of the triple-strand thickness expected from x-ray diffraction [27]. Reasons for this discrepancy are under investigation [17,18]. NCAFM does not yield reliable chain thicknesses measured parallel to the substrate surface, because the chain thickness is convoluted with the thickness of the AFM measuring tip.

Lower molecular weight fractions of native scleroglucan were obtained by ultrasonication and fractional precipitation. The reduced chain lengths present in scleroglucan fraction P3 are clearly visualized in Fig. 6. Samples displaying well-separated molecules are suitable for obtaining information about contour length distributions and mean contour lengths from micrographs like those in Fig. 6. Using the known mass per unit length of scleroglucan/schizophyllan (2140 Da/nm) [34], these features can be compared with experimentally determined mean molar mass and molar mass distributions from dilute solution measurements. The number-average chain length of sample P3 has been determined to be 185 ± 97 nm averaged over hundreds of molecules in many micrographs. The mean chain length corresponds to a number-average molar mass (M_n) of $(3.96 \pm 2.08) \times 10^5$ Da, and the distribution of lengths yields 1.26 for the heterogeneity index M_w/M_n. This shows that scleroglucan fraction P3 contains a relatively narrow distribution of chain lengths.

NCAFM has also been employed to investigate the conversion of

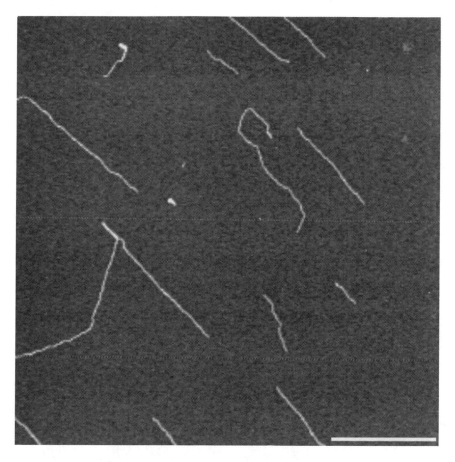

Figure 4 NCAFM image of native schizophyllan. Scale bar = 400 nm.

linear scleroglucan triple helices to cyclic structures prepared by thermal denaturation of the native triple helix and subsequent annealing of the sample below the helix melting temperature [31,35,36]. Figure 7 shows fraction P3 after having been subjected to the denaturation–renaturation procedure described in Section III.B.1. Using NCAFM we have been able to monitor the linear-triple-helix to circular-triple-helix transition in sclero-glucan routinely and reproducibly [17,18]. The contour lengths of the linear and cyclic renatured helices are 96.5 ± 66.7 nm and 100.5 ± 35.8 nm, respectively. The number-average molar mass of all renatured species is $(2.12 \pm 1.01) \times 10^5$ Da. The apparent reduction in molar mass relative to the undenatured material is probably due to thermal degradation of the

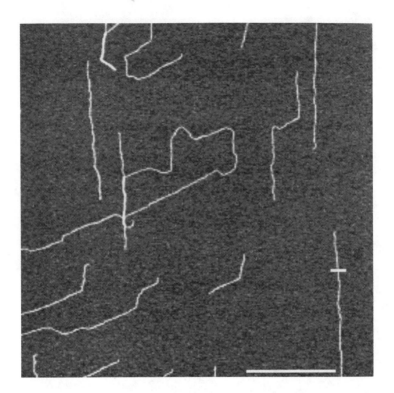

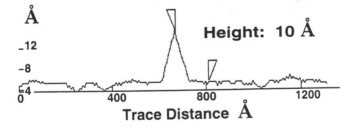

Figure 5 NCAFM image of native scleroglucan. Scale bar = 400 nm. A trace
through one rodlike molecule (indicated by white line) is plotted below the image.

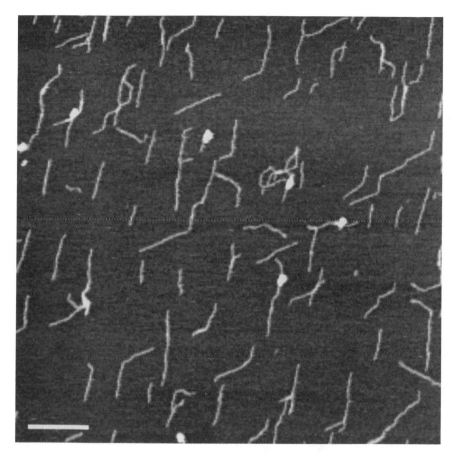

Figure 6 NCAFM image of scleroglucan fraction P3. Scale bar = 200 nm.

chain under the denaturation and/or annealing conditions. The average heights of the renatured linear and cyclic helices are 1.1 ± 0.35 nm and 0.91 ± 0.29 nm, respectively. Explanations for the greater contour length and smaller apparent chain thickness of the cyclic species are being sought through theoretical modeling of the linear-triple-helix to circular-triple-helix equilibrium [18].

B. Gellan

Gellan is a bacterial polysaccharide produced by fermentation of the organism *Pseudomonas elodea* [37]. In contrast to scleroglucan/schizophyllan,

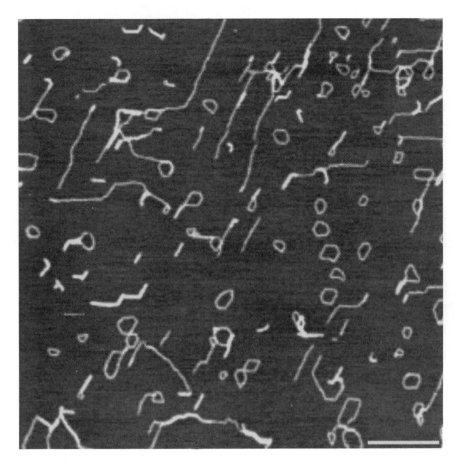

Figure 7   NCAFM image of scleroglucan fraction P3 after denaturation for 15 min in water at 160°C, annealing for 18 h at 70°C, and quenching to room temperature. Scale bar = 200 nm.

gellan is a linear anionic heteropolysaccharide with one carboxylate group per four-sugar repeating unit. Figure 8 is an NCAFM image of a dilute gellan solution in water prepared as previously described. The chains are linear and well separated at the concentration used to prepare the sample imaged in Fig. 8. It has yet to be determined whether or not these well-separated molecules are single strands, the crystallographically determined double strands [29], or higher aggregates.

As the concentration of the deposited solution is increased, the gellan chains form networks, which are composed of filaments of varying thick-

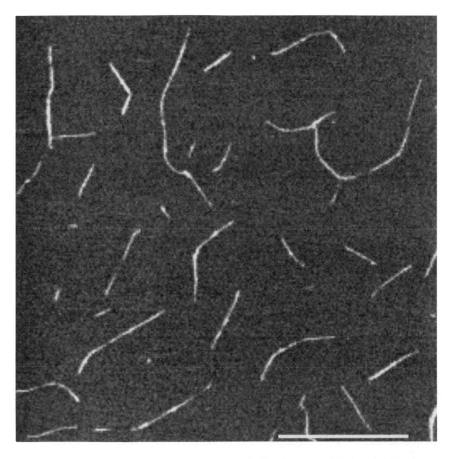

Figure 8 NCAFM image of dilute gellan solution in water (5.33 μg/mL aqueous gellan solution). Scale bar = 400 nm.

ness, as shown in Fig. 9. At still higher polymer concentrations, the chains overlap extensively, it is difficult to find chain ends, and the network pattern that has begun to emerge in Fig. 9 is even more apparent. In the presence of modest amounts of added low molecular weight salt, gellan forms clear, mechanically stable aqueous gels [37]. The network structure seen in Fig. 9 and in micrographs prepared at still higher gellan concentration may well represent two-dimensional projections of the incipient three-dimensional gel network. When micrographs of gellan samples prepared in the presence of ammonium acetate or potassium chloride are examined,

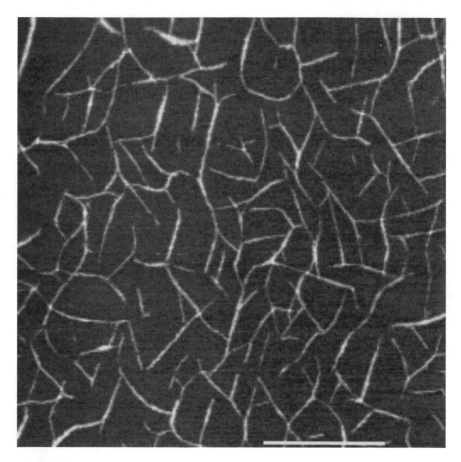

Figure 9 NCAFM image of gellan solution in water (7.99 μg/mL aqueous gellan solution). Scale bar = 400 nm.

they show highly aggregated networks of filaments of varying width and height (data not shown).

C. Xanthan

Xanthan is widely used in the food industry as a thickening and suspending agent [37]. The polysaccharide is produced commercially by fermentation of the bacterium *Xanthomonas campestris*. Its primary structure involves a $(1 \rightarrow 4)$-linked β-D-glucan cellulosic backbone substituted at C-3 on alternating glucose residues with an ionic trisaccharide side chain. The commercial success of xanthan derives in part from the highly pseudoplastic character of its aqueous solutions. They set to weak gels in the absence of shear but are highly sensitive to shear thinning. This behavior is presumably a consequence of a friable network structure stabilized by the same interactions that stabilize the double helical structure [38,39].

The appearance of xanthan chains in noncontact AFM images differs depending on the solvent with which the xanthan stock solution is diluted. Figure 10 shows xanthan, prepared as previously described, diluted with distilled water, deposited on a mica surface, and air dried. The filaments have varying thicknesses and heights, and the chains appear to have under-

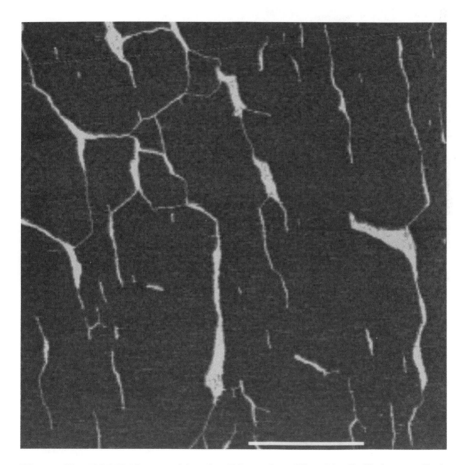

Figure 10 NCAFM image of fraction P2 xanthan diluted in distilled water. Scale bar = 400 nm.

gone lateral aggregation. Some fields of view show much larger aggregated networks consisting of many chains. Other views show the chains aligned side-by-side over long distances (data not shown). The xanthan chains appear well separated and thinner in cross section when diluted with 1 M ammonium acetate, as shown in Fig. 11. To date, no systematic correlation of these NCAFM images of xanthan with other physical properties of the corresponding systems has been carried out, and so the comments that follow should be viewed as highly speculative.

In the complete absence of low molecular weight salt, very dilute xanthan exists close to or above the characteristic order-to-disorder transi-

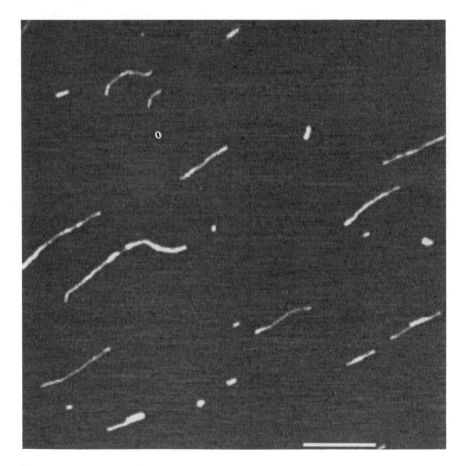

Figure 11   NCAFM image of fraction P2 xanthan diluted in 1 M ammonium acetate. Scale bar = 400 nm.

tion temperature of the double-stranded helical form [40]. Addition of a small amount of salt to such solutions increases the transition temperature significantly, and the ordered double strand becomes the thermodynamically favored form. The image in Fig. 10, prepared by drying a dilute, salt-free xanthan solution may represent the largely disordered aggregation upon dehydration of xanthan chains existing above the conformational transition temperature. The more uniform xanthan chains in Fig. 11 are interpreted in this light as xanthan double helices that were well stabilized in solution by the presence of salt prior to the dehydration process. Their inherent valences for association already having been satisfied by double-strand formation, these chains underwent no further aggregation upon dehydration. Measurements of the thickness of the strands in Fig. 11 normal to the mica substrate are consistent with the interpretation of these molecules as double helical xanthan.

V. CONCLUSIONS

It was shown in this chapter that NCAFM can be used successfully to image a variety of carbohydrate macromolecules quickly, reproducibly, and relatively inexpensively. In some instances, images of well-ordered associated forms, for example, scleroglucan linear and circular triple helices, have been obtained. In other cases, the images appear to be best interpreted as aggregates involving larger numbers of polymer chains, for example, the putative incipient network structure associated with aqueous gellan gels. Imaging of single-stranded chains has also been accomplished, but with less regular success to date. In comparison with conventional contact mode AFM, the NCAFM technique permits imaging without disturbance or displacement of the specimen molecules, because the attractive van der Waals forces of interaction between the sample and probe tip are comparable with or smaller than the forces binding the subject molecules to the underlying substrate. In comparison with TEM techniques, NCAFM images may be obtained without the time, expense, and inherent loss of resolution that accompanies contrast enhancement by heavy metal shadowing. The potential of NCAFM for measurements of polymer chain thickness in the dimension normal to the substrate surface is especially appealing. On the other hand, even the sharpest probe tips currently available have an effective thickness greater than the thickness of the polymer chains, so that dimensions measured parallel to the substrate plane are a convolution of the molecular thickness and the probe thickness. Whereas TEM has been used successfully to assess quantitative measures of inherent polymer chain stiffness [8], alignment artifacts evident in some of the NCAFM images displayed here still present a problem for determination of persistence lengths or other measures of chain stiffness.

VI. ACKNOWLEDGMENTS

This work has been supported by NIH Research Grant GM 33062 to DAB, by NIH Traineeship Grant GM 07311-20 to TMM, and by ONR Research Grant 400X119YIP to Reginald M. Penner, whose generous advice and assistance with access to NCAFM instrumentation has made this work possible. We would like to thank Marco Tortonese at Park Scientific Instruments for donation of some of the cantilevers used in this study. TMM would like to gratefully acknowledge Mike D. Kirk of Park Scientific Instruments for valuable discussions concerning this work and Arthur W. Moore of Advanced Ceramics Corporation for donation of highly oriented pyrolytic graphite used in this work.

ABBREVIATIONS

AFM	atomic force microscopy
EM	electron microscopy
NCAFM	noncontact mode AFM
PSPD	position-sensitive photodiode
STM	scanning tunneling microscopy
TEM	transition electron microscopy

REFERENCES

1. Koehler, J. K. (1973). *Advanced Techniques in Biological Electron Microscopy*. Springer-Verlag, Berlin.
2. Hayat, M. A., ed. (1970–1978). *Principles and Techniques of Electron Microscopy: Biological Applications*, 9 vols. Van Nostrand Reinhold, New York.
3. Helbert, W., and Chanzy, H. (1994). *Carbohydr. Polym.* 24, 119–122.
4. Mazeau, K., Winter, W. T., and Chanzy, H. (1994). *Macromolecules* 27, 7606–7612.
5. Saito, Y., Okano, T., Chanzy, H., and Sugiyama, J. (1995). *J. Struct. Biol.* 114, 218–228.
6. Favier, V., Chanzy, H., and Cavaillé, J. Y. (1995). *Macromolecules* 28, 6365–6367.
7. Stokke, B. T., and Elgsæter, A. (1991). In White, C. A., ed., *Advances in Carbohydrate Analysis*. JAI Press, London, pp. 195–247.
8. Stokke, B. T., and Elgsæter, A. (1994). *Micron* 25, 469–491.

9. Sarid, D. (1994). *Scanning Force Microscopy with Applications to Electric, Magnetic and Atomic Forces.* Oxford University Press, New York.

10. Anselmetti, D., Dreier, M., Lüthi, R., Richmond, T., Meyer, E., Frommer, J., and Güntherodt, H.-J. (1994). *J. Vac. Sci. Technol. B* 12, 1500–1503.

11. Lüthi, R., Meyer, E., Howald, L., Haefke, H., Anselmetti, D., Dreier, M., Rüetschi, M., Bonner, T., Overney, R. M., Frommer, J., and Güntherodt, H.-J. (1994). *J. Vac. Sci. Technol. B* 12, 1673–1676.

12. Wilkins, M. J., Davies, M. C., Jackson, D. E., Mitchell, J. R., Roberts, C. J., Stokke, B. T., and Tendler, S. J. B. (1993). *Ultramicroscopy* 48, 197–210.

13. Wilkins, M. J., Davies, M. C., Jackson, D. E., Roberts, C. J., and Tendler, S. J. B. (1993). *J. Microsc.* 172, 215–221.

14. Lee, I., Atkins, E. D. T., and Miles, M. J. (1992). *Ultramicroscopy* 42–44.

15. Kirby, A. R., Gunning, A. P., and Morris, V. J. (1995). *Carbohydr. Res.* 267, 161–166.

16. Kirby, A. R., Gunning, A. P., Morris, V. J., and Ridout, M. J. (1995). *Biophys. J.* 68, 360–363.

17. McIntire, T. M., Penner, R. M., and Brant, D. A. (1995). *Macromolecules* 28, 6375–6377.

18. Brant, D. A., and McIntire, T. M. (in press). In Semlyen, J. A., ed., *Large Ring Molecules.* Wiley, Oxford.

19. Zhong, Q., Inniss, D., Kjoller, K., and Elings, V. B. (1993). *Surf. Sci. Lett.* 290, L688–L692.

20. AutoProbe CP Manual, Park Scientific Instruments, Sunnyvale, California.

21. Johnson, Jr., J., Kirkwood, S., Misaki, A., Nelson, T. E., Scaletti, J. V., and Smith, F. (1963). *Chemistry and Industry* 820–822.

22. Yanaki, T., Kojima, T., and Norisuye, T. (1981). *Polymer J.* 13, 1135–1143.

23. Jansson, P.-E., Kenne, L., and Lindberg, B. (1975). *Carbohydr. Res.* 45, 275–282.

24. Melton, L. D., Mindt, L., Rees, D. A., and Sanderson, G. R. (1976). *Carbohydr. Res.* 46, 245–257.

25. O'Neill, M. A., Selvendran, R. R., and Morris, V. J. (1983). *Carbohydr. Res.* 124, 123–133.

26. Jansson, P.-E., Lindberg, B., and Sandford, P. A. (1983). *Carbohydr. Res.* 124, 135–139.

27. Bluhm, T. L., Deslandes, Y., Marchessault, R. H., Pérez, S., and Rinaudo, M. (1982). *Carbohydr. Res.* 100, 117–130.

28. Millane, R. P. (1992). In Chandrasekaran, R., ed., *Frontiers in Carbohydrate Research—2.* Elsevier Applied Science Publishers, London, pp. 168–190.

29. Chandrasekaran, R., Millane, R. P., Arnott, S., and Atkins, E. D. T. (1988). *Carbohydr. Res.* 175, 1–15.

30. Kotera, A. (1967). In Cantow, M. J. R., ed., *Polymer Fractionation.* Academic Press, New York, pp. 44–66.

31. Kitamura, S., and Kuge, T. (1989). *Biopolymers* 28, 639–654.

32. Tyler, J. M., and Branton, D. (1980). *J. Ultrastruct. Res.* 71, 95–102.

33. Norisuye, T. (1985). *Makromolekulare Chemie. Suppl.* 14, 105–118.

34. Yanaki, T., and Norisuye, T. (1983). *Polym. J.* 15, 389–396.
35. Stokke, B. T., Elgsæter, A., Brant, D. A., and Kitamura, S. (1991). *Macromolecules* 24, 6349–6351.
36. Stokke, B. T., Elgsæter, A., Brant, D. A., Kuge, T., and Kitamura, S. (1993). *Biopolymers* 33, 193–198.
37. Morris, V. J. (1995). In Stephen, A. M., ed., *Food Polysaccharides and Their Applications*. Marcel Dekker, New York, pp. 341–375.
38. Oviatt, H. W., and Brant, D. A. (1993). *Int. J. Biol. Macromol.* 15, 3–10.
39. Oviatt, H. W., and Brant, D. A. (1994). *Macromolecules* 27, 2402–2408.
40. Paoletti, S., Cesàro, A., and Delben, F. (1983). *Carbohydr. Res.* 123, 173–178.

13

Novel Approach for the Analysis of Interaction Between Sugar Chains and Carbohydrate-Recognizing Molecules Using a Biosensor Based on Surface Plasmon Resonance

Yasuro Shinohara, Hiroyuki Sota, and Yukio Hasegawa
Pharmacia Bioteck K.K., Tokyo, Japan

I. INTRODUCTION

Many kinds of lectins and related molecules have been successfully found in a variety of species [1]. Simultaneously, carbohydrates have been shown to play a role in a multitude of biological processes such as protein folding, clearance, cell–cell interaction, and signal transduction. To clarify the functional roles of these molecules, we must evaluate their sugar binding properties and the mechanisms involved. Many assay systems for analyzing the interactions between lectin and oligosaccharides have been developed, including affinity chromatography [2], microdialysis [3], microcalorimetry [4], electrophoresis [5], and nuclear magnetic resonance (NMR) [6]. However, because these methods usually detect the equilibrium of interactions, information about the on- and off-rate kinetics remains scarce. Recent findings have revealed that the interactions between oligosaccharides and carbohydrate-recognizing molecules are quite complicated, and those interactions should be analyzed in more detail rather than simply observing whether binding occurs. Several important aspects of the interactions have been pointed out, including the clustering effect [7], the importance of transient interactions [8], and the effect of blood flow [9]. For the functional analysis of an oligosaccharide in a binding assay, development of more sophisticated analytical methods is necessary.

A biosensor based on surface plasmon resonance (SPR) has been developed [10,11]. In this technique, one component of the interaction to be studied is immobilized on the sensor surface, while the other interactants are passed over the surface in solution. This technology has made it possible to analyze biomolecular interactions both qualitatively and quantitatively.

To apply this method to the interaction analysis between oligosaccharides and lectins, it was necessary to develop an efficient method for immobilizing a sugar chain on the sensor surface of the biosensor. This was accomplished by a combination of establishing an efficient biotinylation method for sugar chains and biotin–avidin technology. The method reported in this chapter is suitable for analysis of the interactions between lectins and sugar chains because of its simplicity and high sensitivity. The validity of this method was evaluated by measuring the interactions of several lectins, the sugar binding specificities of which are well defined.

II. MATERIALS AND METHODS

A. Instrumentation

A biosensor, BIAcore (Pharmacia Biosensor AB, Uppsala, Sweden), based on surface plasmon resonance (SPR), was used to determine the biomolecular interactions. The SPR response correlates with the changes in mass on the surface of the gold film involving association and dissociation between ligands and analytes, and it is expressed in resonance units (RU). In general, adhesive reactions are detected as an increase in the SPR response, whereas dissociation reactions are detected as a decrease in the SPR response. A response of 1000 RU represents a change in the surface protein concentration of about 1 ng/mm^2.

B. Materials

Sambucus sieboldiana lectin (SSA), *Maackia amurensis* lectin (MAM), *Ricinus communis* agglutinin-120 (RCA$_{120}$), *Datura stramonium* lectin (DSA), and Concanavalin A (ConA) were purchased from Honen Seiyu (Tokyo). *N*-Acetyllactosamine (LacNAc) was obtained from Toronto Research Chemicals (Downsview, Canada), and 6'- and 3'-sialyl LacNAc (6'- and 3'-SLN) were purchased from Oxford GlycoSystems (Abingdon, United Kingdom). The abbreviations and structures of the other oligosaccharides used in this study are shown in Scheme 1. All the oligosaccharides shown in Scheme 1 were purchased from Oxford GlycoSystems. BIAcore sensor chip CM-5, surfactant P20, and chemical activation reagents were obtained from Pharmacia Biosensor AB: 100 mM *N*-hydroxysuccinimide (NHS) in water, 400 mM *N*-ethyl-*N'*-(3-dimethylaminopropyl) carbodi-imide hydrochloride (EDC) in water, and 1 M ethanolamine hydrochloride adjusted to pH 8.5 using NaOH. The HEPES-buffered saline (HBS) comprised 10 mM HEPES, pH 7.4, 150 mM NaCl and 0.05% BIAcore surfactant P20 in

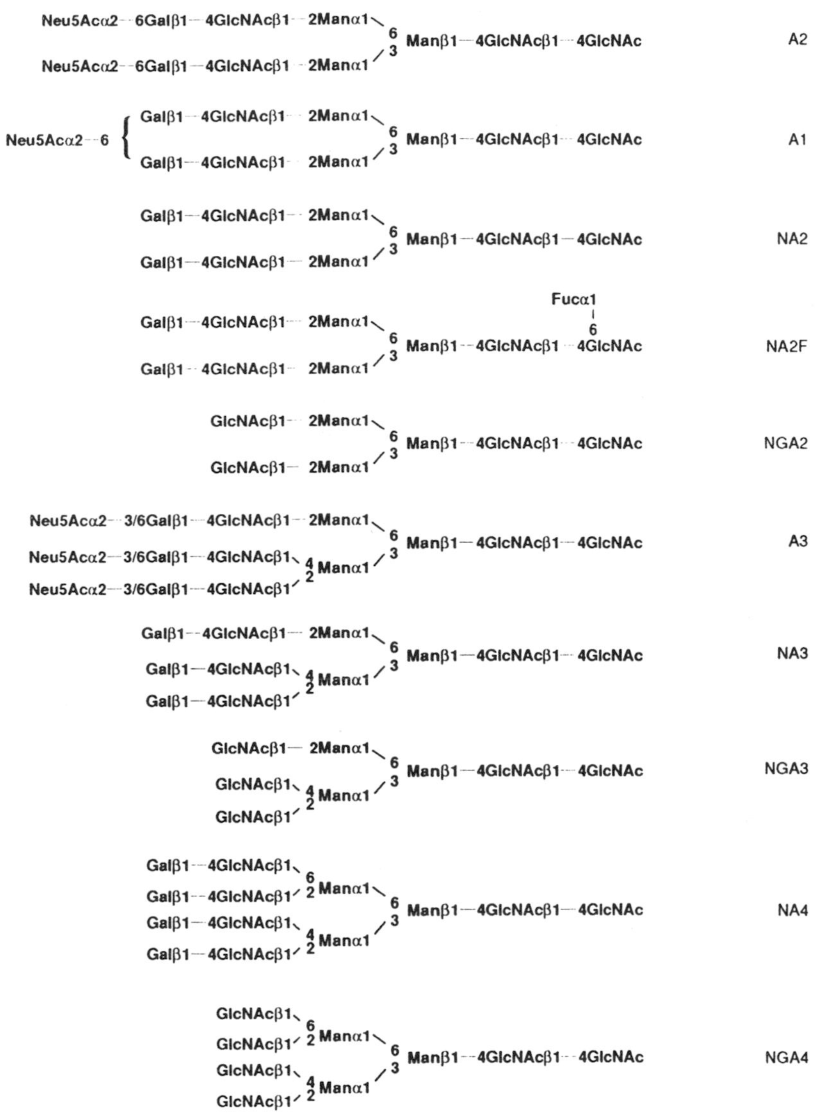

```
Neu5Acα2—6Galβ1—4GlcNAcβ1—2Manα1
                                     \6
                                       Manβ1—4GlcNAcβ1—4GlcNAc        A2
                                     /3
Neu5Acα2—6Galβ1—4GlcNAcβ1—2Manα1


            Galβ1—4GlcNAcβ1—2Manα1
                                     \6
                                       Manβ1—4GlcNAcβ1—4GlcNAc        A1
Neu5Acα2—6 {                         /3
            Galβ1—4GlcNAcβ1—2Manα1


Galβ1—4GlcNAcβ1—2Manα1
                        \6
                          Manβ1—4GlcNAcβ1—4GlcNAc                     NA2
                        /3
Galβ1—4GlcNAcβ1—2Manα1

                                          Fucα1
                                            |
Galβ1—4GlcNAcβ1—2Manα1                       6
                        \6
                          Manβ1—4GlcNAcβ1—4GlcNAc                     NA2F
                        /3
Galβ1—4GlcNAcβ1—2Manα1


GlcNAcβ1—2Manα1
                \6
                  Manβ1—4GlcNAcβ1—4GlcNAc                             NGA2
                /3
GlcNAcβ1—2Manα1


Neu5Acα2—3/6Galβ1—4GlcNAcβ1—2Manα1
                                    \6
Neu5Acα2—3/6Galβ1—4GlcNAcβ1\          Manβ1—4GlcNAcβ1—4GlcNAc        A3
                            4Manα1  /3
Neu5Acα2—3/6Galβ1—4GlcNAcβ1/ 2


Galβ1—4GlcNAcβ1—2Manα1
                        \6
Galβ1—4GlcNAcβ1\          Manβ1—4GlcNAcβ1—4GlcNAc                    NA3
                4Manα1  /3
Galβ1—4GlcNAcβ1/ 2


GlcNAcβ1—2Manα1
                \6
GlcNAcβ1\          Manβ1—4GlcNAcβ1—4GlcNAc                           NGA3
         4Manα1  /3
GlcNAcβ1/ 2


Galβ1—4GlcNAcβ1\
                6
Galβ1—4GlcNAcβ1/ 2Manα1
                        \6
Galβ1—4GlcNAcβ1\          Manβ1—4GlcNAcβ1—4GlcNAc                    NA4
                4Manα1  /3
Galβ1—4GlcNAcβ1/ 2


GlcNAcβ1\
         6
GlcNAcβ1/ 2Manα1
                 \6
GlcNAcβ1\          Manβ1—4GlcNAcβ1—4GlcNAc                           NGA4
         4Manα1  /3
GlcNAcβ1/ 2
```

Scheme 1   Structures and abbreviations of oligosaccharides used in this study.

distilled water. *Arthrobacter ureafaciens* sialidase and Newcastle disease virus sialidase were purchased from Boehringer Mannheim (Mannheim, Germany). 6-Hydrazidohexyl D-biotinamide (BHz–AC5) and 6-(6-hydrazidohexyl) amidohexyl D-biotinamide (BHz–(AC5)$_2$) were obtained from Dojin Do (Tokyo). 4-(Biotinamido) phenylacctylhydrazide (BPH) was synthesized [12]. The HBS buffer was prepared from 10 mM HEPES, pH 7.4, 150 mM NaCl, and 0.05% BIAcore surfactant P20 in distilled water.

C. Preparation of Biotinyl Oligosaccharides

Oligosaccharides (10 pmol–1 nmol, 10 μL) in water were incubated with a 2- to 10-fold molar ratio of BHz–AC5 solutions (10 μL, in 30% acetonitrile) at 90°C for 1 h. This simple derivatization procedure was sufficient to prepare the biotinyl glycan to measure the interaction with lectins qualitatively by the biosensor without any purification steps [13]. In the case where the sample is a mixture of oligosaccharides and purification was required for the binding assay, BPH instead of BHz–AC5 should be used for the labeling reagent, because BPH possesses a phenyl group (Scheme 2). BPH labeling is applicable for highly sensitive detection of oligosaccharides and is advantageous in chromatographic separation. BPH also converted oligosaccharides into their biotin adducts in the same way. Although the biotin adducts were obtained as several stereoisomers, the most stable isomer, cyclic β-glycoside, was selectively obtained if the reaction mixture was further treated with 20 μL of 50 mM formate buffer (pH 3.5) and allowed to stand at 4°C.

D. Detection and Separation of Oligosaccharide–BPH Adducts

For the detection of BPH-labeled oligosaccharide in chromatography, detections were performed at 252 nm. Reversed phase chromatography was conducted on a TSKgel ODS$_{80}$ column (Tosoh, Tokyo, 4.6 × 250 mm).

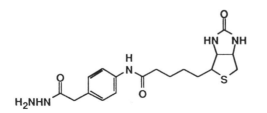

Scheme 2   Structure of 4-(biotinamido) phenylacetylhydrazide (BPH).

Separation of BPH-labeled oligosaccharides was performed at 40°C at a flow rate of 0.5 mL/min. Elution was performed with a linear gradient using solvent A (70 mM phosphate buffer, pH 6.8, containing 10% acetonitrile) and solvent B (70 mM phosphate buffer, pH 6.8, containing 30% acetonitrile). Analysis of BPH adducts of N-linked oligosaccharides was performed using two 30-min linear gradients, from 100% A to 80% A and then, from 80% A to 50% A. Anion-exchange chromatography was performed on a MonoQ column (Pharmacia Biotech, Uppsala, 1.6 × 50 mm). Separation of BPH-labeled oligosaccharides was performed at ambient temperature at a flow rate of 0.05 mL/min. Elution was performed with a linear gradient using solvent A (aqueous ammonia diluted with water, pH 9.0) and solvent B (500 mM ammonium acetate buffer, pH 8.0). After injection, solvent A was passed through the column for 10 min, after which a 25-min linear gradient up to a 50 : 50 ratio of A : B was applied. For gel filtration chromatogrphy, two linked Superdex Peptide columns (Pharmacia Biotech, 3.2 × 300 mm) were used isocratically with 20% acetonitrile at 50°C at a flow rate of 0.4 mL/min.

E.   Interaction Analysis with a Biosensor Based on SPR

The carboxymethyl dextran on the surface of the sensor chip was activated by injection of a mixture (35 μL) of 100 mM NHS and 400 mM EDC. Streptavidin was dissolved in a 10 mM sodium acetate buffer (pH 4.5) at 25 μg/mL and was introduced (35 μL) onto the activated surface of the sensor chip. The immobilized amount of streptavidin was about 2000 RU. After blocking the unreacted NHS ester on the surface with 1 M ethanolamine, the BHz–AC5 or BPH-labeled oligosaccharide (1–10 pmol) was introduced onto the surface. A lectin solution (1–500 μg/mL, in HBS buffer, 10–30 μL) was injected across the surface at a flow rate of 2–10 μL/min. The interaction was monitored at 25°C as the change in the SPR response. After monitoring for 5 min, the HBS buffer was introduced onto the sensor chip in place of the lectin solution to start the dissociation.

The same sensor chip surface, which was immobilized with biotinyl glycan, was used repeatedly to study the interaction with several different lectins. The regeneration was accomplished by washing away the surface-bound lectin with 100 mM H_{3PO4}.

Both the association rate constant (k_a) and dissociation rate constant (k_{diss}) were obtained from the SPR signal binding data and calculated using the BIA-evaluation software (Pharmacia Biosensor AB). For the generalized reaction

$$[A] + [B] \rightleftarrows [AB] \tag{1}$$

the rate of change in the concentration of the complex AB is given by

$$\frac{d[AB]}{dt} = k_a[A][B] - k_{\text{diss}}[AB] \tag{2}$$

A series of time-resolved measurements of the progress of the reaction (measuring concentrations of AB, A, or B) allows the determination of the binding constants for the reaction. At any time t,

$$[B]_t = [B]_0 - [AB]_t \tag{3}$$

Substituting Eq. (3) into (2) gives

$$\frac{d[AB]}{dt} = k_a[A]([B]_0 - [AB]_t) - k_{\text{diss}}[AB]_t \tag{4}$$

In BIAcore, B is the immobilized ligand (oligosaccharide). R_{max} is the total amount of binding sites of ligand B (total concentration of the oligosaccharide sites). $R_{\text{max}} - R_t$ is the number of remaining free binding sites at time t (units RU). The concentration of the added analyte (A, lectin) is C. Substituting these definitions into (4) gives

$$\frac{d[R]}{dt} = k_a C(R_{\text{max}} - R_t) - k_{\text{diss}}R_t \tag{5}$$

When the analyte solution is passed over the surface and replaced by buffer, the concentration of free analyte drops to zero. The derivatives of the response curve reflect the dissociation rate:

$$\frac{dR}{dt} = -k_{\text{diss}}R \tag{6}$$

The k_{diss} was first analyzed by fitting to the integrated form of the rate Eq. (6),

$$Rt = R_0 \exp(-k_{\text{diss}}t) \tag{7}$$

using nonlinear least squares analysis [14,15]: R_0 is the amplitude of the dissociation process. The k_a as well as R_{max} was then analyzed by fitting to the integrated form of Eq. (5),

$$R_t = \frac{CK_a R_{\text{max}}}{Ck_a + k_{\text{diss}}}[1 - \exp\{-(Ck_a + k_{\text{diss}})t\}] \tag{8}$$

The affinity constant, K_a, was then calculated from k_a/k_{diss}.

For the analysis, it was assumed that the binding of RCA $_{120}$ to the surface-bound oligosaccharides fits the generalized reaction of Eq. (1). This is not true, because RCA $_{120}$ is multivalent and the oligosaccharides could also act as multivalent ligands. However, it does allow the comparison of apparent functional affinities with respect to the different ligands and the evaluation of the effect of branching on the apparent affinity of RCA_{120}.

F. In Situ Enzymatic Digestion of BPH-Labeled Oligosaccharide on the Sensor Surface

Sialyn LacNAc (3′-SLN) was used as a model oligosaccharide and was BPH-labeled as previously described and immobilized onto the surface. *Arthrobacter ureafaciens* sialidase (2.5 mU/μL, 10 μL) was introduced onto the surface, and the flow rate was set at zero. After the injection of sialidase, the surface was washed wtih 50 mM phosphoric acid at regular intervals to follow the time course of enzyme digestion by measuring the interaction with MAM and RCA_{120}. The concentration used was 100 μg/mL for MAM, and 1 μg/mL for RCA_{120}.

III. RESULTS AND DISCUSSION

A. Optimization of the Biotinylation Conditions

The conditions for the derivatization with BHz–AC5 were optimized using LacNAc as a model sugar. Using reversed phase chromatography, the yield was estimated as a decrease in added LacNAc by monitoring the peak area. When LacNAc was incubated with a two-fold molar amount of BHz–AC5 relative to the saccharide at 90°C for 1 h, the biotinylation yield was almost quantitative. Positive ion fast-atom bombardment mass spectrometry of the product in thioglycerol/dithiothreitol produced an abundant ion at m/z737 $(M + H)^+$, suggesting that the desired hydrazone was obtained. Care should be taken, in that the yield was remarkably affected by the volume of solvent. The reaction proceeds quantitatively only when the solvent volume was small (ca. 20 μL, under the conditions employed), and almost all the solvent was vaporized in the vial after the reaction. Although the reaction of sugars with substituted hydrazine is thought to proceed by nucleophilic attack of the hydrazine at the hemiacetal carbon atom of a protonated form of the sugar [16], an acid catalyst was not necessary under the conditions we employed. The method was also applicable to sialyl saccharides without any loss of sialyl residues. BPH, a novel biotin hydrazide having UV absorption, also converted LacNAc into its BPH adduct in the same way.

B. Selective Preparation of a Stable β-Glycosidic Form in the Biotinylation

As shown in Fig. 1, the hydrazone could exist in two forms, the acyclic Schiff-base type and the cyclic hydrazino type. Furthermore, *syn/anti* isomers of the acyclic form, anomeric with the cyclic form, might be possible.

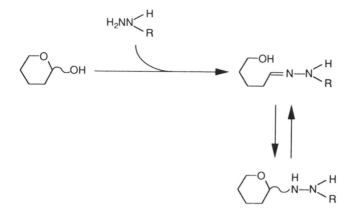

Figure 1 Formation and tautomerism of hydrazones.

The BPH–LacNAc was detected as multiple peaks by reversed phase chromatography. This result was unfavorable because of the inconvenience of purifying the multiple BPH adducts of the same oligosaccharide by reversed phase chromatography and the potential effect of structural heterogeneity at the reducing end of the oligosaccharide on lectin binding.

By further investigating the reaction, we found that one peak became predominant by leaving the mixture under acidic conditions, from the addition of formate buffer (pH 3.5) at 4°C. In the ^{1}H NMR analysis of the peak, the resonances of the anomeric protons of the GlcNAc moiety of LacNAc at 5.24 (α) and 4.51 ppm (β) were lost after the reaction, and a single anomeric resonance was observed at 4.22 ppm (doublet, 1H, $J_{1,2}$ = 9.53 Hz, βH1 proton of the GlcNAc moiety of the product). These results indicate that the other isomers were tautomerized to the β-glycoside under acidic conditions. The reverse tautomerization was not detected when the β-glycoside was allowed to stand under neutral conditions. By prolonging the storage time longer than 5 h, all the peaks including the β-glycoside showed a slight decrease. The peak area of BPH showed a slight increase over this period; apparently, the BPH adducts were gradually decomposed under acidic conditions to give BPH and LacNAc. To minimize the peaks other than the β-glycoside, incubation should be continued for 5 h. The overall yield to produce β-glycoside from LacNAc was estimated to be ca. 70%. To avoid undesirable decomposition of the β-glycoside under acidic conditions, it was recommended to neutralize the reaction mixture or purify the biotin adduct by chromatography. The cyclic β-glycoside was stable at least 2 weeks if it was stored under neutral conditions at 4°C. The limit of detection of labeled LacNAc in reversed phase chromatography was 330 fmol and was linear from 330 fmol to 261 pmol ($r = 0.998$).

Considering the fact that native N-linked oligosaccharides exist as the β-N-glycoside form, this tautomerization procedure may be advantageous for measuring the sugar-binding specificities of lectins and characterizing the function of oligosaccharides.

C. Separation of BPH-Labeled Oligosaccharides by Reversed Phase, Gel Filtration, and Anion-Exchange Chromatographies

The chromatographic behavior of BPH-derivatized oligosaccharides was evaluated using reversed phase, gel filtration, and anion-exchange chromatographies. Chromatograms showing the separation of oligosaccharides by reversed phase chromatography are shown in Fig. 2. Glucose oligomers were well separated, up to 22-mer in 60 min. Typical N-linked complex-type oligosaccharides, including sialylated oligosaccharides, were also well separated. Generally, trimming of the nonreducing end terminii residues and fucose α(1-6) residue at the reducing terminal increased retention times. The BPH-labeled 3′- and 6′-SLN were also well separated under this condition with modification of LacNAc by sialic acid at the 6′ position decreasing retention time.

BPH-labeled oligosaccharides were also well separated by MonoQ according to their charge and by Superdex peptide according to their size (data not shown). When two tandemly linked Superdex peptide columns were used, up to 14-mer glucose oligomers were well separated in 90 min (data not shown).

D. Immobilization of the Biotinyl Glycan onto the Sensor Chip Surface

The biotinyl glycan was injected onto the streptavidin preimmobilized surface and immobilized using a biotin–avidin comlex. Because the molecular mass of the biotinyl glycan is relatively small, it was difficult to detect the immobilized glycans. Whether or not the biotinyl glycan was immobilized should be confirmed by monitoring the interaction with the appropriate lectin. The immobilized amount of streptavidin was set as ca. 2000 RU, which corresponds to ca. 2 ng (33 fmol). Because streptavidin possesses four biding sites, the maximum binding amount of biotinyl glycan would be ca. 130 fmol. The maximum response to RCA_{120} was not affected if ~1 pmol of biotinyl LacNAc was injected. This indicates that at lease 1 pmol of biotinyl glycan per injection was sufficient for the analysis under the described conditions.

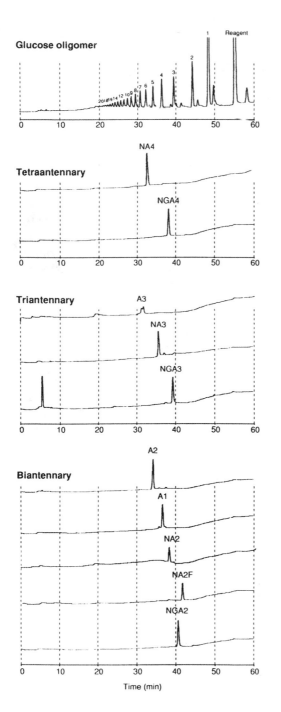

Glucose oligomer

Tetraantennary

Triantennary

Biantennary

Time (min)

It should be noted that the purification of the biotinyl glycan after derivatization was not necessary because of its low reagent/carbohydrate molar ratio (typically a two- to four-fold molar amount of reagent over oligosaccharides was used). It was possible to directly inject the reaction mixture containing biotinyl glycan onto the surface without any purification.

The biotinylated glycans were stable on the sensor surface because more than 90% of the reactivity remained even after 10 repeated injections of lectins and regenerations using 50 mM H_3PO_4. If the tautomerization procedure was performed, the surface-immobilized BPH-labeled LacNAc was more stable, with almost no decrease in the response after 20 repeated injections of RCA_{120}, which involved regeneration wtih 50 mM H_3PO_4. This may be explained by the stable nature of the β-glycoside isomer compared with acyclic Schiff-base type structures.

Glycopeptides or glycoproteins are also readily immobilized onto the sensor surface using the primary amino group of the peptide moiety [17–19]. However, because the coupling efficiency by amine coupling is not very high compared with avidin–biotin coupling, a greater amount of sample is required [17].

E.   Validity of This Methodology for the Analysis of Lectin Specificities

Complex-type sugar chains (NA2, NA3, and NA4) were successfully immobilized onto the surface, and the interactions with several lectins were analyzed. The interactions of each lectin against the surface-bound NA2, NA3, and NA4 are shown in Fig. 3. The increase in RU from the initial baseline is from the binding of injected lectin to the surface-bound oligosaccharide.

The amount of each lectin required for the analysis was from several fmol to 100 pmol per injection. The binding specificities of RCA_{120}, DSA, and ConA have already been well established [20–22]. Similar conclusions could be drawn for RCA_{120} and ConA in this study. Among these surface-bound sugar chains, DSA bound only to NA4. It was unexpected that DSA did not show any binding with NA3. However, by making the linker length

---

Figure 2   Chromatograms showing the separation of BPH-labeled oligosaccharides by reversed phase chromatography. BPH-labeled glucose oligomers, complex-type tetraantennary oligosaccharides, triantennary oligosaccharides, and biantennary oligosaccharides were analyzed by reversed phase chromatography. Each oligosaccharide (ca. 1 nmol) was incubated with fourfold molar amount of BPH at 90°C for 1 h. After tautomerization procedure, one-tenth was injected to high performance liquid chromatography and was analyzed as described in Section II.

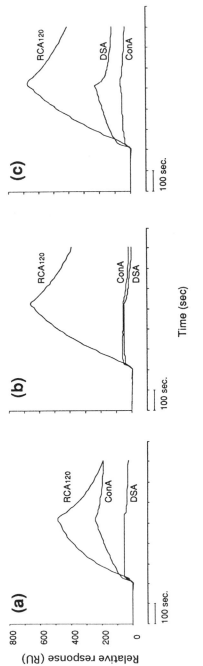
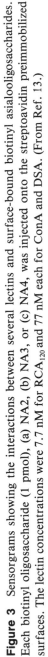

Figure 3 Sensorgrams showing the interactions between several lectins and surface-bound biotinyl asialooligosaccharides. Each biotinyl oligosaccharide (1 pmol), (a) NA2, (b) NA3, or (c) NA4, was injected onto the streptoavidin preimmobilized surfaces. The lectin concentrations were 7.7 nM for RCA_{120} and 77 nM each for ConA and DSA. (From Ref. 13.)

between glycan and biotin longer, DSA acquired reactivity toward NA3. This result suggests that the length between biotin and glycan could make lectins accessible to glycan and affect the interactions between glycan and lectins.

Sialylated complex-type sugar chains were also immobilized on the sensor chip, and the interactions with SSA, MAM and RCA_{120} were analyzed. As shown in Fig. 4a, the monosialylated biantennary oligosaccharide (A1) interacted with SSA and RCA_{120}, but MAM and the disialylated biantennary oligosaccharide (A2) interacted only with SSA (Fig. 4b). Because both A1 and A2 were prepared from human fibrinogen, the linkage of NeuAc to Gal residues in which is exclusively $\alpha(2,6)$ [23], the results obtained here were in agreement to the previously reported sugar-binding specificities of SSA [24] and MAM [25]. The trisialylated triantennary oligosaccharides (A3) prepared from fetuin, the linkage of NeuAc to Gal residues in which is reported to be a mixture of both $\alpha(2,3)$ and $\alpha(2,6)$ [26], interacted with both SSA and MAM (Fig. 4c). When A3 was digested wtih Newcastle disease virus sialidase, which is known to selectively remove the $\alpha(2,3)$-linked NeuAc, and the interactions were measured again, MAM completely lost the ability to bind to A3 (Fig. 5). These results demonstrated that this method is useful for determining carbohydrate-binding specificities of lectins using biotinyl glycans.

F. Enzymatic Digestion of Surface-Bound Oligosaccharides on the Sensor Surface

We further studied exoglycosidase digestion of BPH-labeled oligosaccharides on the sensor surface. Following the BPH labeling of 3'-sialyl N-acetyllactosamine (3'-SLN) and its immobilization onto the sensor surface, digestion wtih *A. ureafaciens* sialidase was performed. After injection of the sialidase solution onto the surface, the flow rate was set at zero. The time course of digestion was monitored by measuring the interaction with MAM and RCA_{120}. As shown in Fig. 6, 4 h after the introduction of sialidase, the response to MAM was drastically decreased with a concomitant increase in the response to RCA_{120}. The abrupt increase and decrease in the response at the beginning and end of the injection of the MAM solution are due to the changes in mass on the surface caused by the injected lectin. From these binding analyses, it was shown that the enzyme digestion was completed 8 h after the enzyme injection. The same responses at 8 h and 20 h after enzyme injection suggests that the labeled LacNAc was stable under the digestion conditions employed. These results indicated the possibilities of other modifications of oligosaccharides on the sensor surface. These may include new approaches to oligosaccharide synthesis and sequential analysis.

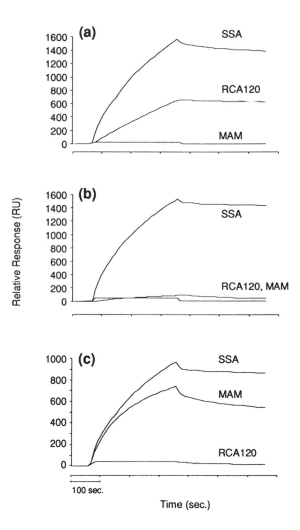

Figure 4 Sensorgrams showing the interactions between lectins and surface-bound biotinyl sialylated-oligosaccharides. Each biotinyl oligosaccharide, (a) A1, (b) A2, and (c) A3, was injected onto the streptavidin preimmobilized surfaces at an amount of 1 pmol. The lectin concentrations used in this study were 7.7 nM for RCA$_{120}$ and 3.9 μM for SSA and MAM.

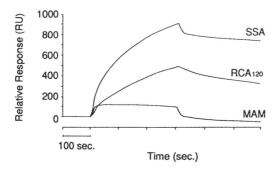

Figure 5 Effect of Newcastle disease virus sialidase digestion on the interaction of trisialylated oligosacchraide (A3) with lectins. The lectin concentration was the same as given in Fig. 4.

G. Kinetic Measurement of Lectin Interactions

Although lectins usually possess broad binding specificities in terms of monosaccharides, they are also known to discriminate between other structural aspects of their ligands. Although *Ricinus communis* agglutinin-120 (RCA$_{120}$) is known to recognize a terminal galactose residue, various galactose-containing sugar chains could be separated into five fractions by RCA$_{120}$-immobilized affinity chromatography [27]. It is interesting to differentiate the recognition mechanism into the inherent affinity of a single

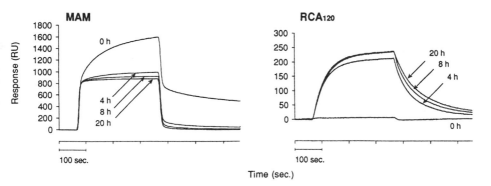

Figure 6 Time course of the sensor surface digestion of sialic acid residues from surface-bound sialyl LacNAc. After injection of the *A. ureafaciens* sialidase solution (2.5 mU/μL) onto the surface, the flow rate was set at zero. The time course of digestion was monitored by measuring the interaction with MAM and RCA$_{120}$. (From Ref. 12.)

nonreducing terminus for a single binding site of lectin and the apparent affinity caused by the multiple potential interactions of a branched glycan.

To evaluate the mechanism of the molecular recognition of RCA_{120}, the interaction of RCA_{120} with NA2, NA3, and NA4 was analyzed kinetically. Following the immobilization of biotinylated NA2, NA3, and NA4 on the surface, RCA_{120} was injected at a concentration of 1 μg/mL and analyzed by nonlinear least squares methods as described in Section II.

The kinetic parameters for each oligosaccharide are summarized in Fig. 7. We found that the parameters for RCA_{120} interactions were different for NA2, NA3, and NA4, although all of these sugar chains possess terminal galactosyl residues. By increasing the number of branches and galactosyl residues at the nonreducing end, both k_a and k_{diss} were reduced. K_a showed the same tendency. The R_{max} values, which reflect the maximum amount of bound lectin, increased with an increase in the galactosyl rsesidues per glycan chain. These results suggested that the number of terminal galactosyl residues or the configuration of residues in the glycan structure affects the kinetic parameters of RCA_{120}. The K_a obtained for RCA_{120} and NA3 was quite similar to the value (1.2×10^8 M^{-1}) reported previously for iodinated RCA_{120} binding to human erythrocytes [28]. The results obtained for k_a and K_a were inconsistent with the previous observation that the more galactosyl residues an oligosaccharide possessed, the more it was retained by an RCA_{120}-immobilized column [27].

Because this study was designed to measure the interaction between surface-bound oligosaccharides and lectins, the multivalency of sugar chains was considered. NA2, NA3, and NA4 could interact with two, three, or four binding sites for RCA_{120}, respectively. The R_{max} value, which reflects the maximum number of bound lectin molecules, increased as the number of terminal galactose residues increased. Because each binding site exists close to one another, the k_a of the third and fourth lectin binding site may be affected by steric hinderance. The experimental value reflects the binding to these independent sites.

Two possible explanations for the reduced k_{diss} are that by increasing the number of branches multivalent binding of RCA_{120} is facilitated, or the increased number of branches may increase the chance for RCA_{120} to rebind. To test the second possibility, the effect of co-injecting of an excess of LacNAc on the dissociation rate was measured. As shown in Fig. 8, increasing amounts of competing LacNAc had a dramatic effect on the dissociation rate. This result suggested that the increased number of branches may fulfill the function of enhancing the apparent affinity by increasing the chance of rebinding. These results also emphasize the importance of kinetic measurement for understanding multi-valent molecular recognition.

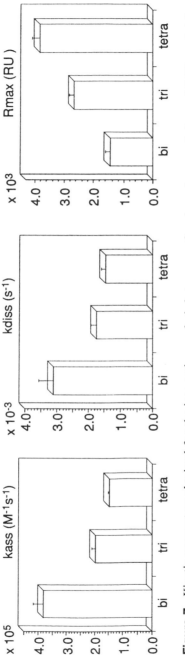

Figure 7 Kinetic parameters obtained for the interaction analysis between surface-bound sugar chains and RCA_{120}. Asialo-biantennary (bi), asialo-triantennary (tri), and asialo-tetraantennary (tetra) complex-type sugar chains were immobilized. Each parameter was determined as described in Section II and is represented as the means ± S.D. of three binding studies.

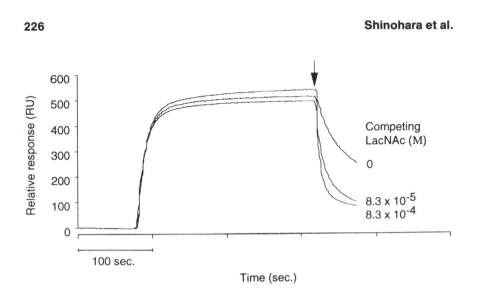

Figure 8   Effect of co-injection of excess *N*-acetyllactosamine (LacNAc) on the k_{diss} of surface-bound NA4. The arrow indicates the end of RCA$_{120}$ injection and the infusion of excess LacNAc.

IV. CONCLUSIONS

We have developed a new method for the analysis of the interaction between lectins and biotin-derivatized oligosaccharides using a biosensor based on surface plasmon resonance. Oligosaccharides, including sialylated glycans, were effectively converted into their biotin adducts as the stable β-glycoside isomer under the optimized conditions. The novel biotin hydrazide derivative was also useful both for highly sensitive detection and for chromatographic purification. The required amounts of sugar chains for the interaction analysis by this method were as low as 1 pmol. Moreover, they were quite stable on the sensor surface of the biosensor during repeated analyses and were also applicable for in situ exoglycosidase digestion. The binding specificities of several lectins were rapidly determined qualitatively by this method. Furthermore, kinetic analysis of the interaction between RCA$_{120}$ and the complex-type asialo bi- , tri-, and tetraantennary oligosaccharides revealed that both the association rate constant (k_a) and the dissociation rate constant (k_{diss}) were reduced by increasing the number of terminal galactosyl residues. Thus, the novel approach to elucidate the interaction provides useful quantitative information and may lead to improved understanding of the physiological functions of these interactions.

ACKNOWLEDGMENT

This work was performed under the management of the Research Association for Biotechnology as part of the R&D Project on Basic Technology for Future Industries supported by NEDO (New Energy and Industrial Technology Development Organization) in the Ministry of International Trade and Industry.

ABBREVIATIONS

BHz–AC5	6-hydrazidohexyl D-biotinamide
BHz–(AC5)$_2$	6-(6-hydrazidohexyl) amidohexyl D-biotinamide
BPH	4-(biotinamido)-phenylacetylhydrazide
ConA	Concanavalin A
DSA	*Datura stramonium* lectin
EDC	*N*-ethyl-*N'*-(3-dimethylaminopropyl) carbodiimide hydrochloride
HBS	HEPES-buffered saline
LacNAc	*N*-acetyllactosamine
MAM	*Maackia amurensis* lectin
NHS	*N*-hydroxysuccinimide
NMR	nuclear magnetic resonance
RCA$_{120}$	*Ricinus communis* agglutinin-120
RU	resonance units
3'(6')-SLN	3'(6')-sialyl LanNAc
SPR	surface plasmon resonance
SSA	*Sambucus sieboldiana* lectin

REFERENCES

1. Varki, A. (1993). *Glycobiology* 3, 97–130.
2. Kobata, A. (1992). *Eur. J. Biochem.* 209, 483–501.
3. Pinckard, R. N. (1978). *Handbook of Experimental Immunology,* Vol. 1. Blackwell, Oxford, Chapter 17.
4. Bains, G., Lee, R. T., Lee, Y. C., and Freire, E. (1992). *Biochemistry* 31, 12624–12628.
5. Honda, S., Taga, A., Suzuki, K., Suzuki, S., and Kakehi, K. (1992). *J. Chromatogr.* 597, 377–382.
6. Jordan, F., Basset, E., and Redwood, W. R. (1977). *Biochem. Biophys. Res. Commun.* 75, 1015–1021.
7. Lee, Y. C. (1992). *FASEB J.* 6, 3193–3200.
8. van der Merwe, P. A., and Barclay A. N. (1994). *TIBS* 19, 354–358.

9. Alon, R., Hammer, D. A., and Springer, T. A. (1995) *Nature* 374, 539–542.
10. Granzow, R., and Reed, R. (1992). *Bio/Technology* 10, 390–393.
11. Schuster, S. C., Swanson, R. V., Alex, L. A., Bourret, R. B., and Simon, M. I. (1993). *Nature* 365, 343–347.
12. Shinohara, Y., Sota, H., Gotoh, M., Hasebe, M., Tosu, M., Nakao, J., and Hasegawa, Y., and Shiga, M. (1996). *Anal. Chem.* 68, 2573–2579.
13. Shinohara, Y., Sota, H., Kim, F., Shimizu, M., Gotoh, M., Tosu, M., and Hasegawa, Y. (1995). *J. Biochem.* 117, 1076–1082.
14. O'Shannessy, D. J., Brigham-Burke, M., Soneson, K. K., Hansley, P., and Brooks, I. (1993). *Anal. Biochem.* 212, 457–468.
15. Karlsson, R., Roos, H., Fagerstam, L., and Persson, B. (1994). *Methods* 6, 99–110.
16. Pigman, W., and Horton, D., eds. (1980). *The Carbohydrates: Chemistry and Biochemistry,* Vol. 1B. Academic Press, New York, p. 932.
17. Shinohara, Y., Kim, F., Shimizu, M., Goto, M., Tosu, M., and Hasegawa, Y. (1994). *Eur. J. Biochem.* 223, 189–194.
18. Yamamoto, K., Ishida, C., Shinohara, Y., Hasegawa, Y., Konami, Y., Osawa, T., and Irimura, T. (1994). *Biochemistry,* 33, 8159–8166.
19. Okazaki, I., Hasegawa, Y., Shinohara, Y., and Kawasaki, Y. (1995). *J. Mol. Recogn.* 8, 95–99.
20. Ogata, S., Muramatsu, T., and Kobata, A. (1975). *J. Biochem.* 78, 687–696.
21. Harada, H., Kamei, M., Tokumoto, Y., Yui, S., Kotama, F., Kochibe, N., Endo, T., and Kobata, A. (1987). *Anal. Biochem.* 164, 374–381.
22. Yamashita, K., Totani, K., Ohkura, T., Takasaki, S., Goldstein, I. J., and Kobata, A. (1987). *J. Biol. Chem.* 262, 1602–1607.
23. Townsend, R. R., Hillker, E., Li, Y.-T., Laine, A., Bell, W. R., and Lee, Y. C. (1982). *J. Biol. Chem.* 257, 9704–9710.
24. Wang, W.-C., and Cummings R. D. (1987). *J. Biol. Chem.* 253, 4576–4585.
25. Shibuya, N., Tazaki, K., Song, Z., Tarr, G. E., Goldstein, I. J., and Peumans W. J. (1989). *J. Biochem.* 106, 1098–1103.
26. Green, E. D., Adelt, G., Baenziger, J. U., Willson, S., and Halbeek, V. (1988). *J. Biol. Chem.* 263, 18253–18268.
27. Merkle, R. K., and Cummings, R. D. (1987). *Methods in Enzymology.* Vol. 138. Academic Press, New York, pp. 232–259.
28. Sandvig, K., Olsnes, S., and Pihl, A. (1976). *J. Biol. Chem.* 251, 3977–3984.

14
Glycoprotein Detection Using the Light-Addressable Potentiometric Sensor

Kilian Dill
Molecular Devices Corporation, Sunnyvale, California

I. INTRODUCTION

The Threshold Immunoassay System is sold commercially as a sensitive, reliable analytical system used for quantitative determination of various solution analytes [1,2]. It is currently utilized by the pharmaceutical industry for the detection of impurities in the production of biopharmaceuticals by companies seeking FDA approval for the release of their products [3–7]. These contaminants include DNA and host cell proteins, which are often present in genetically engineered products. Furthermore, testing for column material such as protein G (or A) used in the monoclonal antibody purification process is also necessary [3,7,8].

Threshold also can be used to detect large molecules by using sandwich immunoassay formats and also may be used to detect small molecules by using competitive immunoassay formats [9]. The system can detect molecules typically at femtomole levels with a dynamic detection range of over two orders of magnitude of analyte concentration.

For analyte detection, Threshold uses the light-addressable potentiometric sensor (LAPS). The LAPS is composed of eight identical etched sites on a silicon chip that can detect pH changes [1,2], simultaneously (Fig. 1). The pH change results from the action of the enzyme urease on the substrate urea; the urease conjugate of the anti-fluorescein antibody (signal generator) is part of the immunocomplex formed (Fig. 2).

In this chapter we discuss the Threshold Immunoassay System, the theory of the LAPS, and the assay formats that can be used. We then show applications to the detection of specific carbohydrate residues of glycoproteins. For glycoprotein detection, lectins as well as specific antibodies were used to form immunocomplexes. Specifically, we have investigated the analysis of two glycoproteins, human α_1-acid glycoprotein (AGP) and calf serum fetuin [10,11], in various forms of glycosylation. Furthermore, we

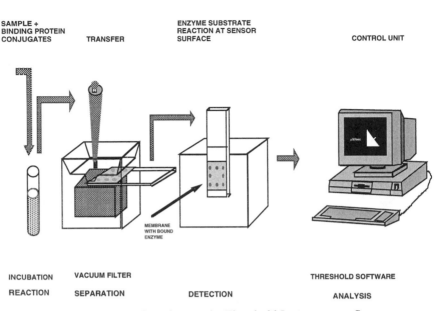

SAMPLE +
BINDING PROTEIN
CONJUGATES   TRANSFER

ENZYME SUBSTRATE
REACTION AT SENSOR
SURFACE

CONTROL UNIT

MEMBRANE
WITH BOUND
ENZYME

INCUBATION   VACUUM FILTER   THRESHOLD SOFTWARE

REACTION   SEPARATION   DETECTION   ANALYSIS

Figure 1   Immunoassay detection on the Threshold Immunoassay System.

obtained binding constants for the interaction of ricin with terminal galactose [10].

II. MATERIALS AND METHODS

A. Threshold System

The system is composed of three parts (Fig. 1). First is a workstation that contains a vacuum manifold for the "filtration capture" of immunocomplexes that have been formed in solution. The capture of the immunocomplexes is made possible by the use of a biotin-coated porous nitrocellulose membrane in the filtration step. Filtration capture also provides a mechanism for concentrating the samples from up to 2 mL down to 0.64 μL assay volume. This step provides increased sensitivity for this system over other immunoassay systems. Second, the LAPS (or Threshold reader) detects minute quantities of immunocomplex (analyte) that are captured on the membrane. Third, Threshold software is available that allows the simultaneous monitoring of eight individual assays sites present on each membrane, and displays the results and quantifies the data.

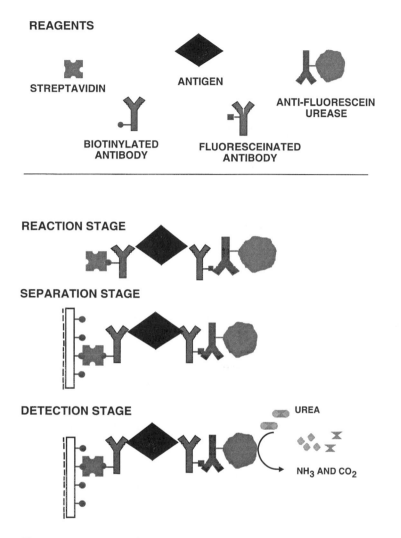

Figure 2 Immuno-Ligand Assay (ILA) stages.

B. Assay Format and Sample Processing

For the quantitative determination of analyte, an immunoassay format is required when using the Threshold Immunoassay System. For the detection of large molecules, a sandwich immunoassay format is used via a four-step process, as shown in Fig. 2. Figure 1 shows the sample processing and data collection and analysis: (1) First, the sample is incubated in a test tube with

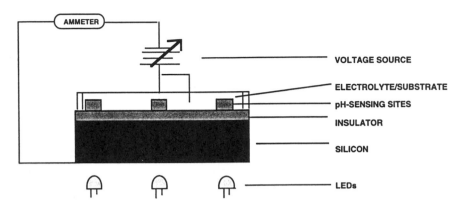

Figure 3   The light-addressable potentiometric sensor (LAPS).

the various assay components, including biotin- and fluorescein-labeled anti-analyte antibodies and the biotin-binding protein, streptavidin (Fig. 2). This allows all molecular associations to be accomplished in solution, thus avoiding artifacts caused by a solid phase. (2) After the end of the incubation period, the sample is filtered (vacuum filtration) through a 100 μm thick biotin-coated nitrocellulose membrane and the immunocomplex is concentrated and captured on the membrane, localized to a 3-mm spot (Fig. 1). Assay volumes can vary from 100 μL to 2 mL. In the next step, a solution containing the urease conjugate of an anti-fluorescein antibody is passed through the membrane and bound to the immunocomplex. (3) Detection of urease-labeled antibody conjugates is accomplished by the enzymatic hydrolysis of urea in the microvolume adjacent to a silicon chip. Eight individual and independent assay sites are present on each immuno-capture membrane (supported on a plastic stick), which can be used to assay different concentrations of an analyte or detect different analytes. Detection occurs via a LAPS when the membrane is placed into the detector and pressed against a silicon chip after the Threshold reader is filled with the enzyme substrate in a buffered solution. All eight assay sites on each stick are detected simultaneously. (4) Data is analyzed with a Molecular Devices' software program.

C.   Detection System

The mainstay of the assay system is the LAPS [12,13]. Figure 3 shows the basics of the sensor. It is a silicon chip with eight identically etched sites [1,2]. The membrane containing the various captured chemistries is aligned and pressed against the chip with a plunger to form tightly sealed 0.64 μL microvolumes. As mentioned earlier, the signal generator in this system is an enzyme, which, upon acting on the substrate, produces an increase in pH within this microvolume. It is this pH change that is monitored by the LAPS.

The exact mechanism by which the pH change is monitored on the chips by the LAPS is as follows: A bias potential is applied to the external circuit that connects the chip to the controlling electrode. The bias potential, together with the pH-dependent surface potential at the pH-sensing sites, determines the magnitude and direction of the electric field in each respective pH-sensing region of the semiconductor adjacent to the dielectric. Illumination of the semiconductor (with LEDs) near a pH-sensing site produces a photopotential, resulting in a transient photocurrent to flow in a circuit external to the semiconductor chip. Changes of pH on the chip surface adjacent to any particular pH-sensing region alters the bias voltage required to produce a given photocurrent when a LED is used to illuminate the semiconductor near that region. In operation, as the pH is altered, the inflection point of the photocurrent–applied voltage curve shifts on the potential axis (see Fig. 4). It is this rate in the shift (μV/sec) that is monitored and is directly related to the change in pH.

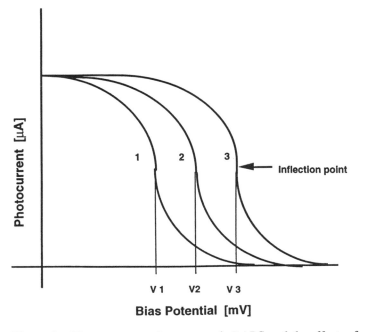

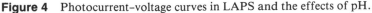

Figure 4   Photocurrent–voltage curves in LAPS and the effects of pH.

D. Glycoproteins and Immunochemical
Sample Preparation

Fetuin (gel filtration purified from fetal calf serum), asialofetuin (Type II from fetal calf serum), biotinylated ricin (B-ricin, 2.2 biotin/ricin), FITC-ricin (F-ricin, 1.7 fluorescein/ricin), human α_1-acid glycoprotein (purified from Cohn fraction VI), polyclonal rabbit anti-human AGP (IgG), and *D*-galactose (ultrapure) were purchased from Sigma Chemical Company (St. Louis). The wash buffers, assay buffers, anti-fluorescein antibody-urease conjugate, capture reagent (streptavidin, SA), and the biotin–bovine serum albumin (BSA)–coated nitrocellulose capture membranes are products of Molecular Devices Corp.

The asialo form of AGP was obtained by treatment of the native material with neuraminidase-coupled agarose beads (Sigma Chemical Company) in phosphate buffer, pH 5.0 [11]. The antibody was biotinylated using biotin–DNP–*N*-hydroxysuccinimide (NHS) reagent (Molecular Devices Corp.) and purified to give a product that contained 3.6 molecules of biotin/antibody. Production of fluoresceinated antibody was accomplished using procedures similar to those described for the biotinylated antibody (using fluorescein–NHS from Molecular Devices Corp.) to give a product that contained 1.0 molecule of fluorescein/antibody. The wash buffers, assay buffers, anti-fluorescein antibody–urease conjugate, and the biotin–BSA coated nitrocellulose capture membranes are components of the Immuno-Ligand Assay (ILA) detection kit (Molecular Devices Corp).

Sample preparations and formation of immunocomplexes were as follows: For the polyclonal antibody (Ab) immunoassay, 150 ng each of biotinylated Ab (B-Ab) and fluoresceinated Ab (F-Ab) per test were mixed with the analyte; streptavidin (2 μg) was added so that the final volume was 250 μL/vial. For the heterosandwich immunoassay, each test contained 50 ng of F-ricin and 50 ng B-Ab (total and not specific), 2 μg of streptavidin, and varying quantities of either AGP, asialo-AGP, or asialofetuin in 250 μL. For the detection of asialofetuin or fetuin using the lectin sandwich immunoassay format, an assay volume of 200 μL was used. Each test contained 20 ng each of B-ricin and F-ricin, and varying quantities of either fetuin or asialofetuin. Finally, 2 μg of streptavidin was added. Mixtures were incubated for 1 h at room temperature. The mixtures were then filtered through the prewashed nitrocellulose membrane mounted on the Threshold workstation. The captured sample was then rinsed with 0.5 mL of wash buffer followed by 200 μL volume of anti-fluorescein antibody-urease conjugate (containing ~2 μg of enzyme conjugate). Lastly, the membrane was rinsed once more with 0.5 mL wash buffer. Typically, for the standard detection curves, three spots were used per analyte concentration used (to obtain data in triplicate).

For the galactose inhibition experiments, the following format was used. Twenty nanograms each of F-ricin and B-ricin were incubated with 16 ng of asialofetuin in 225 μL volume. To each sample, varying amounts of D-galactose were added, and the mixture was allowed to incubate for 1 h. Enough sample was prepared at each galactose concentration so that at least three data points could be obtained. The inhibition curve was analyzed using the equation [10]

$$\text{Sig (Theoretical)}_i = \text{Sig (No Gal)}_i + \frac{\Delta\, 10^{(\text{pK}_I - \text{pK}_{\text{Gal}})}}{1 + 10^{(\text{pK}_I - \text{pK}_{\text{Gal}})}}$$

In this equation, Δ represents the maximum signal difference from a sample in the presence of a large excess of galactose and a sample devoid of galactose. Sig (Theoretical)i represents the theoretical signal based on the specific galactose concentration and pK$_I$ value. Sig (No Gal) is the signal observed when no galactose is present. The term pK$_{\text{Gal}}$ is defined as $-\log$ [Gal]. The percent inhibition was calculated based on the relative signal for that sample concentration of galactose and Δ. The best fit is obtained when

$$\Sigma_i\, [\text{Sig (Theoretical)}_i - \text{Sig (Observed)}_i]^2 \qquad \text{is minimized.}$$

III. RESULTS AND DISCUSSION

The detection of a glycoprotein in a milieu can be accomplished by a number of assay formats involving antibodies and lectins [10,11]. For the simple detection of any glycoprotein, a protein–epitope-based polyclonal immunoassay format was used as shown in Fig. 5. This format is, in principle, independent of the carbohydrate present and will allow quantitation of a specific glycoprotein (all glycoforms). Determination of the presence and quantitation of a specific glycoform(s) may be obtained using a heterosandwich immunoassay as shown in Fig. 6. In this case a labeled lectin may be used in conjunction with an antibody. For probing the terminal carbohydrates of glycoproteins, lectins are attractive because a large assortment are available with different specificities. To determine whether the assay formats shown in Figs. 5 and 6 would allow low level detection of glycoproteins and glycoforms, we used polyclonal rabbit anti-human AGP antibodies and galactose-specific F-ricin to detect human AGP and asialo-AGP.

Alpha-1-acid glycoprotein is found in human serum in various concentrations. This polymorphic glycoprotein contains five N-linked complex carbohydrate chains and displays several glycoforms even at a single site

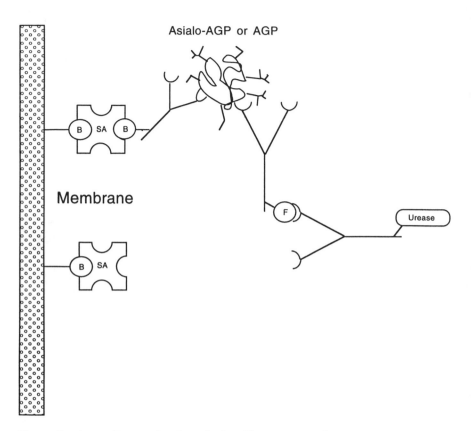

Figure 5   Assay diagram for the polyclonal immunoassay format.

[14,15]. The oligosaccharides present all contain sialic acid at the terminus, which may play a role in the biological function of this glycoprotein. In the native form, ricin binding would be inhibited due to the sialic acid groups present. Asialo-AGP contains terminal galactose residues to which ricin can bind.

Figure 7 shows the standard curve for the detection of AGP and asialo-AGP using the polyclonal antibody protein–based immunoassay procedure as shown in Fig. 5. Both AGP and asialo-AGP elicited similar responses. The lower limits of detection were 2 pg for the glycoprotein. The results indicate that the assay is sensitive and sialic acid independent.

Figure 8 shows the standard curve for the detection of AGP and asialo-AGP using the heterosandwich assay (Fig. 6). An additional glycoprotein, asialofetuin, was also tested. Although this antibody was not spe-

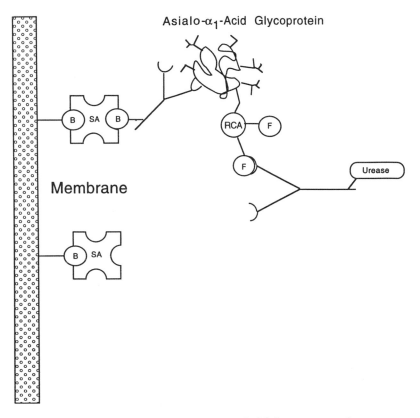

Figure 6   Assay diagram for the heterosandwich immunoassay format.

cific for asialofetuin, we tested this glycoprotein in our assay system because it contains terminal galactose residues to which ricin will bind. Because the B-Ab is not specific for asialofetuin, the ricin–asialofetuin complexes should not be captured on the membrane. Neither AGP nor
asialofetuin elicited a response in the heterosandwich immunoassay. However, asialo-AGP gave a linear response as a function of concentration and
was readily detectable at the nanogram level. The detection limit is about
250 pg for asialo-AGP [11].

A problem that could arise in a complex mixture is the presence of
additional terminal galactose–containing glycoproteins. The heterosandwich assay system would still be usable, but the quantity of ricin used in the
assay would have to be modified. This is because the additional terminal
galactose–containing glycoproteins present would compete with the specific

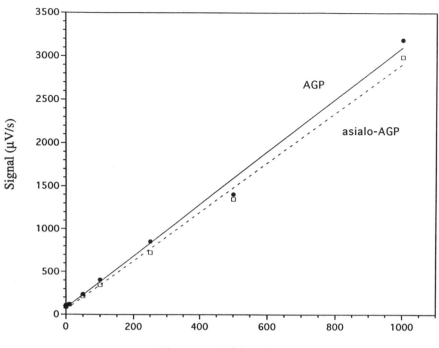

Figure 7 Standard curve for the detection of AGP and asialo-AGP using the format shown in Fig. 5.

glycoprotein (asialo-AGP) for the limited amount of F-ricin. Thus, the assay would have to be augmented by the addition of increased amounts of F-ricin. To determine the quantity of F-ricin needed in the modified assay mixture, a mixture containing fixed amounts of antibody and analyte are titrated with varying amounts of F-ricin until a plateau is reached. The plateau value for ricin (or a value greater than this) can now be used in the modified assay mixture for the detection of a mixture of Gal-terminated glycoproteins. Such a titration is shown in Fig. 9. For this experiment, 10 ng each of asialofetuin and asialo-AGP were incubated with constant amounts of B-Ab and varying quantities of F-ricin. As can be seen, the asialofetuin interferes with the detection of asialo-AGP by binding some of the F-ricin molecules (compare the signal output for the 50-ng level of F-ricin in Fig. 9 with the signal in Fig. 8 at the 10 ng/test for asialo-AGP). Further, addition of F-ricin results in a plateau, and this result can be used to establish the new F-ricin/B-Ab ratio that is required for the heterosand-

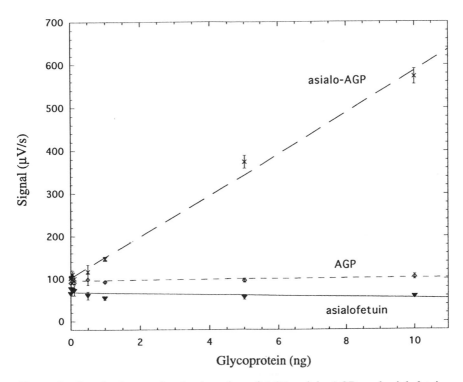

Figure 8 Standard curve for the detection of AGP, asialo-AGP, and asialofetuin using the format shown in Fig. 6.

wich immunoassay. Use of the immunoassay detection system shown in Fig. 6 in conjunction with the modified (corrected F-ricin/B-Ab) heterosandwich immunoassay shown in Fig. 9 would allow detection and quantitation of specific glycoform(s). Note that this approach requires specific standards, and the value obtained for the glycoform(s) would be an "average value"; for instance, the assay cannot distinguish whether AGP molecules have 30% of their sialic acids residues removed (uniformly) or 30% of the AGP molecules have 100% of their sialic acid residues removed.

Another simpler, semispecific immunoassay that could be used to detect glycoproteins is based on the use of only lectins to form the sandwich. Figure 10 shows the format for such a sandwich assay using asialofetuin as the analyte. Both B- and F-ricin are required to form the sandwich with asialofetuin.

We used fetuin as a model glycoprotein in this study because the oligosaccharide structures are well characterized. Fetuin contains six oligosaccharide chains [10] and the asialo forms will be shown presently. Three of the oligosaccharide chains are simple, identical tetrasaccharide linked to Thr/Ser [16,17], also known as the T antigen, which are found on human carcinoma cells:

$$Gal\beta 1 \rightarrow 3GalNAc\alpha \rightarrow Thr (Ser)$$

The asialo *N*-linked oligosaccharide chains are primarily two types of triantennerary structures as shown in Scheme 1 [18]:

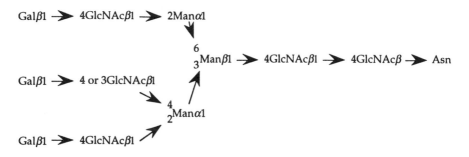

Scheme 1   Two types of triantennerary structures.

As can be seen, the asialo-disaccharide contains only one terminal galactose residue, whereas the triantennary asialo form of the complex carbohydrate contains three terminal galactose residues.

Figure 11 shows the standard curve for the detection of fetuin and asialofetuin using the procedure described in Section II. Fetuin appears to elicit no response in this assay. However, asialofetuin gives a linear response as the quantity of asialofetuin in the assay is increased. Asialofetuin is readily detectable at the nanogram level with a detection limit of 250 pg. As shown earlier, B- and F-ricin molecules are necessary to form the sandwich for capture and detection in the LAPS-based system. Should a glycoprotein have only one terminal galactose available for binding, then a sandwich, in principle, would not be formed, and detection is made impossible.

Information about the lectin/terminal galactose binding may also be obtained using Threshold. It is known that galactose binds to the active site of ricin toxin ($K_a = 1.1 \times 10^4$ M^{-1} for the high affinity site) [19,20] as well as the ricin agglutinin [21]. The lectin appears to have no anomeric preference for the galactose residues [19]. Addition of free galactose to our samples containing B-ricin, F-ricin, and asialofetuin should inhibit formation of the ricin–asialofetuin complex. The inhibition value can be used to deduce a "relative" binding constant for the ricin–fetuin complex. Absolute

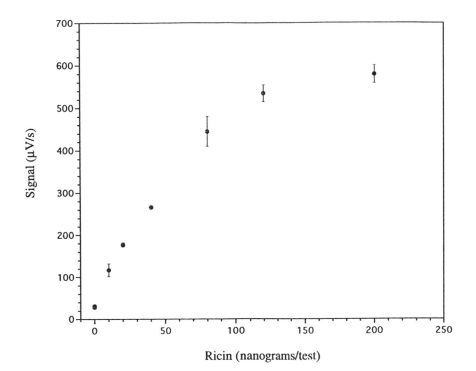

Figure 9 Titration of a 50/50 mixture of asialofetuin and asialo-AGP with a fixed amount of anti-AGP antibody and varying quantitites of F-ricin.

ricin–asialofetuin affinity values could be obtained since the ricin–galactose binding constant is known [10]. A plot of signal inhibition (0 to 100%) versus log of galactose concentration ($-pK_{Gal}$) is shown in Fig. 12. The midpoint of this apparent titration is the pK_I, from which a binding constant can be obtained (see Section II).

To obtain the binding constant for the ricin–fetuin complex, we made the following assumptions and deductions: F- and B-ricin have the same affinity for terminal galactose residues. There are 12 terminal galactose residues present on asialofetuin, and they bind 12 ricin molecules. This is likely correct because recent studies with a complex carbohydrate glycopeptide derived from asialofetuin showed that the three terminal galactose residues were all accessible to the much larger ricin agglutinin (RCA_{120}) [21]. The loss of signal would result from either the displacement of the B- or F-ricin by free galactose. From the data and using the assumptions, we

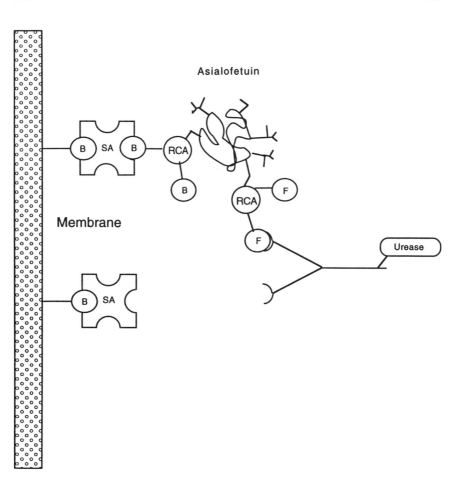

Figure 10   Assay diagram for the lectin immunoassay format.

obtain a K_a value of $3.6 \times 10^8 \, M^{-1}$ for the ricin–asialofetuin complex; this is an average value based on the different terminal galactose sites available. This value is in close agreement with $1.6 \times 10^8 \, M^{-1}$ published by Shinohara et al. [21] for the binding of RCA_{120} to a complex-carbohydrate glycopeptide derived from asialofetuin. Furthermore, the value is also only twice as great as the approximate affinity value obtained by Yamamoto et al. [19] for the binding of ricin to a glycopeptide derived from asialofetuin. The literature values we have cited were obtained using kinetic techniques where the glycopeptide was attached to a solid phase; this means that one binding

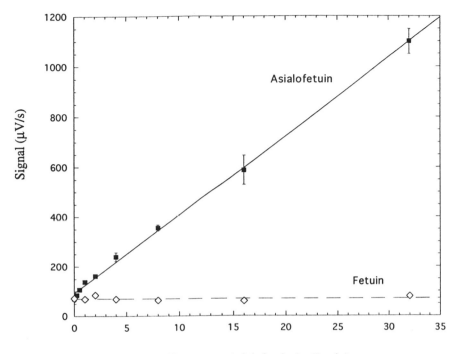

Figure 11 Standard curve for the detection of fetuin and asialofetuin using the format shown in Fig. 10.

component was in the solid phase, whereas the lectin was in the liquid phase. Our immunoassay and detection system is based on solution binding of the various species, and the bound complex is then captured while flowing through the membrane. Our method is not subject to solid phase–liquid phase interactions.

IV. CONCLUSIONS

We have shown that the Threshold System can be used to detect picogram quantities of glycoproteins, and to detect glycoforms. Furthermore, it can be used to assess binding constants of glycoprotein–lectin complexes. The system also has the potential to detect branched carbohydrate structures and even the oligosaccharides present on cells [22].

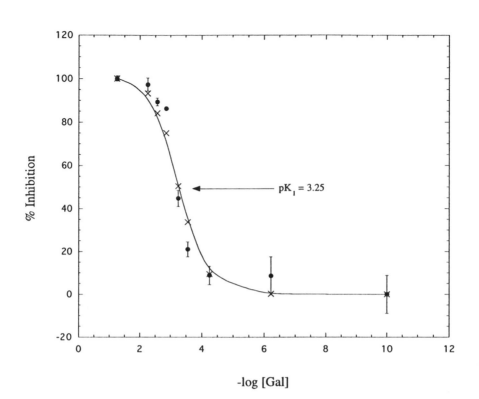

Figure 12 Inhibition of the formation of the lectin complex formed between B-lectin/F-lectin with asialofetuin by D-galactose. The theoretical plot (×) and pK_I of the curve were determined using the equation given in Section II.

ABBREVIATIONS

Ab	antibody(ies)
AGP	human α_1-acid glycoprotein
B-Ab	biotinylated Ab
B-ricin	biotinylated ricin
BSA	bovine serum albumin
DNP	dinitrophenyl
F-Ab	fluoresceinated Ab
FITC	fluorescein isothiocyanate
F-ricin	fluoresceinated-ricin
I_gA	polyclonal rabbit anti-human AGP
ILA	Immuno-Ligand Assay
LAPS	light-addressable potentiometric sensor

NHS   *N*-hydroxysuccinimide
RCA$_{120}$ ricin agglutinin
SA   streptavidin

REFERENCES

1. Merrick, H., and Hawlitschek, G. (1992). *Biotech. Forum Europe* 6, 398–403.
2. Panfili, P. R., Dill, K., and Olson, J. D. (1994). *Curr. Opin. Biotech.* 5, 60–64.
3. McNabb, S., Rupp, R., and Tedesco, J. L. (1989). *Bio/Technology* 17, 343–347.
4. Robinette, R. S. R., and Herbert, W. K. (1993). *J. Immunol. Methods* 159, 229–234.
5. Pcr, S. R., Aversa, C. R., Johnston, P. D., Kopasz, T. L., and Sito, A. F. (1993). *BioPharm.* Nov–Dec., 34–40.
6. Workman, W. E. (1995). *Pharmacop. Forum* 21, 479–484.
7. Eaton, L. C. (1995). *J. Chromatogr. Biomed* 705, 105–114.
8. Briggs, J., and Panfili, P. R. (1991). *Anal. Chem.* 63, 850–859.
9. Dill, K., Lin, M., Poteras, C., Fraser, C., Hafeman, D. G., Owicki, J. C., and Olson, J. D. (1994). *Anal. Biochem.* 217, 128–138.
10. Dill, K., and Olson, J. D. (1995). *Glycoconj. J.* 12, 660–663.
11. Dill, K., and Bearden, D. W. (1996). *Glycoconj. J.* 13, 637–641.
12. Hafeman, D. G., Parce, J. W., and McConnell, H. M. (1988). *Science* 240, 1182–1185.
13. Owicki, J. C., Bousse, L. J., Hafeman, D. G., Kirk, G. L., Olson, J. D., Wada, H. G., and Parce, J. W. (1994). *Ann. Rev. Biophys. Biomol. Struct.* 23, 87–113.
14. Treuheit, M. J., Costello, C. E., and Halsall, H. B. (1992). *Biochem. J.* 283, 105–112.
15. Hermentin, P., Reinhild, W., Doenges, R., Bauer, R., Haupt, H., Patel, T., Parekh, R. B., and Brazel, D. (1992). *Anal. Biochem.* 206, 419–429.
16. Spiro, R. G. (1973). *Adv. Protein Chem.* 27, 349–467.
17. Spiro, R. G., and Bhoyroo, V. D. (1974). *J. Biol. Chem.* 249, 5704–5717.
18. Green, E. D., Adelt, G., Baenziger, J. U., Wilson, S., and Van Halbeek, H. (1988). *J. Biol. Chem.* 263, 18253–18268.
19. Yamamoto, K., Ishida, C., Shinohara, Y., Haegawa, Y., Konami, Y., Osawa, T., and Irimura, T. (1994). *Biochemistry* 33, 8159–8166.
20. Bhattacharyya, L., and Brewer, F. (1988). *Eur. J. Biochem.* 176, 207–212.
21. Shinohara, Y., Kim, F., Shimizu, M., Goto, M., Tosu, M., and Hasegawa, Y. (1994). *Eur. J. Biochem.* 223, 189–194.
22. Dill, K., Song, J. H., Blomdahl, J. A., and Olson, J. O. (1996), submitted to *J. Biochem. Biophys. Methods.*

15

Effects of *O*-Glycosylation on Protein Structure

The Role and Characterization of Heavily O-Glycosylated Mucin Domains

Thomas A. Gerken, Cheryl L. Owens, and Murali Pasumarthy
Case Western Reserve University, Cleveland, Ohio

I. INTRODUCTION

A. *O*-Glycosylation and Protein Structure and Function

Protein *O*-glycosylation is a common post-translational modification of secreted and membrane-associated proteins [1,2]. Many of these glycoproteins have regions in their sequence that contain high percentages ($\geq 30\%$) of Ser and Thr that are heavily *O*-glycosylated with oligosaccharide side chains attached to Ser or Thr via α-*N*-acetylgalactosamine (α-GalNAc). Due to the presence of carbohydrate side chains of 1 to potentially 20 or more residues, these domains can be 50% or more carbohydrate by weight. Little is known of the structural effects of multiple *O*-glycosylation, except that these regions are relatively resistant to proteases and possess extended conformations (Fig.1) [3–5]. These heavily *O*-glycosylated domains are unique structural motifs that can modify the physical properties of proteins and serve as tethers for enzymes and receptors on membrane surfaces. Mucins containing only α-GalNAc side chains have chain dimensions approximately halfway between random coiled peptides and native mucins, whereas fully deglycosylated apomucins behave identical to random coiled peptides [4].

Multiple *O*-glycosylated glycoproteins participate in diverse biological functions. One class of heavily *O*-glycosylated proteins, the mucous glycoproteins, or mucins [7], are responsible for the viscoelastic properties of mucous secretions that play important roles in protecting the internal epithelial surfaces of the digestive, urogenital, and respiratory tracts from physical, chemical, and microbial assault [8,9]. The ease of obtaining large quantities of animal mucins, their high carbohydrate content (50–80%), and

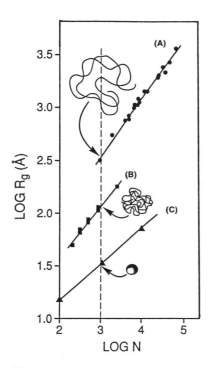

Figure 1 A comparison of the solution size (in terms of radius of gyration, R_g) of heavily O-glycosylated mucins (A), random coiled peptides (B), and globular proteins (C), as a function of the number of peptide residues, N. Relative differences in size are shown schematically adjacent to each curve. For equal-length peptides of 1000 residues, the R_g of a mucin is ~ 370 Å, the R_g of a random coil peptide is ~ 125 Å, and the radius of a spherical globular protein is ~ 33 Å. (Data from Shogren et al. [5]; figure reprinted from Ref. 6.)

the ability to modify their structures has made them ideal model systems for studying the structural roles of multiple O-glycosylation. Mucin-like molecules also line epithelial and endothelial cell surfaces as well as the surface of circulating cells. The cell-surface mucins (for example, Muc1, epiglycanin, epitectin, leukosialin (CD43), macrosialin (CD68), PSGL-1, and CD34 [10–18]) are thought to provide cells with physical and biological protection and may be involved in cellular signaling and modulating cell–cell interactions. The cell-surface O-glycosylated domains of GlyCAM-1, CD34, MAdCAM-1, and PSGL-1 [17,19–21] serve as ligands for selectins, which are involved in the initial attachment stage in the leukocyte adhesion cascade where circulating leukocytes move into tissues at sites of inflammation [22]. The increased expression of the mucin-like glycoproteins, epitectin, and epiglycanin on tumor cell surfaces may also be related to tumorigenicity and metastatic potential, perhaps due to reduced self cell–cell adhesion, increased binding to selectins, and/or increased shielding from the immune system [10,11,23,24].

In another class of cell-surface glycoproteins (e.g., the low density lipoprotein receptor, thrombomodulin, decay accelerating factor, platelet glycoprotein Ib, lactase–phlorizin hydrolase, and sucrase–isomaltose complex [25–30]) mucin-like domains serve as proteolysis-resistant stalks tethering functional globular domains (receptors or enzymes) to the cell surface. These stalks range from ~20 to more than 100 residues and are found in proteins of varied biological functions, from the transport of cholesterol to modulating immune and coagulation responses. Heavily O-glycosylated domains are also found in a small number of nonmucin secreted proteins such as human serum immunoglobulin A1 and D (IgA1, IgD) and κ-casein [31–33], again acting presumably to provide extended structural motifs that may be important for function. Heavily O-glycosylated glycoprotein domains are also common in lower organisms including fungi and slime molds [34,35].

B. Characterization of Heavily O-Linked Glycoproteins

The amino acid composition of the O-glycosylated mucin domains are distinctive: aromatic residues are nearly absent, and 30–40% of the residues are Ser and Thr with another 30–50% of the residues consisting of various combinations of Gly, Ala, and Pro [36]. Nucleotide sequencing of mucins reveals that the mucin O-glycosylated domains are typically composed of multiple tandem repeats (Table 1). Despite their common function, the glycosylated tandem repeats in Table 1 show no sequence homologies among themselves except for the occurrence of clusters of hydroxy amino acids in diad, triad, tetrad, and pentad repeats. These clusters usually occur in numbers higher than predicted statistically from the amino acid composition of the repeat and commonly have neighboring Pro residues. How these varied sequences relate to their properties is currently unknown, although molecular modeling and dynamics studies are beginning to shed some light on this aspect [46,47]. As shown in Table 2, heavily O-glycosylated domains in other nonmucin glycoproteins also possess the same structural features of the glycosylated domains in mucins; they contain similar amino acid profiles, commonly contain tandem repeats, and have a high number of Ser/Thr/Pro clusters. Even the heavily glycosylated domains of IgA1 and IgD are composed of repeats, as are the glycoprotein examples from fungi and slime mold. No strong sequence homologies are observed among these sequences nor with mucins.

Table 1 Selected Mucin Amino Acid Tandem Repeats[a]

Mucin	Amino acid tandem repeat	Source	Reference
PSM	(GAGPGT̂TAŜSVGVT̂ET̂ARPŜVAGŜGT̂TĜTVŜGAŜ GŜTGŜŜGŜPGAT̂GAŜIGQPET̂SRIŜVAGŜSGAPAVŜSGAŜQAAGT̂S)	Porcine salivary	[37]
MUC1	(GŜTAPPAHGVT̂SAPDT̂RPAP)	Human mammary/pancreatic tumor	[12]
MUC2	(PT̂T̂TPIT̂T̂TT̂VT̂PT̂PT̂PT̂GT̂QT̂)	Human intestinal/bronchial	[38]
MUC3	(HŜTPŜFT̂ŜSIT̂T̂TET̂TŜ)	Human small intestinal	[39]
MUC4	(TŜSAŜTGHAT̂PLPVT̂D)	Human tracheobronchial	[40]
MUC5AC	(TT̂ŜTT̂ŜAP), (TT̂VGP/Ŝ)	Human tracheobronchial	[41,42]
MUC5B	(ŜŜTPGT̂AHT̂LT̂VLT̂TT̂AT̂TPT̂AT̂GŜTAT̂P)	Human tracheobronchial	[43]
MUC6	(ŜPFŜŜTGPMT̂AT̂SFQT̂T̂TYPT̂PŜHPQT̂T̂LPT̂HVPPFŜTŜLVT̂ PŜTGT̂VIT̂PT̂HAQMAT̂SAŜIHŜTPT̂GT̂IPPPT̂T̂LKAT̂GŜTHT̂ APPMT̂PT̂ŜGT̂ŜQAHŜSFŜTAKTŜTŜLHŜHTŜSTHHPEVT̂ PT̂ŜTT̂T̂ITPNPT̂ŜTGT̂ŜTPVAHT̂TŜAT̂ŜSRLPT̂PFT̂THŜ PPT̂GŜ)	Human gastric	[44]
MUC7	(TT̂AAPPT̂PŜAT̂TPAPPŜŜAPPE)	Human salivary	[45]

[a]Potential O-glycosylation sites are indicated: Ŝ and T̂.

Table 2 Heavily O-Glycosylated Domains in Selected Membrane and Secreted Glycoproteins[a]

Glycoprotein (Domain)	Sequence	Reference
Human glycophorin (1–43)[b]	-SSTTVAMHTTTSSSVSKSYISSQTNDTHKRDTYAATPRAHEVSEISVRI-	[48,49]
Human LDL receptor (131–172)	-SCLTESESAVTTRGPSTVSSTAVGPKRTASPELTTAESVTMS-	[25]
Human thrombomodulin (516–539)	-SGKVDGGDSGSGEPPSPTPGSTLT-	[27]
Human decay accelerating factor (253–321)	-TSKVPPTVQKPTTVNVPTTEVSPTSQKTTTKTTTPNAQATRSTPVSRTT KHFHETTPNKGSGTTSGTT-	[42]
Human platelet glycoprotein Ib (292–452)	-TRTWKFPTKAHTTPWGLFYSWSTASLDSQMPSSLHPTTQESTKEQT TFPPRNTPNFTLMESITFSKPTKSTTEPTPSPTTSEPVPEPAPNMT TLEPTPSPTTPEPTSEPAPSPTTPEPTIPTIATSPTILVSATS LITPKSTFLTTTKPVSLLESTKKT-$(TTXEPTPXP)_4$	[28]
Rabbit sucrase–isomaltase complex (33–65)	-TKTPAVEEVNPSSSTPTTTSTTTSTSGSVSCPS-	[30]
Human leukosialin (CD43) (1–223)[b,c]	STTAVQTPTSGEPLVSTSEPLSSKMYTTSITSDPKADSTGDDQTSALPPST SINEGSPLWTSIGASTGSPLPEPTTYQEVSIKMSSVPQETPHATSNPAVPIT ANSLGSHTVTGGTITITNSPETSSRTSGAPVTTAASSLETSRGTS GPPLTMATVSLETSKGTSGPPVTNATDSLETGTGTSGPPVTMATDS LETSTGTTGPPVTMTTGSLEPSSGASGPQVSSVKLSTMMSPTTSTNASTVPF-	[14]
Human lamp-1 (163–187)[b,c]	-CEQDRPSPTTAPPAPPSPSPSPVPK-	[50]
Human lamp-2 (159–187)[b,c]	-FLCDKDKTSTVAPTIHTTVPSPTTTPTPK-	[50]
IgA1[b]	-PVPSTPPTPSPSTPPTPSPSC-$(STPPTPSP)_2$ repeat	[31]
IgD[b]	-SPKAQASSVPTAQPQAEGSLAKATTAPATRNTGR-(AXSS/TT) repeat	[32]
Aspergillus glucoamylase (481–515)	-ATGGTTTTATPGSGSVTSTSKTTATASKTSTSTSSTSCTTAV-	[34]
Dictyostelium discoideum (87–124)[b]	-KITATPAPTVTPTVTPTVTPTVTPTPTNPTPSQTSTTTG-$(PTVT)_3$ repeat	[35]

[a]Potential glycosylation sites indicated: T and S.

[b]Experimentally determined glycosylation sites are underlined.

[c]Glycopeptides have not been obtained and sequenced for the entire protein sequences listed; thus, unmarked sites may still be glycosylated.

The extent that the hydroxy amino acid clusters in mucins and other heavily O-glycosylated proteins may be glycosylated is not fully known. Studies on glycopeptides from leukosialin, lamp-1,2, IgA1, IgD, and glycophorin indicate many, but not all, hydroxy amino acid diads, triads, tetrads, and pentads to be glycosylated in vivo (see Table 2 and [51]). Why some sequences are not fully glycosylated is unknown; it may relate to a folded structure of the peptide (i.e., many of these proteins have globular domains), the accessibility of the peptide to the transferase, or even the charge distribution of neighboring residues [52–57]. Failure to detect glycosylation of a site may also reflect that the specific glycosylated glycopeptide may not have been isolated. No true consensus sequences for O-glycosylation have been reported, although several predictive methods have been proposed [52,54,58]. The relationship of acceptor peptide conformation and ability to be O-glycosylated requires further study.

To fully understand the structure, function, and biology of heavily O-linked glycoproteins and their transferases, detailed information regarding the structure and placement of oligosaccharides is desirable. Unfortunately, the heterogeneous nature of the oligosaccharide side chains and their unique physical properties, such as resistance to proteases, make the analytical study of glycoproteins containing these domains very difficult. Adding to this difficulty is the potential for glycosylation pattern heterogeneity (i.e., each Ser or Thr may be partially glycosylated). Thus, attempts to perform classical protein peptide mapping for the characterization of heavily O-glycosylated proteins is fraught with the possibility of obtaining a multitude of partially cleaved glycopeptides with different glycosylation patterns and oligosaccharide structures.

Highly glycosylated residues leave blank cycles in automated Edman sequencing approaches due to the insolubility of the hydrophilic oligosaccharide derivatives in the chlorobutane transfer solvent. Because phosphorylated Ser and Thr residues are also invisible to most automated approaches, the presence of a blank cycle is not necessarily indicative of a glycosylated residue, and the amino acid residue remains undefined. By utilizing trifluoroacetic acid (TFA) as a more polar transfer solvent, Gooley and coworkers [33,35,49,59] have shown that glycosylated amino acids can be efficiently extracted and identified when glycopeptides are covalently bound to a membrane substrate. This improvement is discussed in Chapter 18 of this volume. O-glycosylation sites on isolated glycopeptides can also be determined by collision-induced dissociation tandem mass spectroscopy as shown by Medzihradszky et al. [60,61] and discussed in Chapter 19. These approaches are well suited for determining the structures of homogeneous glycopeptides with relatively homogeneous oligosaccharides. The difficulties associated with cleaving and isolating suitably homogeneous glycopeptides for structural analysis remain.

Another approach for estimating glycosylation patterns is to utilize predictive methods based on amino acid sequence. Unfortunately, unlike N-linked glycosylation, there does not appear to be a specific peptide sequence motif that directs O-glycosylation. As we have discussed, several different algorithms have been proposed [51,54,58] based on known O-glycosylation sites reported in the literature. Because many of the glycoproteins in the database are globular proteins and many contain incomplete glycosylation patterns, the extent that these rules will hold for heavily O-glycosylated proteins is uncertain. As protein O-glycosylation occurs after the folding process, the predictions may more likely reflect the accessibility of surface residues rather than any specific sequence or structure. Studies on the UDP-α-GalNAc:polypeptide-N-acetylgalactosaminyl transferase peptide acceptor specificity show that Thr is more rapidly glycosylated over Ser, that charged residues in the vicinity of the glycosylation site reduce transferase activity, that the presence of Pro, Ser, and Thr tends to enhance glycosylation, and that random coiled acceptor peptides are better substrates [52–54,56]. Elhammer and coworkers discuss their algorithm for predicting mucin type O-glycosylation sites in Chapter 16. An exciting new approach to studying O-glycosylation *in vivo* is presented by Nehrke et al. (Chapter 17).

As a result of the difficulties we have discussed, little is known of the specific glycosylation patterns of heavily O-linked glycoproteins. However, advances in our ability to uniformly modify and selectively deglycosylate heavily O-glycosylated mucins offer one approach by which a variety of methods can be utilized to obtain glycopeptides with homogeneous carbohydrate side chains, from which glycosylation patterns can be readily obtained. These will be discussed in what follows.

C. Conformation and Dynamics of Heavy O-Glycosylated Domains in Mucins

For several years our laboratory has been studying the role of carbohydrate, peptide sequence, and patterns of glycosylation on the properties of mucin glycoproteins. Our work has focused on the physical and spectroscopic properties of native and modified submaxillary gland mucins from sheep and pig (OSM and PSM, respectively), particularly because of their relatively simple carbohydrate side chain structures and because of the ease of obtaining large quantities of glycoprotein. These mucins have been ideal models for perfecting a number of analytical approaches that should have wide applicability to characterizing the glycosylation patterns of heavily O-glycosylated proteins. A review of our techniques and findings follows.

II. RESULTS AND DISCUSSION

A. Measurement of the Radius of Gyration/Molecular Weight Using Light Scattering

These methods allow the determination of the conformational effects of altering mucin carbohydrate side chain length on mucin chain dimensions [4,5]. As shown in Fig. 1, native mucins having two or more carbohydrate residues per oligosaccharide side chain have chain diemsions 3 times as large as random coiled peptides and nearly 10 times as large as globular proteins with similar numbers of amino acids. Mucins with side chains that have been chemically trimmed to monosaccharide α-GalNAc residues have chain dimensions roughly intermediate to those of native mucins and fully deglycosylated apomucin, the chain dimensions of which match those of random coiled peptides [4]. Thus, it appears that maximal mucin chain expansion can be achieved with only two oligosaccharides per side chain on average. These findings suggest that only these residues are sufficiently interacting with the polypeptide core or adjacent oligosaccharides to produce the extended conformation. Rotational isomeric state (RIS) calculations, designed to reproduce the observed mucin chain dimensions, suggested an extended peptide conformation similar to polyproline II ($\phi = -80°$, $\psi = 150°$) for fully and partially deglycosylated mucins [4]. From the analysis it was found that both the glycosylated and nonglycosylated residues of the polypeptide core on average needed to be in this conformation to reproduce the chain dimensions. Additional RIS calculations on PSM [62] suggest that O-glycosylation alters the peptide core conformation over a range of seven consecutive peptide residues.

B. Carbon-13 Nuclear Magnetic Resonance Spectroscopy and Conformational Studies

Having large quantities of mucin available also permits the extensive use of ^{13}C nuclear magnetic resonance (NMR) spectroscopy for quantifying oligosaccharide structure and extent of glycosylation for both native and modified mucins [6,63–65]. Carbon-13 NMR spectroscopy is used in our laboratory extensively to characterize all of our native and modified mucin preparations. One of the major values of ^{13}C NMR is its high chemical shift dispersion and the ability to obtain excellent spectra from very viscous mucin solutions. Representative ^{13}C NMR spectra of native and sequentially deglycosylated OSM, PSM, and human tracheobronchial mucin (HTBM) demonstrating the quantitative removal of carbohydrate are given in Fig. 2.

Note the increase in complexity of the native mucin spectra from OSM

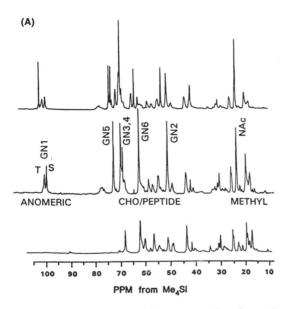

Figure 2 Carbon-13 NMR spectra of native and sequentially deglycosylated mucins showing native mucin oligosaccharide side chain complexity and its simplification after partial deglycosylation. (A) Ovine submaxillary gland mucin (OSM); (B) porcine submaxillary gland mucin (PSM); (C) human tracheobronchial mucin (HTBM). The top spectrum in each panel represents the native mucin; the center spectrum, mucin after partial deglycosylation by treatment with trifluoromethanesulfonic acid (TFMSA) at 0°C, leaving only α-GalNAc side chains (note that HTBM was treated with neuraminidase and twice with TFMSA (see [63]). In (A) and (B), GalNAc carbons are labeled GN, whereas glycosylated Ser and Thr α and β carbons are labeled SOG and TOG, respectively. The bottom spectrum in each panel represents mucin fully deglycosylated by the oxidation–elimination procedure [63] (after 0°C, TFMSA treatments for PSM and HTBM). Peptide carbons are identified in (B). Detailed methods for the TFMSA and oxidation–elimination procedures are described in [63]. Spectra were obtained at 67.9 MHz from 10-mm sample tubes containing 25–150 mg mucin. Spectral accumulation times ranged from ca. 5 h to ca. 2 days.

through HTBM due to longer oligosaccharide side chains and increased structural heterogeneity. OSM is ~56%, PSM ~67%, and HTBM ~85% carbohydrate by weight [63]. The side chains of OSM are the disaccharide α-NeuNAc(2-6) α-GalNAc-*O*-Ser/Thr, whereas those of PSM range from α-GalNAc to the pentasaccharide α-Fuc (1-2)[α-GalNAc(1-3)] β-Gal (2-3) [α-NeuNAc(2-6)] α-GalNAc-*O*-Ser/Thr [64,65]. HTBM oligosaccharide side chains consist of a vast array of structures; see, for example, Refs. 66

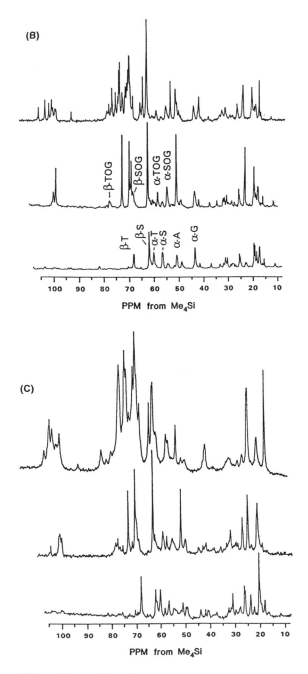

Figure 2   continued

and 67. For OSM and PSM, the oligosaccharide side chain distribution and average side chain length can be readily quantitated from the ^{13}C NMR anomeric carbon resonances (90–100 ppm), as can the extent of glycosylation from the α-carbon resonances of Ser and Thr (53–60 ppm) as described earlier [64,65].

After partial deglycosylation by mild trifluoromethanesulfonic acid (TFMSA) [63] (center spectrum of each panel in Fig. 2) the spectra become similar and much simpler due to the presence of single α-GalNAc side chains (resonances labeled). Note that the Ser/Thr glycosylation differs for each mucin as shown by the α-GalNAc anomeric carbon resonances (which are sensitive to Ser or Thr substitution), and that this represents the Ser/Thr content of the fully deglycosylated mucins (labeled in Fig. 2B, bottom spectrum).

Carbon-13 NMR chemical shift and relaxation studies have also been utilized to monitor mucin conformation and dynamics as a function of carbohydrate content, corroborating the results of our light scattering studies [68]. Subtle changes in chemical shifts of the glycosylated residues confirm that glycosylation causes conformational changes at Ser and Thr. Glycosylation also decreases the peptide core mobility, as shown by changes in the backbone ^{13}C-NT$_1$ (normalized spin-lattice relaxation time) and NOE (nuclear Overhauser enhancement) values. The dynamics of the Ser and Thr α-carbons are furthermore sensitive to carbohydrate side chain length (0 vs. 1 vs. 2 residues) with the Thr residue most sensitive to glycosylation, due to added steric interactions of its methyl group. The finding that glycosylation also affects the mobility of the Gly residue further confirms the longer range effects of O-glycosylation. Carbon-13 NMR relaxation measurements can also monitor oligosaccharide side chain dynamics, revealing differential flexibilities among various glycosidic linkages in OSM and PSM [6,64]. Work is in progress to perform similar studies on a ^{13}C-C-1-glucosamine-labeled human goblet cell mucin containing branched hexasaccharide side chains. Preliminary results suggest that the peptide-linked α-GalNAc residue has considerably reduced mobility compared with those in native OSM and PSM. By sequentially deglycosylating this mucin and monitoring its side chain dynamics by NMR, the origin(s) of the reduced dynamics of the α-GAlNAc residue is expected to be found.

To further explore the origins of the conformational changes induced by mucin O-glycosylation, molecular modeling and molecular dynamics calculations have been performed on a number of short O-linked model glycopeptides containing clusters of Thr. The results of this work suggest that the clustering of hydroxy amino acids serves to stiffen the peptide core from both carbohydrate–carbohydrate and peptide–carbohydrate interactions [46,47], with the possibility that specific hydrogen bonds may stabilize these interactions [46]. Molecular dynamics studies also indicate Pro residues and O-glycosylated Thr residues have similar chain stiffening behaviors, agreeing with the observed polyproline II mucin conformation [47]. Preliminary studies also suggest the pattern of glycosylation may affect glycoprotein conformation and dynamics (unpublished).

C. Isolation of PSM Glycopeptides and Their Characterization

To explore in greater detail the conformational effects of multiple O-glycosylation, we have undertaken the isolation of smaller O-linked glycopeptides from PSM for detailed study by proton NMR. PSM was chosen because it is readily available in gram quantities and its amino acid sequence [37] and carbohydrate structures are known (see Fig. 2 legend and section II. B). This allows the development of a well-directed strategy (Fig. 3) and provides reference sequence criteria for the final isolated glycopeptides. As a result of this work, we have isolated many pure glycopeptides and have determined the glycosylation pattern of nearly the entire 81-residue PSM tandem repeat (data to be reported).

Key aspects of the approach are (1) reducing the O-linked oligosaccharide side chain heterogeneity by mild trifluoromethanesulfonic acid (TFMSA) treatment [63], (2) the subsequent cleavage of the trimmed glycoprotein by specific proteases into smaller glycopeptides, (3) the separation of glycopeptides by reverse phase high performance liquid chromatography (HPLC), and (4) the use of standard amino acid sequencing approaches to detect the α-GalNAc containing Ser and Thr residues.

The first step of the procedure consists of dissociating the PSM molecule into its subunits by reduction and carboxymethylation, followed by extensive trypsinolysis to degrade the nonglycosylated N- and C-terminal regions of the mucin (Fig. 3A,B). Heavily O-glycosylated tandem repeat domains are resistant to trypsin and appear at the void volume of a sephacryl S200 column (Pharmacia Biotech, Uppsala, Sweden) (data not shown). In Fig. 3B,C the carbohydrate side chains are quantitatively trimmed to the peptide α-GalNAc residue using TFMSA/anisole at 0°C as described [63]. At 0°C and for short times (<6 h), the peptide-linked α-GalNAc residues are resistant to TFMSA removal (see Fig. 2 [63]). Thus, oligosaccharide side chain heterogeneity is eliminated while maintaining the original glycosylation pattern. The larger partially deglycosylated domains, containing multiples of the PSM 81-residue tandem repeat, are separated from the resulting lower molecular weight nonglycosylated peptides by S200 gel filtration (Fig. 3B,C). Upon a second digestion with trypsin, the glycosylated domains decrease in size on S200 chromatography (Fig. 3D), giving the

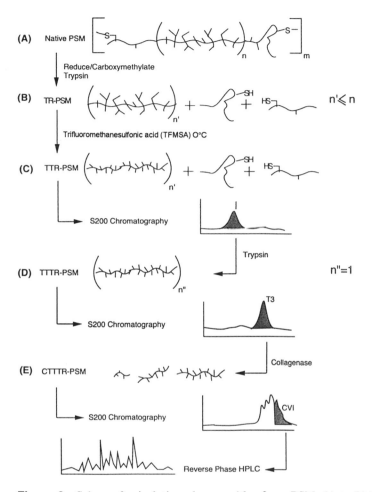

Figure 3 Scheme for isolating glycopeptides from PSM. (A to B) Native PSM is reduced and carboxymethylated in 5 M guanidine hydrochloride and digested for 24 h with 3% trypsin as previously described [69]. (B to C) Mucin is partially deglycosylated with anhydrous TFMSA/anisole (10/3 v/v) at 0°C for 6 h and chromatographed on a sephacryl S200 column [63]. (C to D) Glycosylated domains, I (shaded), are digested with 1% trypsin and rechromatographed on an S200 column, yielding tandem repeat T3. (D to E) Tandem repeat is digested for 24 h with 8% collagenase from *Achromobacter iophagus*, (Boehringer Mannheim, Indianapolis, IN) yielding smaller glycopeptides. Shaded region, CVl, on S200 was subsequently chromatographed by high performance liquid chromatography to obtain glycopeptides as shown in Fig. 4.

glycosylated monomeric PSM tandem repeat. Thus, the single active trypsin cleavage site in the PSM repeat . . . PTSR–|–ISVAG . . . that has neighboring potential glycosylation sites is resistant to trypsin in native mucin but is fully cleavable in the partially deglycosylated mucin.

Digestion of the isolated tandem repeat with collagenase from *Achromobacter iophagus* results in further cleavage, giving a complex pattern of lower molecular weight peaks on S200 (Fig. 3E). This enzyme reportedly cleaves at X–|–Gly-Ala and X–|–Gly-Pro sequences, which are common to heavily *O*-linked domains. However, as shown by the obtained glycopeptides (to be discussed), additional cleavage sites at the N terminal of Ser and Ala are observed for this enzyme. The isolation of individual glycopeptides from the lowest molecular weight peak was performed using preparative reverse phase HPLC on a C_8 column (Fig. 4A) followed by repeated purifications by analytical HPLC on a C_{18} column as described in Fig. 4B,C. For all separations, water/acetonitrile gradients in the presence of 0.1% TFA were utilized. Reproducibility was found to be generally very good. Upon rechromatography of individual fractions, using shallower acetonitrile gradients, many relatively pure glycopeptides have been obtained, as shown, for example, in Fig. 4B,C.

Glycopeptide amino acid sequences were obtained by automated amino acid sequencing, as will be described in Section II. B. For a number of glycopeptides isolated in greater than ∼0.1 mg quantities, their sequences and glycosylation patterns have been confirmed on the basis of 600-MHz TOCSY (total correlation spectroscopy) and NOESY (nuclear Overhauser effect spectroscopy) two-dimensional NMR spectra obtained from 90% H_2O/10% D_2O (see [70]). Thus, high-field NMR approaches will be useful for determining the primary sequence and glycosylation patterns of short *O*-linked glycopeptides (date to be reported).

By focusing on a more hydrophilic region of the HPLC chromatogram (regions 5 and 6 in Fig. 4A), which should contain the more heavily glycosylated and more hydrophilic peptides, we have identified and collected 31 different glycopeptide peaks. Of these, 14 were judged homogeneous enough to be submitted to amino acid sequencing. To date, we have obtained the following glycopeptide sequences: GAṪGAṠIGQPEṪSR, ṠIGQPEṪSR, GAṪGA, SIGQPEṠR, GṠPGAṪGA, GṠTGṠṠṠGṠP#, GAGPGṪṪAṠṠ#, AAGṪṠGAGPGPGṪṪA, GSGṪṪGṪVSGAṠ, and GSGṪṪGṪVṠGAS (where ṡ and ṫ are glycosylated residues, and # indicates a mixture of glycopeptides isolated with variable extents of glycosylation). From the last two glycopeptides, it is clear that glycopeptides with identical sequences and the same number of carbohydrate side chains, but with different sites of glycosylation, can be potentially resolved by reverse phase HPLC. The glycopeptides obtained thus far make up over one-half of the PSM tandem repeat, underlined in the following:

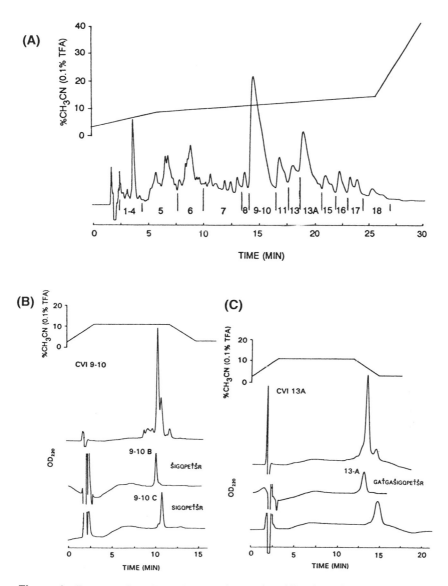

Figure 4   Reverse phase HPLC separation and purification of PSM glycopeptides. (A) Reverse phase chromatography of glycopeptide fraction CVl (Fig. 3E) on preparative 5-μ Spherisorb C$_8$ column (Altech Associates Inc., Deerfield, IL) Column dimensions: 30 × 1 cm. (B,C) Rechromatography of fractions 9–10 and 13A on analytical 3-μ Spherisorb ODS$_2$ reverse phase column (Altech Associates Inc.) yielding purified glycopeptides. Column dimensions: 15 × 0.46 cm. For all separations, water/acetonitrile gradients in the presence of 0.1% TFA were utilized as indicated.

I₁SVAGSSGAPAVSSGASQAAGTSGAGPGTTASSVGVTETARP
SVAGSGTTGTVSGASGSTGSSSGSPGATGASIGQPETSR₈₁

In spite of our ability to obtain and sequence glycopeptides, it remains nearly an impossible task to obtain the detailed glycosylation pattern of the PSM tandem repeat by this approach, sheerly due to the vast number of glycopeptides that would have to be isolated, quantitated, and sequenced from the chromatogram in Fig. 4A. One can readily obtain information on whether a given site is capable of glycosylation from a single isolated glycopeptide, but the relative extent that any given site is glycosylated remains difficult to access by this approach.

After sequencing several glycopeptides the structures of which were confirmed by NMR, it became apparent that we could reasonably quantitate the presence of glycosylated residues (see Section II. E). Thus, the monomeric tandem repeat isolated in step D of Fig. 3 was submitted to sequence analysis. Recall that at this stage the repeat is homogeneous in carbohydrate but heterogeneous regarding the extent that each Ser or Thr may be glycosylated. Therefore, by reverse phase HPLC, the repeat appears as a broad peak (data to be reported). Sequencing of the tandem repeat successfully reveals the extent of glycosylation of each Ser and Thr residue to approximately residue 65. In general, we have found that Thr residues are more highly glycosylated than the Ser residues, with Thr residues nearly fully glycosylated and Ser residues being 25–90% glycosylated. Interestingly, the more poorly glycosylated Ser residues have Ile, Val, Gly, or Ala as N- and C-terminal neighbors. We are currently in the process of repeating the sequencing of the 81-residue repeat to confirm the glycosylation pattern. We have isolated and partially sequenced the C-terminal peptide (residues 38–78) obtained by digesting the tandem repeat with endoprotease GluC, which cleaves the C terminal of Glu and Asp residues (data to be reported). On the basis of this work, the glycosylation patterns of the PSM 81-residue tandem repeat will soon be completely known. With this data at hand, we can begin to compare conformations of sequences that are highly glycosylated with those that are sparsely glycosylated to further determine the acceptor peptide conformational requirements for the α-GalNAc:polypeptide-N-acetylgalactosaminyl transferase.

D. Amino Acid Sequencing of TFMSA-Treated Glycopeptides

Glycopeptides were sequenced on both Applied Biosystms 477A and Applied Biosystems Procise 494 amino acid sequencers (Perkin Elmer, Applied Biosystems Div., Foster City, CA) using standard manufacturer recommended conditions. Unique phenylthiohydantoin (PTH)–amino acid elution patterns are observed for α-GalNAc-O-Thr–PTH and α-GalNAc-O-Ser–PTH derivatives as shown in Fig. 5A,B for cycles 3 and 6 for the glycopeptide GAṪGAṠIGQPEṪṠR, which by NMR is shown to be fully glycosylated. For comparison, an analysis of standard PTH–amino acids is shown in Fig. 5C. As reported earlier [71], α-GalNAc-O-Thr–PTH appears as two peaks: one co-eluting near the position of PTH–Ser (peak T*) and the other eluting between PTH–Thr and PTH–Gly (peak T**). Similarly, we observe that α-GalNAc-O-Ser–PTH appears as two closely spaced peaks (which commonly are not resolved on the Applied Biosystems 477A) that elute between Asp–PTH and Asn–PTH (peaks S* and S**). This doublet nature of the α-GalNAc-Thr/Ser–PTH derivatives arises from the creation of two differently migrating diastereotopic derivatives upon PTH derivation, and it is not due to degradation of the α-GalNAc-O-Thr/Ser as suggested earlier [71]. Thus, the area ratio of the two species is found to remain constant on both sequencers, with T* : T** ratios of ~ 45 : 55 and S* : S** ratios of ~ 1 : 1, based on the sequencing of several purified glycopeptides. Furthermore, the PTH derivatives of authentic α-GalNAc-O-Ser and α-GalNAc-O-Thr (a kind gift of R. Koganty, Biomira Inc., Edmonton, Alberta, Canada) gives identical elution profiles as those in Fig. 5A,B. Relative quantification of the recovery of the PTH derivatives for the sequencing of GAṪGAṠIGQPEṪṠR gives response/recovery factors for α-GalNAc-O-Ser and α-GalNAc-O-Thr of 60–70% of that observed for Ala in this glycopeptide (uncorrected for lag in subsequent cycles).

To determine the relative percent of glycosylation of glycopeptides with heterogeneous glycosylation patterns such as the PSM tandem repeat, the percent of glycosylation can be obtained from the quantitation of the unique PTH peaks in the chromatogram: S and S* + S**, and T and T* + T**. The increased cycle preview and lag, present when sequencing long glycopeptides, results in the presence of PTH–Ser and T* in the same cycle, which overlap. In this situation, T* (and, by difference, PTH–Ser) is quantitated by using the area of the resolved T** peak and the aforementioned T* : T** area ratio. On the Applied Biosystems 477A, Thr–PTH and T** also often co-elute. More elaborate approaches can be utilized to estimate site-specific glycosylation involving the quantitation of free Ser using the dehydro-Ser-PTH peak and the estimation of previous cycle lag. Modification of the HPLC gradients may further help resolve these peaks as needed. Finally, the glycopeptide α-GalNAc-O-Ser/Thr residues are relatively stable with respect to the Edman sequencing and do not appear to be lost even after more than 65 sequencing cycles.

The use of mild TFMSA and subsequent amino acid sequencing of large α-GalNAc-containing glycopeptides to determine glycosylation pat-

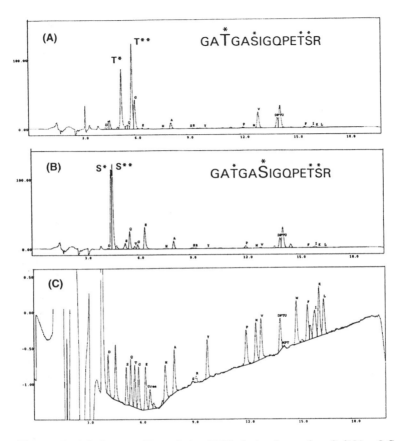

Figure 5 Elution profiles of the PTH derivatives of α-GalNAc-*O*-Ser and α-GalNAc-*O*-Thr on the Applied Biosystems Procise 494 Protein Sequencer. (A,B) Chromatograms of cycles 3 and 6 from the sequencing of GATGASIGQPETSR (ca. 3000 pmol) showing the PTH derivatives of α-GalNAc-*O*-Ser and α-GalNAc-*O*-Thr, respectively. The α-GalNAc-*O*-Ser–PTH derivatives S* and S** elute at 4.51 and 4.58 min, respectively, whereas the α-GalNAc-*O*-Thr–PTH derivatives, T* and T**, elute at 5.24 and 5.84 min, respectively. The elution profiles of the PTH–amino acid standards (5 pmol each) are shown in (C). Retention times for selected PTH derivatives are as follows (in min): Asp(D), 4.22 Asn(N), 4.46; Ser(S), 5.25; Gln(Q), 5.50; Thr(T), 5.73; Gly(G), 5.95; and Glu(E), 6.35. Conditions used for sequencing GATGASIGQPETSR: 15 μL glycopeptide solution (~ 3000 pmol) was dried on a TFA-prewashed glass fiber filter (ABI 401111) spotted with 1.5 mg BioBrene Plus (ABI 400385). Standard pulsed-liquid sequencing cycles were utilized, and PTH derivatives were chromatographed using the unmodified Fast Normal 1 gradient program. Absorbances at OD$_{269}$ are shown.

terns is relatively straightforward compared with alternative approaches to obtain such information. Although information on the individual oligosaccharide side chain structures is lost by this approach, a modification of the approach should be capable of determining the substitution status of the α-GalNAc C-3 hydroxyl (work in progress). This would take advantage of the sensitivity of C-3 unsubstituted α-GalNAc residues to elimination, after periodate treatment and mild alkalinization (see Fig. 2 [63]). By comparison, the quantitation of glycosylation patterns of unmodified glycopeptides with long and heterogeneous oligosaccharide side chains is a considerably more difficult task.

In conclusion, the use of TFMSA to trim O-linked oligosaccharides to single α-GalNAc residues results in glycoproteins that are more susceptible to the action of specific proteases, and produces glycopeptides that can be easily isolated by reverse phase HPLC and that can be readily sequenced using standard sequencing protocols. This approach should be applicable to the characterization of a wide variety of O-linked glycoproteins.

IV. SUMMARY

Many O-linked glycoproteins have regions in their sequence containing high percentages of Ser and Thr that are heavily O-glycosylated with heterogeneous oligosaccharide side chains. These heavily glycosylated domains are resistant to proteases, possess extended conformations, and are present in glycoproteins with diverse biological functions. Detailed knowledge of the role, structure, and glycosylation pattern of these domains is hampered by the difficulty in isolating and characterizing such heterogeneous systems. Light scattering and ^{13}C NMR approaches, however, have yielded important information on the size, conformation, and structure of native and modified mucin glycoproteins. As a result of this work, chemical methods for uniformly modifying the oligosaccharide side chains of mucins have been developed, which now permit the isolation of glycoproteins with homogeneous monosaccharide α-N-acetylgalactosamine (α-GalNAc) side chains. These glycoprotein derivatives are more susceptible to proteases, allowing the isolation of small to moderate size glycopeptides, the glycosylation patterns of which can be readily determined by standard Edman amino acid sequencing methodologies.

ACKNOWLEDGMENTS

This work is supported by NIH grant RO1-DK 39918. Sequencing was performed by the Molecular Biology Core Laboratories of Case Western Reserve University, partially funded by NIH grant P30 CA43703.

ABBREVIATIONS

α-GalNAc	α-N-acetylgalactosamine
CD43	leukosialin
CD68	macrosialin
HPLC	high performance liquid chromatography
HTBM	human tracheobronchial mucin
IgA1	human serum immunoglobulin A1
IgD	human serum immunoglobulin D
NMR	nuclear magnetic resonance
NOE	nuclear Overhauser enhancement
NOESY	2D nuclear Overhauser effect spectroscopy
NT_1	normalized spin-lattice relaxation time
OSM	ovine submaxillary gland mucin
PSM	porcine submaxillary gland mucin
PTH	phenylthiohydantoin
RIS	rotational isomeric state
TFA	trifluoroacetic acid
TFMSA	trifluoromethanesulfonic acid
TOCSY	2D total correlated spectroscopy

REFERENCES

1. Carraway, K., and Hull, S. (1989). *Bioessays* 10, 117–121.
2. Kornfeld, R., and Kornfeld, S. (1980). In Lennarz, W., ed., *The Biochemistry of Glycoproteins and Proteoglycans.* Plenum Press, New York , pp 1–34.
3. Jentoft, N. (1990). *TIBS* 15, 291–296.
4. Shogren, R., Gerken, T., and Jentoft, N. (1989). *Biochemistry* 28, 5525–5536.
5. Shogren, R., Jentoft, N., Gerken, T., Jamieson, A., and Blackwell, J. (1987). *Carbohydr. Res.* 160, 317–327.
6. Gerken, T. A. (1993). *Crit. Rev. Oral Biol. Med.* 4, 261–270.
7. Strous, G., and Dekker, J. (1992). *Crit. Rev. Biochem. Mol. Biol.* 27, 57–92.
8. Kaliner, M. (1991). *Am. Rev. Respir. Dis.* 144, s52–s56.
9. Neutra, M., and Forstner, J. (1987). In Johnson, L., ed., *Physiology of the Gastrointestinal Tract,* 2nd ed. Raven Press, New York, pp. 975–1009.
10. Izumi, Y., Taniuchi, Y., Tsuji, T., Wayne Smith, C., Nakamori, S., Fidler, I., and Irimura, T. (1995). *Exp. Cell. Res.* 216, 215–221.
11. Bhavanandan, V. (1988). *Ind. J. Biochem. Biophys.* 25, 36–42.
12. Gendler, S., Lancaster, C., Taylor-Papadimitriou, J., Duhig, T., Peat, N., Burchell, J., Pemberton, L., Lalani, E., and Wilson, D. (1990). *J. Biol. Chem.* 265, 15286–15293.
13. Slayter, H., and Codington, J. (1973). *J. Biol. Chem.* 248, 3410–3410.
14. Shelly, C., Remold-O'Donnell, E., Davis, A., III, Bruns, A., Rosen, F., Carroll, M., and Whitehead, A. (1989). *Immunology* 86, 2819–2823.

15. Baeckstrom, D., Zhang, K., Asker, N., Ruetschi, U., Ek, M., and Hansson, G. (1995). *J. Biol. Chem.* 270, 13688–13692.
16. Holness, C., Da Silva, R., Fawcett, J., Gordon, S., and Simmons, D. (1993). *J. Biol. Chem.* 268, 9661–9666.
17. Maury, J., Nicoletti, C., Guzzo-Chambraud, L., and Maroux, S. (1994). *Eur. J. Biochem.* 228, 323–331.
18. Shimizu, Y., and Shaw, S. (1993). *Nature (London)* 366, 630–631.
19. Baumhueter, S., Singer, M., Henzel, W., Hemmerich, S., Renz, M., Rosen, S., and Lasky, L. (1993). *Science* 262, 436–438.
20. Norgard, K., Moore, K., Diaz, S., Stults, N., Ushiyama, S., McEver, R., Cummings, R., and Varki, A. (1993). *J. Biol. Chem.* 268, 12764–12774.
21. Lasky, L., Singer, M., Dowbenko, D., Imai, Y., Henzel, W., Grimley, C., Fennie, C., Gillett, N., Watson, S., and Rosen, S. (1992). *Cell* 69, 927–938.
22. Lasky, L. (1992). *Science* 258, 964–969.
23. Hilkens, J. (1988). *Cancer Rev.* 11/12, 25–54.
24. Dilulio, N., Yamakami, K., Washington, S., and Bhavanandan, V. (1994). *Glycosylat. Dis.* 1, 21–30.
25. Russell, D., Schneider, W., Yamamoto, T., Luskey, K. Brown, M., and Goldstein, J. (1984). *Cell* 37, 577–585.
26. Wen, D., Dittman, W., Ye, R., Deaven, L., Majerus, P., and Sadler, J. (1987). *Biochemistry* 26, 4350–4357.
27. Medof, M., Lublin, D., Holers, M., Ayers, D., Getty, R., Leykam, J., Atkinson, J., and Tykocinski, M. (1987). *Proc. Natl. Acad. Sci. USA* 84, 2007–2011.
28. Lopez, J., Chung, D., Fujikawa, K., Hagen, F., Papayannopoulou, T., and Roth, G. (1987). *Proc. Natl. Acad. Sci. USA* 84, 5615–5619.
29. Naim, H., and Lentze, M. (1992). *J. Biol. Chem.* 267, 25494–25504.
30. Cowell, G., Tranum-Jensen, J., Sjostrom, H., and Norgen, O. (1986). *Biochem. J.* 237, 455–461.
31. Baenziger, J., and Kornfeld, S. (1974). *J. Biol. Chem.* 249, 7270–7281.
32. Takayasu, T., Suzuki, S., Kametani, F., Takahashi, N., Shinoda, T., Okuyama, T., and Munekata, E. (1982). *Biochem. Biophys. Res. Commun.* 105, 1066–1071.
33. Pisano, A., Packer, N., Redmond, N., Williams, K., and Gooley, A. (1994). *Glycobiology* 4, 837–844.
34. Williamson, G., Belshaw, N., and Williamson, M. (1992). *Biochem. J.* 282, 423–428.
35. Gooley, A., Marshchalek, R., and Williams, K. (1992). *Genetics* 130, 749–756.
36. Gendler, S., and Spicer, A. (1995). *Ann. Rev. Physiol.* 57, 607–634.
37. Eckhardt, A., Timpte, C., Abernethy, J., Zhao, Y., and Hill, R. (1991). *J. Biol. Chem.* 266, 9678–9686.
38. Gum, J., Byrd, J., Hicks, J., Toribara, N., Lamport, D., and Kim, Y. S. (1989). *J. Biol. Chem.* 264, 6480–6487.
39. Gum, J., Hicks, J., Swallow, D., Lagrace., Byrd, J., Lamport, D., Siddiki, B., and Kim, Y. (1990). *Biochem. Biophys. Res. Commun.* 171, 407–415.

40. Porchet, N., Van Cong, N., Dufosse, J., Audie, J. Guyonnet-Duperat, V., Gross, M., Denis, C., Degand, P., Bernheim, A., and Aubert, J. (1991). *Biochem. Biophys. Res. Commun.* 175, 414–422.

41. Duperat, G., Audie, J., Debailleul, V., Laine, A., Buisine, M., Galiegue-Zouitina, S., Pigny, P., Degand, P., Aubert, J., and Porchet, N. (1995). *Biochem. J.* 305, 211–219.

42. Meerzaman, D., Charles, P., Daskal, E., Polymeropoulos, M., Martin, B., and Rose, M. (1994). *J. Biol. Chem.* 269, 12932–12939.

43. Aubert, J., Porchet, N., Crepin, M., Duterque-Coquillaud, M., Vergnes, G., Mazzuca, M., Debuire, B., Petitprez, D., and Degand, P. (1991). *Am. J. Respir. Cell Mol. Biol.* 5, 178–185.

44. Toribara, N., Robertson, A., Ho, S., Juo, W., Gum, E., Hicks, J., Gum, J., Byrd, J., Siddiki, B., and Kim, Y. (1993). *J. Biol. Chem.* 268, 5879–5885.

45. Bobek, L., Tsai, H., Biesbrock, A., and Levine, M. (1993). *J. Biol. Chem.* 268, 20563–20569.

46. Butenhof, K., and Gerken, T. A. (1993). *Biochemistry* 32, 2650–2663.

47. Gerken, T. A. (1995). *Glycoconj. J.* 12, 393.

48. Tomita, M., and Marchesi, V. (1975). *Proc. Natl. Acad. Sci. USA* 72, 2964–2968.

49. Pisano, A., Redmond, J., Williams, K., Gooley, A. (1993). *Glycobiology* 3, 429–435.

50. Carlsson, S., Lycksell, P., and Fukuda, M. (1993). *Arch. Biochem. Biophys.* 304, 65–73.

51. Hansen, J., Lund, O., Engelbreacht, J., Bohr, H., Nielson, J., Hansen, J., and Brunak, S. (1995). *Biochem. J.* 308, 801–813.

52. Wang, Y., Agrwal, N., Eckhardt, A., Stevens, R., and Hill, R., (1993). *J. Biol. Chem.* 268, 22979–22983.

53. Nishimori, I., Johnson, N., Sanderson, S., Perini, F., Mountjoy, K., Cerny, R., Gross, M., and Hollingsworth, M. (1994). *J. Biol. Chem.* 269, 16123–16130.

54. Elhammer, A., Poorman, R., Brown, E., Maggiora, L., Hoogerheide, J., and Kezdy, F. (1993). *J. Biol. Chem.* 268, 10029–10038.

55. Wilson, I., Gavel, Y., and Heijne, G. (1991). *Biochem. J.* 275, 529–534.

56. O'Connell, B., Hagen, F., and Tabak, L. (1992). *J. Biol. Chem.* 267, 25010–25018.

57. O'Connell, B., Tabak, L. A., and Ramasubbu, N. (1991). *Biochem. Biophys. Res. Commun.* 180, 1024–1030.

58. Chou, K. (1995). *Protein Sci.* 4, 1365–1383.

59. Gooley, A. A., Classon, B. J., Marschalek, R., and Williams, K. L. (1991). *Biochem. Biophys. Res. Commun.* 178,1194–1201.

60. Medzihradszky, K. F., Gillece-Castro, B. L., Settineri, C. A., Townsend, R. R., Masiarz, F. R., and Burlingame, A. L. (1990). *Biomend. Environ. Mass Spectrom.* 19, 777–781.

61. Medzihradszky, K.F., Gillece-Castro, B. L., Townsend, R. R., Burlingame, A. L., and Hardy, M. R. (1996). *J. Am. Soc. Mass Spectrom.* 7, 319–328.

62. Tanpipat, N., and Mattice, W. (1990). *Biopolymers* 29, 377–383.
63. Gerken, T., Gupta, R., and Jentoft, N. (1992). *Biochemsitry* 31, 639–648.
64. Gerken, T., and Jentoft, N. (1987). *Biochemistry* 26, 4689–4699.
65. Gerken, T., and Dearborn, D. (1984). *Biochemistry* 23, 1485–1497.
66. Klein, A., Carnoy, C., Lamblin, G., Roussel, P., van Kuik, A. A., and Vliegenthart, J. F. G. (1993). *Eur. J. Biochem.* 211, 491–500.
67. Lhermitte, M., Rahmoune, H., Lamblin, G., Roussel, P., Strang, A., and Van Halbeek, H. (1991). *Glycobiology* 1, 277–293.
68. Gerken, T., Butenhof, K., and Shogren, R. (1989). *Biochemistry* 28, 5536–5543.
69. Gupta, R., and Jentoft, N. (1989). *Biochemistry* 28, 6114–6121.
70. Clore, G. M., and Gronenborn, A. M. (1994). *Meth. Enzymol.* 239, 349–363.
71. Abernethy, J. L., Wang, Y., Eckhardt, A. E., and Hill, R. L. (1992). In *Techniques in Protein Chemistry III.* R. H. Angeletti, ed. Academic Press, New York, pp. 277–286.

16

Predicting Mucin-Type *O*-Glycosylation Sites

Åke P. Elhammer, Roger A. Poorman, and Ferenc J. Kézdy
Pharmacia & Upjohn Inc., Kalamazoo, Michigan

I. INTRODUCTION

Mucin-type *O*-linked oligosaccharides have been identified in a large number of glycoproteins of diverse origin and function [1,2]. The complexity of these structures ranges from that of a single monosaccharide—*N*-acetylgalactosamine (GalNAc)—to that of large, branched structures composed of more than 20 saccharide units [2–4]. Likewise, the number of glycosylated amino acid residues in any given glycoprotein ranges from one, for example, in the human transferrin receptor [5,6], to the several hundred commonly found in many mucins [3,4,7]. The assembly of mucin-type oligosaccharides occurs by sequential addition of monosaccharide units to the growing oligosaccharide chain, a process invariably initiated by the transfer of an *N*-acetylgalactosamine unit to a serine or threonine residue on the acceptor glycoprotein [8]. Hence, the exact location of mucin-type oligosaccharides on the acceptor polypeptide is largely determined by the specificity of the enzyme catalyzing the transfer of this monosaccharide, namely UDP-GalNAc:polypeptide *N*-acetylgalactosaminyl transferase (E.C. 2.4.1.41), commonly called GalNAc-transferase.

A. Enzymes Catalyzing the Transfer of *N*-Acetylgalactosamine to Acceptor Peptides and Proteins

GalNAc-transferase activities, capable of catalyzing the in vitro transfer of *N*-acetylgalactosamine to acceptor proteins and peptides, have been demonstrated in several tissues and animal species [9–22] and enzyme proteins have been purified to homogeneity from Ascites hepatoma cells, bovine colostrum, and porcine submaxillary glands [12,14,16]. The purified bovine and porcine molecules are generally believed to represent species variations of the same enzyme protein, and cDNA clones containing bovine, human, porcine, and rat sequences encoding this enzyme have been described [23–27]. In vitro studies, using enzymes purified from different tissues as well as recombinant forms of the enzyme, have demonstrated that GalNAc-transferase is a multisubstrate enzyme capable of glycosylating a wide range of acceptor sequences, although with varying efficiencies [10, 15,17,20,21,28–32]. Northern blots showed the presence of this enzyme in a variety of human tissues, including heart, brain, placenta, lung, liver, skeletal muscle, and pancreas [23]. Furthermore, immunolocalization studies have shown that GalNAc-transferase is present at high levels in mucin-producing tissues such as intestinal mucosa, submaxillary gland, and mammary gland (T. Raub and Å. Elhammer, unpublished observations). Taken together, these observations all point to the important role of GalNAc-transferase in the "housekeeping" synthesis of mucin-type O-linked glyco-conjugates in many cell types.

The recent cloning of a second type of GalNAc-transferase suggests that the enzyme discussed above, recently renamed GalNAc-transferase I, may be only one of a family of GalNAc-transferase enzymes [33]. The exact acceptor specificity of the newly discovered enzyme, GalNAc-transferase II, is not yet known, but available data suggest that it overlaps to a large extent that of GalNAc-transferase I [22]. Northern blots and measurements of enzymatic activity suggest that GalNAc-transferase II is expressed in a tissue-specific manner, but it is not yet known to what extent GalNAc-transferase II contributes to the synthesis of O-linked glycoconjugates in a given cell. A third GalNAc-transferase has also been described; this molecule is expressed in a developmentally regulated manner and appears to have a rather narrow acceptor specificity [34,35].

B. The Acceptor Specificity of GalNAc-Transferase

The analysis of the substrate specificity of GalNAc-transferase [31] is based on the postulate that, in mammalian cells, the "housekeeping" biosynthesis of mucin-type O-linked oligosaccharides is carried out by a single type of GalNAc-transferase activity. Thus, even if more than one enzyme is expressed in different tissues, the substrate specificities of all these enzymes should overlap to a large extent. In the following, we will collectively refer to this transferase activity as "GalNAc-transferase," thereby highlighting the enzymological similarity rather than the structural diversity.

Analysis of the sequences surrounding in situ glycosylated serines and threonines revealed that GalNAc-transferase is capable of glycosylating a large number of diverse acceptor sequences (Table 1). This conclusion was further strengthened by in vitro experiments wherein a cloned GalNAc-transferase I readily glycosylated a variety of acceptor peptides as well as intact proteins. These studies also showed that the same enzyme is capable

Table 1 Sequence Around Glycosylation Sites

Peptide	h	Code*	Peptide	h	Code*
PPTP–S–PSTP	1.00	a1hu	TSAV–T–GSEP	0.78	a
TPSP–S–TPPT	1.00	a1hu	DSS–S–KAPP	0.78	kthub
TPPP–T–SGPT	1.00	s03733	TNPA–T–SSAV	0.78	s00842
VTPR–T–PPPS	0.98	b	MASA–S–TTMH	0.77	gfhuc
CPVP–S–TPPT	0.98	a1hue	PSAT–S–PGVM	0.77	gfpge
APAR–S–PSPS	0.97	fqhugm	SSIA–T–VPVT	0.76	s00842
SPSP–S–TQPW	0.97	fqhugm	ARSP–S–PSTQ	0.76	fqhugm
SGEP–T–STPT	0.97	kkbob	PATS–S–AVAS	0.75	s00842
PATW–T–VPPP	0.96	exbo	EPLV–S–TSEP	0.74	d
APPP–S–LPSP	0.96	kthub	TTSI–T–SDPK	0.74	d
NSAP–T–SSST	0.96	ichu2	ESPS–T–SEAL	0.74	s00842
MHTT–T–SSSV	0.95	gfhue	VAVP–T–TSA–	0.74	al6604
TPHA–T–SHPA	0.93	d	TTTS–S–SVSK	0.73	gfhue
SSVP–T–AQPQ	0.93	dhhu	QTPT–S–GEPL	0.73	d
PAPA–T–EPTV	0.93	a16604	VPGG–S–ATPQ	0.73	a
–SKP–T–CPPP	0.92	ghrb	MHTT–T–IAEP	0.73	gfhuc
PGMA–S–ASTT	0.91	gfhuc	GGTI–T–TNSP	0.73	d
TVEP–T–PAPA	0.90	a16604	PGLP–S–TGVS	0.71	a
AMHT–T–TSSS	0.90	gfhue	PVTI–T–NPAT	0.71	s00842
STIT–S–AATP	0.89	gfhoe	KPSA–T–SPGV	0.71	gfpge
QTIA–T–GSPP	0.89	gfhoe	ASAS–T–TMHT	0.71	gfhuc
HTTT–S–SSVS	0.87	gfhue	GGSA–T–PQQP	0.69	a
PPTP–S–PSCC	0.86	a1hu	PSLP–S–PSRL	0.69	kthub
PPTP–S–PSCC	0.86	a1hu	ATAA–T–AATA	0.68	a03191
ESII–T–SPTE	0.85	a16604	TPSP–S–CCHP	0.66	a1hu
ELAP–T–APPE	0.85	plhu	-----T–ETPV	0.66	gfpge
TSAA–T–PTFT	0.84	gfhoe	GVTG–T–SAVT	0.65	a
IITS–T–PETP	0.82	a16604	TAPA–T–TRNT	0.65	dhhu
PLVS–T–SEPL	0.82	d	FTPN–S–ESPS	0.65	s00842
--LA–T–PLPP	0.82	oqhu	NPAT–S–SAVA	0.64	s00842
PSPS–T–QPWE	0.81	fqhugm	SEST–T–QLPG	0.64	a
VQTP–T–SGEP	0.80	d	PVLP–T–QSAH	0.64	bohus
PGAL–S–ESTT	0.79	a	WSTR–S–PNST	0.63	gfhuc
LSTI–T–SAAT	0.79	gfhoe	AQAS–S–VPTA	0.62	dhhu
SESP–S–TSEA	0.79	s00842	PHAT–S–HPAV	0.62	d
PNSE–S–PSTS	0.79	s00842	LPSP–S–RLPG	0.62	kthub
GTIT–T–NSPE	0.78	d	GPVP–T–PPDN	0.61	hchu
SAST–T–MHTT	0.78	gfhuc	-PGG–S–SEPK	0.60	c
AVPT–T–SA--	0.78	a16604	--ST–T–AVQT	0.60	d
SPST–S–EALS	0.78	s00842	AKAT–T–APAT	0.59	dhhu
GIPA–T–PGTS	0.78	a	--MW–S–TRSP	0.59	gfhuc

(*continued*)

Table 1 *Continued*

Peptide	*h*	Code*	Peptide	*h*	Code*
TPTF–T–TEQD	0.59	gfhoe	GTSA–T–ITSD	0.37	*d*
VAMH–T–TTSS	0.58	gfhue	LATG–S–PPIA	0.36	gfhoe
KAQA–S–SVPT	0.58	dhhu	--LS–T–TEVA	0.36	gfhue
STYS–S–IATV	0.57	s00842	-STG–S-----	0.35	*c*
TAVQ–T–PTSG	0.56	*d*	EQPL–T–ENPR	0.35	cthup
STTV–S–LPHS	0.56	kgboh2	ISVR–T–VYPP	0.34	gfhue
DSVV–T–PEAT	0.56	s16604	LSTY–S–SIAT	0.34	s00842
PGST–T–GR--	0.56	*a*	TYAA–T–PRAH	0.34	gfhue
LSES–T–TQLP	0.55	*a*	SGVA–S–DPPV	0.34	s00842
TPGS–T–TGR-	0.55	*a*	GLPG–S–T---	0.33	*a*
APAT–T–RNTG	0.54	dhhu	PDAA–S–AAPL	0.32	zuhu
GLSA–T–LATS	0.54	s00842	ATEP–T–VDSV	0.32	a16604
LPPS–T–SINE	0.54	*d*	VSNA–T–VTAG	0.30	gfpge
MATG–S–LGPS	0.54	s00842	PHQI–S–SKLP	0.29	a05273
GFIS–T–EDPS	0.52	a05273	PNTM–T–MLPF	0.27	s00842
---R–S–SVPG	0.52	*a*	---V–T–LSPK	0.27	nbhua2
TTMH–T–TTIA	0.50	gfhuc	SKMY–T–TSIT	0.26	*d*
PSFN–T–PSTR	0.50	a05273	EVRP–T–SAVA	0.25	lphuc3
---S–T–GS--	0.49	*c*	-LST–T–EVAM	0.25	gfhue
LVST–S–EPLS	0.48	*d*	DNTV–T–SKPL	0.25	riids2
---S–T–TAVQ	0.48	*d*	SEAL–S–TYSS	0.25	s00842
LPGV–T–GTSA	0.47	*a*	SIKM–S–SVPQ	0.24	*d*
PEAT–T–ESII	0.47	a16604	-MWS–T–RSPN	0.24	gfhuc
TTSS–S–VSKS	0.47	gfhue	GPVV–T–AQYE	0.23	a29559
TMHT–T–TIAE	0.47	gfhuc	GPVV–T–AQYE	0.23	kgboh2
GEQG–S–ATPG	0.46	gfpge	AVTG–S–EPGL	0.22	*a*
VTGT–S–AVTG	0.46	*a*	DVNC–S–GPTP	0.21	ahms
KMYT–T–SITS	0.45	*d*	SLGP–S–KETH	0.21	s00842
ATPG–S–TTGR	0.45	*a*	PIAG–T–SDLS	0.21	gfhoe
STGV–S–GPLG	0.45	*a*	---Q–T–IATG	0.20	gfhoe
GLPS–T–GVSG	0.45	*a*	KSYI–S–SQTN	0.19	gfhue
AQPL–T–ENPR	0.44	ctpgp	---L–S–TTEV	0.18	gfhue
LAKA–T–TAPA	0.43	dhhu	DPGM–S–GWPD	0.17	gfhuc
MLPF–T–PNSE	0.41	s00842	ISSQ–T–NDTH	0.17	gfhue
VPQE–T–PHAT	0.41	*d*	HQIS–S–KLPT	0.17	a05273
TSDL–S–TITS	0.40	gfhoe	THGL–S–ATIA	0.16	s00842
VTMA–T–GSLG	0.39	s00842	VSEI–S–VRTV	0.15	gfhue
LPGP–S–DTPI	0.39	kthub	SDLS–T–ITSA	0.14	gfhoe
ETPV–T–GEQG	0.39	gfpge	QVLL–S–NPTS	0.14	JP0105
LPGS–T-----	0.38	*a*	ALSE–S–TTQL	0.13	*a*
SSPL–S–TERM	0.37	plbo	QGSA–T–PGNV	0.13	gfpge
MYTT–S–ITSD	0.37	*d*	AHEV–S–EISV	0.13	gfhue

Table 1 *Continued*

Peptide	h	Code*	Peptide	h	Code*
AWPL–S–LEPD	0.12	gfhuc	IKNT–T–AVVQ	0.04	gfpge
SEPL–S–SKMY	0.11	*d*	-RFS–S–AGIP	0.04	*a*
VSLE–T–SKGT	0.11	*d*	KADS–T–GDQT	0.03	*d*
NATV–T–AGKP	0.10	gfpge	SKLP – T–QAGF	0.03	a05273
IIIP–T–INTI	0.10	A16604	AGVD–S–QQTA	0.03	*a*
---R–S–AGAG	0.10	*a*	LFPK–S–SGVA	0.02	s00842
FVHV–S–ESFP	0.09	s05642	YQEV–S–IKMS	0.02	*d*
--RS–S–VPGG	0.09	α	--RF–S–SAGI	0.02	*a*
DSQQ–T–AR--	0.09	*a*	REDP–S–GTMY	0.02	a05273
AGAG–T–AGVD	0.08	*a*	EEEG–S–GGGQ	0.02	hchu
AGFI–S–TEDP	0.08	a05273	YYNQ–S–EAGS	0.01	hlhub2
REYT–S–ARS-	0.08	xtpgl	RFQD–S–SSSK	0.01	kthub
EPLS–S–KMYT	0.07	*d*	PENF–S–FPDD	0.01	s05642
VQKE–T–GVPE	0.06	gfpge	GFNM–S–LLEN	0.00	s05642
EALS–T–YSSI	0.05	s00842	NVYR–S–HLFF	0.00	xtpgl
NLPN–T–MTML	0.05	s00842			

AV: 0.49 ± 0.30

*NBRF protein sequence database designations:
a03191: bald rockcod antifreeze glycoprotein
a05273: dog glycophorin
a16604: human κ-casein
a29559: bovine HMW I kininogen
a1hu: human Ig α-1 chain C region
ahms: mouse Ig α chain C region
bohus: human sex steroid–binding protein
cthup: human corticotropin–lipotropin (proopiomelanocortin)
ctpgp: pig corticotropin–lipotropin (proopiomelanocortin)
dhhu: human Ig δ chain C region
exbo: bovine coagulation factor Xa
fqhugm: human granulocyte–macrophage colony-stimulating factor (K. Kaushansky et al., 1992)
gfhoe: horse glycophorin
gfhuc: human glycophorin C
gfhue: human glycophorin A
gfpge: pig glycophorin
ghrb: rabbit Ig γ chain C region
hchu: human α-1-microglobulin/inter-α-trypsin inhibitor
hlhub2: human class I histocompatibility antigen HLA-B27
ichu2: human interleukin 2
jp0105: human T-cell surface glycoprotein CD8
kgboh2: bovine kininogen HMW II
kthub: human choriogonadotropin β chain
lphuc3: human apolipoprotein C-III
nbhua2: human leucine-rich α-2-glycoprotein

(*continued*)

Table 1 *Continued*

oqhu: human hemopexin
plbo: bovine plasminogen
plhu: human plasminogen
riids2: channel catfish somatostatin II
s00842: leukosialin
s03733: pig plasmin heavy chain
s05642: human tissue kallikrein
xtpg1: pig acrosin inhibitor
zuhu: human erthyropoietin
[a]Glycosylation site of ovine submaxillary mucin.
[b]Glycosylation site of basic myelin protein.
[c]Taken from J.-P. Aubert et al., 1976.
[d]Glycosylation site of human sialophorin.
Source: Ref. 31.

of glycosylating both threonine and serine. However, catalytic efficiency strongly depends on the chemical structure of the acceptor, and rate differences of as much as three orders of magnitude have been observed between some acceptor sequences [30–32]. As to the site of specificity, experiments with cloned GalNAc-transferase I showed that glycosylatable peptides in which a nonglycosylatable amino acid replaced the acceptor serine or threonine are competitive inhibitors of glycosylation, and their inhibition constants are virtually the same as the K_m of the corresponding acceptor peptides (36). Taken together, these observations establish that GalNAc-transferase I is a multisubstrate enzyme with an extended binding site. In the simplest of models, the interactions of such an enzyme with several amino acid residues of the acceptor substrate are independent of each other; none of them is essential, but each contributes cumulatively to the acceptor efficiency. Of course, the acceptor amino acid is essential; for GalNAc-transferase, this amino acid is invariably serine or threonine. The postulate of independent, noninteractively binding amino acid side chains in the acceptor is borne out by the fact that no statistically significant cross correlation (positive or negative) exists between amino acids at any two sites in the acceptor sequences of Table 1. In other words, no glycosylation motifs are discernable in this collection of peptide sequences. The acceptor specificity of enzymes with a cumulative, noncooperative subsite mechanism can be analyzed quantitatively. For a detailed description and discussion of the analysis of acceptor specificity of GalNAc-transferase, the reader is referred to the work by Elhammer et al. [31].

II. RESULTS AND METHODS

A. Quantitative Analysis of Acceptor Specificity

1. Selection of the Reference Data Set

The analysis of the specificity of GalNAc-transferase was based on a select list of amino acid sequences around serines and threonines of proteins the glycosylation of which had been documented with a reasonable degree of certainty [31]. This list, compiled from the NBRF protein sequence database, does contain some multiple copies of sequences from different proteins; we have kept these duplicates since they merely reflect a high specificity of recognition by the glycosylating enzyme. On the other hand, only one member of any set of highly homologous proteins is included in our list.

2. Computational Model

It is readily apparent that the database, shown in Table 1, contains widely heterologous peptide segments. Thus, O-glycosylation is effected either by a large number of enzymes with different, narrow specificities or, more likely, by a single enzyme with broad specificity. We have analyzed the specificity pattern on the basis of the latter hypothesis; our model embodies a set of assumptions currently taken as fundamental in expressing enzyme specificity: We posited that the enzyme has an "extended specificity site" where individual sections, called subsites, interact independently from each other with individual amino acid side chains of a short segment of the target protein. (The subsites are numbered sequentially toward the N terminal from the glycosylated amino acid as P1, P2, and so forth, and those toward the C terminal as P1′, P2′, and so forth. Ideally, the specificity of an enzyme should be expressed and analyzed quantitatively, using the specificity constants k_{cat}/K_m to access the free energy contribution of the individual subsites. In the absence of such quantitative information, the best one can do for assessing the contribution of the individual subsites is to analyze the experimental data set in terms of probabilities of reaction of a given peptide segment with the enzyme. This subsite contribution to the reactivity should be reflected by the frequency at which a given amino acid side chain is found at a given subsite in our data array. As shown in Table 2, most of the frequencies are indeed far from being random: The abundances differ significantly from those found in globular proteins in general [37]. From these abundances one can then calculate specificity parameters, $s_{i,j}$, for ith amino acid in the jth position (Table 3) and use these parameters to calculate the probability of a serine or threonine in a given peptide segment being glycosylated enzymatically. Since the abundance of the different amino

Table 2 Glycosylated Peptide Segments. Abundance of Amino Acids at Subsites

Amino acid	% S. D. average	# Expected in 196 A. A.	# Found in 196 residues								
			P4	P3	P2	P1	P0	P1′	P2′	P3′	P4′
D	5.7 ± 2.2	11.17 ± 4.31	6	2	3	2	0	2	6	3	6
N	4.4 ± 2.0	8.62 ± 3.92	5	4	6	3	0	4	7	3	4
E	6.4 ± 2.9	12.54 ± 5.68	10	10	8	5	0	15	13	7	11
Q	3.9 ± 1.7	7.64 ± 3.33	4	7	6	5	0	5	8	6	7
S	6.6 ± 2.7	12.94 ± 5.29	21	29	25	28	90	35	33	21	24
G	7.8 ± 3.0	15.29 ± 5.88	15	12	11	13	0	16	14	9	15
H	2.2 ± 1.3	4.31 ± 2.55	2	6	4	2	0	2	3	3	5
R	4.8 ± 2.5	9.41 ± 4.90	3	2	3	7	0	3	7	3	4
T	5.8 ± 2.3	11.37 ± 4.51	20	21	26	27	106	31	21	28	27
A	8.7 ± 3.7	17.05 ± 7.25	20	16	20	24	0	22	20	17	16
P	4.5 ± 2.0	8.82 ± 3.92	24	30	25	30	0	24	25	49	25
Y	3.3 ± 1.9	6.47 ± 3.72	2	3	5	2	0	1	2	3	2
V	7.0 ± 2.5	13.72 ± 4.90	12	12	12	15	0	10	11	12	13
M	2.1 ± 1.3	4.12 ± 2.55	6	4	6	4	0	3	2	5	3
I	5.2 ± 2.3	10.19 ± 4.51	6	5	9	9	0	8	6	6	5
L	8.2 ± 3.2	16.07 ± 6.27	11	10	11	11	0	6	8	7	9
F	3.9 ± 1.9	7.64 ± 3.72	2	4	3	4	0	1	0	3	3
K	6.8 ± 3.3	13.33 ± 6.47	5	4	4	1	0	4	5	2	5
C	1.6 ± 1.5	3.14 ± 2.94	1	0	0	1	0	2	1	1	1
W	1.2 ± 1.0	2.35 ± 1.96	1	1	1	2	0	0	1	1	1
Number of significantly abundant residues			24	80	76	85	196	90	79	77	76
Selectivity of subsites, s_i			0.4	1.1	0.9	1.1	8.1	1.3	1.0	1.6	0.9

Source: Ref. 31.

Table 3 Glycosylation Specificity Parameters, $s_{i,j}$

Amino acid	P4	P3	P2	P1	P0	P1'	P2'	P3'	P4'
D	0.54	0.18	0.27	0.18	0.00	0.18	0.54	0.27	0.54
N	0.58	0.46	0.70	0.35	0.00	0.46	0.81	0.35	0.46
E	0.80	0.80	0.64	0.40	0.00	1.20	1.04	0.56	0.88
Q	0.52	0.92	0.78	0.65	0.00	0.65	1.05	0.78	0.92
S	1.62	2.24	1.93	2.16	6.96	2.71	2.55	1.62	1.86
G	0.98	0.78	0.72	0.85	0.00	1.05	0.92	0.59	0.98
H	0.46	1.39	0.93	0.46	0.00	0.46	0.70	0.70	1.16
R	0.32	0.21	0.32	0.74	0.00	0.32	0.74	0.32	0.43
T	1.76	1.85	2.29	2.38	9.32	2.73	1.85	2.46	2.38
A	1.17	0.94	1.17	1.41	0.00	1.29	1.17	1.00	0.94
P	2.72	3.40	2.83	3.40	0.00	2.72	2.83	5.56	2.83
Y	0.31	0.46	0.77	0.31	0.00	0.15	0.31	0.46	0.31
V	0.87	0.87	0.87	1.09	0.00	0.73	0.80	0.87	0.95
M	1.46	0.97	1.46	0.97	0.00	0.73	0.49	1.21	0.73
I	0.59	0.49	0.88	0.88	0.00	0.78	0.59	0.59	0.49
L	0.68	0.62	0.68	0.68	0.00	0.37	0.50	0.44	0.56
F	0.26	0.52	0.39	0.52	0.00	0.13	0.01	0.39	0.39
K	0.38	0.30	0.30	0.08	0.00	0.30	0.38	0.15	0.38
C	0.32	0.03	0.03	0.32	0.00	0.64	0.32	0.32	0.32
W	0.43	0.43	0.43	0.85	0.00	0.04	0.43	0.43	0.43

Source: Ref. 31.

acids in glycosylated peptides approaches randomness for positions beyond P4 and P4', we opted for a model where the active site interacts with an enneapeptide segment with the reactive alcohol function at the center. The probability of the segment being glycosylated, h, can then be calculated from the equation

$$h = \frac{\prod_{i=1}^{9} s_{i,j}}{56 + \prod_{i=1}^{9} s_{i,j}} \tag{1}$$

where the factor 1/56 is a rough estimate of the relative abundance of glycosylatable serines and threonines among all amino acids in a set of representative globular proteins [31,38]. The values of h calculated from this equation for the reference set are given in Table 1. As a control set we have also compiled a list of serine- and threonine-peptides from proteins that are known to be nonglycosylated: porcine cytochrome C, human apo-lipoprotein A1, the B subunit of human calcineurin, human glucagon, pro-cine elastase, bovine pancreatic trypsin inhibitor, cobra toxin, and A and B chains of human insulin. One should note, however, that not being glyco-sylated need not mean that the peptide could not be recognized by the enzyme, since being buried in a folded protein also could prevent glycosyla-tion. We have calculated h values for these peptides; they were uniformly low with an average of 0.030 compared with the average of 0.49 for the reference set of glycosylated peptides.

3. Acceptor Amino Acid Specificity

The proposed model, where the specificity of the enzyme is due to an extended active site, composed of independent subsites, restricted to inter-act with a short peptide segment of the target protein, is fully consistent with the experimental observations. The high values of h for the reactive peptides and the extremely low values calculated for the nonreactive ones indicate that the computational method based on the extended site model has identified a major underlying cause for the specificity of the enzyme. Also, a detailed correlational analysis indicated the absence of frequent pairing of given amino acids at any two sites. Thus, there are no cooperative interactions between the different subsites. Finally, the segregation of the serine- and threonine-containing reactive peptides into separate specificity sets did not yield significantly different $s_{i,j}$ values. Thus, either a single enzyme is responsible for the glycosylation of both types of residues, or if there is a separate enzyme for serine and threonine glycosylation, then the two enzymes have remarkably similar specificities.

4. Optimization of the Cutoff Value

From a practical point of view, a priori identification of glycosylatable peptide sites is the major application of this aleatoric method. This en-deavor requires the choice of a cutoff value for h that would minimize both overpredictions and underpredictions, according to the criterion of "fraction correct of predicted" [39]. Using the h values calculated for com-parable numbers of reactive and nonreactive peptides, one can estimate the optimal cutoff: For the two sets of peptides considered separately, progres-sive increase of the cutoff value of h results in an increasing number of correct predictions for the nonreactive peptides and a decreasing number of correct predictions for the reactive peptides (Fig. 1). On the other hand, the percentage of correct predictions for the ensemble of reactive and nonreac-tive peptides passes through a broad but nevertheless definite maximum, located at around $h = 0.20$ (see Fig. 1). This value is then taken as the cutoff. It should be reemphasized that the method calculates the probability

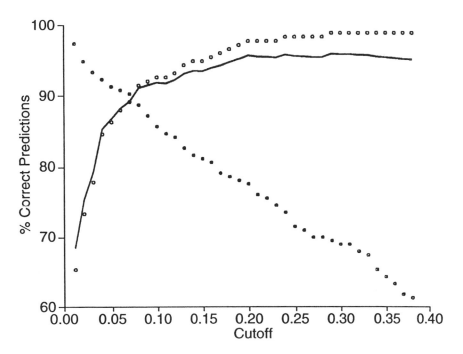

Figure 1 Determination of the cutoff value for h that maximizes the correct predictions for glycosylated and nonglycosylated peptide segments. The percent of correct predictions is shown with solid circles for the set of glycosylated peptides, with open circles for the set of nonglycosylated peptides, and with a solid line for the sum of the two sets.

of glycosylatable sites but not of truly glycosylated sites. Since the tertiary structure of a protein should preclude the reaction of the enzyme with at least a few serines and threonines, the algorithm is expected to be generally overpredictive. The overprediction, however, is not as considerable as one would infer from the distribution of solvation of serine and threonine residues in globular proteins of known tertiary structure [40]. As shown in Fig. 2, in this set of globular proteins, some 15% of all serines and threonines are completely dehydrated, whereas residues with $h > 0.2$ are significantly less abundant in that category.

III. DISCUSSION

A. Application of the Predictive Algorithm

The calculation of h, the probability of glycosylation of a given peptide segment, requires only the multiplication of the $s_{i,j}$ parameters of the nine

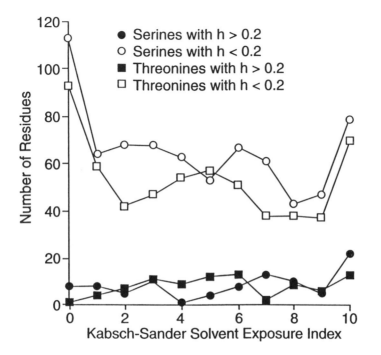

Figure 2 Distribution according to solvent exposure of serine and threonine residues in selected globular proteins of known tertiary structure. The solvent exposure index measures the estimated number of water molecules in contact with the amino acid residue. Amino acids with an index of 10 are in contact with greater than 9 water molecules.

amino acids within the segment and substitution of that product into Eq. (1). The procedure is illustrated for the case of the enneapetide TPTF–T–TEQD of horse glycophorin: From Table 3, we extract $s_{1,T} = 1.76$, $s_{2,P} = 3.40$, $s_{3,T} = 2.29$, $s_{4,F} = 0.52$, $s_{5,T} = 9.32$, $s_{6,T} = 2.73$, $s_{7,E} = 1.04$, $s_{8,Q} = 0.78$, and $s_{9,D} = 0.54$. The product of these nine parameters is 79.42, and therefore $h = 79.42/(56 + 79.42) \approx 0.59$. These calculations are readily incorporated into a simple computational program wherein the amino acid sequence of a protein is scanned by a window of nine amino acids and h is calculated for every segment containing serine or threonine at the center position.

The general usefulness of the predictive algorithm could be illustrated by the examples of several proteins where it was possible to identify domains known to be extensively O-glycosylated [31]. Unfortunately, for most of these proteins only the approximate position of the domains containing the oligosaccharide structures is known, and none of the positions of the O-linked amino acids were experimentally identified. Since the publication of the method, the sequence of a number of O-glycosylated proteins became known. This then provided us with a postive test set — outside the training set of Table 1 — to test the predictive success of the algorithm. Two examples, shown in Fig. 3, illustrate the success in identifying individual glycosylated amino acids: For the bile acid activated lipase (BAL) only one site is a false negative; the nine remaining experimentally identified sites [41] are all correctly predicted (Fig. 3a). For κ-casein, all experimentally identified sites [42] are correctly predicted (Fig. 3b). Thus, underpredictions seem to be the exception rather than the rule. On the other hand, as discussed above and in [31], false positives, that is, overprediction, could be more of a problem, because any glycosylatable sequence must first be exposed on the surface on the acceptor protein before it could actually be glycosylated. O-Glycosylation is a post-transitional modification, likely to occur primarily in the Golgi complex, where the acceptor proteins are already completely folded. Thus, by the time a protein comes within reach of the GalNAc-transferase in situ, only a limited portion of its potential sites may be exposed enough to be glycosylated. This conclusion is supported by the observation that native, unglycosylated proteins containing potential O-glycosylation sites could be glycosylated in vitro, but only after denaturation, reduction, and carboxymethylation [31].

B. Comparison with Other Predictive Methods

Since the publication of the cumulative-specificity model of GalNAc-transferase, methods have been proposed to improve the predictive success of the algorithm, either by setting slightly different $s_{i,j}$ values for the subsets of serines and of the threonines [43], or by using a neural network method wherein, again, the reactive peptides are segregated into four subclasses of presumably different reactivities [44]. These variants obtain marginal gains in predictive performance at the price of a large increase in the number of theoretical parameters. Such a plethora of adjustable parameters cannot be validated by statistical criteria [45]. Furthermore, the partitioning of the data set into subsets would carry with it the unstated assumption that the subsites are not independent of each other and that the specificity of a given subsite depends on the substrate structural elements interacting with other subsites. The present data set does not show any statistically significant cross correlation that would substantiate such a mechanism.

C. Other Serine/Threonine-Modifying Enzymes

GalNAc-transferase is not the only enzyme that targets serines and threonines in native proteins; other glycosyl-transferases and a variety of kinases also modify post-translationally specific serine and/or threonine residues.

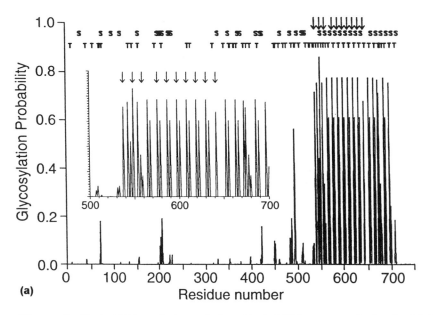

Figure 3 Probabilities of glycosylation. Probabilities were calculated with Eq. (1), using the specificity parameters from Table 3. The sequence position of all serines (S) and threonines (T) is also indicated. Vertical arrows indicate the locations of experimentally identified glycosylated amino acids. (a) Bile acid activated lipase (BAL); the inset is a magnification of the sequence segment between residues 500 and 700. (b) Bovine κ-casein.

Perhaps not surprisingly, the specificity of a cytoplasmic GlcNAc-transferase appears to overlap, at least partially, that of GalNAc-transferase; it selects for residues surrounded by Ser, Thr, and Pro. Still, even though 16 of the 21 identified acceptor sites for this enzyme [46–51] are also predicted for GalNAc-transferase, the specificity overlap of the two types of enzymes is not very broad, as 80% of the sites (in the investigated peptide segments) predicted for GalNAc-transferase have not reacted at all with the cytoplasmic enzyme. In other words, the cytoplasmic GlcNAc-transferase appears to have a somewhat narrower acceptor specificity. In addition, those experimentally observed O-linked GlcNAc sites that are not predicted acceptors for GalNAc-transferase all contain an aspartic acid or arginine in close proximity to the acceptor amino acids. The presence of these amino acids strongly decreases the probability of GalNAc-transferase-catalyzed glycosylation [31]. In contrast to the O-glycosylating enzymes, serine and threonine kinases have, in general, very narrow specificity [52], and the substrate motifs of many of them require the presence of two or more

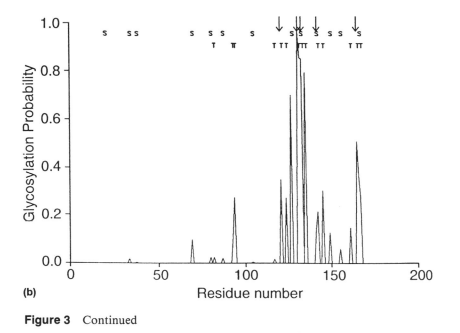

(b)

Figure 3 Continued

arginines and lysines, again amino acids contraindicative of GalNAc-trans-ferase specificity. Thus, there is, in general, very little overlap between predicted glycosylations and phosphorylations. A notable exception is the case of proteins the glycosylation of which controls their phosphorylation or vice versa, for example, in the case of the tau protein [53].

IV. CONCLUDING REMARKS

The predictive method described in this chapter provides a convenient ave-nue for the determination of putative O-glycosylation sites within a known polypeptide sequence. Although the accuracy, in terms of in situ utilized glycosylation sites, may be limited to some extent by the requirement for surface exposure of the hydroxyl groups to be glycosylated—a parameter that currently cannot be predicted—glycosylatable sequences are almost invariably accurately predicted. Hence, sites in protein domains containing few secondary structure features, such as the tandem repeat sequences in mucins or O-linked oligosaccharide cluster domains on many cell-surface molecules, can generally be predicted with excellent accuracy. By inference, knowledge of the three-dimensional structure of the proteins should help to eliminate the false positives. Nevertheless, for each enzyme, an accurate, quantitative predictive method must await the determination of the kinetic specificity parameters for a large number of acceptors.

REFERENCES

1. Sadler, J. E. (1984). In Ginsburg, V., and Robbins, P. W., eds., *Biology of Carbohydrates*, Vol. 2. John Wiley and Sons, New York, pp. 199–213.
2. Schachter, H., and Brockhausen, I. (1992). In Allen, H. J., and Kisialus, E. C., eds., *Glyconjugates*. Marcel Dekker, New York, pp. 263–332.
3. Strous, G. J., and Dekker, J. (1992). *Crit. Rev. Biochem. Mol. Biol.* 27, 57–92.
4. Nieuw Amerongen, A. V., Bolscher, J. G. M., and Veerman, E. C. I. (1995). *Glycobiology* 5, 733–740.
5. Do, S.-I., and Cummings, R. D. (1992). *Glycobiology* 2, 345–353.
6. Hayes, G. R., Enns, C. A., and Lucas, J. J. (1992). *Glycobiology* 2, 355–359.
7. Bansil, R., Stanley, E., and LaMont, J. T. (1995). *Ann. Rev. Physiol.* 57, 635–657.
8. Roseman, S. (1970). *Chem. Phys. Lipids* 5, 270–297.
9. Hagopian, A., Westall, F. C., Whitehead, J. S., and Eylar, E. H. (1971). *J. Biol. Chem.* 245, 2519–2523.
10. Young, J. D., Tsuchiya, D., Sandlin, D. E., and Holroyde, M. J. (1979). *Biochemistry* 18, 4444–4448.
11. Briand, J. P., Andrews, S. P., Jr., Cahill, E., Conway, N. A., and Young, J. D. (1981). *J. Biol. Chem.* 256, 12205–12207.
12. Sugiura, M., Kawasaki, T., and Yamashina, I. (1982). *J. Biol. Chem.* 257, 9501–9507.
13. Cruz, T. F., and Moscarello, M. A. (1983). *Biochem. Biophys. Acta* 760, 403–410.
14. Elhammer, A., and Kornfeld, S. (1986). *J. Biol. Chem.* 261, 5249–5255.
15. Hughes, R. C., Bradbury, A. F., and Smyth, D. G. (1988). *Carbohydr. Res.* 178, 259–269.
16. Wang, Y., Abernethy, J. L., Eckhardt, A. E., and Hill, R. L. (1992). *J. Biol. Chem.* 267, 12709–12716.
17. O'Connel, B. C., and Tabak, L. A. (1993). *J. Dent. Res.* 72, 1554–1558.
18. Tetaert, D., Briand, G., Soudan, B., Richet, C., Demeyer, D., Boersma, A., and Degand, P. (1994). *Anal. Biochem.* 222, 409–416.
19. Nishimori, I., Perini, F., Mountjoy, K. P., Sanderson, S. D., Johnson, N., Cerny, R. L., Gross, M. L., Fontenot, J. D., and Hollingsworth, M. A. (1994). *Cancer Res.* 54, 3738–3744.
20. Nishimori, I., Johnson, N., Sanderson, S. D., Perini, F., Mountjoy, K., Cerny, R. L., Gross, M. L., and Hollingsworth, M. A. (1994). *J. Biol. Chem.* 269, 16123–16130.
21. Stadie, T. R. E., Chai, W., Lawson, A. M., Byfield, P. G. H., and Hanish, F.-G. (1995). *Eur. J. Biochem.* 229, 140–147.

22. Sørensen, T., White, T., Wandall, H. H., Kristensen, A. K., Roepstorff, P., and Clausen, H. (1995). *J. Biol. Chem.* 270, 24166–24173.
23. Homa, F. L., Hollander, T., Lehman, D. J., Thomsen, D. R., and Elhammer, Å. P. (1993). *J. Biol. Chem.* 268, 12609–12616.
24. Hagen, F. K., VanWuyckhuyse, B., and Tabak, L. (1993). *J. Biol. Chem.* 268, 18960–18965.
25. Meurer, J. A., Naylor, J. M., Baker, C. A., Thomsen, D. R., Homa, F. L., and Elhammer, Å. P. (1995). *J. Biochem.* 118, 568–574.
26. Yoshida, A., Hara, T., Ikenaga, H., and Takeuchi, M. (1995). *Glycoconj. J.* 12, 824–828.
27. Hagen, F. K., Gregoire, C. A., and Tabak, L. A. (1995). *Glyconconj. J.* 12, 901–909.
28. O'Connel, B., Tabak, L. A., and Ramasubbu, N. (1991). *Biochem. Biophys. Res. Commun.* 180, 1024–1030.
29. O'Connel, B. C., Hagen, F. K., and Tabak, L. A. (1992). *J. Biol. Chem.* 267, 25010–25018.
30. Wang, Y., Agrwal, N., Eckhardt, A. E., Stevens, R. D., and Hill, R. L. (1993). *J. Biol. Chem.* 268, 22979–22983.
31. Elhammer, Å. P., Poorman, R. A., Brown, E., Maggiora, L. L., Hoogerheide, J. G., and Kézdy, F. J. (1993). *J. Biol. Chem.* 268, 10029–10038.
32. Wragg, S., Hagen, R. K., and Tabak, L. A. (1995). *J. Biol. Chem.* 270, 16947–16954.
33. White, T., Bennett, E. P., Takio, K., Sørensen, T., Bonding, N., and Clausen, H. (1995). *J. Biol. Chem.* 270, 24156–24165.
34. Matsuura, H., Takio, K., Titani, K., Greene, K., Levery, S. B., Salyan, M. E. K., and Hakomori, S. (1988). *J. Biol. Chem.* 263, 3314–3322.
35. Matsuura, H., Greene, T., and Hakomori, S. (1989). *J. Biol. Chem.* 264, 10472–10476.
36. Kurosaka, A., Kézdy, F. J., and Elhammer, Å. P., in preparation.
37. Nakashima, H., Nishikawa, K., and Ooi, T. (1986). *J. Biochem.* 99, 153–162.
38. Poorman, R. A., Tomasselli, A. G., Heinriksen, R. L., and Kézdy, F. J. (1991). *J. Biol. Chem.* 266, 14554–14561.
39. Kabsch, W., and Sander, C. (1983). *FEBS Lett.* 155, 179–182.
40. Kabsch, W., and Sander, C. (1983). *Biopolymers* 22, 2577–2637.
41. Wang, C.-S., Dashi, A., Jackson, K. W., Yeh, J.-C., Cummings, R. D., and Tang, J. (1995). *Biochemistry* 34, 10639–10644.
42. Pisano, A., Packer, N. H., Redmons, J. W., Williams, K. L., and Gooley, A. A. (1994). *Glycobiology* 4, 837–844.
43. Chou, K.-C., Zhang, C.-T., Kézdy, F. J., and Poorman, R. A. (1995). *Proteins: Struct. Funct. Genet.* 21, 118–126.
44. Hansen, J. E., Lund, O., Engelbrecht, J., Bohr, H., Nielsen, J. O., Hansen, J.-E. S., and Brunak, S. (1995). *Biochem. J.* 308, 801–813.
45. Akaike, H. (1974). *IEEE Trans. Automat. Cont.* AC-19, 716–723.
46. D'Onofrio, M., Starr, C. M., Park, M. K., Holt, G. D., Haltiwanger, R. S., Hart, G. W., and Hanover, J. A. (1989). *Proc. Natl. Acad. Sci. USA* 85, 9595–9599.

47. Reason, A. J., Morris, H. R., Panico, M., Marais, R., Treisman, R. H., Haltiwaqnger, R. S., Hart, G. W., Kelly, W. G., and Della, A. (1992). *J. Biol. Chem.* 267, 16911–16921.
48. Kelly, W. G., Dahmus, M. E., and Hart, G. W. (1993). *J. Biol. Chem.* 268, 10416–10424.
49. Dong, D. L.-Y., Xu, Z.-S., Chevrier, M. R., Cotter, R. J., Cleveland, D. W., and Hart, G. W. (1993). *J. Biol. Chem.* 268, 16679–16687.
50. Greis, K. D., Gibson, W., and Hart, G. W. (1994). *J. Virol.* 68, 8339–8349.
51. Chou, T.-Y., Hart, G. W., and Dang, C. V. (1995). *J. Biol. Chem.* 270, 18961–18965.
52. Kemp, B. E., and Pearson, R. B. (1990). *TIBS* 15, 342–346.
53. Morishima-Kawashima, M., Hasagawa, M., Takio, K., Suzuki, M., Yoshida, H., Titani, K., and Ihara, Y. (1995). *J. Biol. Chem.* 270, 823–829.
54. Aubert, J.-P., Biserte, G., and Loucheux-Lefebvre, M.-H. (1976) *Arch. Biochem. Biophys.* 175, 410–418.

17

Experimental Approaches to Studying *O*-Glycosylation In Vivo

Keith Nehrke, Fred K. Hagen, and Lawrence A. Tabak
University of Rochester, Rochester, New York

I. INTRODUCTION

The profound influence of *O*-glycans on the molecular architecture of se-creted hormones, growth factors, and mucin-type glycoproteins [1], and the recognition of *O*-linked glycans of membrane-bound proteins [2] has enhanced interest in predicting sites of *O*-glycosylation from primary amino acid sequence data.

The amino acids that flank reported *O*-glycosylation sites have been examined in great detail to uncover predictive motifs, but no single consen-sus signal has emerged [3–7]. Rather, it is apparent that a wide range of amino acids can be tolerated in the binding site of the enzymes responsible for initiating *O*-glycosylation, UDP-GalNAc polypeptide:*N*-acetylgalacto-saminyl transferase (ppGaNTase). However, there are at least two signifi-cant limitations to the database approach. First, there are the relatively few *O*-glycosylation sites that have been unambiguously mapped [8], particu-larly in clustered glycosylation regions so typical of this class of glycopro-tein. Moreover, advances in solid phase sequencing techniques have pointed to the partial occupancy of specific sites [8], leading to the question of how one treats a partially occupied site in the context of database analysis. Second, all of the database approaches have assumed incorrectly that there is only a single ppGaNTase; it is now clear that there are a family of ppGaNTases that display different substrate specificities as well as biased patterns of cellular and tissue expression [9–11]. Thus, the many in vitro studies using peptide substrates that have been performed using tissue ex-tracts as an enzyme source must be reinterpreted as representing a summa-tion of all surviving ppGaNTase activities. Only recently have in vitro glycosylation studies been conducted with either purified native or recombi-nant ppGaNTase, thus ensuring the effect of a single species (see [12] and references cited therein). Despite these limitations, it can be concluded that the initial rate of GalNAc addition to a synthetic peptide is sensitive to the surrounding amino acid context; charged amino acids at positions -1 and $+3$ relative to a potential glycosylation site appear to be particularly significant in limiting acquisition of O-glycans [3,5,13,14].

We have begun studies designed to relate the initial rate measurements made in vitro with synthetic peptide substrates to glycosylation occurring in vivo. To accomplish this, we have devised a methodology for quickly generating point mutations of a given O-glycosylation site and unambiguously determining the extent to which the sequence can be O-glycosylated in vivo. The specificity of the O-glycosylation reaction can be studied in a variety of cell backgrounds at defined sites using a transfected chimeric fusion protein minigene plasmid construct that encodes the acceptor substrate. We describe a strategy for analysis of single-site O-glycosylation motifs and compare the results obtained in vivo with those generated by analysis of the surrounding primary sequence using the NetOglyc (Center for Biological Sequence Analysis, The Technical University of Denmark, DK-2800, Denmark) O-glycosylation prediction program [6].

II. EXPERIMENTAL PROCEDURES

A. Construction of Vectors Encoding Chimeric Reporters

Previously, we had studied the in vitro glycosylation of a set of peptides based on the parent sequence PHMAQV<u>T</u>VGPGL [13], which was derived from a known O-glycosylation site in human von Willebrand factor (HVF) [15]. To examine the in vivo glycosylation of these substrates, we created vector pKN4 [16], which encoded a recombinant reporter protein (rHVF) containing the parent peptide substrate fused to the insulin secretory signal in the plasmid background of pcDL-SRα296 (DNAX, Research Institute, Palo Alto, CA), a mammalian expression vector. To facilitate purification and detection of the reporter protein, a Ni^{2+} binding site, a FLAG antibody recognition octapeptide (IBI, Kodak Immunology Ventures, Rochester, NY), and a bovine heart muscle kinase site were added (Fig. 1). To create point mutations of the rHVF fusion protein, standard site-directed mutagenesis procedures were employed.

B. Cell Culture and Transfections

COS7 cells were cultured in Dulbecco's modified Eagle's medium (Life Technologies, Inc., Grand Island, NY) supplemented with 10% fetal bovine serum at 37°C in 5% CO_2. Cells were split every 3 to 4 days. Transfections were performed on cells at approximately 70% confluency using DNA prepared with Promega's "midi" prep kit and LipofectAMINE reagent (Life

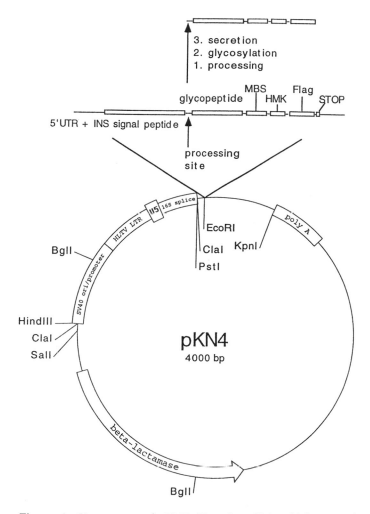

Figure 1 Vector map of pKN4. The plasmid is a high expression level construct that utilizes the SV40 early promoter and contains the SV40 origin of replication for amplification of copy number in transformed hosts such as COS7 cells. The reporter construct contains a 260 bp EcoRI insert encoding and insulin signal peptide, the HVF O-glycosylation site, an immobilized metal binding site (MBS), a heart muscle kinase site (HMK), a FLAG recognition epitope (FLAG), and a termination codon (STOP).

Technologies, Inc.) according to the manufacturer's protocol. At 18 h post-transfection, fresh media was added and then subsequently harvested 72 h post-transfection. The media was centrifuged (13,000 × g; 10 min) prior to immunoprecipitation.

C. Immunoprecipitation, Labeling, and Protein Analysis

To precipitate the reporter protein, 35 μL of 25% anti-FLAG agarose affinity gel (IBI, Kodak Immunology Ventures) was added directly to 0.5 mL of COS7 culture media. The mixture was rocked at 4°C overnight, pelleted, washed three times in phosphate-buffered saline (PBS), 0.5% Tween-20 (buffer A), and then resuspended in 30 μL of heart muscle kinase (HMK) reaction buffer (20 mM HEPES, pH 7.0, 75 mM NaCl, and 15 mM $MgCl_2$) containing 5 μCi of $[\gamma - ^{32}P]$ adenosine triphosphate (New England Nuclear, Boston, MA) and 5 units of bovine heart muscle kinase (Sigma, St. Louis, MO). Following incubation at 30°C for 1 h, the beads were again washed three times in buffer A and brought to a final volume of 15 μL in Tricine gel loading buffer (50 mM Tris pH 7.0, 12% glycerol, 4% sodium dodecyl sulfate (SDS), 2% β-mercaptoethanol, and 0.01% Coomassie G-250). A sample (3 μL) was run on a 16.5% T/6% C resolving phase Tricine polyacrylamide gel electrophoresis minigel containing both stacking and separating phases as described [17]. The gels were dried under vacuum and separated products visualized by autoradiography or with a Molecular Dynamics PhosphorImager.

D. Glycosidase Digestions of Immunopurified Material

Immunopurified reporter proteins were digested with various glycosidases following acid hydrolysis (0.1 NH_2SO_4 for 1 h at 80°C) and neutralization according to the manufacturer's specifications (Oxford Glycosystems, Rosedale, NY) in volumes of 20 μL overnight at 37°C, followed by the addition of 5 μL of 5 times concentrated tricine gel loading buffer and analysis by Tricine SDS-PAGE as previously described.

E. Analysis of O-Glycosylation Sites via the NetOglyc World Wide Web Server

For analysis we included a window of at least eight amino acids on either side of the most N- or C-terminal hydroxy amino acid in the reporter sequence. Queries were sent via the internet to NetOglyc@cbs.dtu.dk [6] (NetOglyc, Center for Biological Sequence Analysis, The Technical University of Denmark, DK-2800, Denmark).

III. RESULTS AND DISCUSSION

In vitro glycosylation assays have demonstrated that GalNAc is readily incorporated into the peptide substrate PHMAQVTVGPGL. Substitution of Glu, Arg, Ala, or Ile at positions +3, −3, or −2 relative to the Thr decreased the initial rate of *O*-glycosylation. Substitution of a charged residue at position −1 led to a marked inhibitory effect [13]. We compared these in vitro observations with predictions based on the neural network NetOglyc server; scores above 0.5 (on a 0 to 1.0 scale) are predictive of glycosylation [6]. The wild-type rHVF sequence was assigned a score of 0.212 and thus predicted to be nonglycosylated. Peptides that resulted from the substitution of Glu at the −1, −1, and +3, or −2, +1, and +3 positions were likewise predicted to not be *O*-glycosylated (Table 1). We next determined if these sequences could be *O*-glycosylated in vivo.

Transient transfection of COS7 cells with the mammalian expression vector pKN4 (Fig. 1) resulted in the expression, *O*-glycosylation, and secretion of the reporter protein rHVF. The secreted product could be immunoprecipitated directly from culture media and radiolabeled using heart muscle kinase to aid in visualization of the product. Facile separation of glycosylated from unglycosylated reporter was achieved with a Tricine SDS-PAGE system (compare lanes 1 and 2 in Fig. 2). Glycosidase digestions were used to verify the *O*-glycosylation of the reporter protein; removal of sialic acid and Galβ (1,3)GalNAc using sequential mild acid hydrolysis and *O*-glycanase digestion resulted in a protein that co-migrated with a nonglycosylated reporter (compare lanes 5 and 2 in Fig. 2). Greater than 95% of the reporter protein was glycosylated in vivo, which was not predicted by NetOglyc analysis (Table 1).

A series of single amino acid changes were made from positions −3 to +3 relative to the potential glycosylation site to determine flanking sequence requirements for in vivo *O*-glycosylation. In contrast to the results obtained in vitro, only when the −1 position was changed to a charged amino acid (either Glu or Arg) was glycosylation reduced in vivo, and even then over 60% of the site was still occupied [16]. The NetOglyc algorithm predicted that this substrate should not be *O*-glycosylated (Table 1). By introducing combinations of charged amino acids into the reporter molecule, we determined that charge distribution was more important than charge density in inhibiting in vivo *O*-glycosylation of single-site substrates. Thus, Glu substitution at the −1 and +3 positions virtually eliminated *O*-glycosylation in vivo, whereas any other combination of positions containing charged residues were fully glycosylated, including a triple Glu substitution at positions −2, +1, and +2. The NetOglyc algorithm predicted that neither of these Glu-substituted substrates would acquire *O*-glycans (Table 1).

Table 1 A Comparison of the Occupancy In Vivo Versus the Prediction of the NetOglyc Neural Network Algorithm for Various Reporter Constructs

	Reporter designation	Peptide sequence and amino acid position	Glycosylation in vivo[a]	NetOglyc prediction[b]
1	rHVF	F–V–N–P–H–M–A–Q–V–T–V–G–P–G–L–L–G–V	>95%	0.212
2	TOG	· · · · · · · · G · · · · · · · · ·	0%	n.d.
3	-1E	· · · · · · · E T · · · · · · · · ·	76% (±3)	0.118
4	−1, +3E	· · · · · · · E T · · E · · · · · ·	11% (±1)	0.036
5	−2, +1, +2E	· · · · · · E · T E E · · · · · · ·	>95%	0.233

[a]The values given in parenthesis for glycosylation in vivo are standard deviations taken from at least three separate experiments.
[b]The assignment of the NetOglyc program is shown for each hydroxy amino acid. An assignment of greater than 0.5 indicates a likelihood of O-glycosylation. (n.d. = not determined.)

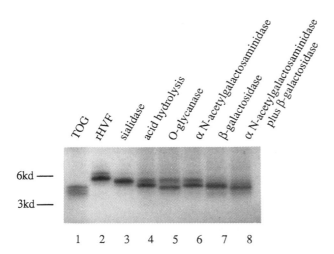

Figure 2 Glycosidase digestions indicate that the chimeric reporter protein rHVF is *O*-glycosylated during secretion from COS7 cells. rHVF was immunoprecipitated from transfected COS7 culture media, labeled with heart muscle kinase, and glycosidase digested prior to analysis by denaturing Tricine SDS-PAGE. Sialidase-digested material migrated more slowly than did acid-hydrolyzed rHVF (lane 4) due to incomplete enzymatic digestion, which may have been caused by the acetylation of a sialic acid residue or by a linkage to Gal or GalNAc that is a poor substrate. A labeled, purified mutant reporter protein containing a Thr-to-Gly substitution (lane 1) migrated similarly to the fully deglycosylated rHVF that had been acid hydrolyzed, then treated with *O*-glycanase or a combination of β-galactosidase and α-*N*-acetylgalactosaminidase (lanes 5 and 8). In a mock transfection, no product the size of either the wild-type rHVF or the Thr-to-Gly mutant was observed (data not shown). (From Ref. 16.)

How universal is the inhibitory effect of charge at positions −1 and +3 on *O*-glycosylation? Work in progress indicates that charge distribution at the −1 and +3 positions can modulate the level of glycosylation of sites derived from six different proteins and in at least four distinct cell backgrounds (K. Nehrke and L. A. Tabak, data to be reported). For the *O*-glycosylation of multiple sequences to be abolished in several distinct cell types, a significant proportion of the ppGaNTase family must be sensitive to charge distribution at the −1 and +3 sites.

Whereas progress has been made in the analysis of single sites of *O*-glycosylation, our understanding of the rules governing the *O*-glycosylation of clustered Thr and Ser is still limited. A major technical problem that must be overcome is the ability to assign the glycosylation occupancy of individual residues within a cluster. This has proven difficult using an SDS-

PAGE system due to the contribution of multiple carbohydrate side chains. Recently, however, a system for the unambiguous assignment of O-glycan occupancy at defined Thr and Ser as part of Edman degradation in solid phase microsequencing has been devised [8] and see the chapter by Gooley et al. Coupling the recombinant expression of defined cluster-type O-glycosylation motifs with solid phase sequencing-based determination of site occupancy should extend the usefulness of both techniques, as well as expand the database of known glycosylated and nonglycosylated sites to increase the robustness of predictive algorithms. Because discrepancies exist between experimental mapping results and predictive algorithms like Net-Oglyc, caution should be applied in assigning sites of O-glycosylation without direct experimental verification.

ACKNOWLEDGMENTS

Original work reported herein was supported in part by National Institutes of Health Grants DE08108 (to L.A.T.). K.N. is supported by training grant T32 DE07202. We thank P. Noonan for her help in preparing this chapter.

ABBREVIATIONS

HMK	heart muscle kinase
HVF	human von Willebrand factor
PAGE	polyacrylamide gel electrophoresis
PBS	phosphate-buffered saline
pKN4	mammalian expression vector
ppGaNTase	UDP-GalNAc polypeptide:N-acetylgalactosaminyl transferase
rHVF	recombinant human von Willebrand factor
SDS	sodium dodecyl sulfate

REFERENCES

1. Jentoft, N. (1990). *Trends Biochem. Sci.* 15, 291–294.
2. Lasky, L. A. (1995). *Ann. Rev. Biochem.* 64, 113–139.
3. Wilson, I. B. H., Gavel, Y., and von Heijne, G. (1991). *Biochem. J.* 275, 529–534.
4. O'Connell, B., Tabak, L. A., and Ramasubbu, N. (1991). *Biochem. Biophys. Res. Commun.* 180, 1024–1030.
5. Elhammer, A. P., Poorman, R. A., Brown, E., Maggiora, L. L., Hoogerheide, J. G., and Kezdy, F. J. (1993). *J. Biol. Chem.* 268, 10029–10038.

6. Hansen, J. E., Lund, O., Engelbrecht, J., Bohr, H., Nielsen, J. O., Hansen, J.-E. S., and Brunak, S. (1995). *Biochem. J.* 308, 801–813.
7. Chou, K.-C., Zhang, C.-T., Kezdy, F. J., and Poorman, R. A. (1995). *Proteins: Struct. Func. Genet.* 21, 118–126.
8. Gooley, A. A., and Williams, K. L. (1994). *Glycobiology* 4, 413–417.
9. Sorenson, T., White, T., Wandall, H. H., Kristensen, A. K., Roepstorff, P., and Clausen, H. (1995). *J. Biol. Chem.* 270, 24166–24173.
10. White, T., Bennet, E. P., Takiol, K., Sorensen, T., Bonding, N., and Clausen, H. (1995). *J. Biol. Chem.* 270, 24156–24165.
11. Hagen, F. K., Gregoire, C. A., and Tabak, L. A. (1995). *Glycoconj. J.* 12, 901–909.
12. Tabak, L. A. (1996). In Baum, B. J., and Cohen, M. M., eds., *Studies in Stomatology and Craniofacial Biology.* IOS Press, in press.
13. O'Connell, B., Hagen, F. K., and Tabak, L. A. (1992). *J. Biol. Chem.* 267, 25010–25018.
14. Wang, Y., Abernethy, J. L., Eckhardt, A. E., and Hill, R. E. (1992). *J. Biol. Chem.* 267, 12709–12716.
15. Titani, K., Kumar, S., Takio, K., Ericsson, L. H., Wade, R. D., Ashida, K., Walsh, K. A., Chopek, M. W., Sadler, J. E., and Fujikawa, K. (1986). *Biochemistry* 25, 3171–3184.
16. Nehrke, K., Hagen, F. K., and Tabak, L. A. (1996). *J. Biol. Chem.*, 271, 7061–7065.
17. Schagger, H., and von Jagow, G. (1987). *Anal. Biochem.* 166, 368–379.

18
Identifying Sites of Glycosylation in Proteins

Anthony Pisano, Daniel R. Jardine, Nicolle H. Packer,
John W. Redmond,* Keith L. Williams, and Andrew A. Gooley
Macquarie University, Sydney, New South Wales, Australia

Vince Farnsworth, Wulf Carson, and Paul K. Cartier III
Beckman Instruments, Inc., Fullerton, California

I. INTRODUCTION

Increasingly, glycosylation on proteins has been found to be important for stability, secretion, biological activity, and cell–cell and cell–extracellular matrix interactions [1]. Despite recognition of the importance of glycosylation, rarely are specific sites of glycosylation assigned, and usually only the pooled oligosaccharides from the protein are characterized [2]. This leads to the loss of positional information such as the identification of the in vivo acceptor site specificity of a particular glycosyltransferase, whether there is site-specific heterogeneity, and how much sugar is at each site of glycosylation.

Mass spectrometry (MS), particularly electrospray ionization MS (ESI/MS), has revolutionized oligosaccharide analysis of glycoproteins. Liquid chromatography coupled to ESI/MS has proved to be a powerful tool for the rapid isolation and characterization of glycopepides and glycoproteins. Carbohydrate-specific ions (m/z) 366 (HexNAc–Hex), m/z 204 (HexNAc) and the m/z 292 (NeuAc) can be generated by collision-induced dissociation [3]. The application of liquid chromatography MS to glycoprotein/glycopeptide analysis has been demonstrated by the characterization of recombinant soluble CD4 [4] and the rapid isolation and characterization of fetuin tryptic glycopeptides [5]. However, the method lacks the capacity to specifically identify individual sites of glycosylation in clustered O-glycosylation domains, such as those found in mucins and mucin-like glycoproteins [6,7].

A novel approach for the identification of site-specific O-glycosylation is solid phase Edman degradation, a chemical method where modified Ser, Thr, and Asn residues are recovered as phenylthiohydantoin

---

*Current affiliation: Research School of Biological Sciences, Australian National University, Canberra, A. C. T., Australia

(PTH)–glycoamino acids during the routine procedure of Edman degradation [8,9]. Solid phase Edman degradation has been used to characterize site-specific glycosylation on the mucin-like glycoprotein glycophorin A [10] and on the macroglycopeptide from bovine κ-casein, where as little as 10% glycosylation was accurately assigned to specific Thr residues [11]. In addition, at the preparative level, solid phase Edman degradation was used to prepare the PTH–glycoamino acid for ESI/MS and compositional analysis by high pH anion-exchange chromatography with pulsed amperometric detection (HPAEC-PAD) [9,12].

Whereas solid phase Edman degradation is a powerful tool for the identification and characterization of glycoamino acids, there no longer exists a commercial instrument specifically designed for this purpose (i.e., the MilliGen ProSequencer is no longer manufactured, and very few remain in service). Here we introduce a prototype instrument, GlycoSite, which is designed to operate in both the absorption and solid phase modes of Edman degradation. GlycoSite has been developed with the covalent attachment chemistries and chomatography suitable for the identification of all three major glycosylated amino acids, PTH–Asn(Sac), PTH–Thr(Sac), and PTH–Ser(Sac), where O-linked structures have a GalNAc core. We have also shown that the GlycoSite chemistry identifies PTH–Thr(GlcNAc) [13] and PTH–Ser(GLc–Xyl–Xyl) (R. Harris and A. Gooley, unpublished data).

II. MATERIALS AND METHODS

Human asialoglycophorin A, type MN, (Cat No. A-9791), bovine κ-casein (Cat. No. C 7278), bovine fetuin (Cat. No. F3004), and N-ethylmorpholine (NEM, Cat No. E-0252) were all obtained from Sigma (Sigma, St. Louis, MO). Acetic acid (Cat No. 45001 3Y), formic acid (Cat No. 45012 2M), acetonitrile (Cat No. 15285 6K), and sodium dodecyl sulfate (SDS, Cat No. 44244 4H) were from BDH (BDH, Poole, UK). Dithiothreitol (DTT, Cat No. 197 777) was from Boehringer Mannheim (Boehringer Mannheim, Mannheim, Germany). Triethylamine (TEA, Cat No. 25108) was from Pierce. 4-Vinylpyridine (Cat No. V320-4) was from Aldrich (Aldrich, Milwaukee, WI). Guanidine hydrochloride (Cat No. 50940) was from Fluka (Fluka, Buchs, Switzerland). Modified trypsin (Cat No. V5122) and endoproteinase Glu-C (Cat No. V5142) were from Promega (Promega, Madison, WI). Chemical reagents for Edman degradation on the GlycoSite sequenator were purchased from Beckman Instruments (Beckman Instruments, Fullerton, CA).

A. Glycoprotein and Glycopeptide Preparation

Asialoglycophorin A (GpA) was used without further preparation and the κ-casein macroglycopeptides were prepared according to Pisano et al. [11].

The *O*-GlcNAc containing *Dictyostelium* peptide obtained from the glyco-protein PsA was prepared according to Zachara et al. [13]. Glycopeptides from bovine fetuin were prepared according to the following procedure: Five milligrams of bovine fetuin was reduced and alkylated with 4-vinylpyridine using the method described by Tarr [14]. The alkylated fetuin was dialyzed into 4 L of 0.1 M *N*-ethylmorpholine acetate buffer, pH 8.3, for 4 h at 4°C and further dialyzed into 4 × 1 L (4 h dialysis for each liter of buffer) of 0.1 M Tris–HCl with 0.02% SDS (w/v), pH 7.8. Alkylated bovine fetuin (0.5 mg) in 0.1 M Tris–HCl with 0.02% (w/v) SDS, pH 7.8, was treated with 10 μg of trypsin at 37°C for 4 h, or 5 μg of Glu-C and incubated at 30°C for 18 h. The incubation reactions were stopped by heating the samples for 2 min at 100°C.

B. Separation and Detection of Proteolytic Glycopeptides by LC/MS

Bovine fetuin tryptic/Glu-C peptides were separated on a Hewlett-Packard 1090 (Hewlett-Packard, Palo Alto, CA) high performance liquid chromato-graph (HPLC) using a Beckman ODS Spherogel (5 μm, 4.6 × 25 mm) C_{18} reversed phase column with a linear gradient over 90 min from 100% sol-vent A (MilliQ water with 0.05% (v/v) formic acid) to 50–70% (v/v) solvent B (acetonitrile) (depending on the digest). The flow rate was 0.7 mL/min. The column eluent was directed through a low volume T piece to the elec-trospray source of a MicroMass VG Quattro triple quadrupole mass spec-trometer. The flow rate into the electrospray was 20 μL/min, while the remaining flow was sent to a Waters programmable fraction collector (set to collect 0.7 mL/tube) via a 214 nm UV detector. Glycopeptides were identified by the characteristic m/z 204 (HexNAc$^+$) ion, m/z 292 (NeuAc$^+$) ion, m/z 366 (HexNAc-Hex$^+$) ions, produced during a mass scan from m/z 70 to 400 at a cone voltage of 100 V. Intact peptides and glycopeptides were observed with a mass scan from m/z 400 to 1950 at a cone voltage of 30 V.

C. Further Fractionation of Bovine Fetuin Peptides by C_8 Reversed Phase Chromatography

Sample fractions were concentrated by vacuum centrifugation on a Speed-Vac (Savant) to 50 μL. Sialylated peptides were desialylated to prevent immobilization via the NeuAc carboxyl by the addition of 50 μL of 0.2 M trifluoroacetic acid (TFA) and incubation at 80°C for 1 h. The desialylated fetuin peptide fractions were then further fractionated by Sephasil C_8 (3 μm, 2 mm × 100 mm) reversed phase microbore chromatography on a SMART system (Pharmacia Biotech, Uppsala, Sweden). Chromatography conditions were MilliQ water with 0.05% (v/v) TFA and a 0–100% linear gradient of solvent B over 75 min, where B is 40–60% (v/v) acetonitrile, depending on the elution position of the LC/MS fetuin peptide fraction. Peptide elution was monitored at 214, 280, and 254 nm for simultaneous detection of peptide bonds, tryptophan and S-β-4 pyridylethyl cysteine, respectively. Fractions (100 μL) were collected and stored at $-20°C$ for Edman sequencing.

D. PTH–Amino Acid Sequence Analysis by Solid Phase Edman Degradation

Asialoglycophorin A (10–200 pmol), *Dictyostelium* PsA glycopeptide (200 pmol), and bovine fetuin peptides (20–500 pmol) were covalently atached to arylamine-derivatized membranes (Millipore, Bedford, MA) following N-ethyl-N'-dimethylaminopropyl carbodiimide (EDC) activation of the α, γ, and δ carboxyl groups at 4°C for 15 min. In brief, the sample (10–500 pmol in up to 100 μL of 20% (v/v) acetonitrile) was dried onto an arylamine-derivatized polyvinylidene difluoride (PVDF) disk in 5 μL aliquots at ~5 min intervals at 55°C. The dried arylamine disk was removed from the heating block and equilibrated to room temperature (RT). One milligram of EDC was dissolved in 100 μL of 0.1 M 4-morpholine ethanesulfonic (MES) acid, pH 5.0, containing 15% (v/v) acetonitrile. Five microliters of the EDC solution was added to the arylamine disk and incubated at 4°C for 15 min.

The immobilized peptides were then subjected to solid phase Edman degradation on the prototype GlycoSite sequenator. The GlycoSite was configured so that one regulator was used for pressurization of both bottles that deliver TFA in the gas phase and liquid phase. The liquid TFA delivery flask had the "pick-up" line extended to the bottom of the flask, and the gas phase delivery flask had the nitrogen "deliver" line extended to the bottom of the flask. These simple adaptations allowed the instrument to operate in both absorption and solid phase modes by a software selection. The solid phase functions for GlycoSite were different from the normal operation of the Beckman LF sequenator range, and we do not recommend modification of existing instruments without consultation with Beckman Instruments Inc.

The on-line System Gold high performance LC (HPLC) (Beckman) consisted of a Model 125s HPLC pump, a model 168 diode array detector set at 269 nm with the reference wavelength set at 293 nm. The column heater was set at 50°C for chromatography on the Shandon reversed phase C_{18} column (3 μm, 2.1 × 250 mm) and at 55°C for the Nova-Pak reversed phase C_{18} column (4 μm, 3.9 × 300 mm or 4 μm, 2 × 300 mm). Two solvent systems were used to separate PTH-(glyco)amino acids. Five millimolar triethylammonium formate (TEAF) at pH 4 was used as solvent A and acetonitrile (HPLC grade) as solvent B for PTH–amino acid analysis on the Nova-Pak (Nova-Pak, Waters, Milford, MA) reversed phase C_{18} (4 μm, 3.9 × 300 mm or 4 μm, 2 × 300 mm) columns [15]. Ammonium formate (14.5 mM, pH 3.9) was used as solvent A and acetonitrile (HPLC grade) as solvent B for PTH–amino acid analysis on the Shandon Hypersil (Shandon, Runcorn, UK) ODS (3μm, 2.1 × 250 mm) column.

E. Corrected Yields for Glycosylated Amino Acids

In an earlier study we found Thr-142, in the κ-casein glycopeptide Val^{139}–Thr^{145} and Ser2 from glycophorin A to be 100% glycosylated [10,11]. The recovered peak areas for κ-casein PTH–Thr-142(Sac) and GpA PTH–Ser2 (Sac) were equal to 0.8 of the equivalent yield for PTH–Val143 (κ-casein) and PTH–Leul (GpA, NN genotype). Hence, the yield for PTH–Asn(Sac), PTH–Ser(Sac), and PTH–Thr(Sac) glycoforms recovered from solid phase Edman degradation is equal to 1.25 multiplied by the recovered pmol area. This is called the corrected yield. Therefore, glycosylation at any particular site (for example, Xaa(Sac)) is quantified by measuring the corrected yield for Xaa(Sac) in pmol and using the following equation: % glycosylation = (corrected yield/expected yield) × 100. The expected yield is the estimated number of pmol from the linear regression calculated from the corrected yield for each amino acid over the entire sequence. It is important to understand that is is not possible to quantitatively recover (unglycosylated) PTH derivatives of Ser and Thr. Both of these hydroxy amino acids undergo β-elimination during Edman degradation, and at best the nonglycosylated form of PTH–Thr is recovered at ~75% and PTH–Ser at ~50%.

III. RESULTS AND DISCUSSION

The development of the solid phase program for the GlycoSite sequenator was achieved using glycopeptides and the glycoprotein human GpA. Most of the glycoforms examined during this study were "mucin-like" type 1 Thr/Ser-GalNAc-Gal. In addition, the glycoform Thr-GlcNAc was also analyzed. GlycoSite analysis of GpA, bovine κ-casein glycopeptide, and the *Dictyostelium discoideum* glycopeptide from recombinant PsA were used to characterize the glycosylated amino acids by their retention times and peak distrbution patterns on reversed phase chromatograms. Desialylation of glycopeptides and proteins was necessary prior to arylamine coupling to the membrane.

Our main criterion for PTH–amino acid chromatography following solid phase Edman degradation was that the PTH–glycoamino acids should each elute separate from the common 20 PTH–amino acids. Previously, we have shown the separation of all common amino acids and glycoamino acids using a 5 mM TEAF (pH 4.0) buffer on the MilliGen ProSequencer [15]. Figure 1 shows the conditions for PTH–amino acid separation using 5

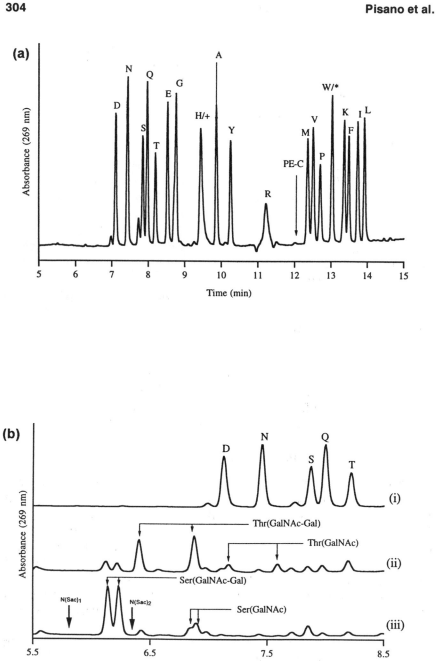

mM TEAF, pH 4.0, on the GlycoSite. One major disadvantage of this system is the high flow rate of 0.7 mL/min. To address this, we have developed an alternative chromatography using 14.5 mM ammonium formate at pH 3.9 and a Shandon ODS narrow-bore column (Fig. 2). Although the elution positions of the glycoamino acids have shifted toward the elution positions of the first of the common amino acids, there still exists a reasonable chromatographic space for the elution of the PTH–Thr(Sac) and PTH–Ser(Sac) glycoforms (Fig. 2b).

GlycoSite solid phase sequencing performs best with peptides (94–96% repetitive yield) rather than proteins (92–94% repetitive yield). Figure 3 shows the corrected yields for sequencing through five glycopeptides and the amino terminus of GpA. One distinct advantage of the GlycoSite solid phase program is the high repetitive yield through proline-rich sequences (Fig. 3a), as well as the increased yield for the recovery of the proline residue. Proline-rich sequences are typically difficult to sequence and require additional acid cleavage to increase the repetitive yield in conventional Edman sequencing. Because many "mucin-like" glycosylated domains, such as the MUC2 gene product [16], contain a high concentration of proline residues, GlycoSite is well suited to characterize proline-rich glycopeptides.

Figure 1 GlycoSite triethylammonium formate chromatography of PTH–amino acids and glycoamino acids. Separation of PTH–glycoamino/amino acids with 5 mM triethylammonium formate (TEAF) buffer. Chromatography was done using a Nova-Pak (Waters, 4μm, 3.9×300 mm) C_{18} reversed phase column with a System Gold HPLC. Flow rate was 0.7 mL/min, and the oven temperature was 55°C. (a) A C_{18}HPLC chromatogram with an elution profile of 19 PTH–amino acid standards (10 pmol) routinely encountered in N-terminal sequence analysis. The PTH–amino acids are (in order of elution): D, N, S, Q, T, E, G, H, A, Y, R, M, V, P, W (which co-elutes with diphenylthiourea), K, F, I, L. The elution position of PE–Cys is indicated on the elution profile by an arrow. (b) An enlargement of the TEAF-derived C_{18} reversed phase chromatograms from the GlycoSite sequenator. (i) PTH-standards chromatogram (10 pmol) from 5.5 to 8.5 min showing the elution position of D, N, S, Q, T. (ii) Glycosylated Thr(GalNAc–Gal) cycle 4 and (iii) Ser(GalNAc–Gal) cycle 2, recovered from solid phase Edman degradation of desialylated human GpA. The two major glycoforms of Ser(GalNAc–Gal) at 6.1 and 6.2 min (iii) are separated from the later eluting two major glycoforms of Thr(GalNAc–Gal), which elute at 6.4 and 6.9 min (ii) and before D at 7 min. The minor peaks at 6.85 min in (iii) represent a minor amount of Ser(GalNAc), which elutes close to the second major Thr(GalNAc–Gal) glycoform at 6.9 min. The elution position for Asn(Sac) glycoforms recovered from GpA (Asn(Sac)26) are indicated by the solid arrows in chromatogram (iii). N(Sac)$_1$ represents the PTH-Asn(HexNAc$_{2-4}$Hex$_{3-5}$) glycoforms (see Fig. 4), and N(Sac)$_2$ represents PTH–Asn(GlcNAc).

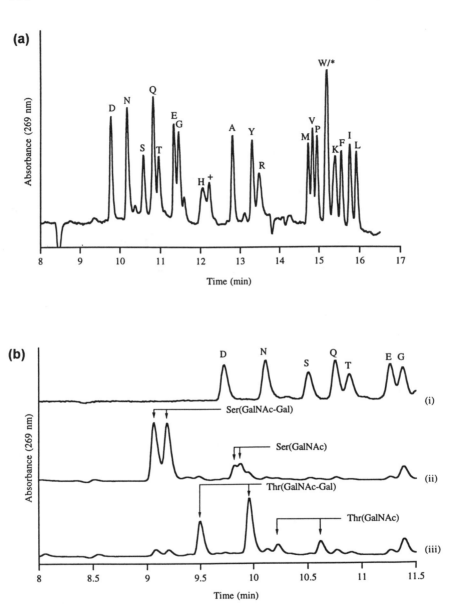

Figure 2 GlycoSite ammonium formate chromatography of PTH–amino acids and glycoamino acids. The PTH–glycoamino/amino acids were separated by on-line System Gold HPLC with a Shandon C_{18} reversed phase (3 μm, 2 × 250 mm) column using 14.5 mM ammonium formate and pH 3.9 and a gradient of acetonitrile. The flow rate was 0.2 mL/min, and the oven temperature was 55°C. (a) Reversed phase chromatogram of the 19 common PTH–amino acid standards (10 pmol). The

The ability to quantitate the degree of O-glycosylation also diminishes the further the sites are from the N terminus. Usually, a glycosylation site identified in the first 16–20 cycles of solid phase Edman degradation can be quantitated accurately, whereas sites occurring between 20 and 30 cycles provide approximate percentage glycosylation values. Those occurring after 30 cycles have little quantitative value, but they still allow assignment of glycosylation.

A. Separation of Fetuin Peptides by LC/MS

Peptide digests were subjected to reversed phase chromatography with on-line electrospray mass spectrometry. Glycopeptides were detected by monitoring the collision-induced fragment ions m/z 366 (HexNAc–Hex$^+$), m/z 204 (HexNAc$^+$), and m/z 292 (NeuAc$^+$). From the full-scan data, the molecular weights of the glycopeptides were determined (Table 1). The Endo Glu-C digest appeared to be only a partial digest with several sites not cleaved (e.g., the N-linked glycopeptides Ser-157–Asn-158(Sac)–Glu-166 or Val-148–Asn-158(Sac)–Glu-166 were never isolated). This may be due to the proximity of Glu to the N-linked glycosylation site at Asn-158.

1. Characterization of the N-Linked Glycopeptides by LC/MS

Six N-linked glycoforms were recovered from the tryptic and Endo Glu-C digests (Table 1). Both digests showed the same two glycoforms for Asn-138 (peptides T1,G5 and T2,G6), which appeared to differ by a single sialylated lactosamine. Three lycoforms were recovered from the Asn-81 site (peptides T4,G2; T5,G3; and G4), although only two of these were recovered from the trypsin digest and all three from the Endo Glu-C digest. The difference was again a single sialylated lactosamine as well as a disialylated lactosamine (G4). Only a single glycoform for Asn-158 was recovered from the

---

Figure 2 *continued*
elution positions of DMPTU is " + ", which elutes close to His, and DPTU is "*", which co-elutes with Trp. (b) An overlay section of the ammonium formate reversed phase chromatogram surrounding elution positions of O-linked glycoamino acids recovered from the solid phase Edman degradation of human GpA. (i) 10 pmol PTH–amino acid standard chromatogram from 8–11.5 min. (ii) Ser(GalNAc–Gal) cycle 2 and (iii) Thr(GalNAc–Gal) cycle 4 recovered from solid phase Edman degradation of desialylated human GpA. The elution positions of the diastereomers of PTH–Ser(GalNAc–Gal) are at 9.1 and 9.2 min and the Edman degradation–induced glycoform PTH–Ser(GalNAc) at 9.8 min (ii). The elution positions for the diastereomers of PTH–Thr(GalNAc–Gal) are at 9.5 and 9.9 min, and the positions of the glycoform induced by Edman degradation, PTH–Thr(GalNAc), are at 10.25 and 10.65 min (iii).

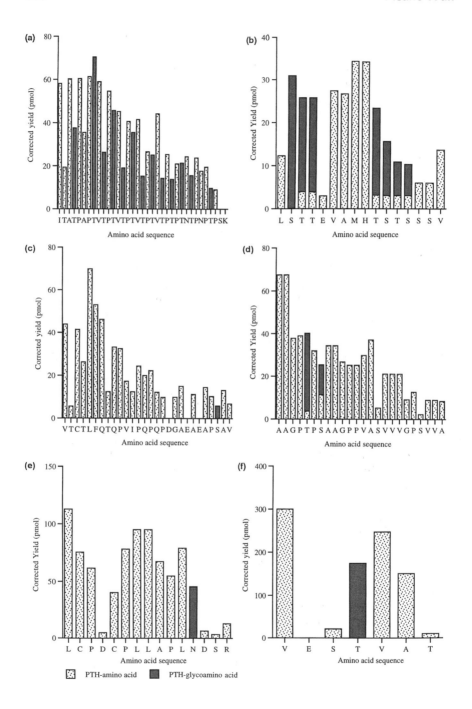

tryptic digest (peptide T12), and as was discussed, no glycoform was recovered for Asn-158 from the Endo Glu-C digest. In addition, the nonglycosylated peptide T12 was recovered at 59.60 min (observed mass 3017.4, data not shown), consistent with previous reports that Asn-158 is only partially modified [18].

2. Characterization of O-Linked Glycopeptides by LC/MS

Multiple O-linked glycoforms were separated from the proteinase digests. In this case, most of the fetuin glycoforms apparently reflect the heterogeneity in the number of glycosylation sites rather than the number of glycoforms at a particular site, although LC/ESI/MS data alone, without CID analysis, cannot determine if glycopeptides have one or more multiple sites. The tryptic digest contained two O-linked glycopeptides, Val^{228}–Arg^{288} (T6–T11), which contained from one to four glycosylation sites of mono- and disialylated disaccharide, and His^{295}–Arg^{330} (T3), which contained a single monosialylated disaccharide. The LC/MS data from the Glu-C digest resolved three glycoforms (G1, G8, G9), representing three of the five potential glycosylation sites from the tryptic digest. However, several of the

---

Figure 3 Corrected yields for glycopeptides/glycoproteins subjected to solid phase sequencing on the GlycoSite. Desialylated glycopeptides were immobilized onto Sequelon-AA. The light shaded bars are PTH–amino acids, and the dark shaded bars are PTH–glycoamino acids. (a) Corrected yields for the tryptic peptide Ile^{88}–Lys^{122} from the *Dictyostelium* recombinant glycoprotein PsA. Every Thr except Thr-89 (cycle 2) was 100% glycosylated. Initial yield (IY) was 20%, and repetitive yield (RY) was ~96%. (b) Corrected yields for the first 16 amino acids from the amino terminus of human glycophorin A. Quantitation of the glycosylation sites (approximate yields are shown in parentheses) are Ser-2 (100%), Thr-3 (80%), Thr-4 (80%), Thr-10 (80%), Ser-11 (70%), Thr-12 (60%), Ser-13 (60%). IY was 30% and the RY was ~93%. (c) Corrected yields for the first 27 cycles of the bovine fetuin tryptic glycopeptide Val^{228}–Arg^{288}. A single glycosylation site at Ser-253 in cycle 25 was recovered with a yield of 50–90%. The IY was 40% and the RY was ~94%. (d) Corrected yields for the first 24 amino acids from the bovine fetuin Endo Gluc-C glycopeptide Ala^{258}–Asp^{292}. Two O-linked glycosylation sites were identified at Thr-262 (cycle 5, 80% glycosylated) and Ser-264 (cycle 7, 50% glycosylated). The IY was 50% and the RY was ~96%. (e) Corrected yields for the N-linked bovine fetuin glycopeptide Leu^{127}–Arg^{141}. The single glycosylated site, Asn138(Sac), was recovered at 100% following 12 cycles of Edman degradation. The RY was 96%. (f) Corrected yields for the bovine κ-casein glycopeptide Val^{139}–Thr^{145}. The single glycosylation site at Thr142(Sac) was recovered with 100% yield. The IY was ~60% and the RY was ~95%.

Table 1 Tryptic (T) and Endo Glu-C (G) Sialylated Bovine Fetuin Glycopeptides Identified by LC/MS

Rt (min)	Bovine fetuin sequence [a]	Observed mass
41.00 **T1**	LBPDBPLLAPLN^{138}DSR + (HexNAc)$_4$ (Hex)$_5$(NeuAc)$_2$	4042.0
41.46 **T2**	LBPDBPLLAPLN^{138}DSR + (HexNAc)$_5$ (Hex)$_6$(NeuAc)$_3$	4698.4
42.38 **T3**	HTFSGVASVESSSGEAFHCGKTPIVGQP S^{323}IPGGPVR + HexNAc–Hex–NeuAc	4233.9
52.86 **T4**	RPTGEVYDIEIDTLETTBHVLDPTPLAN^{81} BSV R + (HexNAc)$_4$(Hex)$_5$(NeuAc)$_2$	5973.8
53.20 **T5**	RPTGEVYDIEIDTLETTBHVLDPTPLAN^{81} BSVR + (HexNAc)$_5$(Hex)$_6$(NeuAc)$_3$	6630.4
53.78 **T6**	VTBTLFQTQPVIPQPQPDGAEAEAP- S^{253}AVP DAAGPT^{262}PS^{264}AAGPPVA SVVVGPS^{278}VVAVPLPLHR + (HexNAc– Hex–NeuAc)$_4$	8691.9
56.31 **T7**	VTBTLFQTQPVIPQPQPDGAEAEAP S^{253}AVPDAAGPT^{262}PS^{264}AAGPPVA SVVVGPS^{278}VVAVPLPLHR + (HexNAc– Hex–NeuAc)$_3$	8033.6
56.54 **T8**	VTBTLFQTQPVIPQPQPDGAEAEAP S^{253}AVPDAAGPT^{262}PS^{264}AAGPPVA SVVVGPS^{278}VVAVPLPLHR + (HexNAc– Hex–NeuAc)$_2$	7377.3
57.02 **T9**	VTBTLFQTQPVIPQPQPDGAEAEAP S^{253}AVPDAA GPT^{262}PS^{264}AAGPPVA SVVVGPS^{278}VVAVPLPLHR + (HexNAc– Hex–Hex–(NeuAc)$_2$) + (HexNAc–Hex– NeuAc)	7668.2
57.02 **T10**	VTBTLFQTQPVIPQPQPDGAEAEAP S^{253}AVPDAAGPT^{262}PS^{264}AAGPPVA SVVVGPS^{278}VVAVPLPLHR + (HexNAc– Hex–NeuAc)	6720.6
57.25 **T11**	VTBTLFQTQPVIPQPQPDGAEAEAP S^{253}AVPDAAGPT^{262}PS^{264}AAGPPVAS VVVGPS^{278}VVAVPLPLHR + (HexNAc– Hex–(NeuAc)$_2$) + (HexNAc–Hex–NeuAc)$_2$	8324.9
60.27 **T12**	VVHAVEVALATFNAESN^{158}GSYLQL VEISR + (HexNAc)$_5$(Hex)$_6$(NeuAc)$_3$	5878.9
25.09 **G1**	AEAPS^{253}AVPD + (HexNac–Hex–NeuAc)	1512.0
32.86 **G2**	TTBHVLDPTPLAN^{81}BSVRQQTQHAVE + (HexNAc)$_4$(Hex)$_5$(NeuAc)$_2$	5164.1

Table 1 Continued

Rt (min)	Bovine fetuin sequence [a]	Observed mass
33.45 **G3**	TTBHVLDPTPLAN[81]BSVRQQTQHAVE + (HexNAc)$_5$(Hex)$_6$(NeuAc)$_3$	5820.8
34.43 **G4**	TTBHVLDPTPLAN[81]BSVRQQTQHAVE + (HexNAc)$_5$(Hex)$_6$(NeuAc)$_4$	6111.5
42.50 **G5**	DVRKLBPDBPLLAPLN[138]D + (HexNAc)$_4$(Hex)$_5$(NeuAc)$_2$	4297.1
42.79 **G6**	DVRKLBPDBPLLAPLN[138]D + (HexNAc)$_5$(Hex)$_6$(NeuAc)$_3$	4953.5
45.05 **G7**	DVRKLBPDBPLLAPLN[138]DSRVVHAVE + (HexNAc)$_4$(Hex)$_5$(NeuAc)$_2$	5832.1
51.74 **G8**	AAGPT[262]PS[264]AAGPPVASVVVGPS[278]V VAVPLPLHRAHYD + (HexNAc–Hex–NeuAc)	4013.5
51.94 **G9**	AAGPT[262]PS[264]AAGPPVASVVVGPS[278]V VAVPLPLHRAHYD + (HexNAc–Hex–NeuAc)$_2$	4669.9

[a]B = S-β-4-pyridylethyl cysteine.
Glycopeptides are ordered according to their elution time and are labeled T1 to T12 and G1 to G9. Each glycosylation site is shown in boldface and its position in the bovine fetuin sequence. Note that the glycosylation sites were not identified by MS but by GlycoSite analysis, except Ser-278, which was identifed by MS according to Medzihradszky et al. [17].

masses were also consistent with the presence of a hexasaccharide, which has peviously been reported by Edge and Spiro [19].

B. Solid Phase Edman Degradatiion of LC/MS Recovered Glycopeptides

1. Identification of N-Linked Glycosylation

All three Asn(Sac) sites (Asn-81, Asn-138, Asn-158) were identified by solid phase Edman degradation of the LC/MS fractions. Figure 4A shows a chromatogram of Asn(Sac) 138 recovered after 12 cycles of solid phase Edman degradation of the pooled peptides T1 and T2; it is typical for the elution profile of Asn residues modified with complex carbohydrate structures. The PTH–Asn(Sac) sites were recovered as one heterogeneous peak that eluted between 4.0 and 4.4 min and another peak that eluted at 4.7 min. The peaks that eluted from 4.0–4.4 min are heterogeneous and consist of the glycoforms PTH–Asn(HexNAc$_{2-4}$Hex$_{3-5}$) as determined by

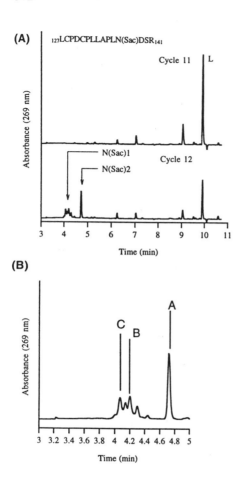

Scan	Mass	Composition
A	453	PTH-Asn-HexNAc
B	1507	PTH-Asn-HexNAc3Hex4
	1345	PTH-Asn-HexNAc3Hex3
C	1871	PTH-Asn-HexNAc4Hex5

Figure 4 Sequencing of bovine fetuin glycopeptide containing Asn-138. (A) Representative chromatogram of a glycosylated PTH–Asn138(Sac) site from bovine fetuin. Approximately 500 pmol of desialylated bovine fetuin glycopeptide Leu[127]–Arg[141] was coupled to Sequelon-AA and sequenced by solid phase Edman degradation on the GlycoSite sequenator. The recovered PTH–glycoamino/amino acids were separated by 5 mM TEAF, pH 4.0, in acetonitile C_{18} reversed phase chroma-

LC/MS analysis of the PTH–Asn 138(Sac) peaks (Fig. 4B). The second major PTH–Asn(Sac) glycoform recovered from solid phase Edman degradation (at 4.7 min) is PTH–Asn(HexNAc) as determined by LC/MS analysis (Fig. 4B). Similar profiles were observed for Asn-81 (peptides T4 and T5) after 28 cycles of solid phase Edman degradation, and for Asn-158 (peptide T12) after 17 cycles of solid phase Edman degradation. The major difference between the three profiles was that the yield of the PTH–Asn (HexNAc) glycoform increased relative to the yield of the PTH–Asn(Hex-NAc$_{2-4}$Hex$_{3-5}$) glycoforms with increasing number of cycles of Edman degradation. This heterogeneity was not observed in the LC/MS data and probably reflects Edman degradation–induced destruction by the exposure of the oligosaccharide to repeated cycles of acid cleavage.

Figure 3e shows the corrected yields for the solid phase Edman degradation of the arylamine-coupled T1 glycopeptide with a repetitive yield of 96%. The PTH–Asn(Sac) combined peak areas represent a recovery of ~ 100% for this site.

2. Identification of O-Linked Glycosylation Sites by Solid Phase Edman Degradation

Solid phase Edman degradation of tryptic and Endo Glu-C glycopeptides separated by LC/MS identified four O-linked sites of glycosylation at Ser-253, Thr-262, Ser-264, and Ser-323. All three Ser(Sac) sites recovered from solid phase Edman degradation were found to have the same elution profile of diastereomers, consistent with the PTH–Ser(Sac) sites recovered from the solid phase Edman degradation of human glycophorin A (Fig. 2b, ii). The only identified Thr(Sac) in bovine fetuin was at Thr-262, which was

---

Figure 4 *Continued*
tography using a System Gold HPLC. Gradient conditions used were those described by Pisano et al. [15]. Flow rate was 0.7 mL/min, and oven temperature was 50°C. Chromatogram (top) shows Leu (from cycle 11) and the glycosylated site Asn(Sac)138 in cycle 12 (bottom). PTH–Asn(Sac) elutes early as a series of peaks: a cluster of peaks that are resolved between 4 and 4.4 min (N(Sac)$_1$) and a single peak at 4.8 min (N(Sac)$_2$). (B) Enlargement of the N(Sac)$_1$ and N(Sac)$_2$ peaks subjected to LC/MS analysis. Mass scans were taken across the cluster of PTH–Asn(Sac) peaks indicated by the lines *A*, *B*, and *C*. Scans *B* and *C* resolved glycoforms with a decreasing lactosamine content. The m/z 1345 ion may result from the Edman degradation–induced hydrolysis of the PTH–Asn(HexNAc$_3$Hex$_4$) glycoform or a MS collision-induced dissociation of the Hex$^+$ from the m/z 1507 ion. The m/z 453 ion is PTH–Asn(HexNAc) resulting from the Edman degradation–induced hydrolysis of the PTH–Asn(HexNAc$_4$Hex$_5$) glycoform.

found to have the same elution profile of diastereomers, consistent with the PTH–Thr(Sac) recovered from the solid phase Edman degradation of human glycophorin A (Fig. 2b, iii). Our understanding of the diastereomers is that during the Edman degradation the Thr(Sac) and Ser(Sac) is racemized at the β position (where the substituted OH group is located). As a result, a mixture of the two Ser and Thr configurations are obtained. These have identical properties in an achiral environment (such as an HPLC column), and one would not expect to separate them. However, when you have a chiral substituent, such as a HexNAc, you have two diastereomers rather than two enantiomers. These now have different properties and resolve on the reversed phase HPLC column.

Medzihradszky et al. [17] have reported MS data that suggests that, in addition to the four O-linked sites we observed on bovine fetuin, Ser-278 is also glycosylated. This site was not glycosylated on the peptide G9 (Ala^{258}–Asp^{292}), as LC/MS detected a mass consistent with only two sites of trisaccharide glycosylation, which were identified by solid phase Edman degradation as Thr-262 and Ser-264. Glycosylation at Ser-278 is consistent with the observed mass of peptide T6 (Val^{228}–Arg^{288}), which could contain four sites of glycosylation at Ser-253, Thr-262, Ser-264, and Ser-278. The mass of T6 (8691.9 Da) is also consistent with the presence of a hexasaccharide (NeuAc(HexNAc-Hex)$_2$), two trisaccharides (NeuAc–HexNAc–Hex), and one unglycosylated site on T6. Three glycosylation sites (Ser-253, Thr-262, or Ser-264) were identified following GlycoSite analysis of peptide T6 (data not shown). We could not confidently assign the glycosylation status of Ser-278 on peptide T6 following 51 cycles of Edman degradation. However, the patterns of PTH–Ser/Thr(HexNAc-Hex) glycoforms recovered from these three sites were only consistent with a mass of NeuAc-HexNAc-Hex glycosylation on all four sites.

C. Quantitation of O-Linked Glycosylation Sites by Solid Phase Edman Degradation

The quantitation of O-linked sites assigned by the solid phase Edman degradation of bovine fetuin glycopeptides is shown in Fig. 3c and d and Fig. 5. Ser(Sac)253 was identified and quantitated as 45% glycosylated from the glycopeptide T6 (Val^{228}–Arg^{288}) after 27 cycles of solid phase Edman degradation (Fig. 3c). Both Thr(Sac)262 and Ser(Sac)264 were quantitated from the glycopeptide G9 (Ala^{258}–Asp^{292}, Fig. 3d). Thr(Sac)262 was found to be 80% glycosylated, whereas the nearby Ser(Sac)264 was found to be only 45% glycosylated. Ser323(Sac), from the glycopeptide Thr^{316}–Arg^{330} (a further trypsin digest of T3, data not shown), was found to have ~5% glycosylation (Fig. 5). The glycosylation status of Ser323(Sac) represents the

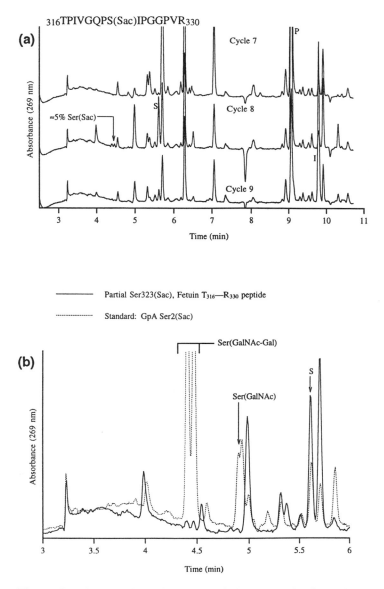

Figure 5 Identification of a new *O*-linked partially glycosylated site in bovine fetuin. (a) Three sequential chromatograms surrounding a partially glycosylated Ser323(Sac) site (with approximately 5% glycosylation) identified by solid phase Edman degradation on the GlycoSite sequenator and 5 mM TEAF, pH 4.0, PTH–amino acid analysis: (top) PTH–Pro-312 in cycle 7; (middle) PTH–Ser(Sac) + PTH–Ser recovered in cycle 8; (bottom) PTH–Ile-314 in cycle 9. PTH–Ser was recovered nonquantitatively in cycle 8 (middle). The peaks eluting at 4 and 5 min in the chromatograms' glycosylation window are phenylthiocarbamyl (PTC)–Ser. (b) Overlay of cycle 8 with PTH–Ser(Sac)2 released after two cycles of solid phase Edman degradation from human GpA (dotted line).

power of solid phase Edman degradation to quantitate partial sites of glycosylation. Previously, we have shown a complex elution pattern for the
same peptide sequence with different amounts of carbohydrate at the same
glycosylation site [11]. This suggests that the glycopeptide and its nonmodified counterparts co-eluted during purification, a phenomenon probably
due to the influence of the sugar on the conformation of the peptide and the
resultant reversed phase interaction. Therefore, quantification of protein
O-glycosylation may require analysis of all peptides and glycopeptides for
each potential glycosylation site.

D. Absorption Phase Identification of Glycoamino Acids

The data reported here have been obtained by solid phase Edman degradation, but there are several reports in the literature on the identification
of PTH–monosaccharide–amino acids recovered from absorption phase
Edman degradation. These include Asn(GlcNAc) and Ser/Thr(GalNAc)
[20–22]. Our studies on the GlycoSite sequenator and the Porton PI2020
have identified the elution position of PTH–Thr(GlcNAc) recovered from
the glycopeptide of the *Dictyostelium* glycoprotein PsA using standard
Beckman sodium acetate/tetrahydrofuran (THF) chromatography (data
not shown). We have also shown that it is possible to recover (nonquantitatively) the disaccharide glycoforms of PTH–Ser(Sac) and PTH–Thr(Sac)
from GpA using absorption phase conditions (data not shown). However,
it is difficult to assign the glycoamino acids using the standard sodium
acetate/THF buffer, as several of the glycoamino acid diastereomers elute
in the void. However, the PTH–glycoamino acids Asn(GlcNAc),
Ser(GalNAc-Gal), Thr(GalNAc-Gal), and Thr(HexNAc) recovered from
gas phase Edman degradation are easily identified using the 5 mM TEAF,
pH 4.0 buffer system.

We have investigated a variety of membrane supports for measuring
the recovery of PTH–glycoamino acids following absorption phase Edman
degradation of both sialylated and desialylated glycopeptides/glycoproteins
(see Table 2). The recovery of sialylated PTH–glycoamno acids was extremely poor, regardless of the membrane support. Figure 6 shows the
corrected yield for absorption phase sequence analysis of asialoglycophorin
A on different membranes using the GlycoSite. The percentage yields for
the absorption phase cycles should be compared with Fig. 3b, which shows
the equivalent sequence using the solid phase chemistry. The yields recovered from the hydrophobic membranes such as PVDF and Hyperbond (Fig.
6c) are sufficient for the purposes of identifying a great majority of sites of
glycosylation found on proteins. Increasing the acidity of the extraction
solvent, ethyl acetate, did not improve the yield of the PTH–glycoamino

Table 2 Absorption Phase Analysis of Glycoproteins and Glycopeptides Using GlycoSite

Protein	Membrane	Glycoamino acid recovery from absorption phase
Sialyoglycophorin A	Electroblotted PVDF (Transblot)	1% PTH–Ser2(GalNAc–Gal)
	[a]Sequelon-DITC	5% PTH–Ser2(GalNAc–Gal)
	Hyperbond	5% PTH–Ser2(GalNAc–Gal)
Asialoglycophorin A	Electroblotted PVDF (Transblot)	30–50% PTH–Ser2(GalNAc–Gal)
		30–50% PTH–Ser2(GalNAc–Gal)
	[a]Sequelon-AA	30–50% PTH–Ser2(GalNAc–Gal)
	Hyperbond	
Dictyostelium PsA	[a]Sequelon-AA	80% PTH–Thr95(GlcNAc)

[a]In these cases the peptide/protein was covalently bound to Sequelon supports but subjected to the gas phase sequencing program.

acids (Fig. 6b). These observations prove that it is possible to sequence electroblotted samples and recover the glycoamino acids, although such experiments are limited to proteins where the glycosylation is in the amino terminal region.

The conclusion from these studies is that GlycoSite offers the glycobiologist the best technolgoy for glycosylation site identification. Solid phase Edman degradation on the GlycoSite quantitatively recovers glycoforms and separates the three common glycoamino acids, Asn(Sac), Ser(Sac), and Thr(Sac). The combined approach of LC/MS and solid phase Edman degradation provides a powerful approach to rapidly identify and recover glycopeptides and glycoamino acids. If there are problems with the recovery and immobilization of glycopeptides using solid phase sequencing, the possibility exists of using the GlycoSite absorption phase mode. However, in many cases the solid phase chemistry is the best choice for proline-rich glycopeptides. We stress that the use of absorption phase sequencing is limited to monosaccharides and disaccharides, and it does not provide the added advantage of the solid-phase chemistry to quantitate glycosylation.

ACKNOWLEDGMENTS

We wish to thank Phill Issacs and Jim Russell of Beckman Instruments Australia for their continuing support of our Glycobiology program. Aspects of this work has been supported by the ARC and NH&MRC grants to AAG, NHP, JWR, and KLW.

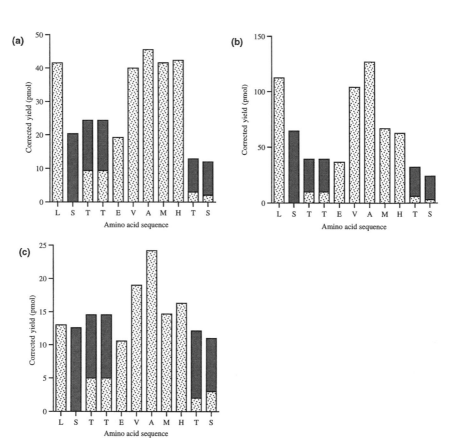

Figure 6 Abosrption phase Edman degradation of glycophorin A. This study only investigated the absorption phase recovery of the disaccharide GalNAc–Gal from the amino terminus of absorbed or electroblotted GpA. Initial yields of GpA vary because different amounts were absorbed onto a variety of supports. (a) Corrected yields for asialoglycophorin A immobilized onto Sequelon-AA and subjected to gas phase program. (b) Corrected yields for asialoglycophorin A immobilized onto Sequelon-AA and subjected to gas phase program that contained a modified extraction solvent of ethyl acetate containing 2% (v/v) TFA. (c) Corrected yields for asialoglycophorin A electroblotted onto the Beckman support Hyperbond.

ABBREVIATIONS

CID	collision induced decomposition
DTT	dithiothreitol
EDC	*N*-ethyl-*N'*-dimethylaminopropyl carbodiimide

ESI/MS	electrospray ionization mass spectrometry
GpA	asialoglycophorin A
HPAEC-PAD	high pH anion-exchange chromatography with pulsed amperometric detection
HPLC	high performance liquid chromatography
LC	liquid chromatography
MES	4-morpholine ethane sulfonic acid
MS	mass spectrometry
NEM	*N*-ethylmorpholine
PTC	phenylthiocarbamyl
PTH	phenylthiohydantoin
PVDF	polyvinylidene difluoride
RT	room temperature
SDS	sodium dodecyl sulfate
TEA	triethylamine
TEAF	triethylammonium formate
TFA	trifluoroacetic acid
THF	tetrahydrofuran

REFERENCES

1. Varki, A. (1993). *Glycobiology* 3, 97–130.
2. Dwek, R. A., Edge, C. J., Harvey, D. J., Wormald, M. R., and Parekh, R. B. (1993). *Ann. Rev. Biochem.* 62, 65–100.
3. Carr, S. A., Hemling, M. E., Bean, M. F., and Roberts, G. D. (1991). *Anal. Biochem.* 63, 2802–2824.
4. Carr, S. A., Hemling, M. E., Folena-Wasserman, G., Sweet, R. W., Anumula, K., Barr, J. R., Huddleston, M. J., and Taylor, P. (1989). *J. Biol. Chem.*, 264, 21286–21295.
5. Carr, S. A., Huddleston, M. J., and Bean, M. F. (1993). *Prot. Sci.* 2, 183–196.
6. Carraway, K. L., and Hull, S. R. (1991). *Glycobiology*, 1, 131–138.
7. Strous, G. J., and Dekker, J. (1992). *Crit. Rev. Biochem. Mol. Biol.* 27, 57–92.
8. Gooley, A. A., Classon, B. J., Marschalek, R., and Williams, K. L. (1991). *Biochem. Biophys. Res. Commun.* 178, 1194.
9. Gooley, A. A., Packer, N. H., Pisano, A., Redmond, J. W., Alewood, P. F., Jones, A., Loughnan, M., Williams, K. L. (1995). In Crabb, J. W., ed., *Techniques in Protein Chemistry VI.* Academic Press, pp. 83–90.
10. Pisano, A., Packer, N. H., Redmond, J. W., Williams, K. L., and Gooley, A. A. (1994). *Glycobiology* 4, 837–844.
11. Pisano, A., Redmond, J. W., Williams, K. L., and Gooley, A. A. (1993). *Glycobiology* 3, 429–435.

12. Gooley, A. A., Pisano, A., Packer, N. H., Redmond, J. W., Williams, K. L. (1994). *Glycoconj. J.* 11, 180–186.
13. Zachara, N. E., Packer, N. H., Temple, M. D., Slade, M. B., Jardine, D. R., Karuso, P., Moss, C. J., Mabbutt, B. C., Curmi, P. M. G., Williams, K. L., and Gooley, A. A. (1996). *Eur. J. Biochem.*, 238, 511–518.
14. Tarr, G. E. (1986). In Shively, J. E., ed., *Methods of Protein Microcharacterization: A Practical Handbook.* Human Press, pp. 162–163.
15. Pisano, A., Packer, N. H., Redmond, J. W., Williams, K. L., and Gooley, A. A. (1996). In Atassi, M. Z., and Appella, E., eds., *Methods in Protein Structural Analysis.* Plenum Press, New York, pp. 69–80.
16. Gum, J. R., Jr., Hicks, J. W., Toribara, N. W., Siddiki, B., and Ki, Y. S. (1994). *J. Biol. Chem.* 269, 2440–2446.
17. Medzihradszky, K. F., Gillece-Castro, B. L., Hardy, M. R., Townsend, R. R., and Burlingame, A. L. (1996). *J. Am. Soc. Mass Spectrom.* 7, 319–328.
18. Medzihradszky, K. F., Maltby, D. A., Hall, S. C., Settineri, C. A., and Burlingame, A. L. (1994). *J. Am. Soc. Mass Spectrom.* 5, 350–358.
19. Edge, A. S. B., and Spiro, R. G. (1987). *J. Biol. Chem.* 262, 16135–16141.
20. Robb, R. J., Kutny, R. M., Panico, M., Morris, H., DeGrado, W. F., and Chowdhry, V. (1983). *Biochem. Biophys. Res. Commun.* 116, 1049–1055.
21. Paxton, R. J., Mooser, G., Pandle, H., Lee, T. P., and Shively, J. E. (1987). *Proc. Natl. Acad. Sci. USA* 84, 920–924.
22. Abernethy, J. L., Wang, Y., Eckhardt, A. E., and Hill, R. L. (1992). In Angeletti, R. H., ed., *Techniques in Protein Chemistry III.* Academic Press, New York, pp. 277–286.

19

High Energy Collision-Induced Dissociation of *O*-Linked Glycopeptides

Katalin F. Medzihradszky
University of California–San Francisco, San Francisco, California

I. INTRODUCTION

Two fundamental questions must be addressed in the structural elucidation of protein glycosylation: (1) What is the structure of the oligosaccharides? and (2) Where are they located on the polypeptide chain? *O*-glycosylation poses unique challenges in both respects. There is no known consensus sequence for this post-translational modification, and *O*-linked carbohydrates usually occur in clusters of amino acid sequences that have no sites for specific proteolytic cleavages. There are a wide variety of *O*-linked structures that occur even in mammalian cells and do not share a common core structure [1–4]. No enzymatic or chemical methods are available to cleave all *O*-linked oligosaccharides without destroying the peptide chain. Modified Edman sequencing can identify Ser and Thr residues bearing the most common mammalian *O*-linked structure, namely, Gal–GalNAc oligosaccharides with high sensitivity [5], but the method fails to address sialylation and other structural aspects of *O*-linked carbohydrate structures.

Mass spectrometry has been used to identify all types of post-translational and other covalent modifications (for review, see [6]). In most cases, proteases of high specificity (e.g., trypsin) are used to generate peptides for mass spectrometric analysis. From the difference between the predicted and observed masses, the chemical structure of even undescribed covalent modifications can be deduced. Fragmentation of the modified species, that is, tandem mass spectrometry experiments, using collision-induced dissociation or post-source decay analysis will reveal more structural detail of the modifying group, identify the sequence of the modified peptide, and locate the site of modification. This chapter illustrates how liquid secondary ion mass spectrometry (LSIMS) with high energy collision-induced dissociation (CID) analysis can be utilized in the structure elucidation of *O*-linked glycopeptides.

II. MATERIALS AND METHODS

A. Isolation of Glycopeptides

Recombinant human PDGF-B was reduced, carboxymethylated, and digested with trypsin as described earlier [7]. The digest components were separated by reversed phase high performance liquid chromatography (HPLC) [7]. Bovine fetuin was digested extensively with Pronase as reported [8]. The isolation and derivatization of these glycopeptides are described in detail in the same paper [8].

B. Mass Spectrometry

High energy CID spectra were recorded on a Kratos (Manchester, UK) Concept IIHH tandem mass spectrometer of EBEB geometry, equipped with an LSIMS source, and a photodiode multichannel array detector (capable of monitoring 4% of the mass range simultaneously) [9]. The collision gas was He, and its pressure was set to reduce the precursor ion abundance by 70%. The collision cell was floated at 4 kV. Glycopeptides (300 pmol–1 nmol) were loaded with 1 μL of liquid matrix (thioglycerol : glycerol, 1 : 1, with 1% trifluoroacetic acid (TFA)).

III. RESULTS AND DISCUSSION

Soft ionization methods suitable for biopolymer analysis usually yield only molecular weight information in the form of protonated molecular ions (positive ion analysis). To gain further structural information about the compounds, fragmentation has to be induced. The energy required for the fragmentation process is usually provided by collision with an inert gas; hence, the method has been termed collision-induced dissociation (CID) analysis. In tandem mass spectrometry experiments, the ion of interest is selected in the first mass spectrometer. The collision cell is situated between the two mass spectrometers. The fragment ions formed within the cell are then detected by the second mass spectrometer. Because only the precursor ion selected will be activated, only fragments reflecting its structure will be recorded during the analysis. Under high energy CID conditions (magnetic sector instruments, 4–8 keV collision energy), usually only the C^{12} isobar of the MH^+ ion is activated; thus, the experiments yield monoisotopic spectra. Glycopeptides described in this chapter were analyzed in positive ion CID. Scheme 1 shows the most commonly observed carbohydrate fragmentation in these experiments [10]. Scheme 2 summarizes the rules and nomenclature of peptide fragmentation (for nomenclature, see [11]; for an overview, see [12]).

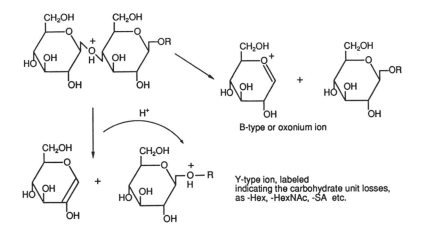

Scheme 1   Glycosidic bond cleavage in carbohydrate structures (reproduced from [12]) [10].

Immonium ions: $NH_2^+=CH-R \rightarrow$ amino acid composition

Sequence information:

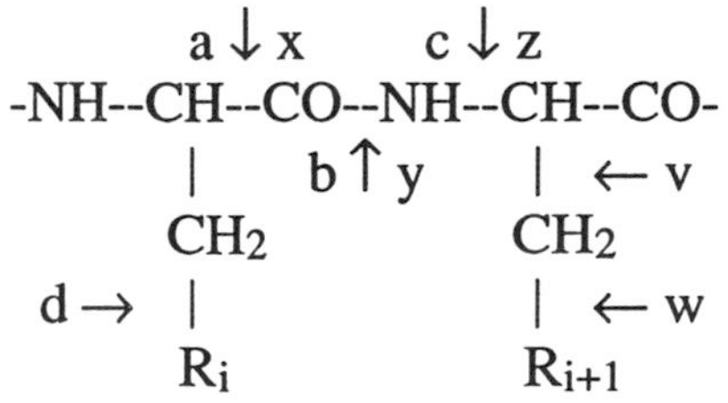

$$
\begin{array}{c}
\mathrm{a}\downarrow\mathrm{x} \qquad \mathrm{c}\downarrow\mathrm{z} \\
\text{-NH--CH--CO--NH--CH--CO-} \\
|\quad \mathrm{b}\uparrow\mathrm{y} \quad | \;\leftarrow\mathrm{v} \\
\mathrm{CH_2} \qquad\qquad \mathrm{CH_2} \\
\mathrm{d}\rightarrow | \qquad\qquad |\;\leftarrow\mathrm{w} \\
\mathrm{R_i} \qquad\qquad \mathrm{R_{i+1}}
\end{array}
$$

a, b, c, x, y, z - sequence ions   d, v, w - side-chain fragments (satellite ions)

a, b, c, d - numbered from the N-terminus
v, w, x, y, z - numbered from the C-terminus

Arg, His, Lys $\rightarrow$ preferential charge retention

Scheme 2   Peptide fragmentation in high energy CID [11,12].

Tryptic peptides of recombinant human PDGF were analyzed by LSIMS. The measured molecular masses were compared with the predicted mass values. A few fractions exhibited molecular weights 162 or 324 Da higher than some of the predicted tryptic peptides. Because the protein was expressed in yeast, one possible explanation for these observations was the presence of O-linked mannosylated glycopeptides. The suspected glycosylated species indeed eluted earlier than the nonmodified counterparts, as expected for a more hydrophilic species. In addition, methylation analysis identified Man residues on the protein. High energy CID analysis was then employed to determine the exact sites of glycosylation. Figure 1 shows the high energy CID spectrum of one of the mannosylated tryptic peptides. From the known amino acid sequence of PDGF and taking the specificity of trypsin into consideration, it was prediced that this species with MH^+ at

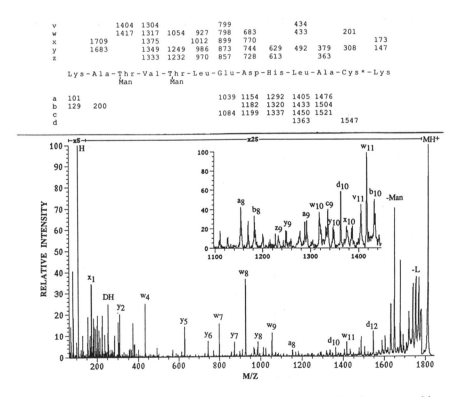

Figure 1 High energy CID spectrum of tryptic glycopeptide from recombinant human PDGF-B. The insert shows an expanded region of the spectrum illustrating the diagnostic ions. All ions identified are included in the ion table above the spectrum. (Reproduced with permission from Ref. 7.)

m/z 1810 represented a doubly mannosylated tryptic peptide, Lys[86]–Lys[98]. High energy CID analysis of this compound revealed two substituents with incremental losses of 162 Da, that is, yielded ions at *m/z* 1648 (-162 Da) and *m/z* 1486 (-2×162 Da), corresponding to the Man losses from the molecular ion. The relatively high abundance of the *m/z* 1648 ion suggests that the cleavage of the glycosidic bond is a preferred fragmentation step under high energy CID conditions. Tryptic peptides usually display ions predominantly with charge retention at the C terminus because of the presence of basic residues, such as Arg or Lys, at this position. When the peptide bond itself is cleaved, the newly formed C-terminal fragment is designated as a **y** ion (see Scheme 2). Masses of all the **y** ions can be predicted for a known sequence. Scheme 3 shows the **y** ions observed for this PDGF glycopeptide. The C-terminal fragment containing Thr-90 indicates the presence of a single hexose residue at this site (162 Da shift). Similarly, the **y** ion formed *via* peptide bond cleavage between Thr-88 and Val-89 shows only a 162 Da shift in comparison with the predicted mass for the nonmodified sequence. Ions including the second Thr residue (y_{11}, y_{12}) exhibit the presence of the second modification (324 Da shift). Thus, it can be concluded that both hydroxy amino acids (Thr-88 and Thr-90) are glycosylated with a single hexose residue. The other ion series can be analyzed similarly. The insert in Fig. 1 shows an expanded region from the spectrum with some of the other diagnostic ions (z_9, w_{10}, x_{10}, v_{11}, w_{11}). All these ions indicate the presence of only one mannose residue. Interestingly, fragments v_{11} and w_{11} are characteristic for the eleventh amino acid from the C terminus (Thr-88), but they are formed *via* side chain cleavages (see Scheme 2 [11]) and thus do not contain the sugar modifying this amino acid.

O-glycosylated mammalian proteins usually contain more complicated carbohydrates. Our model protein, bovine fetuin, contains all of its major O-linked glycosylation sites in a single tryptic peptide of 61 amino acid residues, Val[228]–Arg[288] [13,14]. To isolate glycopeptides containing individual glycosylation sites, the protein was digested with Pronase. During the separation process, the O-linked glycopeptides were identified by carbohydrate composition analysis. The glycopeptides were derivatized with tBOC–Tyr to aid their HPLC separation and to increase their mass spectrometric detection sensitivity. Their molecular weights were measured by LSIMS, and their structures were determined by high energy CID analysis as follows. The CID spectrum of a tBOC–Tyr-derivatized Pronase-digest-derived glycopeptide is shown in Fig. 2. MH$^+$ is at *m/z* 1866.9. The spectrum is dominated by ions formed from glycosidic bond cleavages. Ions at *m/z* 1575, 1413, 1122, and 919 indicate the sequential losses of Neu5Ac, Hex, Neu5Ac, and HexNAc, respectively. The ion pairs that are separated

MH$^+$ at m/z 1810 = Tryptic PDGF peptide [86-98]+ 324 Da

88   90
LysAlaThrValThrLeuGluAspHisLeuAlaCys*Lys
12 11 10 9 8 7 6 5 4 3 2 1

High energy CID analysis → 162 loss from MH$^+$ → -Hex
↓

y_1= 147   Lys
    ↓ + Cys* = + 161 Da (S-carboxymethyl Cys)
y_2 = 308
    ↓ + Ala = +71 Da
y_3 = 379
    ↓ + Leu = + 113 Da
y_4 = 492
    ↓ + His = + 137 Da
y_5 = 629
    ↓ + Asp = + 115 Da
y_6 = 744
    ↓ + Glu = + 129 Da
y_7 = 873
    ↓ + Leu = + 113 Da
y_8 = 986

    ↓ + Thr = + 101 Da
    ↓ + 162 Da (Hex)
y_9= 1249
    ↓ + Val = + 99 Da
y_{10} = 1348

    ↓ + Thr = + 101 Da
    ↓ + 162 Da (Hex)
(y_{11} = 1611 - was not detected)
    ↓ + Ala = + 71 Da
y_{12} = 1682

Scheme 3 High energy CID analysis of a PDGF tryptic peptide.

by 100 Da (e.g., at masses 1575, 1475; 1413, 1313, etc.) are due to the cleavage of the N-terminal tBOC group. Nonreducing terminal fragments were m/z 292 from Neu5Ac, m/z 495 from Neu5Ac–HexNAc, m/z 657 from Neu5Ac–(Hex–HexNAc), and from the intact tetrasaccharide at m/z 948. This ion series was consistent with the observed sequential losses from the molecular ion, and with a tetrasaccharide structure of Neu5Ac–Hex–(Neu5Ac)–HexNAc. The oxonium ion formed by the cleavage of both sialic acids (m/z 366) was not detected. Similarly, the Neu5Ac–Hex oxonium ion was virtually absent. Although this CID spectrum is dominated by ions formed *via* carbohydrate fragmentation, the identity of the peptide also

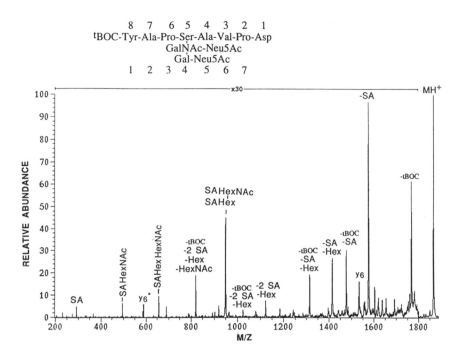

Figure 2 High energy CID spectrum of a fetuin Pronase glycopeptide derivatized with tBOC–Tyr. MH^+ = 1866.9. The Y-type carbohydrate fragments (Scheme 1, [10]) are labeled with the corresponding sugar unit losses. The B-type carbohydrate ions are labeled with the sugar units comprising the oxonium ion. This CID spectrum does not reveal the identity of the hexose or *N*-acetyl hexosamine units. Fragment y_6 is formed *via* peptide bond cleavage between the Ala and Pro residue, with charge retention on the C terminus, and bearing the carbohydrate structure. Fragment y_6^* is the corresponding gas phase deglycosylated ion. (Reproduced with permission from Ref. 8).

could be determined. The molecular mass of the peptide was revealed from the gas phase deprotected ($-$tBOC) and deglycosylated ion at m/z 819. By subtraction of the Tyr residue weight from this fragment, an MH^+ value for the native peptide was calculated to be 656 Da. A search of the fetuin sequence (MacBioSpec, Sciex, Thornhill, Ontario, Canada) gave four matches, two containing Ser-253 ([251]APSAVPD[257] or [252]PSAVPDA[258] with 656.3 Da as MH^+), and two containing Ser-278 ([274]VVGPSV[279] or [275]VVG-PSVV[280] with 656.4 Da as MH^+). Using the aforementioned computer program (or manual calculations) the **a**, **b**, and **y** fragment ions expected for these sequences can be predicted (see a detailed description in [12]). Comparing these tables with the observed peptide fragments, the sequence was determined as ^{251}APSAVPD^{257}. For example, an abundant fragment (y_6) due to a cleavage at the C-terminus of Pro-252 and bearing the complete Neu5Ac–Hex–(Neu5Ac)–HexNAc structure was detected at m/z 1532. The corresponding gas phase partially and completely deglycosylated fragments at m/z 1241 (y_6–Neu5Ac), 1079 (y_6–Neu5Ac–Hex), and 585 (y_6 without the carbohydrate) were also observed. Again, fragments formed *via* elimination of both sialic acids were not detected. Similarly, fragment ion b_6 (not labeled) with the tetrasaccharide was detected at m/z 1536, as well as b_6–Neu5Ac (at m/z 1245), b_6–Neu5Ac–Hex (at m/z 1083), and b_6 deglycosylated (at m/z 589).

A Neu5Acα(2,3)Galβ(1,4)GlcNAcβ(1,6)[Neu5Acα(2,3)Gal β (1,3)]–GalNAc hexasaccharide also has been described in bovine fetuin, but its peptide location has not been determined [15]. Carbohydrate composition analysis indicated the presence of GlcNAc in some of the glycopeptides isolated after pronase and neuraminidase digestion. The GlcNAc-containing, tBOC-Tyr-derivatized glycopeptides were analyzed by LSIMS (MH$^+$ at m/z 1887.9 and 1959.0) and by high energy CID, and they were found to contain only the asialo-hexasaccharide at Thr-262 and the disaccharide at Ser-264 [16]. The peptide sequences were determined as tBOC–Tyr–^{260}GPTPSA^{265} and tBOC–Tyr–^{260}GPTPSAA^{266}, similarly to the method described for the other fetuin glycopeptide. To confirm peptide loci assignments, the shorter glycopeptide was digested with endo-α-N-acetylgalactosaminidase. Only Galβ(1,3)GalNAcα structures are cleaved by this enzyme. Thus, only a single product with a 365 Da lower molecular weight (−Gal-GalNAc) than the untreated glycopeptide was expected from this digestion. To our surprise, two new compounds were isolated from this digest by reversed phase HPLC with molecular weights 162 and 568 Da lower than the undigested glycopeptide, corresponding to a Hex residue loss, and to the anticipated cleavage of a Hex–HexNAc unit plus a Hex residue, respectively. Figure 3 shows the high energy CID spectrum of the glycopeptide with MH$^+$ at m/z 1725.5, that is, 162 Da less than the undigested glycopeptide. This spectrum is again dominated by fragments formed *via* glycosidic bond cleavages. Sugar unit losses from the molecular ion were detected at m/z 1563 (−Hex), 1522 (−HexNAc), 1360 (−Hex-HexNAc), 1198 (−Hex$_2$HexNAc), 1157 (−HexHexNAc$_2$), 995 (−Hex$_2$-HexNAC$_2$), and 792 (−Hex$_2$HexNAc$_3$). The ion pairs 100 Da apart are present due to the cleavage of the N-terminal tBOC group, as previously shown. The ion at m/z 1522, corresponding to the loss of a HexNAc from the molecular ion, indicates that a terminal HexNAc residue has been produced in the endo-α-N-acetylgalactosaminidase digestion. If the intact tetrasaccharide were present, an abundant ion would be seen at m/z 731

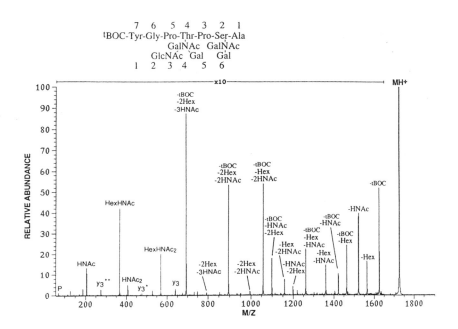

Figure 3 High energy CID spectrum of a tBOC–Tyr-derivatized asialofetuin gly-copeptide after *O*-glycanase digestion. An asialofetuin glycopeptide containing the tetra- and disaccharide was further digested with *O*-glycanase. Two new glycopep-tides were isolated from the digest. MH^+ for this glycopeptide was at m/z 1725.5. The carbohydrate fragment ions are labeled as in Fig. 2. Ion y_3 is a peptide fragment formed *via* peptide bond cleavage between the Thr and Pro residues and with charge retention at the C terminus, bearing the disaccharide. Ions y_3^* and y_3^{**} are the corresponding partially or fully deglycosylated fragments. (Reproduced with permission from Ref. 8).

[16,17]. However, in this spectrum the nonreducing terminal oxonium ions at m/z 204 (HexNAc), 366 (HexHexNAc), 407 (HexNAc$_2$), and 569 (Hex-HexNAc$_2$) suggest the presence of a di- and trisaccharide (Fig 3). The pres-ence of a trisaccharide and the terminal HexNAc residue suggest that the Gal residue linked to the GlcNAc in the original tetrasaccharide was re-moved during the digestion. The endo-α-*N*-acetylgalactosaminidase prepar-tion used was isolated from *Streptococcus pneumoniae*, which is known to contain a β-galactosidase more specific for $\beta(1,4)$ linkages [18]. The hexasaccharide contains the only Gal$\beta(1,4)$ linkage (to the GlcNAc residue) among the *O*-linked carbohydrate structures described for bovine fetuin [15,19]. Fragment ion y_3 (at m/z 639) formed *via* peptide bond cleavage between the Thr and the second Pro residue, with charge retention on the C

terminus, shows a Hex–HexNAc disaccharide on Ser-264. This fragment was also detected as partially and completely gas phase deglycosylated ions at m/z 477 ($-$Hex) and at m/z 274. Figure 4 shows the CID spectrum of the other digestion product with MH^+ at m/z 1360.5. The carbohydrate losses and tBOC group cleavages yielded abundant ions at m/z 1198 ($-$Hex), 1098 ($-$tBOC, $-$Hex), 1157 ($-$HexNAc), 1057 ($-$tBOC, $-$Hex-NAc), 995 ($-$HexHexNAc), 895 ($-$tBOC, $-$HexHexNAc), 792 ($-$Hex-HexNAc$_2$), and 692 ($-$tBOC, $-$HexHexNAc$_2$) and indicated the presence of a trisaccharide. Oxonium ions were detected at m/z 204, 366, 407, and 569, suggesting that this trisaccharide consists of a HexNAc and Hex linked to a HexNAc, consistent with the observation of Hex or HexNAc losses from the molecular ion. The glycosylated $\mathbf{y}_3$ fragments, which were ob-

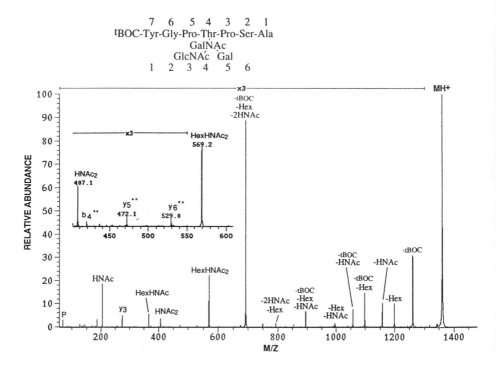

Figure 4 High energy CID spectrum of a tBOC–Tyr-derivatized asialo fetuin glycopeptide after *O*-glycanase digestion. MH^+ = 1360.5. This glycopeptide was prepared as described in Fig. 3. The fragments are labeled as in Fig. 2. The inset shows the lack of glycosylated $\mathbf{y}_3$ ion at m/z 477, which was present in the diglycosylated glycopeptide (see Fig. 3). The asterisks in the peptide fragment labels indicate gas phase deglycosylation and the loss of the tBOC group ($\mathbf{b}_4^{**}$). (Reproduced with permission from Ref. 8)

served prior to the *O*-glycanase treatment [17] and in the diglycosylated species (Fig. 3) at m/z 639 and 477, were not detected for this glycopeptide (Fig. 4). The lack of these fragment ions (see inset in Fig. 4) shows that Ser-264 is not occupied.

IV. CONCLUSIONS

Mass spectrometry reveals unexpected covalent modifications of peptides. Tandem mass spectrometry with CID analysis provides the location and, in many cases, the structural information to characterize covalent modifications. From an ion series of sugar unit losses, the composition and the sequence of the oligosaccharide can be established, as well as the molecular weight of the nonmodified peptide (see peptide fragment). Nonreducing or internal carbohydrate fragments supply additional information on the sequence and branching of the carbohydrate. Unfortunately, neutral sugars usually do not yield these fragments if there is any other structure that can preferentially retain the charge, such as amino acids, HexNAc units, and so forth. In the CID spectrum of *O*-linked glycopeptides, peptide fragments usually are observed as ions of lower abundance and are largely gas phase deglycosylated. These fragments (immonium ions, **y** and **b** ions for "neutral" sequences), in most cases, allow the identification of the peptide. Basic amino acids such as His, Lys, or Arg retain the charge preferentially and may facilitate other bond cleavages (see Scheme 2 and Fig. 1). Carbohydrate-bearing peptide fragments reveal the glycosylation loci. The **y** ions formed at Pro residues are usually abundant; thus, these fragments aid the glycosylation site determination. Most of these phenomena, such as the gas phase deglycosylation of the peptide, sequential carbohydrate losses from the molecular ion, and formation of nonreducing terminal or internal carbohydrate fragments with charge retention at the N-containing carbohydrate units, have also been reported for low energy CID analysis of *O*-linked glycopeptides [20,21].

ACKNOWLEDGMENTS

I would like to thank F. C. Walls for his technical assistance. This work was supported by grants from the National Center for Research Resources, National Institutes of Health (Grant RR01614), and the National Science Foundation (Grant 8700766).

ABBREVIATIONS

CID	collision-induced dissociation
EBEB	electrostatic analyzer (electric sector), magnetic analyzer (magnetic sector)

HPLC   high performance liquid chromatography
LSIMS   liquid secondary ion mass spectrometry
PDGF-B   B-chain of platelet-derived growth factor
tBOC   t-butoxycarbonyl
TFA   trifluoroacetic acid

REFERENCES

1. Schachter, H., and Brockhausen, I. (1992). In Allen, H. J., and Kisailus, E. C., eds., *Glycoconjugates*. Marcel Dekker, New York, pp. 263–332.
2. Lehle, L., and Tanner, W. (1995). In Monteruil, J., et al., eds., *Glycoproteins*, Elsevier Science, New York, pp. 475–509.
3. Hart, G. W., Kelly, W. G., Blomberg, M. A., Roquemore, E. P., Dong, L.-Y. D., Kreppel, L., Chou, T.-Y., Snow, D., and Greis, K. (1994). In *44. Colloquium Mosbach 1993 Glyco- and Cellbiology*. Springer-Verlag, Berlin pp. 91–103.
4. Harris, R. J., and Spellman, M. W. (1993). *Glycobiology* 3, 219–224.
5. Gooley, A. (1997). In Townsend, R. R. and Hotchkiss, A., eds., *Techniques in Glycobiology*.
6. Burliname, A. L., Boyd, R. K., and Gaskell, S. J. (1994). *Anal. Chem.* 66, 634R–683R.
7. Settineri, C. A., Medzihradszky, K. F., Masiarz, F. R., Chu, C., George-Nascimento, C., and Burlingame, A. L. (1990). *Biomed. Environ. Mass Spectrom.* 19, 665–676.
8. Medzihradszky, K. F., Gillece-Castro, B. L., Hardy, M. R., Townsend, R. R., and Burlingame, A. L. (1996). *J. Am. Soc. Mass Spectrom.* 7, 319–328.
9. Walls, F. C., Baldwin, M. A., Falick, A. M., Gibson, B. W., Kaur, S., Maltby, D. A., Gillece-Castro, B. L., Medzihradszky, K. F., Evans, S., and Burlingame, A. L. (1990). In Burlingame, A. L., and McCloskey, J. A., eds., *Biological Mass Spectrometry*. Elsevier, Amsterdam, pp. 197–216.
10. Domon, B., and Costello, C. E. (1988). *Glycoconj. J.* 5, 397–409.
11. Biemann, K. (1990). *Meth. Enzymol.* 193, 886–887.
12. Medzihradszky, K. F., and Burlingame, A. L. (1994). *Methods: Companion to Meth. Enzymol.* 6, 284–303.
13. Carr, S. A., Huddleston, M. J., and Bean, M. F. (1993). *Prot. Sci.* 2, 183–196.
14. Medzihradszky, K. F., Maltby, D. A., Hall, S. C., Settineri, C. A., and Burlingame, A. L. (1994). *J. Am. Soc. Mass Spectrom.* 5, 350–358.
15. Edge, A. S. B., and Spiro, R. G. (1987). *J. Biol. Chem.* 262, 16135–16141.
16. Townsend, R. R., Medzihradszky, K. F., Hardy, M. R., Rohrer, J., and Burlingame, A. L. (1991). Poster Presentation, XI International Symposium on Glycoconjugates, Toronto.
17. Townsend, R. R., Medzihradszky, K. F., and Burlingame, A. L., in preparation.

18. Jacob, G. S., and Scudder, P. (1994). *Meth. Enzymol.* 230, 280–299.
19. Spiro, R. G., and Bhoyhroo, V. D. (1974). *J. Biol. Chem.* 249, 5704–5717.
20. Linsley, K. B., Chan, S. Y., Chan, S., Reinhold, B. B., Lisi, P. J., and Reinhold, V. R. (1994). *Anal. Biochem.* 219, 207–217.
21. Reinhold, B. B., Hauer, C. R., Plummer, T. H., and Reinhold, V. N. (1995). *J. Biol. Chem.* 270, 13197–13203.

20

Analysis of Permethylated Glycoprotein Oligosaccharide Fractions by Gas Chromatography and Gas Chromatography–Mass Spectrometry

**Kristina A. Thomsson, Niclas G. Karlsson, Hasse Karlsson,
and Gunnar C. Hansson**
Göteborg University, Gothenburg, Sweden

I. INTRODUCTION

Oligosaccharides found in nature, bound either to proteins or lipids or in free form, usually occur in very complex mixtures. The variability is not only due to the different constituting monosaccharide building blocks, but also the possibilities of different linkage positions, configurations, branching, and substitutions of the hydroxyl and amine groups with, for example, sulfate or acetyl groups. With few exceptions, the oligosaccharides bound to glycoproteins can be divided into *N*- and *O*-linked oligosaccharides, the former linked to asparagine and the latter linked to either serine or threonine. The large glycoproteins found on mucosal surfaces, the so called mucins, carry a variety of mostly *O*-linked oligosaccharides localized to domains with a high degree of glycosylation [1,2].

Analysis of naturally occurring oligosaccharide mixtures is analytically demanding. The often applied methods for elucidating complex *O*-glycosylation on proteins are based on the release of oligosaccharides from the peptide backbone, purification, and structural characterization of individual oligosaccharides [3]. A first separation step is usually anion-exchange liquid chromatography, separating the oligosaccharides into neutral and acidic components, followed by subfractionation by high performance liquid chromatography (HPLC) using amino bonded or reversed phase columns. Individual oligosaccharides can be isolated and complete structures elucidated with carbohydrate compositional analysis, glycosidase treatments, linkage analysis, mass spectrometry, and nuclear magnetic resonance (NMR). Generally, relatively large amounts of starting material are required for full structural information.

We have chosen another strategy for profiling complex oligosaccharide mixtures, based on an initial separation of the oligosaccharides into three separate groups followed by a relatively fast, but still detailed, patterning of these groups (Fig. 1). After releasing the oligosaccharides, these are separated into fractions of neutral and sialic acid– and sulfate-containing oligosaccharides using a simple on-column procedure [4]. The neutral oligosaccharides are directly eluted, the carboxyl group of sialic acid residues is converted into its corresponding methyl ester and also eluted, after which the sulfate-containing ones are eluted with a salt solu-

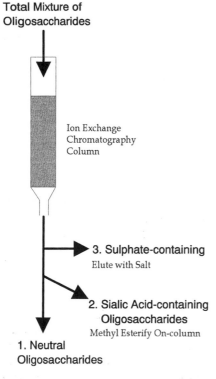

Figure 1 The oligosaccharide mixture is separated into neutral and sialic acid– and sulfate-containing fractions on an anion-exchange column. The neutral oligosaccharides are eluted with dry methanol. The sialic acid–containing species are methyl esterified on-column with methyl iodide in dimethyl sulfoxide (DMSO) and then eluted with methanol. The sulfate-containing oligosaccharides are eluted with pyridinium acetate.

tion. The neutral and sialic acid–containing oligosaccharides are analyzed as their permethylated derivatives utilizing high temperature gas chromatography (GC) and gas chromatography–mass spectrometry (GC/MS) [5–11]. The major advantage of the capillary GC technique is its high resolution, which allows separation of isomeric oligosaccharides. However, despite a substantially increased mass range obtained by high temperature GC, there is an upper restriction in oligosaccharide size to approximately 10 sugar residues. The sequence of individual components can be elucidated from the recorded mass spectra together with some information on linkage positions. The chromatograms can be used to obtain semiquantative information, suitable for comparative studies. Compared with the previously described approaches in which purified individual oligosaccharides are analyzed, this strategy has the advantage of consuming less time and smaller amounts of starting material. The approach presented is suitable for comparing the O-glycosylation of highly purified glycoproteins and altered expression of saccharide epitopes in cells or organs in normal and different disease states.

II. MATERIALS AND METHODS

A. Release of O-Linked Oligosaccharides

The minimum amounts enabling the full procedure with GC and GC/MS are in the low milligram range (1–3 mg) when working with mucin glycoproteins/peptides. The oligosaccharides were released as alditols using a solution of 0.05 M KOH and 1.0 M NaBH$_4$ (1 mL/mg glycopeptide, 45°C for 45 h) [9,12]. The solution was neutralized by adding 2.0 M acetic acid. Desalting was performed on a cation-exchange column (AG 50W-X8, BioRad, Hercules, California) using 0.5 mL resin/mL sample. The sample was eluted with water (5 mL/mL resin), and the water evaporated. Traces of borate were removed by repeated additions (usually four) of methanol and a few drops of glacial acetic acid and evaporation.

B. Oligosaccharide Fractionation

The oligosaccharides were fractionated [4] on a DEAE-Sephadex A-25 anion-exchange column (Pharmacia, Uppsala, Sweden) in its acetate form, using 0.5 mg resin/mg glycopeptide. The applied sample was allowed to equilibrate on the top of the column for a few hours, followed by elution of the neutral oligosaccharides with dry methanol (10 mL/mL resin). The retained sialic acid–containing oligosaccharides were methyl esterified on-column. A mixture of dimethyl sulfoxide (DMSO) and methyl iodide 5/1 v/v (0.6 mL/mL resin) was added to the column in three portions with intervals of 5 min. The neutral sialic acid–containing oligosaccharide methyl esters were collected by washing the column with dry methanol (10 mL/mL resin). The oligosaccharides containing sulfate groups were eluted with 1.0 M pyridinium acetate, pH 5.4 (10 mL/mL resin). Analysis of the sulfated oligosaccharides is not further discussed here. Collected fractions were concentrated and lyophilized.

C. Conversion of Sialic Acid Methyl Esters to Methyl Amides

The sialic acid–containing oligosaccharides were converted to their methyl amide derivatives [4]. This step was necessary due to the relative instability of the methyl esters and the poor yields when permethylating sialylated oligosaccharides containing a reduced GalNAc. The sample was dissolved in dry methanol (1 mL), methyl amine (300 μL, 25% in dry methanol) was added, and the mixture stirred for 10 min. The reaction was stopped by freezing the sample at $-20°C$, and reagents removed by vacuum centrifugation/lyophilization.

D. Permethylation

Neutral and sialic acid–containing oligosaccharides were permethylated according to the method of Ciucanu and Kerek [13], slightly modified [14]. The sample was dried in a stream of nitrogen, then further dried under vacuum for 30 min. Dry DMSO (0.5 mL), methyl iodide (0.1 mL), and powdered dry NaOH (approximately 25 mg) were added, and the mixture stirred with a magnetic bar for 10 min at room temperature. The reaction was stopped by adding 0.1 M HCl (2 mL), followed by chloroform (1 mL). The water phase was removed after agitation and centrifugation; the chloroform phase containing the permethylated species was then further washed five times with 1 mL of glass distilled water. The chloroform phase was moved to a new tube, and the last water phase washed with 0.5 mL chloroform. The pooled chloroform phases were dried in a stream of nitrogen.

E. Evaluating the Oligosaccharide Size Distribution in the Mixture

When the sample mixture contains large amounts of oligosaccharides with more than about 10 sugar residues, these can degrade the chromatography and the column quality. It is recommended that unknown samples are analyzed with fast atomic bombardment- or matrix-assisted laser desorption ionization mass spectrometry to obtain information on the size distribution of the oligosaccharides. Larger permethylated oligosaccharides could be fractionated from smaller ones using gel chromatography (LH-20, Pharmacia, 18 mL resin). Methanol is used as mobile phase at a flow rate of 0.25 mL/min. Fractions of 0.5 mL are collected, concentrated in a stream of nitrogen, and analyzed by GC. The oligosaccharides with a size suitable for GC analysis elute approximately in fractions 15–20.

F. Preparation of Capillary Columns for High Temperature GC

Columns used for high temperature GC have a thin (0.02–0.05 μm) stationary phase (PS 264, Fluka, Buchs, Switzerland). The columns are to our knowledge not commercially available but can be ordered by special request (Chrompack, Middelburg, The Netherlands) or prepared in the laboratory as described [9]. The fused silica column (10 m × 0.25 mm i.d., HT-polyimide coated) is deactivated with D4 (octamethyl tetracyclosiloxane) at 390°C overnight. After extraction with dichloromethane (3 mL) and drying in a GC oven at 150°C with H_2, the column is statically coated using a solution of stationary phase dissolved in dichloromethane/n-pentane 1 : 1. The stationary phase solution should be prepared the day before at a concentration defined according to desired film thickness d_f (d_f [μm] = conc [mg/mL] × d [mm]/4ρ), where d is the inner diameter of the column and ρ the density of the stationary phase ($\rho \approx 1$). The capillary is filled with the solution of stationary phase containing freshly added cross-linking agent (dicumyl peroxide, Merck, Darmstadt, Germany) dissolved in dichloromethane. The cross-linking agent is added at an amount corresponding to 0.5–1% (w/w) of the stationary phase. After coating, the column is flushed with nitrogen, evacuated, sealed, and cross-linked at 145°C for 30 min. The column is extracted with dichloromethane (3 mL), dried with nitrogen, and conditioned using a 5°C /min temperature program from 50°C up to 390°C, kept constant thereafter for 2 h.

G. Chromatography Conditions

A HP5890-II gas chromatograph has been used for the GC and GC/MS analysis. For GC, hydrogen was used as carrier gas (constant pressure mode, 0.5–0.7 bar, linear gas velocity of 80–123 cm/sec at 80°C). An oxygen trap (Oxypurge, Alltech) was installed in the carrier gas line. The flame ionization detector (FID) was kept at 400°C. For GC/MS, helium was used as carrier gas (constant flow mode, vacuum compensation, flow 1.5–1.7 mL/min, linear gas velocity of 73–79 cm/sec at 80°C). Temperature programs used in GC and GC/MS were, for permethylated neutral oligosaccharides: 80°C for 1 min, 80–390°C at 10°C/min, 390°C for 5 min; and for permethylated sialic acid–containing oligosaccharides: 80°C for 1 min, 80–200°C at 50°C/min, 200–390°C at 10°C/min, 390°C for 5 min. The samples were dissolved in ethyl acetate and injected on-column (0.5–1 μL).

H. Mass Spectrometry Conditions

For electron impact mass spectrometry a JEOL SX-102A (Jeol, Tokyo) was used with the following conditions: interface temperature, 380°C; ion source temperature, 370°C; electron energy, 70 eV; trap current, 300 μA; acceleration voltage, +10 kV; mass range scanned, m/z 100–1600; total cycle time, 1.8 sec; resolution, 1400 ($m/\Delta m$, 10% valley definition); ion multiplier, 1.2 kV; pressure in the ion source region, 7 × 10^{-4} Pa. The interface between the gas chromatograph and the mass spectrometer has been slightly modified to eliminate the presence of cold and hot spots.

III. RESULTS AND DISCUSSION

The GC and GC/MS approach presented has been very useful for structural analysis of oligosaccharides with up to about 10 sugar residues. This sets a limit to the oligosaccharide types analyzable; namely, it excludes most Asn-linked oligosaccharides. The oligosaccharides attached to proteins and sometimes also to lipids [15] have to be released. This can be achieved by different methods, but the alkaline borohydride method for O-linked oligosaccharides is the only one described here. This group of saccharides is typical for mucins, the subgroup of glycoproteins being the focus of our group. Typical for these mucin oligosaccharides is their high structural variability, demanding the most advanced analytical tools. However, this is usually not sufficient, and in addition oligosaccharides having different functional groups have to be derivatized and analyzed differently. The procedure with an on-column derivatization, and separation into a neutral fraction, one containing oligosaccharides with carboxylic groups (usually sialic acid), and a fraction with other negatively charged groups (usually sulfate) is simple and fast [4]. This approach is illustrated in Fig. 1. Only the analysis of the neutral and sialic acid–containing oligosaccharides using high temperature GC and GC/MS is presented. The sulfated oligosaccharides can be analyzed by tandem mass spectrometry [16].

An example of a gas chromatogram is illustrated in Fig. 2, showing the permethylated methyl amide derivatives of the sialic acid–containing oligosaccharides from a respiratory mucin glycopeptide. Figure 2 also contains a list of oligosaccharide structures interpreted from analysis of the same fraction by GC/MS. From the gas chromatogram, one can directly obtain an estimation of the relative amount of individual oligosaccharides.

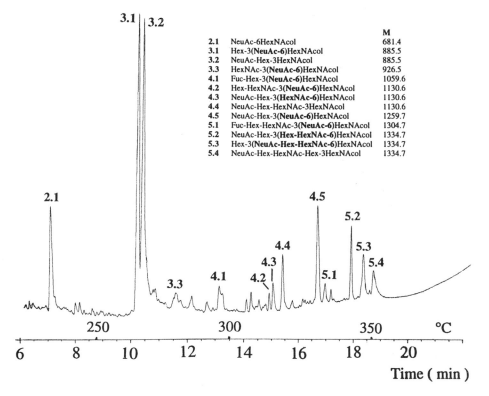

Figure 2 Gas chromatogram of permethylated methyl amide derivatives of sialic acid–containing oligosaccharide alditols released from a human respiratory mucin glycopeptide. The listed oligosaccharide sequences have been determined using GC/MS. The fused silica column (11 m × 0.25 mm i.d.) was coated wtih 0.03 μm of cross-linked PS264. The sample was injected on-column at 80°C, the temperature was kept constant for 1 min and then raised by 50°C/min to 200°C, followed by 10°C/min up to 390°C, and kept constant for 5 min. The carrier gas was hydrogen (constant pressure, 0.7 bar), and the flame ionization detector was kept at 400°C. Abbreviations used are Hex, hexose; HexNAc, N-acetylhexosamine; Fuc, fucose; NeuAc, N-acetylneuraminic acid; HexNAcol, N-acetylhexosaminitol.

However, by comparison with information obtained by matrix-assisted laser desorption–mass spectrometry, we have indications that oligosaccharides with more than 6–7 sugar residues are slightly underestimated by GC, probably due to thermal degradation on the column at high temperature. The fraction presented in Fig. 2 contained oligosaccharides with 5 residues as the largest, but oligosaccharides with 7–8 sugar residues have been ana-

lyzed on this column. Even larger oligosaccharides can be analyzed using
shorter columns (2–5 m) and faster temperature programs [6,11,15].

An example of a GC/MS analysis is illustrated in Figs. 3 and 4, where
the total ion chromatogram of the permethylated neutral oligosaccharides
from a respiratory mucin glycopeptide is shown (Fig. 3). The use of mass
spectrometric detection lowers the resolution slightly, although the range of
oligosaccharides that can be analyzed is the same when an optimized GC/
MS interface is used. One of the mass spectra from the GC/MS analysis in
Fig. 3 is reproduced in Fig. 4, showing a difucosylated hexasaccharide.
The mass spectra of permethylated oligosaccharides are relatively easy to
interpret with fragment ions giving sequence information [6,9]. If the oligosaccharide has more than one substitution on the *N*-acetylhexosaminitol,
their positions are usually easy to assign [6,10]. The mass spectra of the

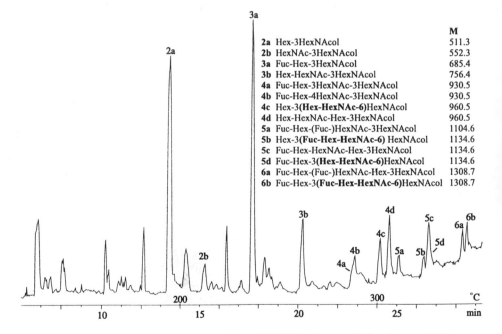

Figure 3 Total ion chromatogram from GC/MS of permethylated neutral oligosaccharides released from a human respiratory mucin glycopeptide. The same GC
column as described in Fig. 2 was used. The sample was injected at 80°C, after 1
min the temperature was raised by 10°C/min to 390°C and was kept constant for 5
min. Helium was used as the carrier gas (constant flow, 1.46 mL/min). The mass
spectrometry conditions are described in the text. For abbreviations used, see legend
of Fig. 2.

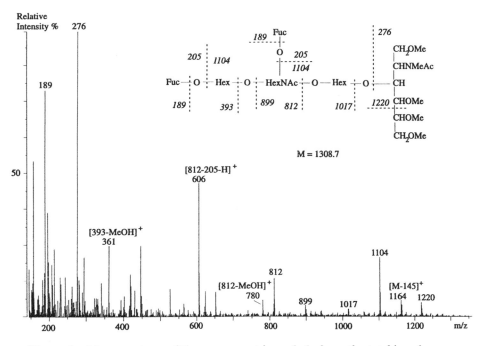

Figure 4 Mass spectrum of the component in peak 6a from the total ion chromatogram of the permethylated neutral oligosaccharide mixture shown in Fig. 3.

permethylated methyl amide derivatives of the sialic acid–containing oligosaccharides have a *N,N*-dimethyl amide side group, easily lost by an α-cleavage resulting in a typical fragment ion [M-72]⁺ [8,11]. This typical α-cleavage was first described by Levery et al. [17].

Collected examples of neutral oligosaccharide sequences identified by GC/MS from different mucins and their relative retention times are listed in Table 1. The relative retention times can often be used to give an initial idea of the type of oligosaccharides present in a mixture. Table 2 contains examples of the sialic acid–containing oligosaccharides as their permethylated *N,N*-dimethyl amide derivative.

Combining the analysis of the neutral and sialic acid–containing fractions with monosaccharide composition analysis [18], one usually can confirm that the deoxyhexose really is Fuc, that the hexose is Gal, and that the *N*-acetylhexosaminitol is GalNAcol. Sometimes the released oligosaccharides contain only GlcNAc, but there is often also GalNAc, making a prediction of GlcNAc or GalNAc more difficult.

Table 1 Sequences of Permethylated Neutral Oligosaccharides Identified by GC/MS and Their Relative Retention Times Obtained with GC

No. of residues	Sequence	t_{rel}[a]	Peak no.[b]
1	HexNAcol[c]	0.25	
2	Hex–3HexNAcol	0.7	2a
2	HexNAc–3HexNAcol	0.83	2b
2	HexNAc–6HexNAcol	0.88	
3	Fuc–Hex–3HexNAcol (**reference**)	1	**3a**
3	Hex–3(**HexNAc–6**)HexNAcol	1.16	
3	Hex–HexNAc–3HexNAcol	1.17	3b
3	HexNAc–Hex–3HexNAcol	1.18	
3	HexNAc–3(**HexNAc–6**)HexNAcol	1.26	
4	Fuc–Hex–HexNAc–3HexNAcol	1.35	4a, 4b
4	HexNAc–(Fuc–)Hex–3HexNAcol	1.37	
4	Fuc–Hex–3(**HexNAc–6**)HexNAcol	1.38	
4	Hex–3(**Hex–HexNAc–6**)HexNAcol	1.45	4c
4	Hex–HexNAc–Hex–3HexNAcol	1.48	4d
4	Hex–HexNAc–3(**HexNAc–6**)HexNAcol	1.52	
4	HexNAc–3(**Hex–HexNAc–6**)HexNAcol	1.54	
4	HexNAc–Hex–HexNAc–3HexNAcol	1.56	
5	Fuc–Hex–(Fuc–)HexNAc–3HexNAcol	1.51	5a
5	Hex–3(**Fuc–Hex–HexNAc–6**)HexNAcol	1.60	5b
5	Fuc–Hex–HexNAc–Hex–3HexNAcol	1.61	5c
5	Fuc–Hex–3(**Hex–HexNAc–6**)HexNAcol	1.62	5d
5	Fuc–Hex–HexNAc–3(**HexNAc–6**)HexNAcol	1.65	
5	HexNAc–3(**Fuc–Hex–HexNAc–6**)HexNAcol	1.66	
5	HexNAc–(Fuc–)Hex–HexNAc–3HexNAcol	1.68	
6	Fuc–Hex–3(**Fuc–Hex–4HexNAc–6**)HexNAcol	1.73	
6	Fuc–Hex–(Fuc–)HexNAc–Hex–3HexNAcol	1.73	6a
6	Fuc–Hex–3(**Fuc–Hex–HexNAc–6**)HexNAcol	1.75	6b
6	Hex–HexNAc–3(**Fuc–Hex–HexNAc–6**)HexNAcol	1.84	
7	Fuc–Hex–(Fuc–)HexNAc–(Fuc–)Hex–3HexNAcol	1.80	
7	Fuc–Hex–3(**Fuc–Hex–(Fuc–)HexNAc–6**)HexNAcol	1.85	
7	Fuc–Hex–HexNAc–3(**Fuc–Hex–HexNAc–6**)HexNAcol	1.95	

[a]Three samples of biological origin have been analyzed on the same GC column. The listed retention times for each oligosaccharide are mean values. The retention time (t_r) of each oligosaccharide has been converted to a relative retention time t_{rel} ($t_{rel} = t_r/t_{ref}$), where t_{ref} is the retention time of the oligosaccharide sequence Fuc–Hex–3HexNAcol.
For GC, a 10.5-m capillary column coated with 0.02-μm stationary phase (PS264) was used. The carrier gas was hydrogen (0.7 bar, constant pressure mode, linear gas velocity of 123 cm/sec at 80°C). The sample was injected at 80°C, the temperature kept at 80°C for 1 min, then raised by 10°C/min up to 390°C for 5 min.
[b]The total ion chromatogram from GC/MS of the marked permethylated oligosaccharides is shown in Fig. 3.
[c]For abbreviations used, see legend of Fig. 2.

Table 2 Sequences of Permethylated Sialic Acid–Containing Oligosaccharides Identified with GC/MS and Their Relative Retention Times Obtained with GC

No. of residues	Sequence	t_{rel}[a]	Peak no.[b]
2	NeuAc-6HexNAcol[c]	0.68	2.1
2	NeuGc-6HexNAcol	0.75	
3	Hex-3(**NeuAc-6**)HexNAcol	0.99	3.1
3	NeuAc-Hex-3HexNAcol (**reference**)	1	**3.2**
3	Hex-3(**NeuGc-6**)HexNAcol	1.01	
3	NeuGc-Hex-3HexNAcol	1.04	
3	HexNAc-3(**NeuAc-6**)HexNAcol	1.11	3.3
3	HexNAc-3(**NeuGc-6**)HexNAcol	1.17	
3	HexNAc-3(**NeuAc-6**)HexNAcol	1.17	
3	HexNAc-3(**NeuGc-6**)HexNAcol	1.22	
4	Fuc-Hex-3(**NeuAc-6**)HexNAcol	1.27	4.1
4	Fuc-Hex-3(**NeuGc-6**)HexNAcol	1.33	
4	HexNAc-(NeuAc-)Hex-3HexNAcol	1.37	
4	HexNAc-(NeuGc-)Hex-3HexNAcol	1.41	
4	Hex-HexNAc-3(**NeuAc-6**)HexNAcol	1.45	4.2
4	NeuAc-Hex-3(**HexNAc-6**)HexNAcol	1.46	4.3
4	Hex-HexNAc-3(**NeuGc-6**)HexNAcol	1.48	
4	NeuAc-Hex-HexNAc-3HexNAcol	1.50	4.4
4	NeuGc-Hex-3(**HexNAc-6**)HexNAcol	1.51	
4	NeuAc-Hex-3(**NeuAc-6**)HexNAcol	1.62	4.5
5	Fuc-Hex-HexNAc-3(**NeuAc-6**)HexNAcol	1.65	5.1
5	Fuc-Hex-HexNAc-3(**NeuGc-6**)HexNAcol	1.68	
5	NeuAc-Hex-3(**Hex-HexNAc-6**)HexNAcol	1.74	5.2
5	Hex-3(**NeuAc-Hex-HexNAc-6**)HexNAcol	1.78	5.3
5	NeuGc-Hex-3(**Hex-HexNAc-6**)HexNAcol	1.78	
5	NeuAc-Hex-HexNAc- Hex-3HexNAcol	1.82	5.4
6	Fuc-Hex-(Fuc-)HexNAc-3(**NeuAc-6**)HexNAcol	1.81	
6	NeuAc-Hex-3(**Fuc-Hex-HexNAc-6**)HexNAcol	1.91	
6	NeuGc-Hex-3(**Fuc-Hex-HexNAc-6**)HexNAcol	1.94	
6	NeuAc-Hex-(Fuc-)HexNAc-Hex-3HexNAcol	1.95	
6	Fuc-Hex-3(**NeuAc-Hex-HexNAc-6**)HexNAcol	1.97	

[a]Three samples of biological origin have been analyzed on the same GC column, and the listed retention times for each oligosaccharide are mean values. The retention time (t_r) of each oligosaccharide has been converted to a relative retention time t_{rel} ($t_{rel} = t_r/t_{ref}$), where t_{ref} is the retention time of the oligosaccharide sequence NeuAc-Hex-3HexNAcol.
For GC, a 10.5-m capillary column coated with 0.02-μm stationary phase (PS264) was used. The carrier gas was hydrogen (0.7 bar, constant pressure mode, linear gas velocity of 123 cm/sec at 80°C). The sample was injected at 80°C, the temperature kept at 80°C for 1 min, then raised by 50°C/min up to 200°C and by 10°C/min up to 390°C for 5 min.
[b]The GC chromatogram of the marked permethylated oligosaccharides is shown in Fig. 2.
[c]For abbreviations used, see legend of Fig. 2; NeuGc, **N**-glycolylneuraminic acid.

ACKNOWLEDGMENTS

We are indebted to Ingemar Carlstedt, Lund, Sweden, for the purified respiratory mucin glycopepides. The work was supported by the Swedish Medical Research Council (grants 7461, 10443, and 3967 for the mass spectrometry) and IngaBritt and Arne Lundbergs Stiftelse.

ABBREVIATIONS

D4	octamethyl tetracyclosiloxane
DMSO	dimethyl sulfoxide
FID	flame ionization detector
GC	gas chromatography
GC/MS	gas chromatography–mass spectrometry
HPLC	high performance liquid chromatography
NMR	nuclear magnetic resonance

REFERENCES

1. Strous, G. J., and Dekker, J. (1992). *Crit. Rev. Biochem. Mol. Biol.* 27, 57–92.
2. Forstner, J. F., and Forstner, G. G. (1994). In Johnson, L. R., ed., *Physiology of the Gastrointestinal Tract*, 3rd ed. Raven Press, New York, pp. 1255–1283.
3. Klein, A., Carnoy, C., Lamblin, G., Roussel, P., van Kuik, J. A., de Waard, P., and Vliegenthart, J. F. G. (1991). *Eur. J. Biochem.* 198, 151–168.
4. Karlsson, N. G., Karlsson, H., and Hansson, G. C. (1995). *Glycoconj. J.* 12, 69–76.
5. Karlsson, H., Carlstedt, I., and Hansson, G. C. (1987). *FEBS Lett.* 226, 23–27.
6. Karlsson, H., Carlstedt, I., and Hansson, G. C. (1989). *Anal. Biochem.* 182, 438–446.
7. Hansson, G. C., and Karlsson, H. (1990). *Meth. Enzymol.* 193, 733–738.
8. Hansson, G. C., Bouhours, J. F., Karlsson, H., and Carlstedt, I. (1991). *Carbohydr. Res.* 221, 179–189.
9. Hansson, G. C., and Karlsson, H. (1993). In Hounsell, E. F., ed., *Methods in Molecular Biology*, Vol. 14. Humana Press, Totowa, New Jersey, pp. 47–54.
10. Carlstedt, I., Herrman, A., Karlsson, H., Sheehan, J., Fransson, L.-Å ., and Hansson, G. C. (1993). *J. Biol. Chem.* 268, 18771–18781.
11. Karlsson, H., Karlsson, N., and Hansson, G. C. (1994). *Mol. Biotech.* 1, 165–180.
12. Carlson, D. M. (1968). *J. Biol. Chem.* 243, 616–626.
13. Ciucanu, I., and Kerek, F. (1984). *Carbohydr. Res.* 131, 209–217.
14. Larson, G., Karlsson, H., Hansson, G. C., and Pimlott, W. (1987). *Carbohydr. Res.* 161, 281–290.

15. Hansson, G. C., Li, Y-T., and Karlsson, H. (1989). *Biochemistry* 28, 6672–6678.
16. Karlsson, N. G., Karlsson, H., and Hansson, G. C. (1996). *J. Mass Spectrom.* 31, 560–572.
17. Levery, S. B., Roberts, C. E., Salyan, M. E. K., Bouchon, B., and Hakomori, S.-i. (1990). *Biomed. Env. Mass Spectrom.* 19, 311–318.
18. Karlsson, N. G., and Hansson, G. C. (1995). *Anal. Biochem.* 224, 538–541.

21

Highly Sensitive Pre-Column Derivatization Procedures for Quantitative Determination of Monosaccharides, Sialic Acids, and Amino Sugar Alcohols of Glycoproteins by Reversed Phase High-Performance Liquid Chromatography

Kalyan Rao Anumula
SmithKline Beecham Pharmaceuticals, King of Prussia, Pennsylvania

I. INTRODUCTION

Carbohydrate composition analysis of glycoproteins is a challenging problem in glycobiology. The ever increasing number of reports dealing with glycoproteins demands simple, highly sensitive methods for characterization of their glycosylation. An accurate determination of the monosaccharide composition of a glycoprotein is a preparatory step in learning about its glycosylation. Furthermore, quantitative determination of monosaccharides is a prerequisite for structural characterization of the isolated oligosaccharides/glycopeptides by various physical, chemical, and enzymatic techniques [1,2]. Methods used for carbohydrate composition analysis have been reviewed in [3].

This chapter describes novel pre-column derivatization procedures for the quantitative determination of monosaccharides typically found in glycoproteins. Specifically, it discusses the use of anthranilic acid (ABA) for derivatizing the reducing monosaccharides, *o*-phenylene diamine (OPD) for sialic acids, and phenylisothiocyanate (PITC) for amino sugar alcohols. The derivatized monosaccharides and sialic acids can be determined with high sensitivity using a fluorescence detector, and the PITC-derivatized amino sugar alcohols are analyzed with a UV detector. All the sugar separations are carried out on C_{18} reversed phase columns using a common high performance liquid chromatography (HPLC) solvent system.

II. EXPERIMENTAL PROCEDURES

A. Hydrolysis of Glycoproteins

Typically, 5–50 μg of glycoproteins were hydrolyzed in 0.25–0.5 mL of 20% trifluoroacetic acid in 1.6-mL conical screw cap freeze vials (polyproplyene with "O" ring seals, Sigma, St. Louis, MO) at 100°C for 6–7 h [4]. The caps on the vials were further sealed with four to five layers of Teflon tape to prevent any accidental evaporation of the sample during hydrolysis. Samples were dried overnight using a vacuum centrifuge evaporator (Savant, Holbrook, NY) without heat.

For hexosamine and hexosaminitol analysis, the glycoproteins were hydrolyzed in 0.05–0.1 mL of 4 N HCl at 100°C for 16 h and dried on the vacuum centrifuge evaporator.

B. Derivatization of Monosaccharides with ABA

The procedure for derivatization of monosaccharides with 2-aminobenzoic acid (anthranilic acid) has been described earlier [4]. A solution of 4% sodium acetate•3H$_2$O and 2% boric acid in methanol was prepared. This solution may be stored at room temperature for several months. The derivatization reagent was prepared by dissolving 30 mg of anthranilic acid (ABA) (Aldrich) and 20 mg of sodium cyanoborohydride (Aldrich) in 1.0 mL of the aforementioned methanol–sodium acetate–borate solution. Dry glycoprotein hydrolysates were dissolved in 1% fresh sodium acetate•3H$_2$O (0.1–0.2 mL), and an aliquot (0.02–0.1 mL) was transferred to a new screw cap freeze vial. Samples were mixed with 0.1 mL of the ABA reagent solution and capped tightly. The vials were heated at 80°C in an oven or heating block (Reacti-Therm, Pierce) for 30–45 min [4]. After cooling the vials to ambient temperature, the volume of the samples was made up to 1.0 mL with HPLC solvent A (see Section II.E) and mixed vigorously by Vortex to expel the hydrogen evolved during the reaction. Duplicate injections of 0.05 mL were made from each vial for analysis. Similarly, monosaccharide standards were derivatized each time with unknown samples, and 20–25 pmol was injected.

C. Derivatization of Sialic Acids with OPD

The procedure for derivatization of sialic acids with OPD has been described earlier [5]. Fifty microliters of the sample containing 0.01 to 0.25 mg of protein was mixed with 50 μL of 0.5 M sodium bisulfate in a 1.6-mL conical screw cap freeze vial. After placing the caps tightly on the vials, they were incubated at 80°C (Reacti-Therm heating module, Pierce) for 20 min. These mild acid hydrolysis conditions were used for the release of sialic acids from the glycoproteins.

The mild acid–released sialic acids were then derivatized with *o*-phenylenediamine•2HCl (OPD, Aldrich). The unknown samples from mild acid hydrolysis and sialic acid standards (1–2 nmol in 0.1 mL), in separate 1.6-mL screw cap freeze vials, were mixed with 0.1 mL of OPD solution (20 mg/mL in 0.25 M NaHSO₄). After placing the caps tightly on the vials, they were incubated at 80°C (Reacti-Therm) for 40 min. After cooling the tubes to room temperature, they were diluted to 1.0 mL with solvent A (see Section A II.E) and mixed vigorously on a Vortex mixer. The vials were centrifuged at maximum speed for 3–5 min in a microcentrifuge to clarify the solution, and the clear supernatant was used for the analysis.

D. Derivatization of Amino Sugar Alcohols with PITC

Amino sugar alcohols from the acid hydrolysis of oligosaccharides released by alkaline–BH₄ can be easily derivatized with PITC, which has been described previously [6]. A 0.02-mL sample containing 20 pmol to 20 nmol in a microcentrifuge tube (1.5 mL) was treated with 0.1 mL of a freshly prepared solution of methanol : triethylamine : PITC (70 : 10 : 5) at room temperature for 20 min. The reaction mixture was diluted with 0.38 mL of 1.5% acetic acid–water, and the excess PITC was extracted into 0.8 mL of methylene chloride by mixing vigorously on a Vortex mixer. The aqueous and organic phases were separated by centrifugation. The aqueous phase containing the derivatized amino sugars can be analyzed at this stage or extracted once more, if required. Typically, 0.05–0.1 mL of the aqueous phase was analyzed by HPLC.

E. HPLC Analysis of Monosaccharides, Sialic Acids, and Amino Sugar Alcohols

ABA–monosaccharides and OPD–sialic acids were separated on a C₁₈ reversed phase HPLC column (Bakerbond, 5 μm, 0.46 × 25 cm, analytical, J. T. Baker (Phillipsburg, NJ), or a Beckman (San Ramon, CA), Ultrasphere-ODS, 0.46 × 25 cm) using a 1-butylamine–phosphoric acid–tetrahydrofuran mobile phase as described previously [4,5]. All the separations were carried out at ambient temperature using a flow rate of 1 mL/min. Solvent A consisted of 0.15–0.3% 1-butylamine, 0.5% phosphoric acid, and 1% tetrahydrofuran (0.025% BHT inhibited, Aldrich) in water, and solvent B consisted of equal parts of solvent A and acetonitrile. For the Bakerbond column, the solvent A contained 0.2% butylamine and the gradient program was isocratic (5% B) for 25 min followed by a linear increase to 15% B by 50 min. For the Beckman column, the solvent A contained

0.15% butylamine and the gradient program was isocratic (5% B) for 30 min followed by a linear increase to 15% B by 50 min. The column was washed for 15 min with 100% of B and equilibrated for 15 min with the initial conditions to ensure reproducibility from run to run. ABA derivatives were detected with an HP 1046A HPLC fluorescence detector (Palo Alto, CA) using 230-nm excitation and 425-nm emission [4].

The sialic acid derivatives were separated on a C_{18} reversed phase column (15 or 25 cm Ultrasphere-ODS, Beckman) at ambient temperature using a flow rate of 1.0 mL/min. The HPLC solvents were prepared as described for the ABA–monosaccharide analysis. Solvent A for this column contained 0.15% butylamine. The common sialic acid derivatives were eluted isocratically (10–13%) with solvent B, for 15 min followed by 10 min wash with 95% of solvent B and equilibration with 10–13% solvent B for 10 min as described previously [5].

PITC-derivatized amino sugar alcohols were also separated on a C_{18} reversed phase column (15 cm Beckman Ultrasphere-ODS) using the HPLC and the solvent systems described previously for the ABA–monosaccharide separation. Separations were carried out with a gradient consisting of initial 0% B for 2 min followed by linear increase to 10% B over 20 min. The column was washed with 95% B for 10 min followed by equilibration with 0% B for 10 min before the next injection. The derivatized amino sugar alcohols were detected with a UV detector operating at 254 nm.

III. RESULTS AND DISCUSSION

Initially, 2-aminobenzoic acid (ABA, anthranilic acid) was introduced for labeling carbohydrates and their detection with high sensitivity using a fluorescence detector [4]. ABA is being promoted by Oxford Glycosystems as 2-AA, and therefore, the reader should keep in mind that ABA and 2-AA are the same molecule, 2-aminobenzoic acid [7]. Fluorescence detection of the derivatized sugars is the most sensitive way of quantitation. With the fluorescent tags of ABA and OPD, ~0.1 pmol of the monosaccharides and <2 pmol of the sialic acids can be easily determined. Reactions for the derivatization of monosaccharides and sialic acids with ABA and OPD are shown in Schemes 1 and 2, respectively. A typical separation of the ABA–monosaccharides is shown in Fig. 1. The separation of the common sialic acid standard derivatives, OPD-N-acetylneuraminic acid and OPD-N-glycolylneuraminic acid is shown in Fig. 2. A baseline separation of all the sugar derivatives was obtained in both analyses. These procedures are highly reproducible with standard deviations <3.0%. In addition, analyses of various glycoproteins containing 2–18% carbohydrate (w/w) indicates that these procedures are highly reproducible, and the compositions deter-

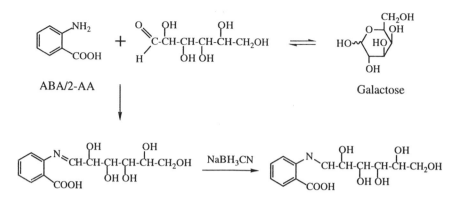

Scheme 1 Reaction scheme for labeling carbohydrates with anthranilic acid as shown by the example of galactose.

mined were in good agreement with reported/calculated values. Chromatograms obtained with fetuin and α_1-acid glycoprotein are shown in Fig. 3. Most of the problems in the determination of carbohydrate composition using these procedures appear to reside in the measurement of the protein concentration by various methods.

One advantage of the current procedures with fluorescence detection is that all these sugars can be conveniently determined using the same

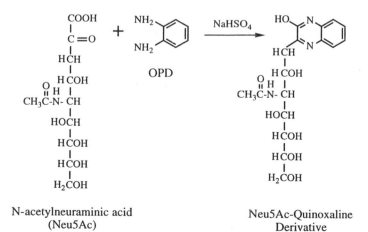

Scheme 2 Reaction scheme for derivatization of sialic acids with o-phenylene diamine as shown by the example of N-acetylneuraminic acid.

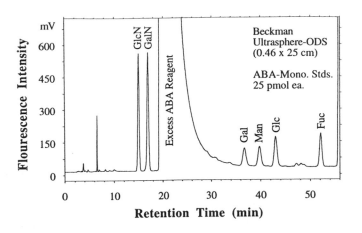

Figure 1　HPLC separation of the 2-aminobenzoic acid (ABA)–derivatized mono-saccharide standards.

HPLC system, column solvent systems, and detector settings. It is advisable to use a separate column (15 cm, Beckman Ultrasphere-ODS) for sialic acid estimations in order to save time in cleaning the column (4–5 blank runs) that has been previously used for the monosaccharide analysis. Experience with this type of setup indicates that the composition of glycoproteins can be determined conveniently and efficiently.

It has been noted that the fluorescence of ABA/2-AA derivatives was

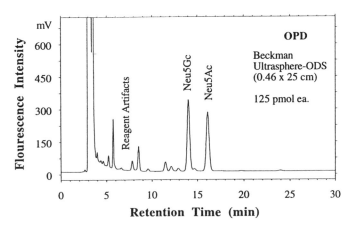

Figure 2　HPLC separation of the common sialic acid standards derivatized with OPD on a C_{18} reversed phase column.

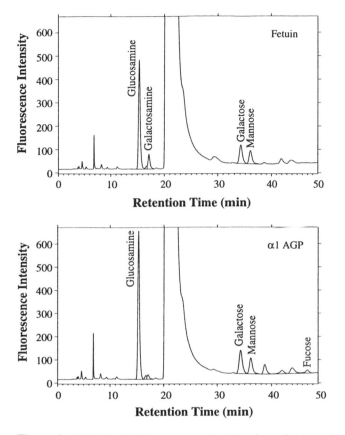

Figure 3 HPLC of ABA–monosaccharides from fetuin and α_1-acid glycoprotein. The separations were carried out on a C_{18} Bakerbond column [4].

about 10 times higher with Waters 474 fluorescence detector (Milford, MA) at 230-nm excitation than the HP 1046A detector, whereas the fluorescence of OPD–sialic acids was the same in both HP 1046A and Waters 474 detectors.

For characterization of Ser/Thr-linked oligosaccharides, glycoproteins are usually treated with mild alkaline–borohydride to release specifically the O-linked oligosaccharides. Thus, the released oligosaccharides contain reduced GalNAc ends and, therefore, cannot be derivatized by reductive alkylation. However, the amino sugar alcohols after de-N-acetylation by acid hydrolysis can be easily derivatized with PITC, and these derivatized amino sugar alcohols can be easily separated using the same HPLC, column, and the solvent systems, as described previously. Quantitative detection with a UV detector is understandably not as sensitive as analysis with fluorescence. In the separation using the solvent system for the ABA–monosaccharides, the PITC–amino sugar alcohols are well resolved and the PITC-derivatized amino acids did not interfere with the amino sugar alcohols, as shown in Fig. 4. It has been noticed that glucosaminitol undergoes transformation in diluted solutions to yield a peak that elutes in front of the $GlcNH_2$ and is not well resolved from the $GlcNH_2$ peak.

IV. CONCLUSIONS

A comprehensive approach for the analysis of carbohydrates present in glycoproteins has been described in which different novel procedures use the same HPLC system, column, and detector conditions except for the UV absorbing, PITC-derivatized amino sugar alcohols. In addition, a useful derivatization procedure has been described for the nonreducing amino sugar alcohols for characterization of isolated *O*-linked oligosaccharides using an alkaline–borohydride method. All the methods including the

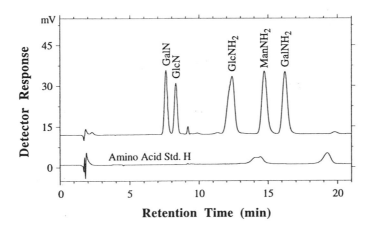

Figure 4 HPLC of PITC-derivatized amino sugars and amino sugar alcohols. The HPLC separation was performed on a 15-cm Beckman Ultrasphere-ODS column using the HPLC and the solvent systems used for the separation of ABA–monosaccharides. The derivatives were detected with an Applied Biosystems 757 UV detector (Foster City, CA) at 254 nm. Approximately 1.5 nmol of each amino sugar was injected. The amino acid standard mixture H was from Pierce Chemical Co. $GlcNH_2$, $MaNH_2$, and $GalNH_2$ represent glucosaminitol, mannosaminitol, and galactosaminitol, respectively.

PITC-derivatized amino sugar alcohols use common HPLC and solvent systems for convenience and efficiency in determining the carbohydrate composition of glycoproteins. In addition, these procedures are highly reproducible because the sample preparations involve simple steps without any sample cleanup for the HPLC analyses. The ABA labeling of monosaccharides with the current procedure allows quantitation of ~100 fmol (with a standard analytical setup), and therefore carbohydrate composition analyses can be performed on less than 1.0 μg of glycoprotein. The sensitivity of these methods can be increased further by simple adaptation of these procedures to a narrow-bore column and using a highly sensitive fluorescence detector such as Waters 474. It is estimated that <0.1 μg of glycoprotein can be easily analyzed with a narrow-bore HPLC system. Therefore, current procedures should be useful in determining the carbohydrate composition of glycoproteins and isolated oligosaccharides/glycopeptides available in small amounts more easily and efficiently.

ACKNOWLEDGMENT

I would like to thank Ms. M. B. Ebert for critical reading.

ABBREVIATIONS

2-AA	alternative name for ABA
ABA	anthranilic acid (2-aminobenzoic acid)
HPLC	high performance liquid chromatography
OPD	*o*-phenylene diamine
PITC	phenylisothiocyanate

REFERENCES

1. Anumula, K. R. (1995). In Atassi, Z. M., and Apella, E., eds., *Methods in Protein Structure Analysis*. Plenum, New York, pp 195–206.
2. Anumula, K. R., and Taylor, P. B. (1991). *Eur. J. Biochem.* 195, 269–280.
3. Townsend, R. R. (1993). *Am. Chem. Soc. Sym. Ser.* 529, 86–101.
4. Anumula, K. R. (1994). *Anal. Biochem.* 220, 275–283.
5. Anumula, K. R. (1995). *Anal. Biochem.* 230, 24–30.
6. Anumula, K. R., and Taylor, P. B. (1991). *Anal. Biochem.* 197, 113–120.
7. Bigge, J. C., Patel, T. P., Bruce, J. A., Goulding, P. N., Charles, S. M., and Parekh, R. B. (1995). *Anal. Biochem.* 230, 229–238.

22
Glycan Labeling with the Fluorophores 2-Aminobenzamide and Anthranilic Acid

Mark R. Hardy*

Oxford GlycoSystems, Inc. Bedford, Massachusetts

I. INTRODUCTION

The science of glycobiology often requires the structural analysis of small quantities of purified materials (e.g., glycoproteins) to permit meaningful structure–function correlations. There is thus a need for highly sensitive, selective, and reproducible methods for the analysis and characterization of oligosaccharides. Methods and strategies for glycan (mono- and oligosaccharide) analysis in glycobiology have been reviewed [1,2]. Historically, chromatographic methods, both gas and liquid chromatography (see [3] for review), have been widely employed. More recently, capillary electrophoresis and polyacrylamide gel electrophoresis have gained in popularity for carbohydrate analysis [2].

Most mono- and oligosaccharides lack an intrinsic chromophore suitable for sensitive and selective detection in chromatographic analyses. One popular liquid chromatographic (LC) method for carbohydrate analysis, high pH anion-exchange chromatography with pulsed amperometric detection (HPAEC-PAD, see [4] for review), exploits the electrochemical properties of carbohydrates to provide reasonably specific and selective detection. However, the electrochemical response of different sugars can vary significantly (and rather unpredictably), and so it is often only possible to quantify any given carbohydrate structure analyzed by HPAEC-PAD with respect to an identical external oligosaccharide standard sample.

To introduce a chromophore, to modulate the selectivity of various chromatographic media, and also to improve and normalize quantitative analysis, many researchers have turned to methods for preanalysis derivatization of carbohydrates. Because reducing carbohydrates possess an aldehyde (or ketone) functionality, reductive amination [5] with a suitable amino compound has become a popular method for selective derivatization

---

*Current affiliation: Genetics Institute, Andover, Massachusetts

of carbohydrates. In reductive amination, the reducing terminus of a carbohydrate reacts with an amine-containing derivatization reagent to form a
Schiff's base. The Schiff's base is then selectively reduced with an appropriate reducing agent, typically NaCNBH$_3$ [5] or an amine–borane complex
[6], to form a stable secondary amine product. Many methods and reagents
have been described for reductive amination of carbohydrates, but it is
clear that the best reagents for this purpose are arylamines, due to the low
basicity of the aryl-amino group. The use of arylamines permits the reductive amination to be done at mild pH. Use of extremes of pH for the
derivatization reaction can result in artifactual loss of labile groups from
the carbohydrate (e.g., desialylation at acid pH, de-*O*-acylation at alkaline
pH) or epimerization (at alkaline pH). Arylamines generally also have the
advantage of being strong UV-chromophores. Many are also fluorophores
or fluorogeneic. The aromatic character of the reductive amination products prepared from arylamines has an additional benefit of increasing the
retention of derivatized carbohydrates on many commonly used LC media
(e.g., C$_{18}$ for reversed phase high performance LC [HPLC]).

Bigge and coworkers have described the use of the arylamine (aniline)
derivatives 2-aminobenzamide (2-AB) and anthranilic acid (2-aminobenzoic
acid, 2-AA) as reagents for the labeling of oligosaccharides by reductive
amination [7]. They have shown that it is possible to label reducing oligosaccharides reproducibly and in high yield with 2-AB and with 2-AA. The
derivatization conditions that were described (dimethyl sulfoxide [DMSO]/
acetic acid solvent system, NaCNBH$_3$ as reducing agent) were shown to
effect minimal degradation (desialylation, epimerization) of a number of
standard oligosaccharides. The 2-AA label was shown to be useful for
separation of oligosaccharides (both neutral and charged species) by polyacrylamide gel electrophoresis (PAGE).

This chapter provides some further examples of the utility of the 2-AA
and 2-AB labels for chromatographic and electrophoretic separation of
mono- and oligosaccharides. The 2-AA label is shown to be useful for
sensitive qualitative monosaccharide analysis of acid hydrolysates of glycopeptides fractionated by reversed phase HPLC as well as PAGE analysis of
oligosaccharides. The utility of the 2-AB label for highly selective oligosaccharide profiling is illustrated using a new stationary phase [8].

II. MATERIALS AND METHODS

A. Materials

Bovine fetuin (Pedersen method and Spiro method) was from Life Technologies (Grand Island, NY). Monosaccharide standards were from Pfanstiehl
(Waukegan, IL) and were Reference Grade materials whenever available.

Reagent kits for the derivatization of carbohydrates with 2-AB (kit K-404) or 2-AA (K-402), oligosaccharide standards and glycoprotein oligosaccharide libraries, prototype GlycoSep R (reversed phase octadecylsilica), and GlycoSep N (normal phase) HPLC columns were from Oxford GlycoSystems (Bedford, MA). GlycoGel polyacrylamide gels for analysis of 2-AA-labeled glycans by PAGE were also from Oxford GlycoSystems. Hydrochloric and trifluoroacetic acids, tetrahydrofuran, and *n*-butylamine were from Sigma Chemical Co. (St. Louis). Other chemicals and reagents were obtained from VWR Scientific (Boston).

B. Preparation of Monosaccharide Samples

Glycoprotein and glycopeptide samples were hydrolyzed in a final volume of 400 μL of 4 N trifluoracetic acid (TFA) or 6 N HCl for 4 h at 100°C [9] in 500-μL screw cap microcentrifuge tubes (Sarstedt, Newton, NC).

C. Derivatization of Mono- and Oligosaccharides with 2-AB and 2-AA

Dried, salt-free oligosaccharide samples were derivatized with 2-AB and purified according to the K-404 kit protocol. Samples for monosaccharide analysis were derivatized with 2-AA according to the K-402 kit protocol. Purified 2-AB-labeled glycans (in ca. 1.5 mL water) were dried in a Savant SpeedVac (Holbrook, NY) vacuum centrifuge and redissolved in an appropriate solvent for subsequent analysis.

2-AA-labeled monosaccharide samples were purified on "GlycoClean" cartridges by a modified version of the K-404 protocol. GlycoClean cartridges were prepared by sequential washes with 1 mL each of water, 30% (v/v) acetic acid, and finally acetonitrile. Each 2-AA-labeling reaction was then spotted onto a GlycoClean disk and allowed to stand for 15 min at room temperature. Each reaction vessel (500-μL microcentrifuge tube) was then rinsed with 100 μL of acetonitrile, and the rinse was applied to the GlycoClean disk. Each GlycoClean cartridge was then washed with 3 × 1 mL of acetonitrile, and the 2-AA-labeled monosaccharides were eluted with 3 × 0.5 mL of water, dried as previously described, and dissolved in water for HPLC analysis.

Samples to be labeled with 2-AA for PAGE analysis were prepared as described in the K-402 protocol.

D. HPLC Analysis of Labeled Glycan Samples

HPLC analyses were performed on a Waters Corporation (Milford, MA) dual-pump, high pressure mixing HPLC system equipped with Waters WISP autosampler and controlled by Waters Millenium software (PC-

Table 1 Gradient Conditions for 2-AA-
Monosaccharide Analysis

Gradient Mono-1
Eluant A: 0.2% (v/v) *n*-butylamine, 0.5% (v/v)
          phosphoric acid, 1% tetrahydrofuran
Eluant B: 1 : 1 (v/v) mixture of eluant A and aceto-
          nitrile

Time (min)	%A	%B
0	95	5
25	95	5
50	85	15
51	0	100
66	0	100
67	95	5

Gradient Mono-2
Eluant A: 0.2% (v/v) triethylamine, 0.5% (v/v)
          phosphoric acid, 1% tetrahydrofuran
Eluant B: 1 : 1 (v/v) mixture of eluant A and aceto-
          nitrile

Time (min)	%A	%B
0	100	0
2	100	0
30	95	5
40	0	100
55	0	100
56	100	0

compatible computer-based). All system stainless steel tubing was replaced
with 0.007 inch i.d. polyetheretherketone (PEEK) tubing. Some HPLC
separations were done using a Waters M625 low pressure mixing, quaternary gradient HPLC pump and controller equipped with a Rheodyne (Co-
tati, CA) 9125 PEEK injector valve.

A prototype GlycoSep R (100 × 4.6 mm) column was used for analysis of 2-AA-labeled monosaccharides, essentially as described by Anumula
[10]. The gradient conditions used for 2-AA-monosaccharide analysis are
given in Table 1. The flow rate was 0.7 mL/min. In some experiments,

the column was thermostatted at 25 °C using a column oven, and eluant temperatures were maintained at 25 °C using a temperature-controlled water bath. Detection of the eluted 2-AA-sugars was by fluorescence (Hitachi, San Jose, CA, L-7480 dual monochromator fluorimeter) using $\lambda_{ex} = 230$ nm and $\lambda_{em} = 400$ nm [10].

Separation of 2-AB-labeled oligosaccharides was performed using a prototype GlycoSep N (250 × 4.6 mm) column [8]. The flow rate used was 0.4 mL/min, at ambient temperature. Detection was by fluorescence, $\lambda_{ex} = 330$ nm, $\lambda_{em} = 420$ nm. The gradient conditions used for 2-AB-oligosaccharide separation on GlycoSep N are given in Table 2.

E. Reversed Phase-HPLC Glycopeptide Separation

Fetuin glycopeptides (trisialylated [FI] fraction [11]), lot number AA019, supplied by ImmunoGen, Inc. (Cambridge, MA), were separated using the Waters M625 quaternary solvent gradient HPLC system. The column used was a 3.9 × 300 mm Waters μ-Bondapak C_{18}. The flow rate was 0.5 mL/min. Column temperature was maintained at 35 °C with an Eppendorf (Madison, WI) column heater. Detection was at 220 nm (Waters 486 detector). The gradient used to fractionate the glycopeptides is given in Table 3. One-minute fractions were collected using an Isco Foxy Jr. fraction collector. The fractions were collected, evaporated to dryness, redissolved in 400 μL of 6 N HCl, and hydrolyzed as previously described before derivatization with 2-AA.

F. Polyacrylamide Gel Electrophoresis of 2-AA-Labeled Oligosaccharides

Aliquots of 2-AA-labeled oligosaccharides were applied to a GlycoGel polyacrylamide gel and electrophoresed using Tris-TAPS (N-tris[Hydroxy-

Table 2 Gradient Conditions for 2-AB-Oligosaccharide Separations on GlycoSep N

Eluant A: 250 mM formic acid, pH 4.5 (ammonia)
Eluant B: 90 : 10 (v/v) acetonitrile/water

Time (min)	%A	%B
0	27	73
70	45	55
71	100	0
76	100	0
110	27	73

Table 3 Gradient Conditions for Separation of Fetuin FI Glycopeptides

Eluant A: 0.08% (v/v) trifluoracetic acid

Eluant B: 90 : 10 (v/v) acetonitrile/water

Time (min)	%A	%B
0	98	2
5	98	2
45	70	30
50	70	30
51	98	2

methyl]methyl-3-aminopropane sulfonic acid) buffer according to the supplied protocol manual. Gels were placed onto ice immediately after electrophoresis and then imaged as quickly as possible. Image acquisition was either by photography of the gel on a UV transilluminator or by use of a digital image analyzer.

III.   RESULTS AND DISCUSSION

A.   Separation of 2-AB-Labeled Oligosaccharides Using a GlycoSep N Column

Figure 1 shows the separation of the *N*- and *O*-linked oligosaccharides released by hydrazinolysis [12] from bovine fetuin. The GlycoSep N column is capable of separating oligosaccharides with a high degree of selectivity, depending on both the size and charge of the oligosaccharides [8]. Guile and coworkers have developed rules of retention of neutral and sialylated oligosaccharides on GlycoSep N, and they have adapted the use of a (fluorescently labeled) partially hydrolyzed dextran "ladder" as a reference standard to normalize oligosaccharide elution data between runs, HPLC systems, and laboratories [8]. The use of a volatile buffer system for separations on GlycoSep N permits easy recovery of the eluted oligosaccharides for further purification by orthogonal HPLC techniques or for structural studies such as mass spectrometry or exoglycosidase treatment.

B.   Labeling of Monosaccharides with 2-AA and 2-AB

Preliminary experiments showed that 2-AB efficiently labeled the neutral monosaccharides galactose, glucose, and mannose (Gal, Glc, and Man) as well as the *N*-acetylated sugars GalNAc and GlcNAc (data not shown). Conversely, little or no labeling of glucosamine, galactosamine, or fucose

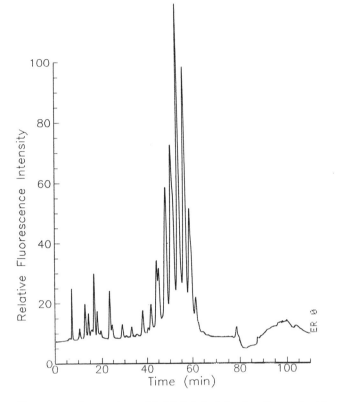

Figure 1 Separation of 2-AB-labeled fetuin oligosacharides on GlycoSep N. Approximately 25 nmol of total fetuin *N*- and *O*-linked oligosaccharides, released by hydrazinolysis, were labeled with 2-AB. The purified, labeled oligosaccharides were dried, redissolved in 1 mL of water, and a 5-μL aliquot was injected onto a prototype GlycoSep N HPLC column. Chromatographic conditions are given in Section II. The gradient conditions are given in Table 2.

(GlcN, GalN, or Fuc) was observed using the standard K-404 protocol (data not shown). However, Anumula [10] has shown efficient labeling of monosaccharides using 2-AA in a methanol–borate–acetate solvent system with NaCNBH₃ as reductant. We found that 2-AA also labels all of the monosaccharides previously mentioned using the K-402 labeling protocol (DMSO/acetate solvent system).

The separation of an admixture of 2-AA-labeled monosaccharide standards (20 pmol each) labeled using the K-402 protocol, cleaned up as described in Section II and separated according to reported gradient conditions ([10]; see Table 1, Gradient Mono-1) is shown in Fig. 2A. Use of

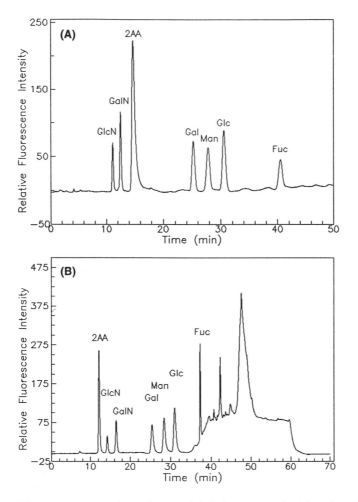

Figure 2 Separation of 2-AA-labeled monosaccharides. An admixture (2 nmol each) of monosacharides was labeled with 2-AA and prepared using GlycoClean cartridges as described in Section II. The labeled monosaccharides were dried and redissolved in 1 mL of water. A 10-μL aliquot (20 pmol per monosaccharide) was then analyzed by reversed phase HPLC on a prototype GlycoSep R column. (A) The separation using gradient Mono-1 (Table 1); (B) the separation using gradient Mono-2 (Table 1).

slightly different elution conditions than those described by Anumula (Table 1, Gradient Mono-2) resulted in elution of the GlcN and GalN derivatives after the residual free 2-AA peak, but there was no change in the relative elution order of the 2-AA-monosaccharides (Fig. 2B). In the present experiment, only the GlcN and GalN standards were used for the quantitative evaluation of sample hydrolysates.

We did not evaluate the absolute yield or recovery of monosaccharides labeled using the K-402 kit, but the relative yields of the six different monosaccharide derivatives are similar (Fig. 2). The calculated response factors for 2-AA-aminosugar standards ($n = 2$) were GlcN, 42,817 relative fluorescence intensity (RFI-sec/pmol $\pm$ 2.0%; GalN, 96,171 RFI-sec/pmol $\pm$ 0.1%.)

Anumula [10] used no sample cleanup in his experiments, leaving a large free 2-AA peak in his samples. Monosaccharide samples labeled using the K-402 protocol, omitting the cleanup step described in Section II, also display a large 2-AA peak that trails into the aminosugar peak region (data not shown) when using the Mono-2 gradient (Table 1). The small residual amount of free 2-AA dye present in standards was observed to disappear with time. After ca. 48 h at room temperature, the 2-AA peak was completely gone (data not shown). There was no change observed in the peak areas for the 2-AA-monosaccharide standards.

C. Reversed Phase HPLC Separation of Fetuin FI Glycopeptides

Baenziger and Fiete [11] described the fractionation of sialylated glycopeptides isolated from bovine fetuin using anion-exchange chromatography. The FI fraction was shown to contain primarily trisialyltriantennary N-linked complex glycopeptides, as well as some O-linked glcopeptides of similar charge to mass ratios.

The glycopeptide map of FI was used as a model to evaluate the usefulness of the 2-AA-labeling method as a sensitive, selective qualitative assay for N- and O-linked glycopeptides in a peptide map. In the case of bovine fetuin, all of the N-acetylgalactosamine (GalNAc) residues occur at the O-glycosylation sites of the protein, and almost all of the N-acetylglucosamine (GlcNAc) residues occur at the three N-glycosylation sites. Thus, to a first approximation, the presence of GalNAc versus GlcNAc can be used as a diagnostic for the presence of N- versus O-glycosylation sites in a glycoprotein peptide map. The present FI fraction is enriched in trisialylated, triantennary N-glycopeptides, but it was observed by monosaccharide compositional analysis to contain also some GalN and thus some O-glycopeptides (Fig. 3). The aminosugar content of AA019 FI

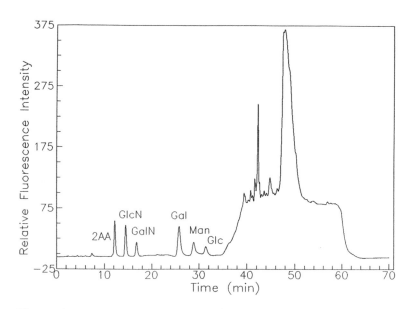

Figure 3 2-AA-monosaccharide analysis of AA019 FI HCl hydrolysate. An aliquot (5 μL) of AA019 FI (0.56 mM) was hydrolyzed in 6 N HCl, dried, labeled with 2AA, and purified. The labeled hydrolysate was dried and redissolved in 1 mL of water. A 5-μL aliquot was then analyzed by HPLC as described in Section II.

measured with the present method was 1024 μM [GlcN] and 206 μM [GalN], compared with an assayed glycopeptide content of 560 μM as [triantennary] by phenol–sulfuric acid assay [13].

Fig. 4 shows the HPLC fractionation of a commercial preparation of fetuin FI glycopeptides [14] (50-μL injection of a 560-μM solution) on a C_{18} column. The fractions were dried, hydrolyzed, and labeled with 2-AA as described in Section II. Each dried, 2-AA-labeled hydrolysate was redissolved in 500 μL of water for reversed phase HPLC monosaccharide analysis.

The aminosugar content of the first 34 fractions of the collected FI reversed phase-HPLC chromatogram was determined and quantified with respect to external 2-AA-labeled standards. Fractions that lacked 220-nm peaks in Fig. 4 were observed to contain no GalN or GlcN (data not shown). Detection of the aminosugars in the reversed phase-HPLC peak fractions is illustrated with the analysis of 100-μL aliquots (i.e., 20% of the HCl hydrolyzates) of fraction 8 (a breakthrough fraction) and fraction 20 (major retained A_{220} peak) in Fig. 5A and B, respectively.

The calculated aminosugar content measured for the AA019 FI re-

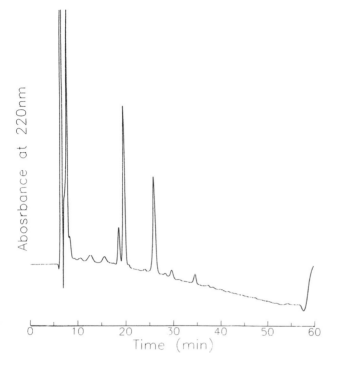

Figure 4 Reversed phase HPLC separation of AA019 FI glycopeptides. A 50-µL aliquqot of AA019 FI (0.56 mM) was separated as described in Section II. One-minute fractions were collected for aminosugar analysis.

versed phase-HPLC fractions is shown graphically in Fig. 6. It is clear from these data that the N-glycopeptides present in AA019 FI were unretained on the µ-Bondapak column, as virtually all the GlcN was observed in break-through fractions 7–9. The retained A_{220} peaks in Fig. 4 all contained GalN, and thus are presumably O-glycopeptides. The small amount of GlcN seen in some retained peaks (e.g., fraction 20) may represent minor amounts of a GlcNAc-containing O-linked sugar reported in bovine fetuin [15].

D. PAGE Analysis of 2-AA-Labeled Oligosaccharides

Figure 7 shows the separation of a variety of 2-AA-labeled oligosaccharides by PAGE using a Tris-TAPS buffer system. The low charge : mass ratio of 2-AA (contributing a single net negative charge at the pH of the running buffer) and this buffer system affords high resolution of both neutral and charged (sialylated) oligosaccharides. Lanes 4 and 5, containing total glycan

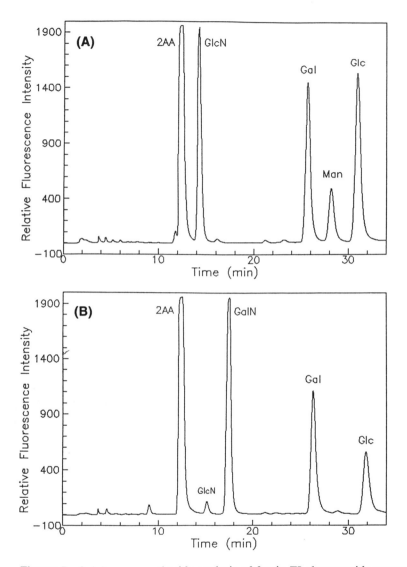

Figure 5 2-AA monosaccharide analysis of fetuin FI glycopeptide reversed phase HPLC fractions. (A) The analysis (Table 1, gradient Mono-2) of 20% (100/500 μL) of the 6 N HCl hydrolysate from fraction 8 of the preparative reversed phase HPLC fractionation of AA019 FI glycopeptides. (B) The analysis of a 20% aliquot of the 6 N HCl hydrolysate from fraction 20.

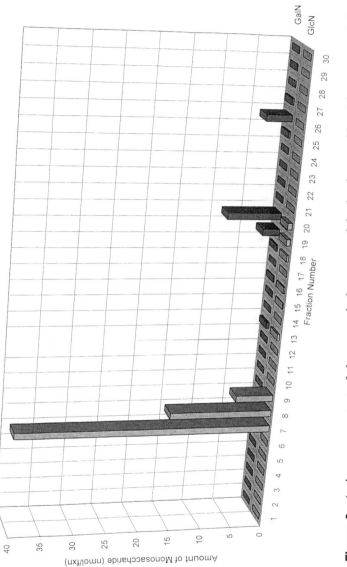

Figure 6 Aminosugar content of the reversed phase map of fetuin glycopeptides. The amount of the aminosugars GlcN and GalN was measured for each of the thirty 1-min reversed phase HPLC fractions collected from the preparative reversed phase HPLC of AA019 FI glycopeptides (Fig. 4). The quantity of 2-AA-monosaccharide was calculated using an external standard.

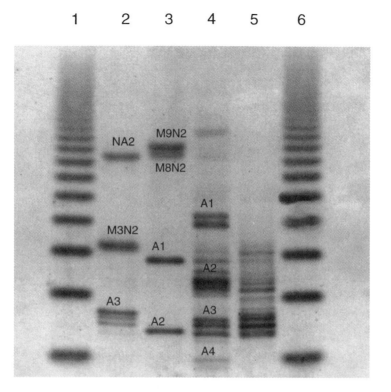

Figure 7 Polyacrylaminde gel electrophoresis of 2-AA-labeled oligosaccharides. Samples of oligosaccharide standards or hydrazinolyzed glycoprotein oligosaccharide libraries were labeled with 2-AA and analyzed by PAGE using a Tris-TAPS running buffer system. Sample identities were as follows: lanes 1 and 6, partially hydrolyzed dextran ladder (isomaltooligosaccharides); lane 2, admixture of labeled asialobiantennary (NA2), trimannosyl-core oligosaccharide (M3N2), and trisialyltriantennary (A3, mixture of sialyl-isomers); lane 3, admixture of high-mannose oligosaccharides (M9N2 and M8N2), monosialylbiantennary (A1) and disialylbiantennary (A2); lane 4, total (N + O)-glycan library from bovine fetuin; lane 5, total glycan library from human α-1-acid glycoprotein. Lanes 1 and 6 contain 200 ng of total carbohydrate. Lanes 2 and 3 contain 60 pmol of each oligosaccharide standard. Lanes 4 and 5 contain glycan libraries equivalent to 4 μg of each glycoprotein. The gel was visualized and photographed on a UV transilluminator.

libraries from bovine fetuin and human α-1-acid glycoprotein, respectively, show separation of the N-glycans into classes based on total charge (sialylation). Within a charge group, multiple bands are observed, presumably due to sialic acid linkage isomerism within a charge group (e.g., fetuin, lane 4).

Also note that the size of an oligosaccharide also influences its migration on the gel (compare, e.g., the migration of standard A3 (trisialo*tri*antennary) in lane 2 to the migration of standard disialo*bi*antennary (A2) in lane 3).

E. Perspective on the Use of Fluorescent Labels for Carbohydrate Analysis

The use of fluorophores or fluorogens to label the reducing termini of mono- or oligosaccharides affords a sensitive, specific mode of detection for these traditionally difficult analytes. Even better, the use of hydrophobic, aromatic labels such as 2-AA and 2-AB can be exploited to improve chromatographic selectivity for sugars. The presence of a formal charge in 2-AA also enables its application as a fluorescent tag for carbohydrate analysis by gel electrophoresis.

A significant drawback to the use of 2-AA, 2-AB, or other fluorophores for carbohydrate analysis is the necessity of performing a chemical modification (derivitization) of the sample to be analyzed. The conditions used for some fluorescent derivatization protocols involves rather harsh conditions (extremes of pH and/or high temperature), which may cause degradation of the sample before analysis. Bigge and coworkers [7] have carefully optimized the reaction conditions using a wide array of pure, standard oligosaccharides for 2-AA and 2-AB. I have confirmed that these conditions for oligosaccharide labeling permit a high and reproducible degree of fluorescent labeling of most samples with minimal degradation such as desialylation. Anumula [10] carefully documented conditions for 2-AA monosaccharide labeling using a methanolic borate/acetate solvent system. It is also important to realize that the derivatization of biological samples can be complicated by the presence of noncarbohydrate contaminants. Thus, the labeling methods described here will work best when samples are as pure and salt-free as possible.

Researchers accustomed to using absorbance measurements in analytical methods should use caution in the application of fluorescent assays. Generally, the linear range of fluorescent assays is narrower than for absorbance-based assays, with the best linearity at the low end of the standard curve (i.e., with very low sample concentration). The presence of contaminants in the sample that absorb light at the excitation wavelength used for the assay can interfere with quantitative analysis. Finally, the measurement of fluorescence intensity on commercial HPLC fluorimeters is done on a relative scale. Thus, it is difficult to compare data between detectors or laboratories without the use of standard calibration compounds. Despite these caveats, the use of fluorescent tags for carbohydrates affords considerable improvement in the sensitivity and in the selectivity (i.e., background reduction) of detection of carbohydrates. The modern availability of extremely sensitive analytical techniques such as mass spectrometry is complemented nicely by the application of fluorescence-based separation methods for carbohydrates.

ACKNOWLEDGMENTS

The author thanks Paul Goulding and Mary Lopez for their invaluable development work on the GlycoGel PAGE system, and Andrew Ventom for his assistance with 2-AA-monosaccharide method development.

ABBREVIATIONS

2-AA	anthranilic acid (2-aminobenzoic acid)
2-AB	2-aminobenzamide
DMSO	dimethyl sulfoxide
FI	trisialylated fetuin glycopeptide fraction
GalNAc	N-acetylgalactosamine
GlcNAc	N-acetylglucosamine
HPAEC-PAD	high pH anion-exchange chromatography with pulsed amperometric detection
HPLC	high performance liquid chromatography
LC	liquid chromatography
PAGE	polyacrylamide gel electrophoresis
PEEK	polyetheretherketone
RFI	relative fluorescence intensity
TAPS	N-tris[Hydroxymethyl]methyl-3-aminopropane sulfonic acid
TFA	trifluoroacetic acid

REFERENCES

1. Townsend, R. R. (1993). *Am. Chem. Soc. Symp. 529*, 86–101.
2. Hardy, M. R. (1994). *ABRF Newslett*. 5(1), 12–19.
3. Honda, S. (1984). *Anal. Biochem*. 140, 435–442.
4. Townsend, R. R., and Hardy, M. R. (1991). *Glycobiology* 1, 139–147.
5. Schwartz, B. A., and Gray, G. R. (1977). *Anal. Biochem*. 181, 542–549.
6. Wong, W. S. D., Osuga, D. T., and Feeney, R. E. (1984). *Anal. Biochem*. 139, 58–67.
7. Bigge, J. C., Patel, T. P., Bruce, J. A., Goulding, P. A., Charles, S. M., and Parekh, R. B. (1995). *Anal. Biochem*. 230, 229–238.

8. Guile, G. R., Rudd, P. M., Wing, D. R., Prime, S. B., and Dwek, R. A., (1996) *Anal. Biochem. 240*, 210–226.

9. Hardy, M. R., Townsend, R. R., and Lee, Y. C. (1988). *Anal. Biochem.* 170, 54–62.

10. Anumula, K. R. (1994). *Anal. Biochem.* 220, 275–283.

11. Baenziger, J. U., and Fiete, D. (1979). *J. Biol. Chem.* 254, 789–795.

12. Patel, T., Bruce, J., Merry, A., Bigge, C., Wormald, M., Jaques, A., and Parekh, R. (1993). *Biochemistry* 32, 679–693.

13. Ashwell, G. (1966). *Meth. Enzymol.* 8, 85–95.

14. Lambert, J. M., McIntyre, G., Gauthier, M. N., Zullo, D., Rao, V., Steeves, R. M., Goldmacher, V. S., and Blättler, W. A. (1991). *Biochemistry* 30, 3234–3247.

15. Edge, A. S. B., and Spiro, R. G. (1987). *J. Biol. Chem.* 262, 16135–16141.

23

Capillary Gel Electrophoresis of 8-Aminopyrene-3,6,8-trisulfonate-Labeled Oligosaccharides

András Guttman

Beckman Instruments, Inc., Fullerton, California

I. INTRODUCTION

The emergence of glycoproteins as pharmaceutical products made it necessary to develop fast, high resolution, and reproducible methods for the analysis of complex carbohydrates [1,2]. Due to the increasing evidence that carbohydrate moieties of glycoproteins are important as recognition factors in receptor–ligand or cell–cell interactions, in immunogenicity modulation, in the folding/unfolding process of protein molecules, and in the regulation of protein bioactivity, understanding the role of glycoproteins in the function of normal and abnormal cells is increasingly important [3]. Even small changes in the oligosaccharide structures and/or site occupancy can significantly influence the biological activity of glycoproteins. The wide variety of similar structures of oligosaccharides in glycoproteins have made it difficult to obtain all the necessary information to address the foregoing problems. With the advent of high performance capillary gel electrophoresis (HPCGE) in conjunction with the ultrasensitive detection capability of the laser-induced fluorescence (LIF) systems, faster and higher resolution separations enable the tentative identification of small quantities of closely related complex oligosaccharides [4–8]. Using this novel methodology, individual oligosaccharides can be quantified to obtain molar ratios, and to detect changes in the extent and nature of glycosylation [9,10].

II. MATERIALS AND METHODS

A. Chemicals

The standard ladder of glucose oligomers (maltooligosaccharides) M040 (Grain Processing Co., Muscatine, Iowa) labeled by trisodium 8-amino-1,3,6-pyrene trisulfonate (APTS) (Beckman Instruments, Inc., Fullerton,

California) was used as a reference standard in the separations. The disialy-lated-biantennary, asialo-triantennary and asialo-agalacto-tetra-antennary structures were purchased from Oxford Glycosystems (Rosedale, New York). The trisialylated triantennary [2 × α(2,6)] and trisialylated trianten-nary [2 × α(2,3)] glycans were from Dionex (Sunnyvale, California). Ribo-nuclease B and fetuin were purchased from Sigma (St. Louis). All other chemicals were from Aldrich (Milwaukee). The APTS-derivatized oligosac-charide samples were used directly after derivatization or stored at −20°C. All buffer solutions were filtered through a 0.45 μm pore size filter and vacuum degassed at 100 mbar.

B. Procedures

The flow chart in Fig. 1 [11] depicts the necessary steps of the N-linked oligosaccharide profiling protocol (eCAP N-linked Oligosaccharide Profil-ing Kit, Beckman Instruments, Inc., Fullerton, California): Step I is release of the N-linked oligosaccharides from the glycoprotein, performed enzy-matically. Step II is labeling of the mixture of released oligosaccharides with a charged fluorescent tag. Step III is separation of the fluorophore-labeled oligosaccharides by capillary gel electrophoresis. The distribution of the resulting peaks in the electropherogram obtained after capillary elec-trophoresis separation represents an N-linked oligosaccharide profile of the glycoprotein. This pattern can be thought of as a N-glycosylation "finger-print" with individual peaks representing unique oligosaccharides released from the glycoprotein.

1. Release of N-Linked Oligosaccharides from Glycoproteins

The first step was to isolate glycoproteins according to reported procedures [12]. The glycoprotein should be prepared in a non-Tris-containing buffer and contain a minimum amount of salt. The glycoprotein was dried in a 0.5-mL microcentrifuge tube in a centrifugal vacuum evaporator (CVE). Generally, 20–200 μg of glycoprotein was used for analysis. The actual amount of glycoprotein required will depend on the size of the protein and the extent of glycosylation. At least 0.1 nmol of glycoprotein should be digested for analytical work; a greater amount is required for the prepara-tion of purified oligosaccharides. The dried glycoprotein was dissolved in 45 μL of 50 mM phosphate buffer, pH 7.5. Sodium dodecyl sulfate (SDS) was required in most instances to denature the glycoprotein prior to enzy-matic digestion. SDS and β-mercaptoethanol (β-ME) were added to a final concentration of ~0.1% and 50 mM, respectively. The mixture was boiled for 5 min, then cooled to room temperature, followed by the addition of Nonidet P-40 to a concentration of 0.75% and mixed with finger flicks. PNGase F (peptide-N^4-(acetyl-β-D-glucosylaminyl)–asparagine amide, from

Purified Glycoprotein

- Boiling with SDS
- PNGase F digestion
- Ethanol precipitation
  and centrifugation

5 min
2 hours
15 min

Free Oligosaccharides
(can be stored frozen for later use)

- Dry in CVE
- Label with APTS

30-60 min
90 min at 55°C
16 hours* at 37°C

Labeled Oligosaccharides
(can be stored frozen for later use)

- Capillary gel
  electrophoresis

10-20 min

Separated Oligosaccharides

Figure 1 N-linked oligosaccharide profiling protocol. (*16 hours at 37°C is recommended with sensitive (e.g. sialylated) structures.)

Glyko Inc., Novato, CA) (1 mU) was added to the sample, which was mixed and centrifuged for 5 sec. The reaction was incubated for 2 h at 37°C. Some proteins, such as several immunoglobulins, precipitated when boiled. The following procedure was used for precipitated proteins. After the addition of SDS and β-ME, standing for 5 min at room temperature, and addition of Nonidet P-40 as just described, PNGase F (1 mU) was added and the mixture incubated overnight at 37°C. After the release of oligosaccharides, the protein was precipitated by the addition of 3 volumes of cold (-20°C) ethanol. The samples were kept on ice for 10 min, followed by centrifugation for 5 min at 10,000 rpm. The supernatant, which contained the released oligosaccharides, was transferred to a clean 0.5-mL microcentrifuge tube. Maltose (2 nmol) was added to the supernatant as an internal labeling control.

If a large amount of glycoprotein is digested (>250 μg), 5–10% of the released oligosaccharides may be trapped in the pellet. The recovery of these oligosaccharides was accomplished by drying the pellet in a CVE or lyophilizer followed by the addition of CE grade water. The protein was then precipitated by the addition of 150 μL of cold ethanol and centrifuged. The combined supernatants were dried in a CVE or lyophilizer to translucent pellet. Samples may be stored at $-20°$C for 2 weeks or labeled with the fluorophore as described in the following section.

2. Fluorophore Labeling Reaction

An APTS (Beckman Instruments, Inc.) solution (0.2M) was prepared in 15% acetic acid, and mixed by vortexing until the dye was dissolved. The APTS solution (2 μL) was added to each dried oligosaccharide pellet in 0.5-mL microfuge tubes and mixed well until the pellet was dissolved. Two microliters of 1 M sodium cyanoborohydride in tetrahydrofuran (Aldrich) was added, and the mixture was vortexed. The samples were incubated at 55°C for 90 min in a water bath after closing the vial caps tightly to avoid evaporation (if necessary, use parafilm). (Note: Although greater than 95% of the oligosaccharides are labeled in 90 min at 55°C, which is recommended for most profiling applications, to avoid desialylation samples can be incubated overnight at 37°C for 16 h. The reaction was stopped by the addition of 46 μL of CE grade water. After appropriate dilution (usually 1 : 20), the samples were injected onto the column and separated by capillary electrophoresis.

3. Capillary Electrophoresis Separation of the Labeled Glycans

Capillary electrophoresis was performed on a P/ACE 5000 system (Beckman Instruments, Inc.) with the cathode on the injection side and the anode on the detection side (i.e., reversed polarity). The negatively charged, APTS-labeled carbohydrate molecules migrate toward the anode under the influence of the electric field and under conditions that of zero or very low electroosmotic flow. The separations were monitored on-column with a Beckman laser-induced fluorescence (LIF) detection system using a 4-mW argon ion laser with an excitation wavelength of 488 nm and emission wavelength filter of 520 nm. The temperature of the capillary in the P/ACE instrument was controlled at 20 $\pm$ 0.1°C. The electropherograms were acquired and stored on an IBM 486/66 computer using the System Gold software package (Beckman Instruments, Inc.).

A coated capillary column (50 μM i.d.) with a 40 cm effective length (47 cm total length) was used (eCAP PVA-1, Coated Capillary, Beckman Instruments, Inc.). For all separations, 25 mM acetate buffer (pH = 4.75) with 0.4% polymeric additive used as a running buffer. This buffer resulted in high relative migration time reproducibility ($RSD_{RMT} < 0.1\%$) and a low stable current ($< 20 \mu A$). The labeled samples were diluted 500 times (2500 times for the M040 ladder) in the running buffer and injected by the pressure injection mode of the system, typically for 5–10 sec at 0.5 psi (3447 Pa).

III. RESULTS AND DISCUSSION

The principle of the carbohydrate analysis system described in this chapter is based on the use of capillary electrophoresis to separate and quantify intact oligosaccharides released from glycoproteins. As the flow chart in Fig. 1 shows, the analysis begins with the release of N-linked oligosaccharides from glycoproteins using recombinant PNGase F. The released oligosaccharides are then labeled with 8-aminopyrene-3,6,8-trisulfonate (APTS) at the reducing terminal by reductive amination [13,14] as shown in Fig. 2. The stoichiometry of labeling is one mole of fluorophore for each mole of oligosaccharide.

When labeling 5 nmol or less of total sugar using the reagents and labeling conditions described in Section II, the labeling efficiency was $> 97\%$. Figure 3 shows the effect of temperature on the yield of coupling APTS to a trisialylated triantennary oligosaccharide [$2 \times \alpha(2,3)$] (Table 1) [15]. As the lower section ($///$) of each column depicts, the derivatization yield increased with increased temperature, and up to approximately 55°C, no significant desialylation occurred. Applying temperatures higher than 55°C, desialylation increased, as the higher section (xxx) of each column in Fig. 3 indicates. At higher temperatures (> 65°C), the increasing desialylation resulted in a decrease in the amount of labeled trisialylated structure, as shown by a decline of the lower section ($///$) of each column. Labeling > 5 nmol sometimes resulted in reduced labeling efficiency; therefore, a quantitation control should be added to the reaction mixture as an internal labeling control.

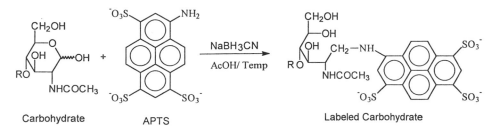

Figure 2 N-linked glycan labeling reaction with 8-aminopyrene-3,6,8-trisulfonate (ATPS).

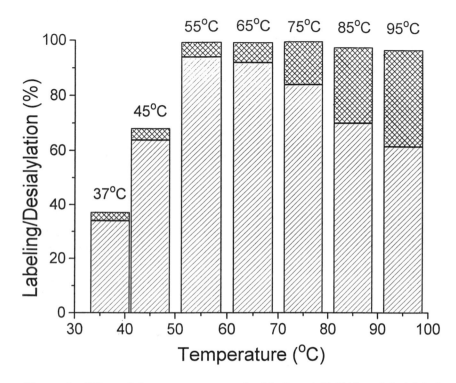

Figure 3 Effect of the temperature on the labeling yield (///) and desialylation (xxx) of a trisialylated-triantennary [2 × α(2,3)] glycan (structure in Table 1).

After derivatization, the fluorescently labeled oligosaccharides were separated and quantified by capillary gel electrophoresis using LIF detection. The separation of the Glc ladder standard, consisting of a mixture of various chain length Glc oligomers (Glc_1 to $>Glc_{20}$) should appear at the completion of the capillary gel electrophoresis separation (Fig. 4). The area of the Glc_5 peak in the standard mixture is less than the adjacent peaks, allowing for easy identification of this peak and further peak assignments.

Figure 5 compares the separations of the Glc ladder with individual, commercially available oligosaccharides, disialylated biantennary (A), asialo-agalacto-tetraantennary (B), and asialo-tetraantennary (C). The structures are shown in Table 1. Based on the relative migration positions of the standards, the contribution of each individual sugar unit to the resulting relative migration positions of the complex structures was calculated from the corresponding Glc unit (GU) values.

The profiles of the APTS-labeled *N*-linked oligosaccharides, enzymat-

Table 1 Structure of a Trisialylated-Triantennary [2 × α(2,3)] Glycan

Complex-Type Structures

Disialylated biantennary

NeuAcα(2,6)Galβ(1,4)GlcNAcβ(1,2)Manα(1,6)

|

Manβ(1,4)GlcNAcβ(1,4)GlcNAc

|

NeuAcα(2,6)Galβ(1,4)GlcNAcβ(1,2)Manα(1,3)

Trisialylated-triantennary − 2 × α(2,6) (F3):

NeuAcα(2,6)Galβ(1,4)GlcNAcβ(1,2)Manα(1,6)

|

Manβ(1,4)GlcNAcβ(1,4)GlcNAc

|

NeuAcα(2,6)Galβ(1,4)GlcNAcβ(1,2)Manα(1,3)

|

NeuAcα(2,3)Galβ(1,4)GlcNAcβ(1,4)

Trisialylated-triantennary − 2 × α(2,3) (F4)

NeuAcα(2,3)Galβ(1,4)GlcNAcβ(1,2)Manα(1,6)

|

Manβ(1,4)GlcNAcβ(1,4)GlcNAc

|

NeuAcα(2,6)Galβ(1,4)GlcNAcβ(1,2)Manα(1,3)

|

NeuAcα(2,3)Galβ(1,3)GlcNAcβ(1,4)

Tetrasialylated-triantennary − 2 × α(2,6) (F1)

NeuAcα(2,6)Galβ(1,4)GlcNAcβ(1,2)Manα(1,6)

|

Manβ(1,4)GlcNAcβ(1,4)GlcNAc

|

NeuAcα(2,6)Galβ(1,4)GlcNAcβ(1,2)Manα(1,3)

|

NeuAcα(2,3)Galβ(1,3)GlcNAcβ(1,4)

|

NeuAcα(2,6)

Tetrasialylated-triantennary − 2 × α(2,3) (F2)

NeuAcα(2,3)Galβ(1,4)GlcNAcβ(1,2)Manα(1,6)

|

Manβ(1,4)GlcNAcβ(1,4)GlcNAc

|

NeuAcα(2,6)Galβ(1,4)GlcNAcβ(1,2)Manα(1,3)

|

NeuAcα(2,3)Galβ(1,3)GlcNAcβ(1,4)

|

NeuAcα(2,6)

Table 1 *Continued*

Neutral Complex-Type Structures

Asialo-agalacto-tetraantennary

```
    GlcNAcβ(1,6)
         |
  GlcNAcβ(1,2)Manα(1,6)
                 |
                     Manβ(1,4)GlcNAcβ(1,4)GlcNAc
                 |
  GlcNAcβ(1,2)Manα(1,3)
             |
       GlcNAcβ(1,4)
```

Asialo-galacto tetraantennary

```
    Galβ(1,4)GlcNAcβ(1,6)
              |
  Galβ(1,4)GlcNAcβ(1,2)Manα(1,6)
                          |
                              Manβ(1,4)GlcNAcβ(1,4)GlcNAc
                          |
  Galβ(1,4)GlcNAcβ(1,2)Manα(1,3)
                      |
  Galβ(1,4)GlcNAcβ(1,4)
```

High-Mannose Structures

Man 5

```
    Manα(1,6)
         |
             Manα(1,6)
         |       |
    Manα(1,3)        Manβ(1,4)GlcNAcβ(1,4)GlcNAc
                         |
             Manα(1,3)
```

Man 6

```
    Manα(1,6)
         |
             Manα(1,6)
         |       |
    Manα(1,3)        Manβ(1,4)GlcNAcβ(1,4)GlcNAc
                         |
    Manα(1,2)Manα(1,3)
```

Man 7

```
                 Manα(1,6)
                      |
                          Manα(1,6)
    Manα(1,2)             |       |
                     Manα(1,3)        Manβ(1,4)GlcNAcβ(1,4)GlcNAc
                                          |
                     Manα(1,2)Manα(1,3)
```

Table 1 *Continued*

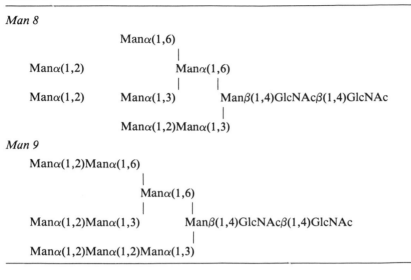

Man 8

Manα(1,6)
|
Manα(1,2)   Manα(1,6)
| |
Manα(1,2)   Manα(1,3)   Manβ(1,4)GlcNAcβ(1,4)GlcNAc
|
Manα(1,2)Manα(1,3)

Man 9

Manα(1,2)Manα(1,6)
|
Manα(1,6)
| |
Manα(1,2)Manα(1,3)   Manβ(1,4)GlcNAcβ(1,4)GlcNAc
|
Manα(1,2)Manα(1,2)Manα(1,3)

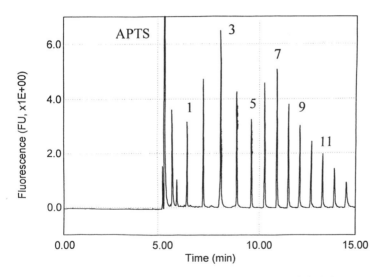

Figure 4 Capillary gel electrophoresis of the APTS-labeled glucose ladder standard test mixture. Numbers on peaks correspond to the degree of polymerization of the glucose oligomers. Separation conditions: 40-cm eCAP PVA-1 coated column (47 cm total length), i.d. 50 μm; LIF detection: argon-ion laser, excitation 488 nm, emission 520 nm; separation buffer: 25 mM acetate buffer, pH 4.75, $E = 500$ V/cm; $i = 19$ μA, 20°C.

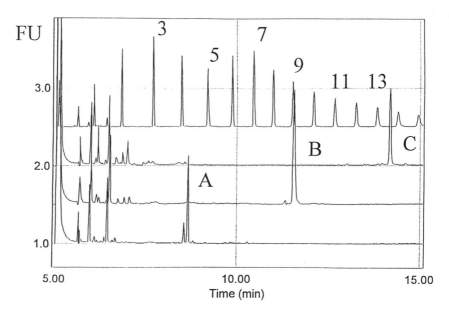

Figure 5 Capillary electrophoresis of the glucose ladder standard (upper trace) and oligosaccharide structures of disialylated-biantennary (A), asialo-agalacto-triantennary (B), and asialo-galacto-tetraantennary structures (C) (structures in Table 1). Conditions are the same as described in Fig. 4.

ically released from bovine fetuin and ribonuclease B (outlined in Fig. 1), are shown in Fig. 6. The trace in Fig. 6A depicts the separation of the APTS-labeled N-linked glycan pool of bovine fetuin. Two well-separated doublets (F1/F2 and F3/F4), migrating between 7 and 9 min, were observed. Peaks F3 and F4 were spiked with purified, individual standards and were assumed to be trisialylated triantennary structures with different proportions of $\alpha(2,3)$- and $\alpha(2,6)$-linked sialic acid residues. The other two large peaks, F1 and F2, were postulated to be tetrasialylated triantennary structures, based on calculations from their corresponding GU values [16]. The proposed structures of these complex carbohydrates are given in Table 1. At lower pH (2.5), only two major peaks, corresponding to the tri- and tetrasialylated structures, were resolved; the differences in sialic acid linkages were not resolved. Increasing the pH of the buffer to 4.75 resulted in complete separation of the two doublets (F3/F4 and F1/F2) and several minor peaks (Fig. 6A). The effects of gel concentration, electric field strength, and temperature on the migration behavior of peaks F1, F2, F3, and F4 were also examined. The data suggest that capillary gel electrophoresis separation of multisialylated, branched oligosaccharides with different

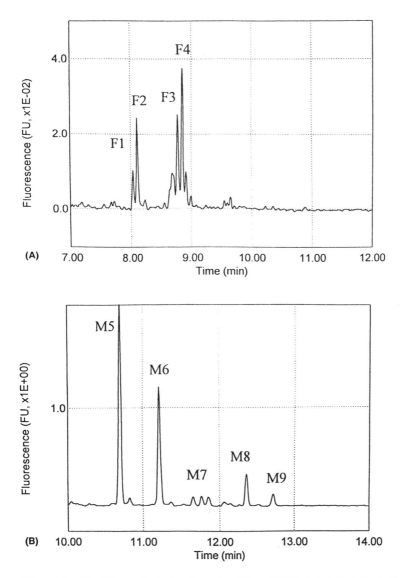

Figure 6 Capillary gel electrophoresis of the APTS labeled *N*-linked oligosaccharides of fetuin (A) and ribonuclease B (B). Peak assignments in the electropherograms correspond to the structures shown in Table 1. Conditions are the same as described in Fig. 4.

proportions of differently linked sialic acid residues depends on the degree of sialylation and hydrodynamic volumes.

High resolution separation of APTS-labeled ribonuclease B glycans is shown in Fig. 6B. Here, in addition to the complete separation of the high-mannose structures according to size [Man_5 (M5) to Man_9 (M9)], baseline separation of the positional isomers of Man_7 (M7) and Man_8 (M8) structures was also obtained (for structures, see Table 1). The proportion of the peak areas agreed with those previously reported using ^{1}H nuclear magnetic resonance (NMR) [17] and high performance liquid chromatography (HPLC) [18]. The M7 triplet showed a distribution of 38 : 37 : 25 and 31 : 33 : 35 by NMR and capillary gel electrophoresis, respectively. The NMR data also revealed three positional isomers in the M8 fraction with proportions of 10 : 6 : 84, in agreement with data from capillary gel electrophoresis, 17 : 4 : 78. Run-to-run variation of the proportions was less than 2%. Co-elution of most of the ribonuclease B glycan peaks was achieved with individual standards, and peaks were identified as follows: Man_5 to M5; Man_6 to M6; Man_7D3 to M7, middle peak; Man_7D1 to M7, last peak; $Man_8D1,3$ to M8, last peak; and Man_9 to M9 [19]. Based on our previous results [20] and other studies [17] on ribonuclease B glycans, the first peak of the M7 triplet and the first two peaks of the M8 triplet were assigned to Man_7D2 and $Man_8D1,2$ and $Man_8D2,3$, respectively.

IV. CONCLUSION

N-linked oligosaccharides released from glycoproteins can be tagged with a charged, fluorescent label for subsequent analysis, using reductive amination chemistry. The labeled oligosaccharides can then be rapidly profiled with capillary gel electrophoresis, and the individual oligosaccharides in the mixture can be quantified into the low femtomole range by the laser-induced fluorescence detection system. It has been demonstrated that capillary gel electrophoresis provides an excellent carbohydrate profiling tool: it can resolve differences in oligosaccharide size, net charge, and linkage positions of the individual monosaccharide units.

ACKNOWLEDGMENTS

The author gratefully acknowledges Dr. Nelson Cooke for his support. The help of the Beckman Research Library is also highly appreciated.

ABBREVIATIONS

APTS   8-aminopyrene-3,6,8-trisulfonate
CE   capillary electrophoresis

CVE	centrifugal vacuum evaporator
GU	Glc unit(s)
HPCGE	high performance capillary gel electrophoresis
HPLC	high performance liquid chromatography
LIF	laser-induced fluorescence
β-ME	β-mercaptoethanol
NMR	nuclear magnetic resonance
PNGase F	peptide-N^4(-acetyl-β-D-glucosylaminyl)-asparagine amide
SDS	sodium dodecyl sulfate

REFERENCES

1. Hardy, M. R., and Townsend, R. R. (1988). *Proc. Natl. Sci. USA* 85, 3289–3293.
2. El Rassi, Z., ed. (1995). *Carbohydrate Analysis*. Elsevier, Amsterdam.
3. Varki, A. (1993). *Glycobiology* 3, 97–130.
4. Honda, S., Iwase, S., Makino, A., and Fujiwara, S. (1989). *Anal. Biochem.* 176, 72–77.
5. Liu, J., Shirota, O., Wiesler, D., and Novotny, M. (1991). *Proc. Natl. Acad. Sci. USA* 88, 2302.
6. Chiesa, C., and Horváth, Cs. (1993). *J. Chromatogr.* 645, 337–352.
7. Linhardt, R. J. (1994). *Meth. Enzymol.* 230, 265–280.
8. Paulus, A., and Klockow, A. (1995). *J. Chromatogr.* 716, 245–257.
9. Starr, C., Striepeke, S., and Guttman, A. (1994). *Biomedical Products*, Nov., 22–25.
10. Oefner, P. J., and Chiesa, C. (1994). *Glycobiology*, 4, 397–412.
11. eCAP N-linked Oligosaccharide Profiling Kit, Care and Use Instructions (#725805), Beckman Instruments, Inc., Fullerton, California.
12. Merkle, R. K., and Cummings, R. D. (1987). *Meth. Enzymol.* 138, 232–345.
13. Jackson, P. (1994). *Anal. Biochem.* 216, 243–252.
14. Evangelista, R., Liu, M. S., and Chen, F. T. A. (1995). *Anal. Chem.* 67, 2239–2245.
15. Guttman, A., Chen, F. T. A., Evangelista, R., and Cooke, N. (1995). *Anal. Biochem.* 233, 234–242.
16. Guttman, A., Chen, F. T. A., and Evangelista, R. (1996). *Electrophoresis* 17, 412–417.
17. Fu, D., Chenm, L., and O'Neil, R. A. (1994). *Carbohydr. Res.* 261, 173–186.
18. Parekh, R., and Ventom, A. (1995). *Amer. Biotech. Lab.* 13, 8–9.
19. Guttman, A., and Herrick, S. (1996). *Anal. Biochem.* 235, 236–239.
20. Guttman, A., and Pritchett, T. (1995). *Electrophoresis* 16, 1906–1911.

24
Evaluation of Glycosylpyrazole Derivatives in Carbohydrate Analysis

Warren C. Kett and Michael Batley
Macquarie University, Sydney, New South Wales, Australia

John W. Redmond
Research School of Biological Sciences, Australian National University, Canberra, A. C. T., Australia

I. INTRODUCTION

The methods for the derivatization of sugars described elsewhere in this volume are based on reductive amination of reducing sugars or their condensation with suitable pyrazolones. The preliminary account in this chapter describes an alternative technique in which UV-absorbing derivatives are formed from glycosylhydrazones under mild conditions. The method could be particularly useful for the analysis of the products of hydrazinolysis of glycoproteins, because it avoids the hydrolysis of the glycan hydrazones and any associated incompleteness of recovery. This method can also be used for reducing sugars, because they can be converted quantitatively to the hydrazones by treatment with hydrazine hydrate at room temperature.

II. MATERIALS AND METHODS

A. Preparation of 1-Glycosyldimethylpyrazoles

Monosubstituted hydrazines condense readily with β-diketones to form 1-substituted pyrazoles [1] (Fig. 1, top scheme). Similarly, because the cyclic forms of glycan hydrazones correspond to monosubstituted hydrazines, the hydrazones can be also converted to 1-glycosylpyrazoles. In the present study, glycosylpyrazoles (Fig. 1, bottom scheme) were prepared using 2,4-pentanedione (acetylacetone). Other diketones can, in principle be used, but it is important to use a symmetrical diketone ($R^2 = R^3$ in Fig. 1, top scheme) in order to avoid the formation of a mixture of isomeric pyrazoles. A practical advantage of acetylacetone (Acac) is its good solubility in water and its easy removal by evaporation. The method was applied successfully

β diketone

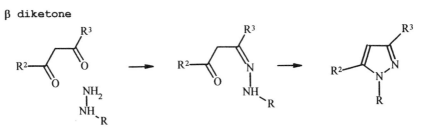

Monosubstituted hydrazine   1-substituted pyrazole

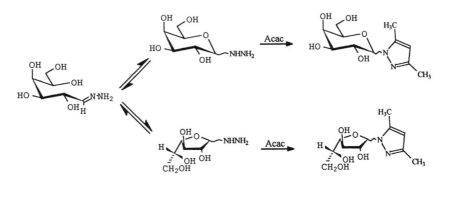

Hydrazone   Glycosylhydrazine   1-Glycosylpyrazole

Figure 1 Top scheme: The general synthesis of 1-substituted pyrazoles with β-diketones. Bottom scheme: The formation of 1-galactosyl-3,5-dimethylpyrazoles from galactose hydrazone with 2,4 pentanedione (acetylacetone).

to a range of monosaccharides including aldohexoses, aldopentoses, ketoses, aminodeoxy sugars, and uronic acids. The sugars were incubated for 4–16 h in hydrazine hydrate, and evaporated to dryness in vacuo. The residues were dissolved in a *minimum* of water, treated with a 10-fold molar excess of Acac, allowed to stand for 2 h, and again evaporated to dryness. The conversion of monosaccharide hydrazones to 1-glycosyl-3,5-dimethylpyrazoles occurred with an efficiency of greater than 95% as indicated by ^{1}H nuclear magnetic resonance (NMR).

The 1-glycosylpyrazoles are stable and resist hydrolysis in 2 M trifluoroacetic acid at 100°C for 4 h. They have a UV absorption maximum at 220 nm with a molar extinction coefficient of approximately 4000 L mol^{-1} cm^{-1}. This UV absorption of the heterocyclic ring is modest, but useful for detection in liquid chromatography and capillary electrophoresis. The UV peak areas obtained for a standard mixture of maltooligosaccharides (degree of polymerization (dp) = 2, 4, 6) analyzed by high performance liquid chromatography (HPLC) showed that the efficiency of derivatization was independent of chain length (data not shown). Serial dilution of the same sample before derivatization was used to demonstrate that the efficiency was independent of the concentration of the sample, down to the method detection limit.

B. Characterization of 1-Glycosylpyrazoles

A shortcoming of 1-glycosylpyrazoles for the analysis of sugars is the formation of a mixture of isomeric products, corresponding to α- and β-pyranosyl- and α- and β-furanosylpyrazoles. The proportions of the different isomers vary for the different monosaccharides, and they are a sensitive function of the stereochemistry of the sugars. For example, in an oligosaccharide with a 1-4 linkage to the reducing-terminal residue, the formation of furanose forms is in fact impossible, because there is no free hydroxyl group at the 4 position. The isomeric products were fractionated by preparative reversed phase HPLC and characterized by ^{1}H, ^{13}C, and ^{1}H/^{13}C heteronuclear correlation NMR spectroscopy. Examples of the isomeric distributions are shown in Table 1. The relative amounts of the 1-glycosylpyrazoles from a given monosaccharide were altered slightly by preequilibrating the hydrazone in the buffer solution, but it was not possible to obtain more

Table 1 The Isomeric Distribution of 1-Glycosyl-3,5-Dimethyl Pyrazoles Expressed as a Percentage of the Total Yield of 1-Glycosylpyrazoles

Monosaccharide	α-Furanosyl	β-Furanosyl	α-Pyranosyl	β-Pyranosyl
Glucose	4	6	7	83
Xylose	13	26	12	48
Galactosamine	10	45	5	40
Glucuronic acid	2	6	12	80
Galacturonic acid	12	4	50	34
Fucose	3	12	3	80
Maltose	—	—	12	88
Isomaltose	6	6	7	81

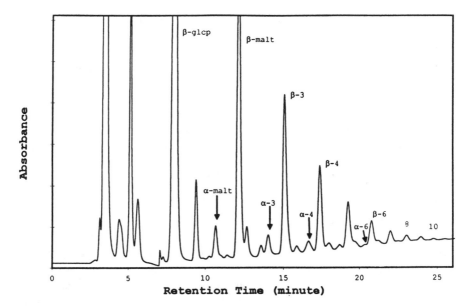

Figure 2 Normal phase HPLC separation of the pyrazole-labeled maltooligosaccharides from corn syrup. The labels at the peak apices indicate the configuration
of the pyrazole-labeled terminal sugar and the degree of polymerization. The column used was a Rainin (Woburn, Massachusetts) NH_2 8 μm (250 × 4.6 mm).
Gradient elution was with acetonitrile/water at 1 mL/min. Twenty percent water
initially for 4 min with a linear increase to 80% water by 34 min. Detection was by
UV absorbance at 220 nm.

than 85% of a single isomer. Morever, caution should be exercised when
attempting equilibration at pH values below 7.0, because significant hydrolysis of the hydrazones is likely to occur. Once the pyrazole ring is formed,
no further equilibration between the isomers is possible.

III. RESULTS AND DISCUSSION

A. Normal Phase HPLC

A mixture of 1-glycosylpyrazoles was prepared from the maltooligosaccharides in corn syrup and separated by normal phase HPLC on an aminopro-
pylsilica column eluted with a gradient of water–acetonitrile. Oligomers of
up to 10 glucose units were resolved without peak distortion (Fig. 2). The
maltooligosaccharides with dp = 2, 3, 4, and 6 were assigned using standards. The remaining peaks formed a regular series that converged to the

same limit. One of the advantages of using UV detection is that gradient elution was more practicable. The instrument detection limit was a few picomoles of each component. The retention times for the pyrazole derivatives were only slightly shorter than those for underivatized sugars. The α-anomers eluted before the corresponding β-anomers, and the difference decreased with increasing chain length. The anomers were not resolved when the chain length was greater than 8 glucose units.

B. Size Exclusion

The elution times of the derivatized components of corn syrup using Toyopearl (Tosoh Corp. Tokyo, Japan) with water as the eluent (Fig. 3) obeyed the usual semilogarithmic relationship with respect to degree of polymerization. Unsubstituted 3,5-dimethylpyrazole, resulting from reaction of Acac with traces of residual hydrazine, eluted well after the pyrazole derivatives of glucose. The use of the inorganic salts, sodium chloride and ammonium chloride, in the eluent did not affect the elution times of the glycosylpyra-

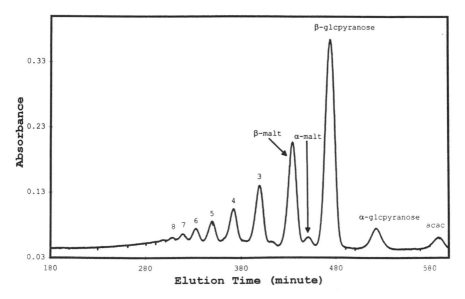

Figure 3 Size exclusion chromatography using Toyopearl HW 40-S of the pyrazole-labeled maltooligosaccharides from corn syrup. The labels at the peak apices indicate the configuration of the pyrazole-labeled terminal sugar and the degree of polymerization. The column dimensions were 95 × 1.5 cm. Water was the eluent at 0.3 mL/min. Detection was by UV absorbance at 220 nm.

zoles, but it did have the desirable effect of reducing the elution time of the contaminating 3,5-dimethylpyrazole.

C. Capillary Electrophoresis

The basicity of pyrazoles can be exploited for sugar analysis by capillary electrophoresis. At low pH, the peaks for α- and β-anomers of the maltooligosaccharides are widely separated, and the two series of peaks barely overlap (Fig. 4). Within each series, the migration times obey a semilogarithmic relationship with molecular size. The difference in mobilities presumably reflects the relative basicities of the two anomeric forms. Isomaltose derivatives, however, have mobilities that differ from those of the maltose pyrazoles, indicating that hydrodynamic volume also influences mobility.

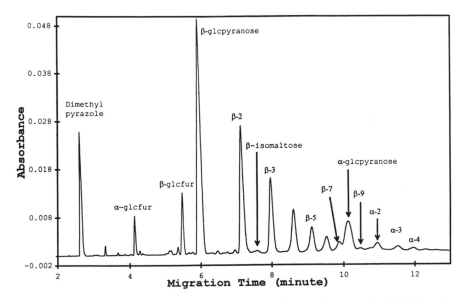

Figure 4   Low pH capillary electrophoresis separation of the pyrazole-labeled maltooligosaccharides in corn syrup. The labels at the peak apices indicate the configuration of the pyrazole-labeled terminal sugar and the degree of polymerization. The buffer was 0.1 M H_3PO_4 (approximate pH 1.5) with a capillary length of 20 cm (13 cm to window). The applied voltage was 8 kV, current $= 130 \ \mu A$. Detection was by UV absorbance at 226 nm. A PACE 5000 (Beckman Instruments, Fullerton, CA) instrument was used.

D. Electrospray Mass Spectrometry

The pyrazole moiety enhances the ionization efficiency in positive ion electrospray mass spectrometry. Using the solvent systems employed in normal phase and hydrophilic interaction chromatography, an instrument detection limit of approximately 2 pmol was obtained for the positive pseudomolecular ion of maltose pyrazole. The advantage of using the 1-glycosylpyrazoles is that, at low orifice potential, the only significant ion in the mass spectrum corresponds to the protonated 1-glycosylpyrazole, allowing the determination of molecular mass at very low levels. Negative ion detection is the more sensitive mode for observing underivatized maltose. This has a detection limit of 10 pmol, which is five times as great as the limit for the pyrazole derivative in positive ion mode.

E. Chromatographic Properties of Monosaccharide Derivatives

The resolution of the isomeric forms complicates reversed phase HPLC analysis of monosaccharide mixtures and limits the use of pyrazole derivatives to samples containing only a few monosaccharides. The method becomes practical, however, when used for end-group analysis of oligosaccharides. In this approach, the linkage of the reducing-terminal residue to the pyrazole substituent is stable under the conditions used to hydrolyze all the internal glycosyl linkages. It is, therefore, possible to establish the terminal residue by identification of the single 1-glycosylpyrazole in the hydrolysate. This can be done by either reversed phase HPLC or gas chromatography of the TMS or acetyl derivatives of the 1-glycosylpyrazole. In addition, a nitrogen–phosphorus detector could provide sensitive and selective detection of the pyrazole derivatives.

The acetyl and TMS derivatives of the 1-glycosylpyrazoles are less volatile than the corresponding sugar derivatives, and when gas chromatographic (GC) analysis of mixtures was performed, the pyrazoles were well separated from sugars without a pyrazole substituent (and from the alditols and methylglycosides). The electron impact mass spectra of the TMS ethers are similar to those of the TMS ethers of the corresponding monosaccharides and the usual rules for determining ring size [2] can be used.

IV. SUMMARY

The conversion of reducing glycans to 1-glycosylpyrazoles under mild nonreducing conditions offers a useful alternative to derivatization by reductive amination. The incorporation of a UV-absorbing chromophore is useful in the analysis of glycans, but its absorbance at 220 nm still precludes its use for some sample matrices and some solvent systems. The development of alternative chromophores is under investigation.

The modest absorbance at 220 nm, 4000 L mol^{-1} cm^{-1}, still confers significantly greater sensitivity than detection of underivatized sugars by direct UV absorbance at below 200 nm, indirect UV absorption, or refractive index (RI) detection [3,4]. The sensitivity is inferior to that obtained with fluorescent labeling or pulsed amperometric detection, but the constant molar response avoids the calibration steps required to allow for the variable response of a pulsed amperometric detector.

The pyrazole derivatives are extremely stable, but the formation of isomeric products is a disadvantage. For a mixture of related polysaccharides, where the objective is determination of the molecular size distribution, they may still be valuable. The presence of isomeric pyrazoles is more vexatious for oligosaccharides, such as those derived from glycoproteins, for which more detailed molecular characterization is necessary.

Although these difficulties preclude the use of 1-glycosylpyrazoles in a general method for analyzing the hydrazinolysis products from glycoproteins, the pyrazoles may be useful for certain aspects of such analyses. They can provide improved electrospray ionization mass spectrometry (ESI/MS) sensitivity and may be valuable for end-group analysis. The method is simple, robust, and inexpensive, and the derivatives are chemically stable. It may, therefore, have a place in a suite of methods for the analysis of glycans.

ABBREVIATIONS

Acac	acetylacetone
dp	degree of polymerization
ESI/MS	electrospray ionization mass spectrometry
GC	gas chromatography
HPLC	high performance liquid chromatography
NMR	nuclear magnetic resonance
RI	refractive index
TMS	Trimethylsilyl

REFERENCES

1. Elguero, J. (1984). In Katritzky, A. R., ed., *Comprehensive Heterocyclic Chemistry*, Vol. 5. Pergamon Press, Oxford, pp. 167–303.
2. Lonngren, J., and Svensson, S. (1974). *Adv. Carb. Chem. Biochem.* 29, 42–106.
3. Bruno, A. E., and Krattiger, B. (1995). In El Rassi, Z., ed., *Carbohydrate*

Analysis—High Performance Liquid Chromatography and Capillary Electrophoresis, Journal of Chromatography Library, Vol. 58. Elsevier, Amsterdam, pp. 431–446.

4. El Rassi, Z., and Smith, J. T. (1995). In El Rassi, Z., ed., *Carbohydrate Analysis-High Performance Liquid Chromatography and Capillary Electrophoresis*, Journal of Chromatography Library, Vol. 58. Elsevier, Amsterdam, pp. 607–640.

25

Discovery and Uses of Novel Glycosidases

Sharon T. Wong-Madden, David Landry, and Ellen P. Guthrie
New England Biolabs, Beverly, Massachusetts

I. INTRODUCTION

There is an ever expanding number of techniques used to identify and sequence oligosaccharides. Many of the sequencing techniques require exoglycosidases to sequentially remove sugar residues from an oligosaccharide. Because of this requirement, the more exoglycosidases available that are highly specific for either a particular sugar residue or glycosidic linkage, the more powerful these techniques become. Screening for novel specific exoglycosidases from bacterial species was initiated to try to augment the number of exoglycosidases available to the research community. Until this project was started, screening for exoglycosidases was primarily performed using chromogenic substrates such as p-nitrophenyl-glycosides. Though this technique is simple and rapid, requiring only a color change upon incubation of the substrate with the extract to detect exoglycosidase activity, the results give no information about the enzyme's linkage specificity or its ability to cleave an oligosaccharide substrate. To overcome these limitations and to detect exoglycosidases that do not cleave these synthetic substrates, a screening technique was developed that would use oligosaccharides as substrates for exoglycosidase digestion. This chapter describes the technique used at New England Biolabs for screening, purification, and characterization of novel exoglycosidases using fluorescently labeled oligosaccharides as the substrate for glycosidase digestion and thin-layer chromatography (TLC) for analysis of the cleavage products. The exoglycosidases isolated and characterized as a result of this screening project will be described as well as a demonstration of the uses of these enzymes to confirm the sequence of an oligosaccharide.

II. MATERIALS AND METHODS

A. Fluorescent Labeling of Oligosaccharides

Oligosaccharide substrates were purchased from Accurate Chemical and Scientific Corp. (Westbury, NY), Pfanstiehl Labs (Waukegan, IL), and

V-Labs Inc. (Covington, LA), or isolated from natural sources [1]. Oligosaccharides were labeled by reductive amination with the fluorophore 7-amino-4-methylcoumarin (AMC) as described by Prakash and Vijay [2]. Oligosaccharides (0.1–10 μmol) to be labeled were dissolved in 100 μL H_2O. The aqueous carbohydrate solution was added to a solution containing 300 μL methanol, 20 mg AMC (Eastman Kodak, Rochester, NY), 35 mg NaCNBH$_3$, and 41 μL glacial acetic acid. The mixture was sealed into a screw cap microfuge tube and heated in a dry block at 85°C for 45 min. The fluorescently labeled carbohydrate was separated from the unreacted AMC by loading the reaction onto a Sephadex G$_{25}$ column (2 × 50 cm) that had been equilibrated with H_2O. The fluorescent peak containing the labeled oligosaccharide was eluted with H_2O, and 1-mL fractions were collected. After assaying fractions for purity by TLC as will be described, appropriate fractions were pooled and vacuum concentrated to 0.1–1 μmol/mL before storage at −20°C.

B.   Analyzing AMC-Labeled Oligosaccharides Using TLC

Two to three microliters of the reaction containing the AMC-labeled oligosaccharide was spotted in a thin tight band on a silica gel 60 w/o F glass-backed TLC plate (EM Science, Gibbstown, NJ). The bands were dried completely with a hot-air gun using care that the temperature of the plate did not exceed 70°C. Once dried, the plate was developed in a tank containing isopropanol : ethanol : H_2O until the solvent front moved 10 cm from the bottom of the plate. The bands were visualized with a 314-nm UV lamp. This absorption chromatography separates the AMC-labeled oligosaccharides by the number of hydroxyl residues present on the carbohydrate, which generally corresponds to the size of the oligosaccharide. Smaller oligosaccharides migrate faster and further up the TLC plate than the larger oligosaccharides, which remain closer to the origin. Changing the ratio of isopropanol to water in the solvent system allows for the separation of different size ranges of oligosaccharides (Fig. 1). By decreasing the amount of isopropanol relative to the amount of water present in the solvent mixture, the separation of larger oligosaccharides up to 20 sugar residues in length can be achieved.

C.   Screening for Glycosidase Activity

Cell pellets from each strain to be tested were suspended in three volumes of buffer containing 20 mM Tris-HCl (pH 7.5), 50 mM NaCl, and 1 mM Na$_2$EDTA. After a brief sonication, the cell extracts were centrifuged at 14,000 rpm for 10 min at 4°C in an Eppendorf microcentrifuge model 5415c, (Brinkman Instruments Inc., Westbury, NY). Cell supernatants were

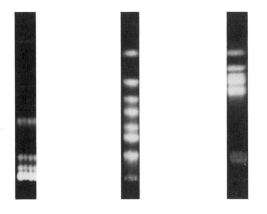

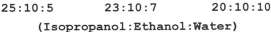

25:10:5   23:10:7   20:10:10

(Isopropanol:Ethanol:Water)

Figure 1 Comparison of the TLC migration of identical sets of AMC-labeled oligosaccharides (monomer to septamer and a decamer) under different solvent conditions. The oligosaccharides used were GlcNAc–AMC, GlcNAcβ(1-4)GlcNAc–AMC, Manβ(1-4)GlcNAcβ(1-4)GlcNAc–AMC, Manα(1-6)Manβ(1-4)GlcNAcβ(1-4)GlcNAc–AMC, Manα(1-3)(Manα(1-6))Manβ(1-4)GlcNAcβ(1-4)GlcNAc–AMC, Manα(1-3)(GlcNAcβ(1-4))(Manα(1-6))Manβ(1-4)GlcNAcβ(1-4)GlcNAc–AMC, GlcNAcβ(1-2)Manα(1-3)(GlcNAcβ(1-2)Manα(1-6))Manβ(1-4)GlcNAcβ(1-4)GlcNAc–AMC, and Manα(1-2)Manα(1-2)Manα(1-3)[Manα(1-2)Manα(1-3)(Manα(1-2)Manα(1-6))Manα(1-6)]Manβ(1-4)GlcNAc–AMC. The ratios of isopropanol : ethanol : water used as the solvent system for each TLC are indicated below each of the lanes.

tants were assayed for glycosidase activity by adding 5 μL of the supernatant to a 10-μL mixture containing 1 nmol of AMC-labeled oligosaccharide in 50 mM sodium citrate (pH 6.0), 5 mM CaCl$_2$. After incubation at 37°C for 15 min and 2 h, the reactions were analyzed using TLC as previously described.

D. Characterizing the β-Galactosidase Specificity

The specificity of the purified β-galactosidase isolated from *Bacteroides fragilis* was determined by incubating the enzyme preparation with the AMC-labeled oligosaccharides shown in Fig. 2. For each reaction, 4 units of the β-galactosidase was incubated with 1 nmol of each substrate in 50 mM sodium citrate pH 6.0 with one unit defined as the amount of enzyme required to cleave >95% of the β-D-galactose from 1 nmol of substrate E

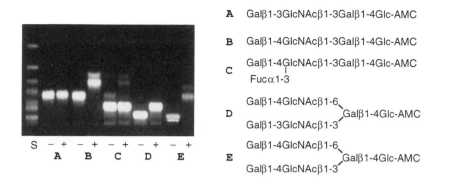

A   Galβ1-3GlcNAcβ1-3Galβ1-4Glc-AMC

B   Galβ1-4GlcNAcβ1-3Galβ1-4Glc-AMC

C   Galβ1-4GlcNAcβ1-3Galβ1-4Glc-AMC
     |
     Fucα1-3

D   Galβ1-4GlcNAcβ1-6
                        Galβ1-4Glc-AMC
     Galβ1-3GlcNAcβ1-3

E   Galβ1-4GlcNAcβ1-6
                        Galβ1-4Glc-AMC
     Galβ1-4GlcNAcβ1-3

Figure 2   Determination of the specificity of the β-galactosidase purified from *B. fragilis*. Lane S contains the oligosaccharide markers, monomer to septamer, described in Fig. 1. The other lanes are in pairs with the oligsoaccharides A through E (as indicated) incubated without (−) or with (+) the β-galactosidase for 4 h at 37°C and at pH 6.0. The reactions were spotted on a TLC plate and run in a solvent system of isopropanol : ethanol : water (23:10:7). The structures of oligosaccharides A through E are shown to the right of the TLC plate.

(Fig. 2) in a 10-µL reaction in 1 h at 37°C and pH 6.0. Following incubation at 37°C for 18 h, the cleavage products were analyzed using TLC.

III.   RESULTS AND DISCUSSION

The branched oligosaccharides D and E (see Fig. 2 for the structures) labeled with AMC were used to screen for a β-galactosidase, which was specific for 1-4 linkages and capable of cleaving branched oligosaccharides. Crude extracts from 165 bacterial strains from the New England Biolabs culture collection were incubated with 1 nmol of either oligosaccharide D or E in a buffer of pH 6.0 in 15 min at 37°C. An extract from one strain, *B. fragilis* (NEB 668), contained an enzyme that was capable of removing the terminal β(1-4)-galactose from both substrates. This β(1-4)-galactosidase was purified from *B. fragilis* cell extracts using oligosaccharide E as a substrate to monitor the enzyme activity through the purification. Once purified, different AMC-labeled substrates were used to characterize the specificity of the β(1-4)-galactosidase (Fig. 2). These experiments showed that this β-galactosidase purified from *B. fragilis* cleaved terminal β(1-4)-galactose from branched and linear oligosaccharides (Fig. 2, substrates B, D, and E) unless an α(1-3)-linked fucose was adjacent to the terminal galactose (Fig. 2, substrate C). The *B. fragilis* β-galactosidase did not cleave any terminal β(1-3)-galactose linkages tested (Fig. 2, substrates A and D).

The additional bands observed upon digestion of substrates B and E with the enzyme preparation were due to the presence of a contaminating β-hexosaminidase. Once fully purified, the *B. fragilis* β-galactosidase should be useful for identifying most of those oligosaccharides containing a terminal $\beta(1-4)$-linked galactose.

Using AMC-labeled oligosaccharides, it has been possible to screen, purify, and characterize a wide variety of exoglycosidases (Table 1, Ref. 3, and unpublished data). These exoglycosidases, when used in combination or alternately, can be used to modify as well as sequence oligosaccharides. Combinations of the enzymes listed in Table 1 can show the presence of GlcNAc versus GalNAc residues, or the identification of specific sialic acid, α-fucose, α-mannose, and β-galactose linkages. For example the $\beta(1-4)$-galactosidase previously described can be used in combination with a β-galactosidase from *Xanthomonas manihotis*, which has a preference for $\beta(1-3)$ linkages, to determine the galactose linkages present on an oligosaccharide as shown in the following illustration.

To demonstrate that these exoglycosidases could be used to identify specific linkages, combinations of exoglycosidases were incubated with *N*-linked oligosaccharides isolated from bovine fetuin. Several papers have been published on the carbohydrate structures of the *N*-linked glycans found on bovine fetuin. Early reports suggested that only galactose-linked

Table 1   Exoglycosidases Discovered and Available from New England Biolabs

Enzyme	Specificity[a]	Source	Cloned[b]	Reference
β-Galactosidase[c]	$\beta1$-4	*Bacteroides*	No	This work
β-Galactosidase	$\beta1$-3 > 6 > 4	*Xanthomonas*	Yes	[3]
β-*N*-Acetylglucosaminidase	$\beta1$-2,3,4,6	*Xanthomonas*	No	[3]
β-*N*-Acetylhexosaminidase	$\beta1$-2,3,4,6	*Streptomyces*	Yes	[4]
α-Neuraminidase	$\alpha2$-3 > 6,8	*Salmonella*	Yes	[5]
α-Neuraminidase	$\alpha2$-3,6,8	*Clostridium*	Yes	[6]
α-Fucosidase	$\alpha1$-2	*Xanthomonas*	No	[3]
α-Fucosidase	$\alpha1$-3,4	*Xanthomonas*	No	[3]
α-Mannosidase	$\alpha1$-2,3	*Xanthomonas*	Yes	[3]
α-Mannosidase	$\alpha1$-6	*Xanthomonas*	Yes	[3]
α-Galactosidase	$\alpha1$-3,6	*Xanthomonas*	No	[3]
β-Xylosidase[c]	$\beta1$-2,3,4,6	*Xanthomonas*	No	[7]
β-Mannosidase	$\beta1$-2,3,4,6	*Xanthomonas*	No	[7]

[a]Specificity of the enzyme: > indicates that the enzyme will cleave the first linkage with a least a 100-fold preference over successive linkage.
[b]Indicates whether the enzyme is sold from a recombinant source.
[c]Enzymes soon to be released for sale.

$\beta(1-4)$-was present on these oligosaccharides [8,9]. However, more recently, a study using nuclear magnetic resonance (NMR) [10] and another study using sequential glycosidase digestion [11] have shown that some of the oligosaccharides contain $\beta(1-3)$-linked galactose. For this experiment, the oligosaccharides were isolated from fetuin by hydrazinolysis [1], purified on a G_{25} Sephadex column, and then labeled with AMC (see Section II) prior to exoglycosidase treatment. The results of the exoglycosidase digestion are shown in Fig. 3. Removal of the sialic acids from the oligosaccharide, accomplished with a general neuraminidase from *Clostridium perfringens* [6], caused a shift down of the labeled oligosaccharide on the TLC plate, (Fig. 3, lane 2). It has been observed previously [12] that in a solvent system of isopropanol : ethanol : water the presence of $\alpha(2-3)$-linked sialic acid on an oligosaccharide caused the carbohydrate to run faster (i.e., appear smaller) by two sugar residues on a TLC plate than an oligosaccharide of similar size that contained no sialic acid residues (possibly due to the interaction of the sialic acid with neighboring sugar residues). Incubation of the oligosaccharides with the *B. fragilis* $\beta(1-4)$-galactosidase (previously described) and the neuraminidase gave an upward shift of the labeled oligosaccharides on the TLC plate to give one broad band (Fig. 3, lane 3). When the β-*N*-acetylglucosaminidase from *X. manihotis*, which cleaves all terminal β-GlcNAc residues from oligosaccharides [3], was added in addi-

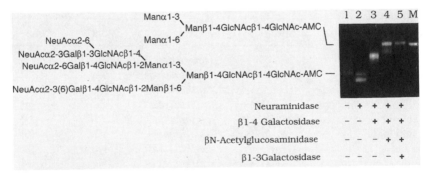

Figure 3 Digestion of the *N*-linked glycans purified from bovine fetuin and labeled with AMC. The enzyme(s) incubated with the AMC-labeled fetuin oligosaccharides are indicated below each lane. The lower oligosaccharide structure shows the sequence containing the $\beta(1-3)$-galactose linkage of interest. The top oligosaccharide structure shows the sequence of the trimannosyl core that should be produced from digestion of the *N*-linked fetuin oligosaccharides with the four exoglycosidases listed. Lines from each oligosaccharide structure indicate where these oligosaccharides migrate on the TLC plate. Lane M is an internal standard containing only the purified trimannosyl core oligosaccharide labeled with AMC.

tion to the neuraminidase and the β(1–4)-galactosidase, two bands were observed by TLC (Fig. 3, lane 4). One of the bands co-migrated with the AMC-labeled trimannosyl core marker (Fig. 3, lane M). The other band ran slower than the trimannosyl core by approximately two sugar residues compared with the size standards (standards not shown). Adding the *X. manihotis* β-galactosidase, which cleaves β(1–3)-galactose linkages preferentially [3], in addition to the aforementioned enzymes resulted in the slower band seen in lane 4 moving up to form a single band (Fig. 3, lane 5) that aligns to the trimannosyl core marker. Because one single band the size of the trimannosyl core was observed only after the addition of the β(1–3)-galactosidase, this indicated that a certain percentage of the *N*-linked oligosaccharides that were present on bovine fetuin contained a β(1–3)-galactose linkage and that this linkage could be detected using specific exoglycosidases in combination with AMC labeling and TLC analysis.

IV. CONCLUSIONS

The use of AMC-labeled oligosaccharides greatly assisted in the screening, purification, and characterization of novel exoglycosidases. At present, 13 exoglcosidases are available to modify or sequence oligosaccharides. Work is presently under way to show that these exoglycosidases transglycosylate sugar components of glycoproteins and glycolipids. Discovery of additional exoglycosidases with other novel specificities should aid workers in glycobiology to further address the roles that sugars play in various aspects of biology.

ABBREVIATIONS

AMC   7-amino-4-methylcoumarin
NMR   nuclear magnetic resonance
TLC   thin-layer chromatography

REFERENCES

1. Montrevil, J., Bouquelet, S., Debray, H., Bournet, B., Spik, G., and Strecker, G. In Chaplin, M. F., and Kennedy, J. F., eds (1986). *Carbohydrate Analysis: A Practical Approach*. IRL Press, Washington, D.C., pp. 143–204.
2. Prakash, C., and Vijay, I. K. (1983). *Anal. Biochem.* 128, 41–46.
3. Wong-Madden, S. T., and Landry, D. (1995). *Glycobiology* 5, 19–28.
4. Robbins, P., Overbye, K., Albright, C., Benfield, B., and Pero, J. (1992). *Gene* 111, 69–76.

5. Hoyer, L. L., Roggentin, P., Schauer, R., and Vimr, E. R. (1991). *J. Biochem*. 110, 462–467.
6. Roggentin, P., Rothe, B., Lottspeich, F., and Schauer, R. (1988). *FEBS Lett*. 238, 31–34.
7. Wong-Madden, S. T., and Landry, D., unpublished results.
8. Baenzinger, J. U., and Fiete, D. (1979). *J. Biol Chem*. 254, 789–795.
9. Nilsson, B., Norden, N. E., and Svensson, S. (1979). *J. Biol. Chem*. 254, 4545–4553.
10. Townsend, R. R., Hardy, M. R., Wong, T. C., and Lee, Y. C. (1986). *Biochemistry* 25, 5716–5725.
11. Takasaki, S., and Kobata, A. (1986). *Biochemistry* 25, 5709–5715.
12. Landry, D., unpublished results.

26

Resolution of Oligosaccharides in Capillary Electrophoresis

Morgan Stefansson
Uppsala University, Uppsala, Sweden

Milos Novotny
Indiana University, Bloomington, Indiana

I. INTRODUCTION

In the rapidly developing field of glycoanalysis, capillary electrophoresis (CE) has recently assumed a significant position alongside techniques such as high-field nuclear magnetic resonance (NMR) spectroscopy, mass spectrometry, and high performance liquid chromatography with pulsed amperometric detection. This is primarily due to its high resolving power, which is needed to separate extremely complex glycoconjugate mixtures. Second, in combination with fluorescent derivatization, capillary electrophoresis with laser-induced fluorescence detection gives considerable promise that ultra-sensitive determinations in glycobiology will become readily available. Third, with the new developments in combining capillary electrophoresis with nanoelectrospray and laser desorption ionization mass spectrometry, structural elucidation capabilities should be enhanced in the near future. The advances in detection methodologies, in turn, necessitate the development of better CE separations. For glycoconjugates the resolution and selectivity can be maximized through a better understanding of electromigration phenomena.

In CE, oligosaccharide separations are affected by a variety of parameters, originating both from the sample characteristics and the experimental design. A comprehensive and universal separation methodology, which solves all the problems associated with the analysis of complex glycoconjugate mixtures, has not been developed. However, remarkable progress has been made [1–3]. Some knowledge of the physicochemical properties and solution behavior of the sample components is beneficial in the optimization procedures. Alternatively, the best separation may be established empirically by a systematic variation of the separation parameters such as pH, buffer composition, and additives. Due to the complexity of oligosaccharide structures and the presence of closely related structures in various proportions (often in trace levels), the separation power is a major consideration for an analytical technique to be useful in resolving all species for their structural determination. The latter is particularly important due to a lack of available standards.

In this chapter, the processes affecting the separation of oligosaccharide and polysaccharide mixtures by capillary electrophoresis will be discussed. Different separation conditions will be examined and some guidelines provided, in the context of current progress in the field.

II. RESULTS AND DISCUSSION

A. Basic Concepts

CE separations are typically performed in fused silica capillaries of internal diameters of 10–100 mm and effective lengths of 10–100 cm, utilizing electric field strengths up to 1000 V/cm. The presence of acidic silanol groups on the capillary inner surface results in the formation of an electric double layer generating an electroosmotic flow, with the mobility μ_{eo}, in the direction of the cathode when an external voltage is applied. Through a covalent or dynamic chemical treatment of the surface, this bulk flow can be totally suppressed [1]. Furthermore, an ionic species (sample) present in the electrolyte will exhibit an electrophoretic mobility, μ_{ep}, dependent on the charge-to-friction ($\approx$ mass) ratio, which is negative for anions.

Assuming that the only contribution to band broadening is longitudinal diffusion, the efficiency of a separation is determined by the number of theoretical plates, N, which is expressed as

$$N = \frac{(\mu_{eo} + \mu_{ep})\,V}{2D} \tag{1}$$

where V and D are the applied voltage and diffusion coefficient of the solute, respectively. Equation (1) predicts an increase in peak performance with increasing voltage and overall mobility. Hence, to utilize the applied voltage to improve separation, the conductivity of a buffer electrolyte must be as low as possible (to avoid temperature effects from Joule heating). Procedures to increase the electophoretic mobility will be discussed later.

Compared with a cation, an anionic component (i.e., an acidic carbohydrate constituent) needs a relatively higher electrophoretic mobility to display similar peak efficiencies as neutral species. However, cations frequently interact with the negatively charged surface, thereby severely impairing peak shape and performance [4]. In addition, other components present in a sample, namely, proteins or peptides, might adsorb to the capillary surface and give rise to nonreproducible migration times. In the case of an ideally coated capillary (when μ_{eo} is diminished), all these effects are avoided. Surface coating has proven essential for high separation efficiencies with plate numbers over one million per meter [5]. Nevertheless, most reported carbohydrate separations are performed in uncoated capillaries.

The resolution, R_S, of two adjacent peaks is defined as

$$R_s = \frac{\Delta\mu_{ep} N^{1/2}}{4(\mu_{eo} + \mu_{ep})} \qquad (2)$$

where $\Delta\mu_{ep}$ is the difference in electrophoretic mobilities. For uncoated columns exhibiting an electroosmotic flow, maximum resolution is obtained where $\mu_{eo} = -\mu_{ep}$. Unfortunately, the analysis times approach infinity under such conditions. Increasing the capillary length has little influence ($N^{1/2}$) on resolution and is moreover counteracted by increased longitudinal diffusion. For coated capillaries, μ_{eo} becomes vanishingly small.

B. Assessment of Variables

1. Pre-Column Treatment and Derivatization

Oligosaccharide structures, covalently linked to proteins or lipids, have a variety of biological functions. Isolation is usually a difficult and cumbersome task, especially when oligosaccharides are present in minute quantities. Differences in chain length and branching, isomeric forms, and modifications (e.g., acetylation, methylation, phosphorylation, and esterifications) are often encountered. Low contamination levels might be detrimental when universal detection methods are used. Relatively little work has been published regarding this problem, and thus far there is no report on the direct coupling of on-line sample treatment procedures to capillary electrophoresis of oligosaccharides, an area which might prove fruitful in the future. Ultimately, selective enzymatic hydrolysis and sequencing schemes could be developed, analogous to protein sequencing.

The introduction of fluorescent pre-column derivatization [1,6] has revolutioned glycoanalysis in both its sensitivity and selectivity. It has become an important part of the fundamental strategy in carbohydrate analysis. The reactions involve a Schiff-base formation between a primary aromatic amine and the aldehyde group of the open-chain form of reducing sugar, followed by reduction with cyanoborohydride. Alternatively, the carbohydrate can be derivatized through the amino group with an amine-reactive probe, as exemplified by 3-(4-carboxybenzoyl)-2-quinoline carbox-aldehyde (CBQCA) [7], with detection limits at the attomole level. Derivatization of the analytes also enhances the electrophoretic properties of carbohydrates. In a recent study [5], the influence of the number of negative charges (zero, one, and three, respectively) on the separation selectivity and analysis time was investigated. The electropherograms are displayed in Fig. 1. Clearly, multiple charges are beneficial; several papers have now been published using 8-aminonaphthalene-1,3,6-trisulfonate (ANTS) or 8-aminopyrene-1,3,6-trisulfonate (APTS) as labels [6]. A fluorogenic reagent that is specific for sialic acids, 4,5-dinitrocatechol-O-O-diacetic acid, has also been used [8]. Introducing additional functional groups to enable efficient sample cleanup and to promote separation selectivity is an important area of development.

The migration order of negatively charged κ-carrageenan oligomers has been shown to be governed by the label [9]. The charge-to-friction ratio of the oligosaccharides was increased by the use of negatively charged ANTS moiety, resulting in a migration order from smaller to larger oligomers. The selectivity was further improved by the addition of linear polyacrylamide as a sieving medium in the electrolyte buffer. The migration order could be entirely reversed by using the positively charged 6-aminoquinoline derivative and, thus, altering the friction coefficient (Fig. 2). The electrophoretic mobilities experimentally obtained in free solution were found to be in excellent agreement with a recently reported theoretical model [10]. For samples with constant charge-to-friction ratio, as observed with many complex polysaccharide and other biopolymer mixtures, the number of resolved components is presumed to be greatly improved by an appropriate design of the end label. The highest resolution between adjacent oligomers will be obtained using a label of opposite charge to the analytes. Another major advantage of this approach is that the unreacted label itself migrates in the opposite direction and does not interfere in the electropherogram, facilitating sample cleanup.

2. Columns

As was discussed, the choice of column surface is an important factor for CE separations. For simple mixtures (containing small and neutral sugars), adequate selectivity is often obtained in uncoated capillaries using borate buffers. However, as the degree of polymerization (dp) and complexitiy increases, the charge-to-friction ratio decreases and the electroosmotic flow finally dominates the migration. Hence, separation can be severely compromised [11]. This is exemplified in Fig. 3. A reversed migration order is obtained with suppressed electroosmotic flow, making the separation of

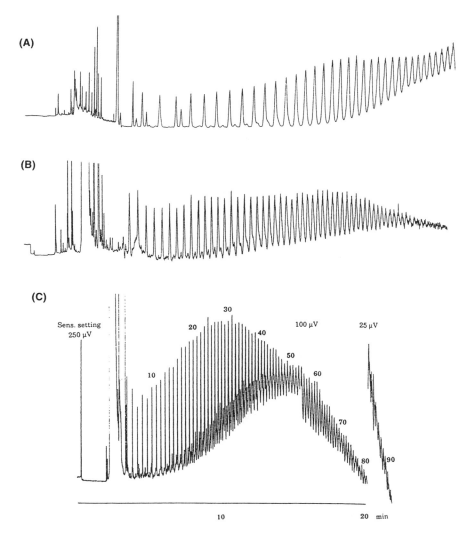

Figure 1 Influence of the fluorescent tag on the separation of a dextran standard with an average molecular weight of 18,300. The reagents used were (A) 2-aminopyridine, (B) 5-aminoaphthalene-2-sulfonate, and (C) 8-aminonaphthalene-1,3,6-trisulfonate. Conditions: -500 V/cm (10 μA) using 0.1 M Tris-borate, pH 8.65, as the electrolyte. The effective length of the capillary was 35 cm (50 μM i.d.). (From Ref. 5.)

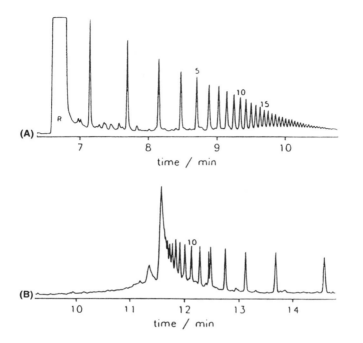

Figure 2 Separation of the (A) ANTS- and (B) 6-aminoquinoline-derivatized oligosaccharides derived from partially hydrolyzed κ-carrageenan. Numbers 5 and 10
are the elution positions of the deca- and eicosasaccharide, respectively. Capillary:
60 cm effective length (70 cm total); 50 μm i.d. (363 μm o.d.); coated with a linear
polyacrylamide. Buffer: 25 mM sodium citrate, pH = 3.0; 1% linear polyacrylamide (M_W = 5–6 × 10^6); electrokinetic injection, 10 sec at 100 V/cm. Applied
voltage: 350 V/cm. (From Ref. 9.)

larger oligomers feasible. Although a similar migration order can be obtained with uncoated capillaries at low pH [12], we have found that surface
coatings are indispensable in the resolution of complex oligo- and polysaccharide mixtures [5]. Furthermore, the use of borate buffers at alkaline pH
can be employed in the absence of a high electroosmotic flow, resulting in a
larger "separation window," where subtle differences among larger oligomers can be distinguished. These conditions are also characterized by
shorter analysis times. This is demonstrated with the CE of corn amylose
oligomers with resolution occurring up to degrees of polymerization of ~70
[13] and taking in less than 15 min (Fig. 4). Moreover, satellite oligomer
peaks were observed in the electropherograms. These were later shown,
through selective enzymatic treatment with isoamylase, to be branched species withα(1 → 6) glycosidic bonds.

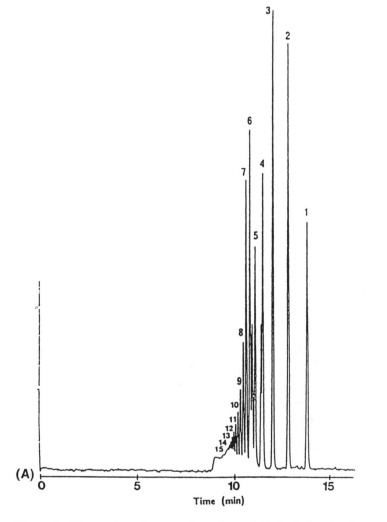

Figure 3 Electrophoretic separation of a partially hydrolyzed polysaccharide (Dextrin 15) in the (A) open-tubular format (58 cm effective length × 50 μm i.d.; buffer, 10 mM Na_2HPO_4–10 mM $Na_2B_4O_7 \cdot 10H_2O$ (pH 9.4); applied field, 227 V/cm) and (B) in a polyacrylamide gel-filled column (gel concentration, 10% T, 3% C, 19 cm effective length × 50 μm i.d.; buffer, 0.1 M Tris–0.25 M borate–7 M urea (pH 8.33); applied field 269 V/cm). (From Ref. 11.)

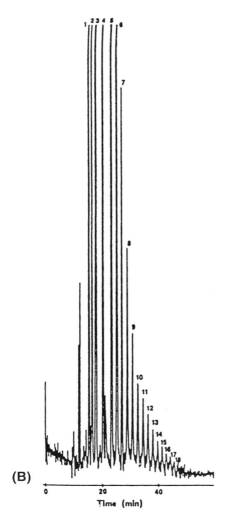

(B)

Figure 3 Continued

Separation by degree of polymerization of highly charged polymers, consisting of monomer units with constant and unfavorable charge-to-friction ratio, is complicated and analogous to the situation found with the uniformly charged nucleotides. Through the introduction of an additional separation mechanism, sieving according to the hydrodynmic radius ($\propto$ molecular weight) in highly concentrated polyacrylamide gel-filled capillaries, hydrolyzed dextrins, chrondroitin sulfate, and hyaluronate oligosaccharides were successfully analyzed [11]. An example of a hydrolyzed polygala-

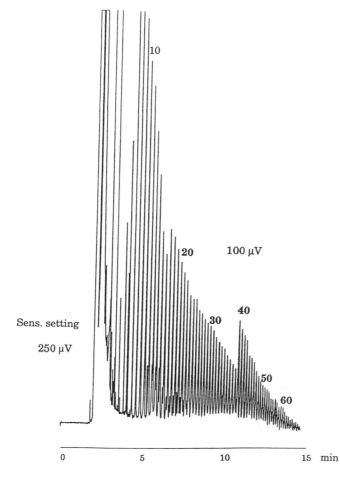

Figure 4 Separation of corn amylose oligomers. Conditions: 0.2 M Tris-borate, pH 8.65; −500 V/cm (18 μA) and 35 cm effective capillary length (50 μm i.d.). The numbers refer to the degree of polymerization. (From Ref. 13.)

cturonate sample is shown in Fig. 5 [14]. However, preparation of the gel-filled columns was not straightforward, and a unique polymerization technique had to be developed [15].

C.   Additives and "Entangled" Polymer Solutions

1.   Oligosaccharides
Electrophoretic separations are based on the differences in analyte mobilities, as discussed, and one way to enhance resolution is the use of a suitable

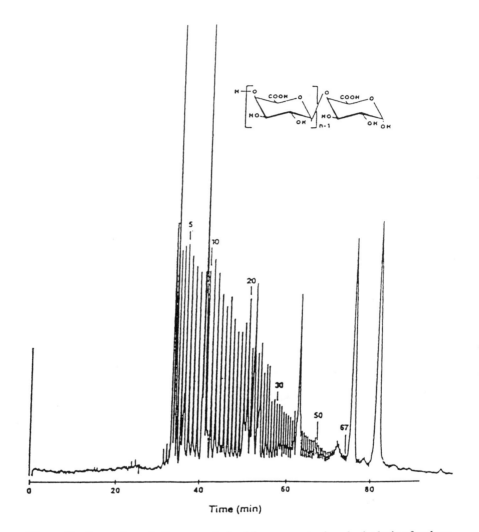

Figure 5 Separation of oligomers derived from an autoclave hydrolysis of polyga-lacturonic acid. The numbers indicate the estimated degree of polymerization. Capillary: 50 μm i.d. × 32 cm (23 cm effective length). Polyacrylamide gel concentration: 18% T, 3% C. Buffer: 0.1 M Tris–0.25 M borate–2 mM EDTA (pH = 8.48). Electromigration injection: 5 kV, 25 sec. Applied electric field: 234 V/cm (6.5 μA). (From Ref. 14.)

derivative. Introduction of charges through derivatization is in some cases the predominate strategy to generate electrophoretic mobility with a sample molecule. There is, nevertheless, an upper limit to the oligomer size where the charge-to-mass ratio becomes too small, regardless of the number of charges on the label, and the solute will not migrate through the capillary on a reasonable time scale. An alternative strategy for neutral as well as highly charged samples is the use of secondary equilibria that are induced by certain additives to the electrolyte solution [16]. The complex formation between the additive and sample compounds may include interactions such as ion pairing or ion exchange [16], inclusion phenomena, hydrophobic adsorption to detergents, and hydrogen bonding. The borate complexation with polyols is probably the most common example of the additive approach. For example, various monosaccharides have been separated as their 2-aminopyridyl derivatives [17] in 200 mM borate buffer at pH 10.5.

The main feature of this approach is the possibility of selectively altering the mobilities for separation optimization. The key parameters are the type and the concentration of the additive, which may result in either an increase or a decrease in mobilities. Furthermore, several agents can be used simultaneously, sometimes resulting in a separation of the analytes with minor structural differences. An illustrative example is the separation of a complex, derivatized monosaccharide enantiomer mixture shown in Fig. 6 [18]. For compounds containing one or several chiral centers, the enantiomeric forms are the nonsuperimposable mirror images having exactly the same physicochemical properties in an achiral environment. However, when one enantiomeric form of a chiral additive was included in the buffer (such as β-cyclodextrin), diasteromeric complexes were formed between the respective solute enantiomer and the additive. These complexes were shown by NMR spectroscopy [18] to be held together by borate as a complexing and bridging ion, with chiral recognition through the interaction of the C-3 hydroxyl group of the β-cyclodextrin molecule and the C-2 hydroxyl group of the sugar derivative. The naphthyl moiety of the reagent was inserted into the cyclodextrin cavity. Due to a difference in the stability constants for the ternary complexes of the D and L forms, a baseline separation could easily be obtained for all of the 14 pairs tested (see Table III in Ref. 18). In addition to the concentrations of borate and chiral reagent, the type of a fluorescent tag and a chiral selector were additional means available to optimize such separations. Incorporation of the small amounts of tetrahydrofuran as organic solvent was used in the optimization of the eight-pair enantiomeric separation (Fig. 6). Moreover, linear maltooligomers displayed chiral selectivity when used as the chiral additives, due to their ability to form helices around certain organic molecules [18]. A striking correlation between the chain length of the selector molecules and chiral

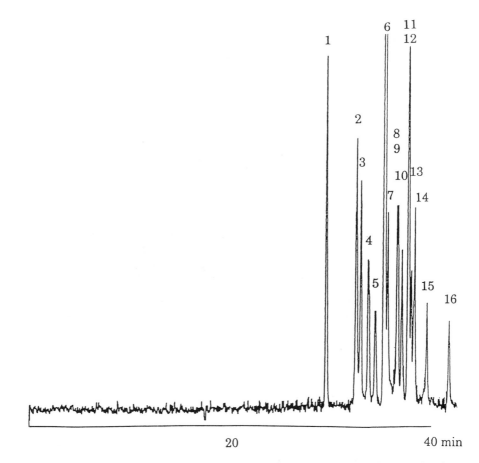

Figure 6   High efficiency separation of a complex enantiomeric mixture. The electrolyte was 12.5 mM β-CD (β-cyclodextin), 2% tetrahydrofuran, and 0.5 M borate (pH 8.2). The separation was performed in a 85/100 cm fused silica capillary, and at 25 kV and 38 μA. Peak assignments: 1. D-ribose; 2. D-xylose; 3. L-arabinose; 4. D-fucose; 5. D-glucose; 6. reagent 5-aminonaphthalene-2-sulfonate (ANA); 7. L-ribose; 8. D-galactose; 9. L-mannose; 10. D-lyxose; 11. L-xylose; 12. D-mannose; 13. L-glucose; 14. D-arabinose; 15. L-fucose; and 16. L-galactose. (From Ref. 18.)

discrimination was observed. It is believed that hydrogen bonding in combination with steric factors are responsible for the high selectivity. Borate complexation was found crucial for both the carbohydrate–carbohydrate recognition mechanisms, and we propose that future developments in borate chemistry, such as synthesis of tailor-made reagents, could result in separation advances for glycoconjugates.

As shown by Mechref and El Rassi [19], carboxylated carbohydrates can be converted into UV-absorbing of fluorescent glycoconjugates. The derivatization scheme involved a condensation reaction between the carboxyl group of the sugar and the amino group of the reagent in the presence of a water-soluble carbodiimide. This procedure was successfully utilized in the derivatization and then the separation of ganglioside isomers [20]. Gangliosides are sialic acid–containing glycosphingolipids present in the plasma membranes of all vertebrates. They act as receptor binding sites and interact with toxins, viruses, antibodies, and growth factors. Some of the gangliosides are closely related, consisting of a hydrophilic sialooligosaccharide chain connected to a hydrophobic ceramide tail (Fig. 7). Due to the amphiphilic nature of these compounds, aggregations occur within the low

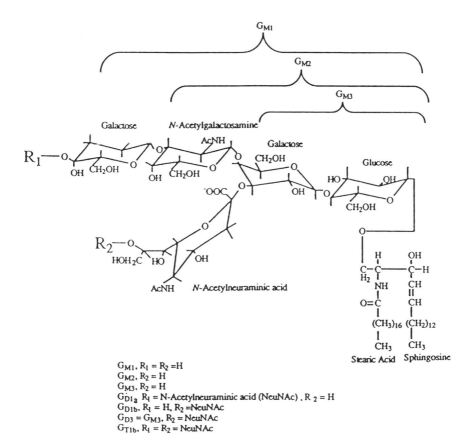

G_{M1}, $R_1 = R_2 = H$
G_{M2}, $R_2 = H$
G_{M3}, $R_2 = H$
G_{D1a} $R_1 = $ N-Acetylneuraminic acid (NeuNAc) , $R_2 = H$
G_{D1b}, $R_1 = H$, $R_2 = $NeuNAc
$G_{D3} = G_{M3}$, $R_2 = $ NeuNAc
G_{T1b}, $R_1 = R_2 = $ NeuNAc

Figure 7 Structures of the gangliosides. (From Ref. 19.)

nanomolar range, resulting in the formation of mixed micelles. This severely hinders efficient separations of the respective monomeric species. The additives used to dissociate such micellar aggregates include organic solvents and complexing agents such as α-cyclodextrins. The separation of GD1a, GD1b, and GD3 (peak number 1, 2, and 3, respectively) is shown in Fig. 8. An in situ micellar system, based on complexation of the alkylglycoside surfactant, decanoyl-N-methylglucamide, (MEGA) and borate, was used to control the micellar charge density, enabling the separation of GD1a and GD1b (Fig. 8). Nashabeh and El Rassi [21] have also reported the use of tetrabutylammonium bromide as an additive in the separation of linear and branched 2-aminopyridyl derivatives of the homologous series of N-acetylchitooligosaccharides, xyloglucan oligosaccharides from cotton cell walls, and high-mannose glycans of ribonuclease B, using a fused silica tube with polyether-interlocked, coated inner walls.

The divalent cation diaminopentane has been utilized in the resolution of $\alpha(2,3)$- and $\alpha(2,6)$-linked isomers of sialooligosaccharides [22]. The analog diaminobutane was also employed in the separation of the glycoforms of recombinant human erythropoetin [23], as well as the oligosaccharides released by hydrazinolysis from α_1-acid glycoprotein [24]. An improved resolution of the sialylated N-linked oligosaccharides, derived from recombinant tissue plasminogen activator, was accomplished by increasing the "separation window" with sodium dodecyl sulfate (SDS) micelles in the presence of either calcium or magnesium [25]. Camilleri et al. [26] describe the separation of 2-aminoacridone derivatives of branched oligosaccharides released from various glycoproteins using taurodeoxycholate micelles as an electrolyte additive.

2. Polysaccharides and Enzymatic Modification

Celluloses are perhaps the most abundant biopolymers in nature (wood, plants, seaweed, etc.). They are highly polydisperse materials, characterized by different degrees of polymerization and low water solubility. After chemical modification, a variety of chemical and physical properties can be imposed on the cellulose molecules: water solubility, gel strength, viscosity, and stability toward enzymatic degradation. Various modified celluloses are important materials for the food and pharmaceutical industries. Due to the lack of hydroxyl groups, chemically modified celluloses do not complex readily with borate, with consequent minimal migration in electric fields. Through the incorporation of charged detergents (S), such as SDS and dodecylamine, into the electrolyte buffer system, electrophoretic mobility can be induced [16] via association of the detergent molecules with the cellulose molecules (C). This can be described by the equation, C + S = CS, and the apparent stability constant:

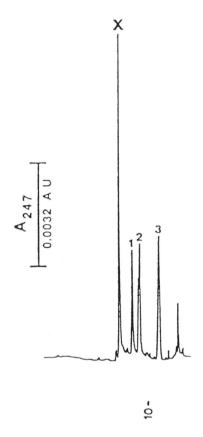

Figure 8 Electropherogram of 7-aminonaphthalene-1,3-disulfonate-disialogangliosides. Running electrolyte: 50 mM borate (pH6.0) containing 5.0 mM MEGA surfactant and 15 mM α-CD (α-cyclodextin). Peaks: 1 = GD1a; 2 = GD1b; 3 = GD3; and x = by-product. Capillary: 80 cm (50 cm effective length) × 50 μm i.d.; running voltage, 18 kV. (From Ref. 19.)

$$K_{app} = \frac{[CS]}{[C]\,[S]} \tag{3}$$

The electrophoretic migration velocity, μ_{ep}, is expressed in a linear equation:

$$\frac{1}{\mu_{ep}} = \frac{1}{\mu_{max}} + \frac{1}{\mu_{max}K_{app}[S]} \tag{4}$$

where μmax, the electrophoretic mobility when saturated with detergent, and Kapp can be obtained from the intercept and the slope, respectively.

If more than one molecule of S is complexed/adsorbed to the same site (i.e., a multilayer formation is achieved), the expression for μ_{ep} becomes

$$\log \frac{\mu_{ep}}{\mu_{max} - \mu_{ep}} = \log K_n + n \log [S] \tag{5}$$

The influence of SDS in increasing the electrophoretic mobility of chemically modified celluloses mimics Langmuirian adsoption phenomena. Through various migration models, several types of adsorption sites could be identified, and by modification of the buffer medium an optimized separation of a hydroxypropyl–methylcellulose sample was obtained [16], revealing an expected complexity (Fig. 9). A similar approach was applied to the highly charged heparins, utilizing multivalent cations for a charge reduction through complex formation.

This concept has since been extended using aminated dextrans for the separation of anionically charged polysaccharides such as hyaluronate and polygalacturonate and their enzymatic digests [27]. A strong and cooperative complexation, while being an equilibrium reaction, has to be carefully adjusted for a given sample type. Linear and short-chain polyacrylamide polymers have often been used as sieving media. The primary advantage of these systems is the fast analysis times compared with those for separations in concentrated and covalently cross-linked gels, as is demonstrated in Fig. 10 for an enzymatic digest of hyaluronate from the rooster comb. In addition to the expected distribution of oligomers, differing by one disaccharide unit, satellite peaks were apparent (Fig. 10). The origin of these peaks has not been determined, but it is suspected that they might be chondroitin sulfate impurities present in the original sample.

Structural characterization of polysaccharides is essential for correlating their composition with their physicochemical properties and biological activity. The enzymes that cleave at specific sites are presently more useful tools than methods employing chemical degradation. The oligosaccharide maps from laminarin [5], amyloses, amylopectins, pullulan, lichenan [13], chondroitin sulfate A, and hyaluronate [11] demonstrate the importance of highly selective and sensitive separations for the assessment of glycosidase specificities for residues, anomericity, and other structural features. In comparative studies on the compositional analysis of heparin, heparan sulfate, and chodroitin/dermatan sulfate, a number of papers employing enzymatic degradation have been reported, as reviewed by Linhardt and Pervin [28].

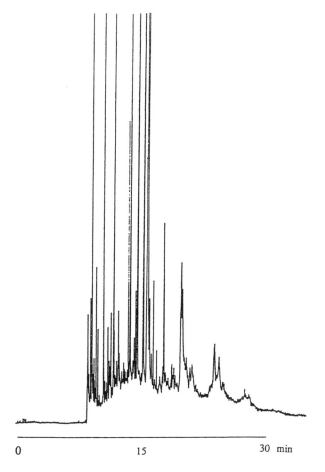

0   15   30 min

Figure 9 Optimized separation of a hydroxypropyl–methylcellulose sample. 25 mM morpholino-ethanesulfonate, 12.5 mM Tris, 20 mM SDS, 1.0 mM spermine, and 5% tetrahydrofuran (pH 6.0). The effective capillary (coated with linear poly-acrylamide) length was 65 cm (50 μm i.d.) using −400 V/cm (14 μA). (From Ref. 26.)

3. Capillary Affinity Electrophoresis

Affinity electrophoresis [29] has developed into a group of techniques for measuring the interactions between biological molecules. An important feature of the capillary format is that minute quantities of sample are used (femto- to picomoles) for the determination of stoichiometry and affinity constants. This methodology involves varying the concentrations of one

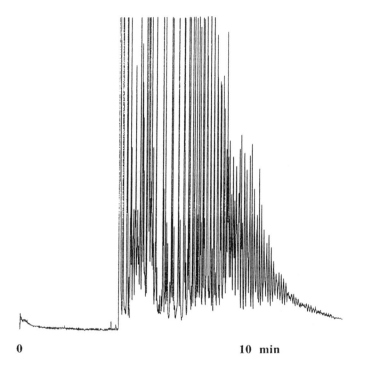

0   10 min

Figure 10   Partial enzymatic digest of hyaluronic acid (rooster comb) using hyaluronidase. Capillary: 60 cm (50 cm effective length; 50 μm i.d.); running buffer, 25 mM citrate, 25 mM Tris, 4 M urea, aminodextran 0.6 mg/mL, and 2% linear polyacrylamide (M_W = 0.7–1 $\times$ 10^6); electrokinetic injection, 2 sec at 500 V/cm. Applied voltage: 500 V/cm (12 kV). (From Ref. 27.)

reactant in the running buffer and measuring the corresponding changes in electrophoretic mobilities of the analyte. There is now pharmaceutical interest in studying the binding of glycoconjugates and oligosaccharides to carbohydrate-binding proteins, for example, in the discovery of drugs to inhibit cell adhesion. In contrast to most techniques, large association constants can be measured with high precision. Affinity differences as small as 5% in dissociation constants can be differentiated by two-dimensional affinity electrophoresis. However, as pointed out by Winzor [30], binding to equivalent and independent sites should be described by an intrinsic binding constant, as presented in the theoretical model.

D.   Pulsed Field Separations

The separation of polymer chains in gel eletrophoresis is conceptualized as a wormlike motion through gel pores created by a theoretical tube. Along this curvilinear contour the mobility is restricted, producing separations according to molecular size. However, for higher molecular weights or electric fields, the dependence of mobility on chain length diminishes, due to molecular stretching of the solutes, and a constant friction per length segment of the analytes (biased reptation with stretching) is obtained. This results in co-migration regardless of differences in molecular mass. The reptation concept was originally introduced by de Gennes [31] and Doi and Edwards [32] for polymer melts.

Pulsed field capillary gel electrophoresis is an emerging technique for the efficient separation of very large biopolymers, extending to molecular weights of several hundred million. Furthermore, entangled polymer matrices as sieving media provide the means for a higher degree of reproducibility. Theoretical studies of the electrophoretic transport of solutes through polymer matrices have been published by Grossman and Soane [33] and Viovy and Duke [34]. Both theories find pore size and its fluctuations crucial for the sieving effect as a chain is restricted along a curvilinear path of its backbone by constraints of the entangled polymer matrix. Except for Brownian motion, which creates fluctuations of end sections by a stochastic process (tube theory), the central chain sometimes develops "bulges" and "hernias."

Small molecules maintain their Gaussian conformation, and hence, mobility is inversely proportional to molecular weight. Larger molecules stretch in the field and migrate independently of their molecular weight, that is, through reptation-by-stretching. Besides the Gaussian and extended conformation, U- or V-shaped conformations with an overall zero migration distance are possible. Ultimately, Brownian motion pushes one end ahead. The self-trapping phenomenon can give rise to "band inversion" effects where smaller molecules migrate slower in a gel than the large ones. A recently developed technique [35–37] involves the use of biased sinusoidal field gel electrophoresis (BSFGE) where band inversion is avoided. The obstruction experienced by a separation medium is the main parameter [38], except for the electric field and temperature.

Thus far, the report by Sudor and Novotny [39] on the pulsed field separation of polysaccharides is the only one in the literature, whereas DNA molecules are studied more frequently. Using dextran standards with the molecuar weights from 8800 to 2×10^6 Da, separations according to molecular mass were achieved in an entangled linear polyacrylamide matrix (4% T; 0% C). The pulse regime of 3 Hz with a forward-to-reverse ratio of 2 at 345 V/cm resulted in an efficient separation, as shown in Fig. 11. The reptation behavior was manipulated by the pulsed fields and was found to occur at vastly different molecular weight ranges compared with those of DNA. This was attributed to the generally more rigid structures of polysaccharides, possibly amplified by the borate complexation. Pulsed field sepa-

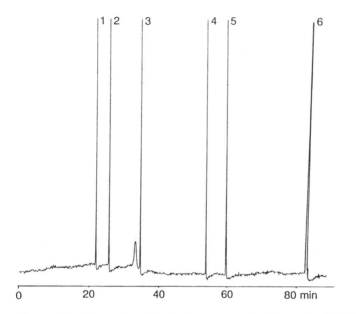

Figure 11 Separation of polydextran standards by pulsed field capillary electrophoresis. Peak assignments are 1. 8800 Da; 2. 39,100 Da; 3. 70,000 Da; 4. 503,000 Da; 5. 667,800 Da; 6. 2,000,000 Da; linear polyacrylamide (4% T; 0% C). Electrophoretic conditions: capillary 50 μm i.d., 29 cm total length (20 cm effective length), coated with linear polyacrylamide, and filled with linear polyacrylamide (5% T; 0% C); buffer: 50 mM sodium borate–50 mM boric acid–100 mM Tris, pH 8.81; electromigration injection, 5 sec (1 kV); 2:1, applied voltage, ±10kV; and frequency, 3 Hz, ratio of forward-to-reverse times, 2:1. (From Ref. 39.)

ration methodologies can become useful in fine-tuning difficult separations of polysaccharide mixtures, with a prerequisite being the molecular flexibility of polymer chains during the relaxation process when alternating the electric field.

III. FUTURE PROSPECTS

Compared with the overwhelming number of applications in the DNA and protein areas, capillary electrophoresis remains a relative newcomer to the field of glycobiology. However, with increasing knowledge of sample derivatization, modern electromigration techniques for complex carbohydrate samples can be applied. Understanding the electromigration phenomena is clearly central to improving the separations of glycoconjugates. While there are examples indicating high resolution of glycoconjugates is feasible, additional aspects should be explored: (1) the role of fluorescent tags in migration mechanisms (including design of superior reagents); (2) optimization of separation media, including the exploration of novel entangled matrices; and (3) development of integrated schemes that combine effective sample treatment, including wider utilization of enzymes, fractionation of extremely complex mixtures, and separations by capillary electrophoresis. The extant separations of complex oligosaccharide mixtures provide considerable incentive for the future of ultrasensitive analysis of glycoproteins as well as the challenging investigations into the chemical nature of the "molecular giants" of glycobiology—proteoglycans and large polysaccharide molecules.

ACKNOWLEDGMENTS

This study was financially supported by the Swedish Natural Science Research Council (to M.S.) and Grant No. GM24349 from the National Science Foundation (to M. N.). The grant-in-aid from Astra Hässle AB is gratefully acknowledged.

ABBREVIATIONS

ANTS	8-aminonaphthalene-1,3,6-trisulfonate
APTS	8-aminopyrene-1,3,6-trisulfonate
BSFGE	biased sinusoidal field gel electrophoresis
CBQCA	3-(4-carboxybenzoyl)-2-quinolinecarboxaldehyde
CE	capillary electrophoresis
dp	degree of polymerization
MEGA	decanoyl-N-methylglucamide
NMR	nuclear magnetic resonance
SDS	sodium dodecyl sulfate

REFERENCES

1. Novotny, M., and Sudor, J. (1993). *Electrophoresis* 14, 373–389.
2. Oefner, J. P., and Chiesa, C. (1994). *Glycobiology* 4, 397–412.
3. Kakehi, K., and Honda, S. (1996). *J. Chromatogr. A* 720, 377–393.
4. Hjertén, S. (1985). *J. Chromatogr.* 347, 191–198.
5. Stefansson, M., and Novotny, M. (1994). *Anal. Chem.* 66, 1134–1140.
6. Paulus, A., and Klockow, A. (1996). *J. Chromatogr. A* 720, 353–376.
7. Liu, J., Shirota, O., Wiesler, D., and Novotny, M. (1991). *Proc. Natl. Acad. Sci. USA* 8, 2302–2306.

8. Shirota, O. (1992). Doctoral thesis, Indiana University, Bloomington, Indiana.

9. Sudor, J., and Novotny, M. (1995). *Anal. Chem.* 67, 4205–4209.

10. Mayer, P., Slater, G. W., and Drouin, G. (1994). *Anal. Chem.* 66, 1777–1780.

11. Liu, J., Shirota, O., and Novotny, M. (1991). *J. Chromatogr.* 559, 223–235.

12. Honda, S., Makino, A., Suzuki, S., and Kakehi, K. (1989). *Anal. Biochem.* 176, 72–77.

13. Stefansson, M., and Novotny, M. (1994). *Carbohydr. Res.* 258, 1–9.

14. Liu, J. Shirota, O., and Novotny, M. (1992). *Anal. Chem.* 64, 973–975.

15. Dolnik, V., Cobb, K. A., and Novotny, M. (1991). *J. Microcol. Sep.* 3, 155–159.

16. Stefansson, M., and Novotny, M. (1994). *Anal. Chem.* 66, 3466–3471.

17. Honda, S., Iwase, S., Makino, A., and Fujiwara, S. (1989). *Anal. Biochem.* 176, 72–77.

18. Stefansson, M., and Novotny, M. (1993). *J. Am. Chem. Soc.* 115, 11573–11580.

19. Mechref, Y., and El Rassi, Z. (1994). *Electrophoresis* 15, 627–634.

20. Mechref, Y., Ostrander, G. K., and El Rassi, Z. (1995). *J. Chromatogr. A* 695, 83–95.

21. Nashabeh, W., and El Rassi, Z. (1992). *J. Chromatogr.* 600, 279–287.

22. Hermentin, P., Witzel, R., Doengers, R., Bauer, R., Haupt, H., Patel, T., Parekh, R., and Brazel, D. (1992). *Anal. Biochem.* 206, 419–429.

23. Watson, E., and Yao, F. (1993). *Anal. Biochem.* 210, 389–393.

24. Hermentin, P., Doenges, R., Witzel, R., Hokke, C. H., Vliegenthardt, J. F. G., Kamerling, J. P., Conradt, H. S., Nimtz, M., and Brazel, D. (1994). *Anal. Biochem.* 221, 29–41.

25. Taverna, M., Baillet, A., and Ferrier, D. B. (1993). *Chromatographia* 37, 415–422.

26. Camilleri, P., Harland, G. B., and Okafo, G. (1995). *Anal. Biochem.* 230, 115–122.

27. Stefansson, M., Hong, M., Sudor, J., Chmelikova, J., Chmelik, J., and Novotny, M., submitted.

28. Linhardt, R., and Pervin, A. (1996). *J. Chromatogr. A* 720, 323–335.

29. Chu, Y.-H., Avila, L. Z., Gao, J., and Whitesides, G. M. (1995). *Acc. Chem. Res.* 28, 461–468.

30. Winzor, D. J. (1995). *J. Chromatogr. A* 696, 160–163.

31. De Gennes, P. G. (1977). *J. Chem. Phys.* 55, 572–586.

32. Doi, M., and Edwards, S. F. (1978). *J. Chem. Soc. Faraday Trans. 2* 74, 1789–1798.

33. Grossman, P. D., and Soane, D. S. (1991). *Biopolymers* 31, 1221–1228.

34. Viovy, J. L., and Duke, T. A. (1993). *Electrophoresis* 14, 322–329.

35. Shikata, T., and Kotaka, T. (1991). *Biopolymers* 31, 253–254.

36. Shikata, T., and Kotaka, T. (1991). *Macromolecules* 24, 4868–4873.

37. Kotaka, T., Adachi, S., and Shikata, T. (1993). *Electrophoresis* 14, 313–321.

38. Noolandi, J., Rousseau, J., Slater, G., Turmel, C., and Lalande, M. (1987). *Phys. Rev. Lett.* 58, 2428–2431.

39. Sudor, J., and Novotny, M. (1993). *Proc. Natl. Acad. Sci. USA* 90, 9451–9455.

27

A Comparison of Oligosaccharide
Profiling Methods

Elizabeth Higgins and Richard Bernasconi
Genzyme Corporation, Framingham, Massachusetts

I. INTRODUCTION

To monitor routinely the consistency of the glycosylation of recombinant glycoproteins, one must have an oligosaccharide profiling method available. Within the last five years, several new options for profiling have become available. To evaluate some of these methods, we have tested two different derivitizations (all with commercially available labeling kits) and four different separations.

Two N-glycosylated recombinant glycoproteins were used in these experiments, antithrombin III (ATIII) and thyroid stimulating hormone (TSH). ATIII has 4 N-linked sites and a molecular weight of 49 kDa, and TSH is a dimer with a molecular weight of 24 kDa and 3 N-linked sites. The ATIII sample was produced transgenically in the milk of a goat and is glycosylated with complex (mostly biantennary), oligomannose, and hybrid structures. This protein is sialylated with approximately 15% N-glycolylneuraminic acid (NeuGc), and 85% N-acetylneuraminic acid (NeuAc) (unpublished data). TSH was produced in Chinese Hamster Ovary cells [1] and is glycosylated with a mixture of bi-, tri-, and tetraantennary structures [2].

Our goal was to evaluate the resolution and sensitivity of profiling methods. Because sialylation affects the activity and clearance of TSH and the clearance of ATIII [3,4], the ability of these methods to resolve differences in sialylation was evaluated. Although it was not possible to evaluate the reproducibility of these methods within the scope of this study, this is also an important factor in choosing a profiling method. Finally, because the peaks/bands in a profile must be identified, it must be possible to isolate oligosaccharide structures from the profile and characterize them by other methods, such as glycosidase sequencing and mass spectrometry.

II. MATERIALS AND METHODS

A. Release of Oligosaccharides

Oligosaccharides were released from ATIII and TSH by hydrazinolysis using the N + O program on the Glycoprep 1000 from Oxford Glycosystems (Abingdon, U.K.). Recovery of oligosaccharides from these two proteins was 10% for ATIII and 37% for TSH, as estimated from fluorophore-assisted carbohydrate electrophoresis (FACE) profiles of released oligosaccharides. Profiles were similar when oligosaccharides were released by PN-Gase F or hydrazinolysis (data not shown). Subsequently, we have been able to release greater than 90% of the oligosaccharides from TSH using the N* program on the Glycoprep 1000; however, we have not been able to release greater than 30–40% of the oligosaccharides from ATIII using either program.

B. High pH Anion-Exchange Chromatography

Oligosaccharides released from ATIII and TSH were separated by high pH anion-exchange chromatography (HPAEC) using a Dionex BioLC and detected by a solvent-compatible pulsed amperometric detection (PAD) cell with a gold electrode (Dionex, Sunnyvale, CA). Conditions used for chromatography and detection were those described by Hermentin et al. [5]. Briefly, the CarboPac PA100 column with a PA100 guard column (Dionex) was equilibrated in 100% of buffer A (0.1 M NaOH). After holding for 5 min in 100% of buffer A, the oligosaccharides were eluted with a linear gradient from 0 to 35% of buffer B (0.6 M sodium acetate in 0.1 M NaOH) over 45 min, then a linear gradient to 70% of buffer B over 7 min. The flow rate was 1.0 mL/min. Pulse potentials and durations were 0.05V for 300 msec, 0.60 V for 120 msec. and −0.60 V for 60 msec. Internal standards were added to our samples to normalize retention times between runs. The internal standards used were 2′-fucosyllactose and lacto-N-neo-tetraose (Oxford Glycosystems) and (NeuAc)$_3$ (E-Y Laboratories, San Mateo, California).

C. Fluorophore-Assisted Carbohydrate Electrophoresis

Oligosaccharides released from 100 μg of ATIII and TSH were labeled with 8-aminonaphthalene-1,3,6-trisulfonate (ANTS) using the fluorophore-assisted carbohydrate eletrophoresis (FACE) oligosaccharide labeling reagent kit from Glyko (Novato, California). Briefly, oligosaccharide samples were dried in a SpeedVac, and then 5 μL of labeling dye and 5 μL of reducing agent were added to the samples. Samples were incubated at 37°C overnight and loaded onto gels without any further sample preparation. Oligoaccharides were also desialylated at 80°C for 1 h in 0.5 M formic acid (Baker) before labeling. Labeled oligosaccharides were separated on N-linked oligosaccharide profiling gels (Glyko) for 1 h and 20 min at 15 mA/gel, and images were acquired for 0.5–1.5 sec using an SC1000 FACE imager (Glyko).

D. Anion-Exchange Separation of 2-Aminobenzamide-Labeled Oligosaccharides

Oligosaccharides released from 500 μg of ATIII and TSH and oligosaccharide standards (Oxford Glycosystems) were labeled with 2-aminobenzamide (2-AB) using the Signal labeling kit from Oxford Glycosystems. Briefly, 5 μL of the labeling reagent (containing reducing reagent) was added to each sample, and each was incubated at 65°C for 2 h. Sample cleanup was performed by paper chromatography. Samples were spotted onto filter paper and allowed to dry. Filter paper was then placed into a chromatography tank that had been equilibrated during the labeling reaction, with the spotted samples at the bottom. Solvent was allowed to move up the filter paper until it was 1 cm from the end of the paper and then removed from the tank and allowed to dry. The sample spot was cut out of the filter paper and allowed to elute into water for 5 min. The sample was then filtered before injecting onto the high performance liquid chromatograph (HPLC). Oxford Glycosystems has since replaced the paper chromatography step in their kit with a cleanup step where samples are spotted onto a filter unit (supplied in the kit) and the filter then is washed to remove unreacted 2-AB. Both methods effectively remove unreacted label from the samples.

Oligosaccharides were separated on a Glycosep C column (Oxford Glycosystems) using a Hewlett-Packard 1090 HPLC. We used UV detection at 330 nm for these profiles simply because we needed an HPLC capable of pumping three different buffers instead of an HPLC connected to a fluorescent detector. This does demonstrate that even with UV detection 2-AB is a useful label. Chromatograms were generated using a gradient suggested by Oxford Glycosystems in the instruction booklet for the column. This gradient is designed to separate neutral oligosaccharides first using the hydrophilic nature of the column, and then to separate oligosaccharides according to negative charge using anion-exchange chromatography. The column was equilibrated in 75% buffer A (acetonitrile), 25% buffer C (water) and then held at these conditions for 5 min. Neutral oligosaccharides were eluted with a linear gradient from 25 to 40% of water over 25 min, and then negatively charged oligosaccharides were eluted with a linear gradient from 0 to 60% buffer B (100 mM ammonium formate, pH 6.5) over 20 min. The flow rate was 0.3 mL/min.

E. Reversed Phase Separation of 2-AB-Labeled Oligosaccharides

Oligosaccharides released from ATIII and TSH were labeled with 2-AB, as previously described, separated using a Beckman System Gold HPLC (San Ramon, CA), and detected using a fluorescent detector (ABI Spectroflow 980, Foster City, CA). Oligosaccharides were detected using an excitation wavelength of 330 nm, and emissions were filtered through a 389-nm cutoff filter. A Glycosep-H column (Oxford Glycosystems) was equilibrated in 100% of buffer A (5% acetonitrile, 0.1% trifluoroacetic acid) and held under these conditions for 5 min. Oligosaccharides were then eluted with a linear gradient from 0 to 20% of buffer B (acetonitrile) over 5 min and then a linear gradient from 20 to 40% of buffer B over 20 min. The flow rate was 0.3 mL/min.

III. RESULTS AND DISCUSSION

HPAEC-PAD profiles of TSH and ATIII are shown in Fig. 1. This method resolved more oligosaccharide structures from both the ATIII and TSH samples than the other methods examined. The shortcomings of this method are that it is not quantitative, because oligosaccharides do not respond equally to PAD [6,7] and subsequent characterization of peaks separated by HPAEC is made difficult by the nature of the eluants. As shown in Fig. 1, there was variability in the retention time of the $(NeuAc)_3$ internal standard (Table 1) between the two runs.

The oligosaccharides in the TSH profile (Fig. 1A) are separated into groups according to negative charge, or the number of sialic acid residues. This relationship between negative charge and the number of sialic acid residues did not apply to the ATIII oligosaccharides, because they are sialylated with both NeuAc and NeuGc. NeuGc carries two negative charges at high pH [5,8], whereas NeuAc has only one. This causes structures sialylated with NeuGc to elute later under HPAEC conditions than structures sialylated with NeuAc.

FACE profiles of ATIII and TSH before and after desialylation are shown in Fig. 2. On FACE gels, the separation of structures is primarily due to the charge/mass ratio [9]. Thus, structures with a similar charge/mass ratio, such as monosialylated biantennary structures and trisialylated triantennary structures, will co-migrate. It is difficult to evaluate the degree of sialylation from FACE profiles, because the oligosaccharides do not separate into groups according to negative charge. Once desialylated, oligosaccharides separate on the basis of size (unless they carry other charged groups such as phosphate or sulfate). Because of the heterogeneity in antennarity, fucosylation, and degree of sialylation, three of the four major

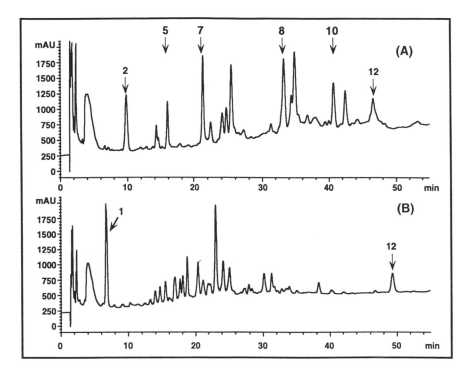

Figure 1 HPAEC-PAD profiles of TSH (A) and ATIII (B). Oligosaccharides released from 50 μg of TSH and 200 μg of ATIII were separated by HPAEC and detected by PAD, as described in Section II. The elution positions of oligosaccharide standards, as well as internal standards (Table 1, 2, and 12) added to these samples using this method are indicated.

bands in the TSH profile are composed of two or even three structures (data not shown). The desialylated profile still gives an excellent determination of the relative size of the oligosaccharide structures on this protein.

The dp (degree of polymerization) values for the sialylated TSH bands from Fig. 2 are shown in Table 2. A dp value for each band was determined by comparing its electrophoretic migration with the dextran ladder. The dp values were determined relative to the dextran ladder standard using FACE analytic software (Glyko). These dp values represent the average of three gels. The dp values are very reproducible; but they are not all significantly different, and therefore, an isolated band cannot be identified by its dp value.

In the ATIII profile there is some overlap between the biantennary complex structures and the oligomannose structures. For example, the oli-

Table 1 Oligosaccharide Standards

Key	
■ GlcNAc	○ Man
● Gal	◇ SA
△ Fuc	□ Glc

gomannose 6 and monosialo bi + core fucose structures have the same electrophoretic mobility. Once desialylated, oligomannose and complex oligosaccharides did not co-migrate. This method works well with exoglycosidases [10] and can also be used to separate mannose-6-phosphorylated structures [11]. It is also relatively tolerant of salts and detergents in the applied sample.

Profiles of oligosaccharides labeled with 2-AB are shown in Figs. 3 and 4. Separation of 2-AB-oligosaccharides by anion exchange (Fig. 3) separates oligosaccharides into charge groups, as seen with HPAEC; however, it lacks the resolution within charge groups of the HPAEC separation. This type of chromatography has traditionally been used to separate oligosaccharides according to negative charge as a first step in oligosaccharide purification. Profiles of 2-AB-labeled oligosaccharides separated by re-

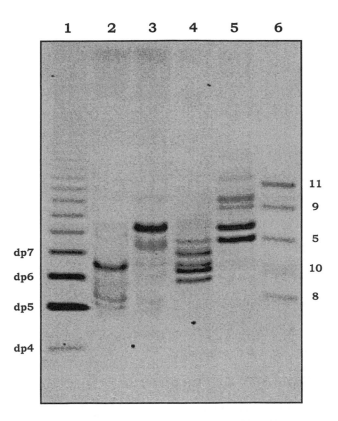

Figure 2 FACE profiles of oligosaccharides released from TSH and ATIII. Oligosaccharides were released, labeled with ANTS, and separated as described in Section II. Lane 1, ANTS-labeled dextran ladder; lane 2, oligosaccharides released from 20 μg of ATIII; lane 3, desialylated oligosaccharides released from 20 μg of ATIII; lane 4, oligosaccharides released from 5 μg of TSH; lane 5, desialylated oligosaccharides released from 5 μg of TSH; lane 6, oligosaccharide standards (Table 1) labeled with ANTS (50 pmol of each standard).

versed phase HPLC are shown in Fig. 4. This method separates the oligosaccharides from ATIII and TSH into a limited number of peaks; however, we have not yet identified the oligosaccharide structures with these peaks. Like FACE, it is difficult to evaluate the degree of sialylation from these profiles because oligosaccharides are not separated into groups according to charge.

A summary of the advantages and disadvantages of the different techniques is shown in Table 3. The anion-exchange separation gives insuf-

Table 2 Values of dp for Five
TSH FACE Bands

Band	Mean dp[a]
1	7.25 ± 0.22
2	6.79 ± 0.23
3	6.46 ± 0.12
4	6.07 ± 0.15
5	5.72 ± 0.16

[a]Values represent the average of three
gels ± standard deviations.

ficient resolution for a profiling method, whereas we have yet to determine
whether the resolution of the reversed phase profile will be sufficient. The
reversed phase HPLC separation is a very robust method and can be run
with volatile salts, making the cleanup of isolated oligosaccharides simple.
Further characterization is required to determine the true value of the reversed phase HPLC separation for profiling oligosaccharides from these
proteins.

The 2-AB label has been demonstrated to be compatible with both
matrix-assisted laser desorption ionization mass spectrometry [12–14] and
electrospray ionization mass spectrometry [15].

Although we concentrated on the labeling and separation methods in
this study, the approach to release the oligosaccharides from the glycoprotein is a critical part of any profiling method. A representative pool of
oligosaccharides must be released from the glycoprotein in order to monitor
effectively glycosylation. Automated hydrazinolysis using the Glycoprep
1000 is a simple method, although sample preparation is crucial. The sample must be completely desalted and free of detergents and heavy metals.
This method also requires an expensive piece of equipment, utilizes hazard-
ous chemicals, and is only able to process two samples at a time. A major
advantage of this technique is that the oligosaccharides are desalted and
separated from the protein during the process. Enzymatic release of oligosaccharides requires that the enzyme has access to all of the glycosylation
sites and can nonselectively remove the oligosaccharides. This may be accomplished using detergents, reducing agents, and sometimes proteases.
These additives, as well as the salts in the digestion buffer, can interfere
with subsequent analysis. The ease with which oligosaccharides are enzymatically released can vary from glycoprotein to glycoprotein.

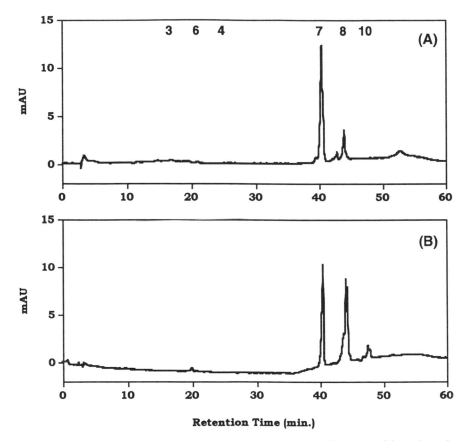

Figure 3 Anion-exchange chromatography profile of oligosaccharides released from 100 µg of ATIII (A) and TSH (B), labeled wtih 2-AB. Oligosaccharides were released, labeled with 2-AB, and separated by anion exchange, as described in Section II. Oligosaccharides were detected by UV absorbance at 330 nm. The elution positions of oligosaccharide standards (Table 1), labeled with 2-AB are indicated by numbers above the profiles.

IV. CONCLUSIONS

When choosing a profiling method there are three areas that must be considered: sensitivity, resolution, and characterization of the profile. If the protein samples are only available in small amounts, then derivitization with a fluorophore is required, whereas if milligram quantities of the protein are available, derivitization can be avoided by using HPAEC. It is helpful to understand what oligosaccharide structures need to be resolved

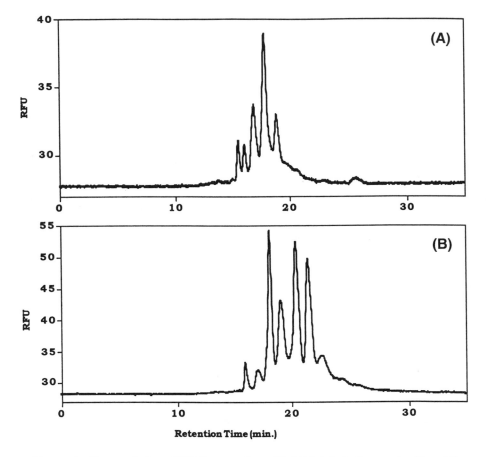

Figure 4   Reversed phase HPLC separation of 2-AB-labeled oligosaccharides. Oligosaccharides labeled with 2-AB, released from 100 μg of ATIII (A) or TSH (B) were separated as described in Section II.

by the method. For example, if the separation of different oligosaccharide isomers is necessary, HPAEC would be a good method, whereas if the degree of sialylation is most critical, then FACE or an anion-exchange separation could provide the necessary information while avoiding the complexity of the HPAEC profiles. Finally, a profiling method is much more useful if the oligosaccharide structures present in the peaks/bands are identified. This requires isolation and characterization. If mass spectrometry will be used, the reversed phase HPLC separation would be a good candidate for the profiling method, whereas FACE works very well with glycosi-

Table 3 Advantages and Disadvantages of the Different Methods

Method	Advantages	Disadvantages
HPAEC-PAD	Excellent resolution	Not quantitative
	Derivitization not necessary	High salt buffers required
	Sensitive	Epimerization of reducing GlcNAc possible
	Separation into charge groups	
FACE	Very senstive due to fluorescence	Requires labeling
	Multiple samples run on one gel	
	Compatible with glycosidase sequencing	
	Relatively tolerant of salts and detergents	
	Quantitative	
Reversed phase-HPLC	Salt-free buffer system	Requires labeling
Anion exchange	Separation into charge groups	Requires labeling
		Poor resolution

dase sequencing. Overall, with the profiling methods now available routine monitoring of protein glycosylation is now possible. As these new methods are used in more laboratories, other strengths and weaknesses will become more evident.

ABBREVIATIONS

2-AB	2-aminobenzamide
ANTS	8-aminonaphthalene-1,3,6-trisulfonate
ATIII	antithrombin III
dp	degree of polymerization
FACE	fluorophore-assisted carbohydrate electrophoresis
HPAEC	high pH anion-exchange chromatography
HPLC	high performance liquid chromatography
NeuAc	N-acetylneuraminic acid
NeuGc	N-glycolylneuraminic acid
PAD	pulsed amperometric detection
TSH	thyroid stimulating hormone

REFERENCES

1. Cole, E. S., Lee, K., Lauziere, K., Kelton, C., Chappel, S., Weintraub, B., Ferrara, D., Peterson, P., Bernasconi, R., Edmunds, T., Richards, S., Dickrell, L., Kleeman, J. M., McPherson, J. M., and Pratt, B. M. (1993). *Bio/Technol.* 11, 1014–1024.
2. Higgins, E. (1995). *Glycoconj. J.* 12, 407.
3. Szkudlinski, M. W., Thotakura, N. R., Bucci, I., Joshi, L. R., Tsai, A., East-Palmer, J., Shiloach, J., and Weintraub, B. D. (1993). *Endocrinology* 133, 1490–1503.
4. Higgins, E. A., Pollock, J., Li, T., Albee, K., Hanson, E., Logvinenko, K., DiTullio, P., Bernasconi, R., Garone, L., and Friedman, B. (1994). *XVIIth International. Carbohydr. Symp.* 508.
5. Hermentin, P., Witzel, R., Vliegenthart, J. F. G., Kamerling, J. P., Nimtz, M., and Conradt, H. S. (1992). *Anal. Biochem.* 203, 281–289.
6. Townsend, R. R., Hardy, M. R., Hindsgaul, O., and Lee, Y. C. (1988). *Anal. Biochem.* 174, 459–470.
7. Townsend, R. R., Hardy, M. R., Cumming, D. A., Carver, J. P., and Bendiak, B. (1989). *Anal. Biochem.* 182, 1–8.
8. Monica, T. J., Williams, S. B., Goochee, C. F., and Maiorella, B. L. (1995). *Glycobiology* 5, 175–185.
9. Jackson, P. (1993). *Biochem. Soc. Trans.* 21, 121–125.
10. Jackson, P., and Williams, G. R. (1991). *Electrophoresis* 12, 94–96.
11. Friedman, Y., and Higgins, E. A. (1995). *Anal. Biochem.* 228, 221–225.
12. Nuck, R., Gohlke, M., Kannicht, C., Prime, S., and Reutter, W. (1995). *Glycoconj. J.* 12, 410.
13. Townsend, R. R., Lipniunas, P. H., Bigge, C., Ventom, A., and Parekh, R. (1995). *Glycobiology* 5, 721.
14. Bigge, J. C., Patel, T. P., Bruce, J. A., Goulding, P. N., Charles, S. M., and Parekh, R. B. (1995). *Anal. Biochem.* 230, 229–238.
15. Lipniunas, P. H., and Townsend, R. R. (1995). *Glycobiology* 5, 710.

28

High-Performance Liquid Chromatographic Mapping of Oligosaccharides Using High pH Anion-Exchange Chromatography

Improvements from Sample Preparation, Reduction, and Fluorometric Detection

Peter H. Lipniunas,* Lorri Reinders, and R. Reid Townsend
University of California–San Francisco, San Francisco, California

James Bruce, Collette Bigge, and Raj Parekh
Oxford GlycoSystems Limited, Abingdon, Oxon, England

I. INTRODUCTION

High performance liquid chromatographic (HPLC) separations of complex carbohydrates are now sufficiently developed that isomeric species in complicated mixtures can be resolved as single peaks (for review, see Ref. 1). The term *oligosaccharide mapping* has been applied to such higher resolution chromatographic analyses and different HPLC methods, singly or in combination, are currently employed [2–8]. High pH anion-exchange chromatography (HPAEC) with pulsed amperometric detection (PAD) has often been used for "mapping" complex carbohydrates from glycoproteins [2–4,6–8]. The method has the distinct advantage that not only can oligosaccharides be separated according to size, charge, and linkage isomerism, but also that low nanomole amounts of carbohydrates can be detected in column effluents without derivatization (for review, see Refs. 9 and 10). However, artifactual peaks occur from several sources in HPAEC-PAD chromatograms. A duplicity of peaks results from base-catalyzed epimerization of the reducing GlcNAc residue of N-linked oligosaccharides [11]. Oligosaccharides with 3-O-substituents on the reducing end residue readily undergo base-catalyzed elimination from the β carbon [12,13], which can be avoided by preparation of the alditol form before chromatography [11,14,

---

*Current affiliation: Astra Draco AB, Lund, Sweden.

15]. Although the electrochemical wave forms that have been optimized for carbohydrates are relatively specific, other compounds (e.g., alcohols, glycols, amino acids, and organo-sulfur compounds) can also be detected, albeit to a lesser extent, under these conditions [16]. We report, herein, microscale methods to minimize artifactual peaks during the HPAEC analyses of reducing oligosaccharides, which have been released from glycoproteins either enzymatically or by automated hydrazinolysis. These include (1) a simplified, nondegradative method for the preparation of oligosaccharide alditols to prevent base-catalyzed degradation; (2) a method to remove noncarbohydrate, electrochemically active contaminants; and (3) fluorescent detection after derivatization of oligosaccharides with 2-aminobenzamide.

II. METHODS

A. Materials

Human Immunoglobulin G (IgG), ovalbumin, and Sephadex G_{50} were obtained from Sigma Chemical Company (St. Louis, MO). Bovine fetuin, prepared according to the method of Spiro [17], was from GIBCO (Grand Island, NY). Peptide-N^4-(N-acetyl-β-D-glucosaminyl)-asparaginase (PNGase F) was a gift from Tony Tarentino, New York State Department of Health (Albany, NY). A milliunit (MU) of PNGase F activity was defined as the amount of enzyme required to hydrolyze 1 nmole of a pentaglycopeptide from fetuin [18]. Biantennary oligosaccharides with (C2-024301) and without a core Fuc residue (C2-024300), the glycan adsorption matrix, and the Signal Kit were from Oxford GlycoSystems Ltd. (Abingdon, England). The monosaccharide standards were a gift from Dionex Corporation (Sunnyvale, CA). Labeling of the reducing oligosaccharides with 2-aminobenzamide was performed with the Signal Kit (Oxford GlycoSystems). Controlled porous glass beads were from CPG Inc. (Fairfield, NJ). Microspin cartridges (1.5 mL) with nylon membranes (0.45-μm pore size) were obtained from Alltech Assoc. Inc. (Deerfield, IL).

B. Preparation of Oligosaccharides from Glycoproteins

Fetuin (25 mg) was dissolved in 5 mL of 25 mM sodium phosphate buffer, pH 7.4. PNGase F (32 mU) was added, and the solution was incubated at 37°C for 3 h, followed by an equivalent addition of the enzyme and an overnight incubation. The digest was then applied to a Sephadex G_{50} column (1.6 × 94 cm) equilibrated in 1% acetic acid. The flow rate was maintained at 0.34 mL/min using a Ranin Rabbit-plus peristaltic pump (Woburn, MA). The effluent from the first 230 min was sent to waste, and then 3-min fractions were collected. Fractions 12–26 were pooled and lyophilized. Oligosaccharides were released from human IgG and ovalbumin by hydrazinolysis [19] in a GlycoPrep 1000 (Oxford GlycoSystems, Ltd.). The protein (2 mg) was lyophilized overnight in reaction vials for the instrument, and the oligosaccharides were released with the instrument in the "N + O" mode. The resulting solution of oligosaccharides was dried within 6 h after release from the protein in a SpeedVac equipped with a liquid nitrogen cold trap and a Varian vacuum pump, Model CD200 (Palo Alto, CA). The oligosaccharides were then dissolved in 1 mL of water.

C. Preparation of Oligosaccharide Alditols

An aliquot (~5 nmole) was dried in the SpeedVac apparatus and then dissolved in 70 μL of borate buffer, pH 11.0. A 100-fold molar excess of sodium borohydride was added, and the reaction was incubated at 30°C for 30 min. The reaction was terminated by drop addition of 10 μL of 4 M acetic acid. The sample was then mixed with a 2-mL layer of butanol over a column (0.5 × 4 cm) of glycan adsorption matrix (Oxford GlycoSystems, Ltd.), which had previously been washed sequentially with 4 mL of water, twice, 2 mL of ethanol, and 5 mL of butanol. The reaction vessel was extracted with 30 μL of water and then with 50 μL of butanol. Both extractions were applied to the column. The column was then washed sequentially with 1 mL of butanol, twice, 1 mL of a butanol–ethanol mixture (2/1, v/v), four times, and finally twice with 9 mL of ethanol. At each step, all the solvent was allowed to flow completely through the column. The glycans were eluted with water (5 × 0.5 mL). All the water washes were collected and dried in the SpeedVac.

D. Preparation of [³H-Gal]-Biantennary Oligosaccharides

The agalacto-biantennary oligosaccharide, C2-004300 (~1 μg), was dissolved in 100 μL of reaction buffer (20 mM HEPES containing 0.05% Triton X-100, 2.5 mM AMP, 10 mM Gal, 5 mM $MnCl_2$, and 34 μCi of UDP-6[³H]-Gal (17.2 Ci/mole)). β-Galactosyl transferase (10 mU) was added, and the reaction was placed at room temperature for 3 h. The tritiated glycans were purified by application to a four-sectioned column consisting of 0.25 mL each of Chelex 100 (Na^+ form), AG50X 12 (H^+ form), AG3X 4A (OH^- form), and QAE Sephadex. A final purification was performed on the GlycoMap 1000.

E. Sample Preparation Using Controlled Porous Glass Beads

Aliquots (200 μL) of oligosaccharides, after automated hydrazinolysis and drying, were transferred to a 1.5-mL microcentrifuge tube and dried in the SpeedVac-type apparatus. A 50% suspension of controlled porous glass

(CPG) beads in water was prepared, and 50 μL was added to the dried oligosaccharide sample. The suspension was evaporated to dryness under a stream of nitrogen. Butanol (200 μL) was then added, and the suspension was transferred to a (MicroSpin) apparatus. The insert, containing its 0.45-μm nylon filter, was placed into the carrier microfuge tube and centrifuged for 2 min at ~1500 × g. The same volume of butanol was added to the insert, followed by centrifugation. This organic wash cycle was repeated three times. After replacing the used carrier tube, the oligosaccharides were then eluted twice with water (200 μL each). The dried oligosaccharides were dissolved in 100 μL of water for analysis.

F. Derivatization of Oligosaccharides with 2-Aminobenzamide

Reducing oligosaccharides were coupled to 2-aminobenzamide (2-AB) as previously described [20]. Briefly, glycans were rendered particle-free and salt-free prior to conjugation. The oligosaccharides were dissolved in 5 μL of a solution of 2-AB (0.35 M) in dimethyl sulfoxide/glacial acetic acid (30% v/v)/sodium cyanoborohydride (1 M). The glycan solution was then incubated at 65°C for 2 h. The reaction mixture was applied to a cellulose disk (1 cm in diameter) in a glass holder. The disk was washed five times with 1 mL of acetonitrile to remove unreacted dye and noncarbohydrate reactants. Labeled glycans were then eluted using two washes (0.5 mL) of water and then filtered (0.2 μM) prior to analysis.

G. Monosaccharide and Oligosaccharide Analyses

Monosaccharide analysis by HPAEC with PAD was used to either quantify oligosaccharide preparations or carbohydrate in column effluents that had been desalted "in-line" with a Dionex anion micromembrane suppressor (100 mM H_2SO_4 at a counterflow rate of 20 mL/min) [21]. Samples were hydrolyzed using trifluoroacetic acid (2 N) at 100°C for 3 h in Chromacol (Trumbull, Connecticut) glass vials (Cat. No. 1.1-STVG) with Teflon cap liners (Cat. Nos. 8-SCY and 8-TST-1). We have found that different lots of plastic microfuge tubes, which were recommended for hydrolyses in our previously described method [22], varied considerably in Glc content. The hydrolysates were dried in a SpeedVac apparatus and analyzed using the earlier described chromatograph and column conditions for monosaccharides [22]. The concentration of fetuin oligosaccharides was calculated based on GlcN content (5 mol GlcN/mol of glycan). For oligosaccharide analyses, the sodium acetate eluant (0.5 M, pH 5.5) was prepared from glacial acetic acid [7]. For fluorometric detection, a Gilson fluorometer (Model 121, Middleton, WI) was connected after the electrochemical detector using polyetheretherketone tubing (40 cm of 0.01 in. i.d.). The detector was attenuated at 0.1 relative fluorescent units. Both electrochemical and fluorometric data were collected using Dionex AI450 software through a Dionex ACI convertor, and the chromatographic profiles were exported as "X,Y" data files using the "Optimize" module in the software.

III. RESULTS AND DISCUSSION

A. HPAEC After Microscale Oligosaccharide Alditol Preparation

The usefulness of oligosaccharide mapping is predicated on each peak representing a single carbohydrate structure from the parent glycoconjugate. From reversible epimerization of the reducing end GlcNAc, singular N-linked oligosaccharide species elute as a major (~ 80%) and a minor peak during high pH anion-exchange chromatography (HPAEC) [23]. These two forms of the same oligosaccharide are separated by ~ 1.5 min with the ManNAc terminal oligosaccharide eluting later. This duplicity of peaks can be prevented by reducing the oligosaccharides to their alditol forms [22]. The preparation of alditols has most often involved treatment with large excesses of sodium borohydride in the presence of millimolar concentrations of NaOH, making recovery of small quantities of glycans difficult. Further, these reaction conditions can also result in the C-2 epimerization of the reducing GlcNAc terminal, which, under some HPLC conditions, elutes as two peaks for each oligosaccharide. Epimerization during reduction can be avoided by reacting the oligosaccharides in a buffer (pH $\approx$ 10) [24]. The removal of salt after the reduction reaction has been accomplished using either gel filtration [15] or ion-exchange chromatography [15,25,26], followed by successive evaporations in the presence of methanol to remove borate. We investigated the usefulness of the glycan adsorption matrix for the one-step desalting and recovery of alditols after treatment with buffered sodium borohydride. Oligosaccharides were purified from bovine fetuin after release with PNGase F, reduced in buffered sodium borohydride, desalted on the glycan adsorption matrix, and analyzed using HPAEC-PAD. Figure 1 shows the HPAE chromatograms of the reducing (top trace) and the alditol forms of fetuin sialylated (lower trace). The oligosaccharide alditols eluted ~ 10 min earlier, due to the conversion of the relatively acidic hemiacetal hydroxyl at the reducing end residue to an alcohol group [27]. There was no significant change in the resolution within the complexes. The inset of Fig. 1 details the time interval of the chromatogram where the trisialylated complexes eluted (t = 34–45 min). The two smaller peaks trailing the major components (Fig. 1 inset, top trace) were replaced with a

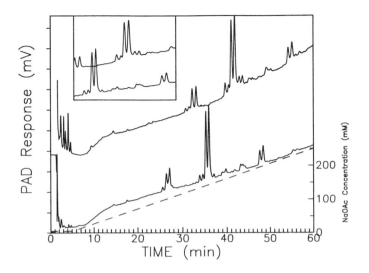

Figure 1   HPAEC with PAD of reducing bovine fetuin oligosaccharides and their alditols. Oligosaccharides from bovine fetuin were prepared and quantified by monosaccharide analysis after release by PNGase F as described in Section II. Approximately 250 pmol were injected into the described chromatograph. The scales for the reducing oligosaccharides (top trace) and alditols (lower trace) were 800 mV and 1 V, respectively. Inset details $t = 34–45$ min, during which the trisialylated complexes eluted.

single peak after reduction (Fig. 1 inset, lower trace). From our previous studies [11], we concluded that these trailing peaks contained the same oligosaccharides as the two major peaks except with ManNAc reducing termini and that the minor trailing peak, which remained after reduction, was another oligosaccharide species.

The single-step adsorption method avoids subjecting oligosaccharides to the acidic forms of strong anion-exchange resins to remove sodium ions. This method of desalting can result in desialylation (data not shown) and desulfation [15]. To determine if desialylation or selective adsorption of different oligosaccharide species to the glycan matrix occurred, the peaks in the di- ($t = 25.6–27.1$ and $30.7–33.1$ min), tri- ($t = 33.9–37.1$ and $39.7–43.6$ min), and tetrasialylated ($t = 47.6–48.4$ and $54.3–57.0$ min) regions of the chromatograms from alditol and reducing oligosaccharides, respectively, were integrated. The ratios of the peak areas of the di- : tri- : tetrasialylated species were similar: $1.0 : 4.3 : 0.93$ and $1.0 : 3.9 : 0.80$ for the alditols and reducing oligosaccharides, respectively. The peaks that eluted near the void in the reducing oligosaccharides and did not contain carbohydrate (Townsend et al., unpublished results) were also partially removed during reduction and sample preparation. The overall recovery (reduction and desalting) of ~1 nmol of either a neutral trisaccharide, Manα(1 → 3)Manα(1 → 4)GlcNAc, or a sialylated tetrasaccharide, Neu5Acα(2 → 3) Galβ(1 → 3)[Neu5Acα(2 → 6)] GalNAc, was ~80%.

B. Sample Preparation for HPAEC After Automated Hydrazinolysis

Hydrazinolysis and the associated preparation steps have recently been automated in the GlycoPrep 1000. The upper traces in Fig. 2 show the HPAEC-PAD analysis of oligosaccharides released from IgG (A) and ovalbumin (B), after automated hydrazinolysis and prior to any sample preparation. An aliquot of oligosaccharides from each glycoprotein was treated with CPG beads and reanalyzed using HPAEC-PAD. The lower chromatographic profiles in Fig. 2 (A and B) show that this treatment removed the broad peak with a retention time (t_r) of ~5 min and left smaller, sharper peaks with t_r < 10 min. Significantly, a peak present in both preparations (t_r ~ 15), and that would otherwise be considered to correspond to an oligosaccharide component, was removed after treatment with CPG beads. The radio of peak areas with t_r > 15 min remained the same after CPG treatment, indicating no selective loss during sample preparation. The inset of (Fig. 2B) shows that the HPAEC-PAD profiles from t_r = 0 to 10 min of the CPG-treated oligosaccharides were identical for both glycoproteins, suggesting that these early-eluting peaks did not contain carbohydrate and providing one measure of the reproducibility of automated hydrazinolysis. We collected the first 10 min of the column effluent, after injection of the CPG-treated oligosaccharides (~1 nmol in GlcN) from both IgG and ovalbumin. The separate effluents were lyophilized, hydrolyzed with 2 N trifluoroacetic acid (TFA), and analyzed for monosaccharides. No detectable GlcN was found in either the IgG or the ovalbumin oligosaccharide preparations, indicating that less than 1% of the total GlcN eluted in this region of the chromatograms (t_r = 0–10 min).

The recovery of low-level amounts of glycans from CPG beads was determined using a [^{3}H-Gal]-biantennary oligosaccharide. Aliquots containing 10 pmol–1 nmol were treated with CPG beads, transferred to microspin cartridges, and treated sequentially with butanol and water as was done for the IgG and ovalbumin oligosaccharides. Table 1 shows that 83–96% of the counts were recovered after the final wash of the CPG beads with water. Most of the losses (4–17%) were attributed to the physical transfer of the beads to the microspin cartridges, because only 1.6–5% of the radioactivity was found in the butanol washes. We concluded that CPG

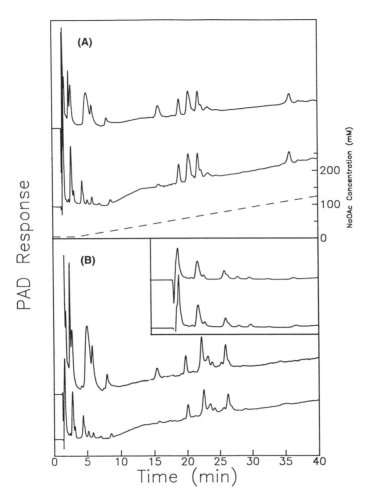

Figure 2 Treatment of oligosaccharides after automated hydrazinolysis with controlled porous glass beads. Oligosaccharides from either human IgG (A) or ovalbumin (B) were prepared by automated hydrazinolysis as described in Section II. An aliquot of oligosaccharides (~1 nmol) with (lower traces) or without (upper traces) treatment with CPG beads was injected into the described chromatograph. The CarboPac PA100 column (4.6 × 250 mm) was equilibrated in 100 mM NaOH and 5 mM sodium acetate at 1 mL/min. Five minutes after sample injection, the oligosaccharides were eluted with a linear acetate gradient over 55 min to a limit concentration of 125 mM, as indicated by the dashed line in (A).

Table 1 Recovery of [³H]-Gal-Asialo-Biantennary Oligosaccharide from CPG Beads

Amount treated[a]	Cpm in water phase[b]	Cpm in butanol phase[b]
12	88	1.6
22	86	3.1
62	94	3.9
112	83	5.2
512	90	2.8
1012	96	2.0

[a]Approximately 12 pmol of the radiolabeled oligosaccharide (C-024300) was added to each tube.
[b]Percent from averaging the results of two experiments.

was suitable for the microscale (< 1 nmol) sample preparation of oligosaccharides.

C. HPAEC with Fluorescent Detection

Reductive amination has been used extensively to couple the reducing termini of oligosaccharides to chromophores [28] and fluorophores [6] for both separation and detection during HPLC. Another potential benefit of this procedure is the removal of contaminants due to both the selectivity of the reductive amination reaction and the steps to remove reaction contaminants. A method to couple fluorophores quantitatively to neutral and sialylated oligosaccharides has been developed [20]. The ability of this approach to eliminate the artifactual peaks in HPAEC-PAD oligosaccharide maps was investigated by first analyzing standard reducing and 2-AB-labeled oligosaccharides using HPAEC and either PAD or fluorescent detection (FD). Figure 3A shows the HPAEC-PAD chromatogram of a reducing, core fucosylated biantennary oligosaccharide (C2-024301). The characteristic artifacts are apparent in the profile: (1) non-carbohydrate-containing peaks eluting near the void; (2) a peak associated with the onset of the acetate gradient ($t_r \sim$ 10 min); (3) a minor peak trailing the major one ($t_r \sim$ 30 min) from C-2 epimerization of the reducing end GlcNAc residue; and (4) baseline elevation between the major and minor oligosaccharide peaks. Figure 3B shows the HPAEC-FD analysis of the same oligosaccharide after derivatization with 2-AB. All the artifactual peaks were eliminated, and the 2-AB-derivative eluted earlier ($t_r \sim$ 5 min) as a single, symmetrical peak.

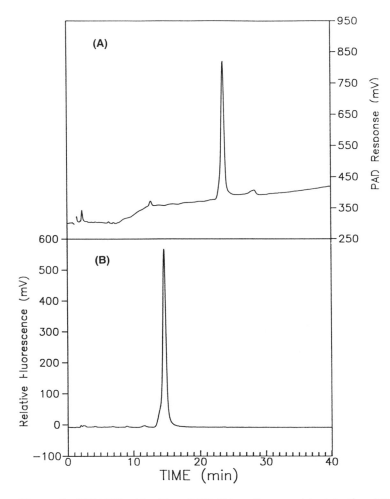

Figure 3 HPAEC with either PAD (A) or fluorometric detection (B) of a 2-AB-derivatized biantennary oligosaccharide (C2-024301). For electrochemical detection, ~ 500 pmol of the oligosaccharide in 100 μL of water was injected into the chromatograph as described in Section II. The full scale attenuation of the pulsed amperometric detector was 300 nA. For fluorescent detection, approximately 90 pmol of the 2-AB-modified biantennary oligosaccharide was analyzed (B). The full scale attenuation of the fluorimeter was 0.1 relative fluorescent units. Both analyses were performed with a CarboPac PA100 column (4 × 250 mm) equilibrated in 5 mM sodium acetate, and the oligosaccharides were eluted with the gradient described in the legend of Fig. 2.

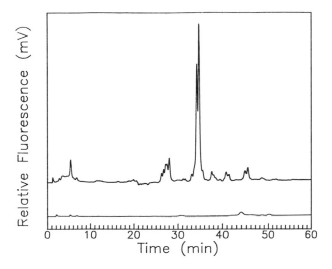

Figure 4 HPAEC with fluorometric detection of oligosaccharides from bovine fetuin. Oligosaccharides (~1 nmol), released either by automated hydrazinolysis (top trace) or by PNGase F treatment (middle trace), were derivatized with 2-AB, and the dried samples were dissolved in 200 μL of water. An equivalent aliquot of water was also carried through the derivatization protocol (lowest trace). The 2-AB-labeled oligosaccharides and the water blank were analyzed (20 μL) using the described chromatograph with FD. The oligosaccharides were eluted with the same gradient described in the legend of Fig. 1. The full scale attenuation of the fluorimeter was 0.1 relative fluorescent units.

The usefulness of the 2-AB-labeling method for HPAEC profiling of sialylated, lactosamine-type oligosaccharides was next determined. Figure 4 shows the 2-AB-labeled oligosaccharides that were released from bovine fetuin by PNGase F (top trace). The characteristic tetra-, tri-, and disialylated peak complexes, as were seen in the unlabeled oligosaccharides (Fig. 1), were observed. The small differences in the profiles of the disialylated oligosaccharides from that of the underivatized oligosaccharides (Fig. 1) may be due to the small amount of desialylation (<5%) that occurs during 2-AB coupling [20]. Only minor peaks were observed in the water blank (Fig. 4, lower trace). We concluded that HPAEC with FD, after 2-AB-labeling [20], is a useful method for mapping sialylated oligosaccharides.

IV. SUMMARY

We have developed simplified, microscale methods to remove artifactual peaks from HPAE chromatograms. A reduction protocol with a single desalting step was developed to prevent the formation of additional peaks from base-catalyzed C-2 epimerization of the reducing GlcNAc residue of N-linked oligosaccharides and β-elimination of oligosaccharides with 3-O-substituted reducing termini. There was neither selective adsorption of oligosaccharides nor desialylation during the desalting step, and the yields were ~80% for even a neutral trisaccharide or a disialylated tetrasaccharide. Non-carbohydrate-containing, electrochemically active peaks from automated hydrazinolysis were removed using a protocol with controlled porous glass beads. The recovery of a biantennary oligosaccharides (10 pmol up to 1 nmol) from the glass beads was 83–96%. Derivatization of oligosaccharides with 2-AB and fluorometric detection also eliminated the major artifacts in HPAEC oligosaccharide maps. The usefulness of 2-AB labeling for mapping sialylated oligosaccharides by HPAEC-FD was demonstrated.

ACKNOWLEDGMENTS

This work was supported in part by the Biomedical Research Technology Program of the National Center for Research Resources, NIH NCRR BRTP 01614, and a grant from the Wenner-Gren Foundation (to PHL).

ABBREVIATIONS

2-AB	2-aminobenzamide
AMP	adenosine monophosphate
CPG	controlled porous glass
FD	fluorescent detection
IgG	immunoglobulin G
HPAEC	high pH anion-exchange chromatography
HPLC	high performance liquid chromatography
PAD	pulsed amperometric detection
PNGase F	peptide-N^4-(N-acetyl-β-D-glucosaminyl)-asparaginase
TFA	trifluoroacetic acid

REFERENCES

1. El Rassi, Z., ed. (1995). *Carbohydrate Analysis: High Performance Liquid Chromatography and Capillary Electrophoresis*. Elsevier, New York.
2. Kumarasamy, R. (1990). *J. Chromatogr.* 512, 149–155.
3. Hermentin, P., Witzel, R., Vliegenthart, J. F. G., Kamerling, J. P., Nimtz, M., and Conradt, H. S. (1992). *Anal. Biochem.* 203, 281–289.

4. Rice, K. G., Takahashi, N., Namiki, Y., Tran, A. D., Lisi, P. J., and Lee, Y. C. (1992). *Anal. Biochem.* 206, 278–287.
5. Takahashi, N., Wada, Y., Awaya, J., Kurono, M., and Tomiya, N. (1993). *Anal. Biochem.* 208, 96–109.
6. Hase, S. (1994). *Meth. Enzymol.* 230, 225–237.
7. Anumula, K., and Taylor, P. B. (1991). *Eur. J. Biochem.* 195, 269–280.
8. Barr, J. R., Anumula, K. R., Vettese, M. B., Taylor, P. B., and Carr, S. A. (1991). *Anal. Biochem.* 192, 181–192.
9. Spellman, M. W. (1990). *Anal. Chem.* 62, 1714–1722.
10. Townsend, R. R. (1995). In Rassi, E. L., ed., *Carbohydrate Analysis: High Performance Liquid Chromatography and Capillary Electrophoresis*. Elsevier, New York, pp. 181–209.
11. Townsend, R. R., and Parekh, R. (1994). Application Note IAN-06, Oxford GlycoSystems Ltd, Abingdon, United Kingdom.
12. Lloyd, K. O., and Kabat, E. A. (1969). *Carbohydr. Res.* 9, 41–48.
13. Mayo, J. W., and Carlson, D. M. (1970). *Carbohydr. Res.* 15, 300–303.
14. Shibata, S., Midura, R. J., and Hascall, V. C. (1992). *J. Biol. Chem.* 267, 6548–6555.
15. Midura, R. J., Salustri, A., Calabro, A., Yanagishita, M., and Hascall, V. C. (1994). *Glycobiology* 4, 343–350.
16. Johnson, D. C., and Lacourse, W. R. (1994). In Rassi, E. L., ed., *Carbohydrate Analysis: High Performance Liquid Chromatography and Capillary Electrophoresis*. Elsevier, New York, pp. 391–430.
17. Spiro, R. G. (1960). *J. Biol. Chem.* 235, 2860–2869.
18. Plummer, T. H., Phelan, A. W., and Tarentino, A. L. (1987). *Eur. J. Biochem.* 163, 167–173.
19. Patel, T., Bruce, J., Merry, A., Bigge, C., Wormald, M., Jaques, A., and Parekh, R. (1993). *Biochemistry* 32, 679–693.
20. Bigge, J. C., Patel, T. P., Bruce, J. A., Goulding, P. N., Charles, S. N., and Parekh, R. B. (1995). *Anal. Biochem.* 230, 229–238.
21. Townsend, R. R. (1991). In Conradt, H. S., ed., *Protein Glycosylation: Cellular, Biotechnological and Analytical Aspects*. VCH, New York, pp. 147–160.
22. Hardy, M. R., and Townsend, R. R. (1994). *Meth. Enzymol.* 230, 208–225.
23. Hardy, M. R., Townsend, R. R., and Lee, Y. C. (1987). *Proc. 16th Annu. Meet., Soc. Complex Carbohydr.*, Washington, D.C., Abstr. No. 12.
24. Mellis, S. J., and Baenziger, J. U. (1981). *Anal. Biochem.* 114, 276–280.
25. Gabriel, O., and Ashwell, G. (1992). *Glycobiology* 2, 437–443.
26. Takeuchi, M., Takasaki, S., Inoue, N., and Kobata, A. (1987). *J. Chromatogr.* 400, 207–213.
27. Rendleman Jr., J. A. (1974). In Isbell, H. S., ed., *Carbohydrates in Solution*. American Chemical Society, Washington, D.C., pp. 70–87.
28. Wang, W. T., LeDonne, N. C., Jr., Ackerman, B., and Sweeley, C. C. (1984). *Anal. Biochem.* 141, 366–381.

29

Monosaccharide Analysis of Recombinant Glycoproteins by High pH Anion-Exchange Chromatography/High-Performance Liquid Chromatography Using a Refractive Index Detector

M. Janardhan Rao, Yong K. Cho, and Roderic P. Kwok
Baxter Biotech/Hyland Division, Duarte, California

I. INTRODUCTION

The glycosylation of a protein involves co- and post-translational reactions that have significant impact on expression levels, structural integrity, specific activity, immunogenicity, and solubility [1–3]. Mammalian cells produce a wide range of oligosaccharide structures, and the type of structures depends on the species [4]. Common sources of heterogencity include variation in the sites of attachment of oligosaccharides, differences in branching, and terminal modification with different sialic acid residues. The polypeptide chains appear to direct glycosylation at individual sites, unique to a single protein or class of proteins. Each of these sources of oligosaccharide heterogencity can have significant effects on biological properties [5]. The structure of glycoprotein glycans can be influenced not only by cell culture methodology, but also by culture time [6,7].

Proteins without "appropriate" carbohydrate structures may show altered pharmacokinetics and may also cause adverse effects [8]. The Anti-Hemophilic Factor (AHF) is mainly used to correct the bleeding disorder in hemophilia A. When the human AHF is expressed in Chinese hamster ovary (CHO) cells, its glycosylation is 30% by weight [9]. This recombinant AHF (rAHF) plays a vital role in the blood coagulation cascade process as a co-factor of factor X activation by factor IXa. The sequence of the cDNA clone comprises 2351 amino acids. The primary structure of factor VIII (FVIII) exhibits three distinct structural domains, and they are arranged in the order A1–A2–B–A3–C1–C2. The AHF possesses two types of glycosylation, O-linked and N-linked. The large B domain is extensively glycosylated (19 asparagine-linked glycosylation sites out of 25 potential sites). The possible influence of glycans on the structural integrity of glycoproteins led us to study the consistency of monosaccharide content in different preparations of rAHF.

Many laboratories have developed and adopted different methods to evaluate protein glycosylation by performing monosaccharide analysis. Methods employed to estimate the monosaccharide composition of a glycoprotein include high pH anion-exchange chromatography (HPAEC) with pulsed amperometric detector (PAD) [10–13], high performance capillary electrophoresis (HPCE) [14–16], mass spectrometry (MS) [17], and gel electrophoresis (GE) [18,19]. In this chapter, we describe a high pressure liquid anion-exchange chromatographic method using refractive index detection (HPAEC-RI) and potassium phosphate buffer as the mobile phase to quantify the anionic sugars (sialic acid) from a glycoprotein. The amino sugars [galactosamine (Gal.Am) and glucosamine (Glu.Am)] and neutral sugars [fucose (Fuc), galactose (Gal), and mannose (Man)] were also well resolved using the same chromatographic system, except dilute NaOH was used as an eluant. This method is fast, simple, and reliable. It does not require any pre- or post-column derivatization. We demonstrate here the utility of this analytical tool in estimating the monosaccharide content of rAHF. The validity of this chromatographic method was established on the basis of its selectivity, linearity, precision, and accuracy based on recovery.

II. MATERIALS AND METHODS

A. Preparation of Buffers and Eluants

A 10 mM phosphate-buffered saline (PBS), pH 7.4, containing 138 mM NaCl and 2.7 mM KCl, was prepared by dissolving one packet of PBS (Sigma) in 1L of glass-distilled water. Before use, the buffer was filtered with a 0.22 μm pore size nylon filter. The buffer was stored at room temperature and prepared fresh every week. Trifluoroacetic acid (TFA) (4 M), 0.46 mL of HPLC grade TFA (B & J) diluted to a final volume of 1.5 mL, was prepared with glass-distilled water and stored at room temperature. A fresh acid solution was prepared at least once a week.

The HPLC eluants were prepared as follows. For eluant A (25 mM NaOH), 1997 mL of glass-distilled water was placed in a 2.0-L plastic container and degassed by sparging with helium gas for 30 min. Sodium hydroxide was pipetted, 2.6 mL of a 50% (w/w) solution (19.3 M) (J. T. Baker), from the center, not from the edges, of the bottle. The bottle was not stirred or shaken. Degassing was continued for at least 30 min more to ensure complete mixing of the NaOH solution. Eluant B (200 mM NaOH) was prepared by placing 1980 mL of glass-distilled water into a 2.0 L-plastic container, degassing with sparging helium gas for 30 min, and adding 20.8 mL of 50% (w/w) (19.3 M) NaOH. Eluant C (25 mM potassium phosphate buffer, pH 4.5) was prepared by weighing 6.8 g of monobasic potassium phosphate crystals (Mallinkrodt) and adding 2.0 L of glass-distilled water. The buffer was vacuum filtered through a 0.22-μm pore size nylon filter. The pH of the buffer should be 4.5. The eluants (phosphate buffer and NaOH) were kept under helium gas (5 to 7 psi) at all times, unless not used for more than a week.

B. Desalting of rAHF

The most commonly encountered problems in the analysis of proteins are sample matrices, salt, and detergent. To eliminate these interferences during hydrolysis and/or analysis, recombinant glycoprotein, such as rAHF samples, were desalted or buffer exchanged using concentrators. A schematic diagram of sample preparation and monosaccharide analysis is shown in Fig. 1.

The Centricon-10 (2 mL volume, molecular weight (MW) cutoff 10,000) concentrators were washed with 1 mL of 10 mM PBS buffer by spinning at 5000 rpm for 20 min at 5 $\pm$ 1°C. The protein in 1 mL (0.5 mg/mL) volume was concentrated to approximately 100 μL under similar conditions as described for the washing of the Centricon-10. Two milliliters of 10 mM PBS buffer was added to the protein and concentrated for 80–100 min. Similar steps were repeated two more times. Finally, approximately 100 μL of the protein concentrate was collected. Equal aliquots were placed into two 1.5-mL microfuge tubes. The protein recovery was confirmed by absorbance at 280 nm. The protein was stored at -70 ± 2°C until further use.

C. Hydrolysis of rAHF

Acid hydrolysis was used to release the monosaccharides from the rAHF protein [14]. The sugars were identified and analyzed by HPLC. The protocol for acid hydrolysis often depended on the method of analysis and the amount of carbohydrate available. The neutral sugars were released by strong acid hydrolysis, and milder conditions were used to release sialic acids (Fig. 1). Hydrolysis of glycosidic linkages involving hexosamines required more vigorous conditions. For this purpose, hydrochloric acid (HCl) was used (Fig. 1). Highly purified acid was used. Only highly purified and glass-distilled water should be used to dilute the acid, because metallic impurities may lead to the destruction of amino sugars. A lower concentra-

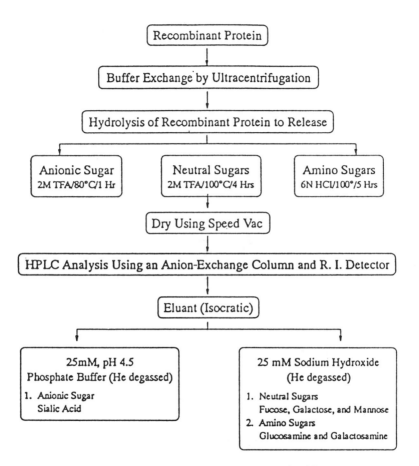

Figure 1 Schematic presentation of the monosaccharide assay.

tion of glycoprotein (< 1 mg/mL) during hydrolysis was used to decrease the destruction of sugars, which presumably occurs as result of the reaction of sugars with amino acids.

1. Hydrolysis of rAHF to Release Sialic Acid

An aliquot containing 0.2 mg of desalted rAHF was added to 200 μL of 4 M TFA. The concentrations of TFA and protein were adjusted to 2 M and 0.5 mg/mL, respectively, by adding the required amount of water to the cocktail. This solution was hydrolyzed at 80 $\pm$ 2°C for 1 h. After hydrolysis, the microfuge tubes containing protein were removed from the incubator and cooled at room temperature for 5 min. The tubes were centrifuged for 5 min to return the condensate to the liquid. The samples were then evaporated at room temperature. The SpeedVac was operated under a strong vacuum with a cold trap ($-110°C$) and chemical acid trap (Savant, SpeedVac-110 with VP 100 vacuum pump). The protein samples were dried completely in 4 to 5 h. The hydrolyzed and dried protein samples were stored at $-70 \pm 2°C$ until analyzed by HPLC.

2. Hydrolysis of rAHF to Release Fucose, Galactose, and Mannose

The same protocol used for sialic acid was employed for neutral sugars, except that the protein cocktail was incubated at $100 \pm 2°C$ for 4 h. After hydrolysis, the tubes were prepared for evaporation as described previously.

3. Hydrolysis of rAHF to Release Galactosamine and Glucosamine

The protein cocktail was prepared in the same way as for sialic acid hydrolysis, except a different acid and incubation time were used. Concentrated HCl was used to release amino sugars. The protein and HCl concentrations were 0.5 mg/mL and 6 M, respectively. The protein cocktail was hydrolyzed at $100 \pm 2°C$ for 5 h. The microfuge tubes containing protein were removed from the heating block and cooled at room temperature for 5 min. The tubes were centrifuged to bring down the condensate to liquid in the tubes. For all hydrolysates, microfuge tube caps were removed and covered with parafilm. Holes were made in the parafilm, and the microfuge tubes were placed into a SpeedVac concentrator. The hydrolyzed proteins were evaporated to dryness at room temperature.

D. HPLC Analysis of Hydrolyzed rAHF

1. Chromatographic Apparatus

Due to the strong alkaline nature of the eluants, an eluant degassing module (Shimadzu, HDU-1) was used to sparge and to pressurize the eluants with helium. This also minimized absorption of atmospheric CO_2. Eluant delivery was performed with a Shimadzu Solvent Control Module (Shimadzu, FCV-10AL) and a double-plunger pump (Shimadzu, 10-AD). An anion-exchange column (Hamilton, RCX-30) was used to separate the monosaccharides. A guard column was used to protect the anion-exchange column against deterioration. The elution of monosaccharides was monitored with a refractive index detector (RID) (Shimadzu, RID-6A). The temperature of the column and RID was maintained at $25 \pm 1°C$. This was achieved by keeping the room temperature below ambient and the column in a temperature-controlled heating module (Eppendorf TC-50). Fluctuation in column temperature results in drift of the RID baseline, as it is sensitive to changes in temperature, ionic strength, and flow rate. The RI cell temperature was set to 25°C and allowed to stabilize at 25°C for at least 24 h. The RID was not turned off unless it was not used for 4 to 5 days. The reference cell (R cell) of the RID was filled with mobile phase. If, even after filling the R Cell with a mobile phase, the "BALANCE" indication did not stabilize to "0.000," the slit image knob was adjusted until the "BALANCE" reading was "0.000" or close to "0.000." Whenever the concentration or flow rate of the eluant was changed, the "BALANCE" indicated was readjusted to zero.

The standard and the protein hydrolysate were injected using an autosampler (Shimadzu, SIL-10A) using the system controller (Shimadzu, SCL-10A). To withstand the alkalinity of the eluants, the rotor seal (stainless steel Rheodyne injection valve) was replaced with a high pH–resistant (Shimadzu) seal. Tefzel or tubing capable of withstanding high pressure was used to connect the pump, autoinjector, column, and detector. To use efficiently the limited recombinant glycoprotein sample, disposable plastic conical insert vials (0.1 mL volume) with Teflon septa were used for processing the multiple samples. The autoinjector was rinsed and purged with glass-distilled water before and after use.

2. Column Equilibration

The column was equilibrated with eluant A (25 mM NaOH) for at least 30 to 60 min or until the baseline stabilized. The flow rate (1.5 mL/min) was the same during the column equilibration and the standard sugar or the glycoprotein analysis. Any change in flow rate altered the baseline stability. The column was regenerated after every four injections with 30 mL of 200 mM NaOH to maintain the retention times of sugars. If freshly prepared and continuously degassed with helium, eluants were used for a week. The pump was always purged with water before and after the run.

3. Preparation of a Monosaccharide Standard Solution

Monosaccharides (Fuc, Gal, Man, Glu.Am, Gal.Am, and sialic acid) are hygroscopic and should be stored in a vacuum desiccator containing NaOH pellets. To prepare standards, monosaccharides were transferred to dry scintillation vials with a loose cap and stored in a desiccator under continuous vacuum for a week. An exact amount of dried sugar was weighed in a dry vial, and the required amount of glass-distilled water was added to the vial for a 100 μM solution. This primary standard solution was stored at -20°C. A working standard solution was prepared by 100-fold dilution. This was achieved by pipetting 100 μL (electronic pipettes were used, Rainin, EDP-2) of each primary standard sugar solution to a final volume of 10 mL and mixing thoroughly. Aliquots, sufficient for one day of analysis, were prepared.

4. Calibration Curves

Before using a new column for monosaccharide analysis, we calibrated it with a five-point standard curve. The protein sample was preceded by a single standard sugar run. Standard sugar runs were also repeated with two different concentrations (one above and one below) to confirm a 1 to 1 linear dose–response relationship. The concentration of standard/control was maintained so that it fell within the established standard curve range. To minimize error and to keep the same volume load on the column, the same volume of standard and test protein samples were injected.

5. Preparation and HPLC Analysis of Hydrolyzed and Dried Glycoprotein

The hydrolyzed and dried protein from the $-70°C$ freezer was thawed at room temperature. Glass-distilled water (80 μL) (for the hydrolysate of amino and neutral sugars) or 80 μL of 25 mM phosphate buffer (for the hydrolysate for sialic acid) was used to dissolve the hydrolysate of the glycoprotein (0.2 mg). The volume of water or buffer depended on the initial amount of glycoprotein used for hydrolysis. After vortexing, the hydrolysates were transferred to microfilterfuge tubes (containing 0.45-μM Nylon-66 filters, Rainin) and spun for 5 min at $10,000 \times g$ using a microcentrifuge. The hydrolysates were pipetted from the receiver of the microfilterfuge tube to HPLC sample 0.1-mL vials. The hydrolyzed protein solution was analyzed on the same day or stored at $-70°C$. At this stage, protein samples were treated as standard mixtures. Standard sugar and protein sugar results fell within the established standard curve range. If not, a more concentrated or diluted sample was prepared. Amino and neutral sugars were analyzed using 25 mM NaOH as the mobile phase, at a flow rate of 1.5 mL/min. Sialic acid was eluted using 25 mM phosphate buffer, at a flow rate of 1.0 mL/min. In both cases, mobile phase was used isocratically.

III. RESULTS AND DISCUSSION

High pH anion-exchange chromatography (HPAEC) is a routine method for the rapid analysis of monosaccharides from a glycoprotein. Due to the development of new ion-exchange materials of small particle size and narrower size ranges, the stability at higher pressures, and improvements in the column efficiency, reductions as well as reproducible retention times have become possible. In the present assay, monosaccharides from a recombinant glycoprotein, rAHF, were analyzed at high and low pH by anion-exchange chromatography using a refractive index detector. The anion-exchange column, quaternary ammonium ion on a column of a highly cross-linked anion-exchange resin (polystyrene–divinyl benzene trimethyl ammonium), was employed to separate the monosaccharides. The anionic sugar, sialic acid, was separated on the basis of ionic interaction of the sugar with the charges on the stationary phase (reversible adsorption of sample anions onto a charged surface). It is well known that negatively charged glycans bind to the column with weak interactions. The pK_a of sialic acid is 2.6. At pH 4.6, it is negatively charged. To elute it from the column, potassium phosphate buffer of pH 4.5 was used. Because the pK_a's of neutral (Fuc, Gal, and Man) and amino (Gal.Am and Glu.Am) sugars are in the range of 12 to 13, sodium hydroxide (pH ~ 13) was used to elute these sugars from the column. Retention times and the selectivity of each sugar was controlled by varying the ionic strength of the eluant. A typical chromatogram showing the elution pattern of five standard monosaccharides is shown in Fig. 2A. A change in column temperature or flow rate altered the baseline, as depicted in Fig. 2B. The representative HPAEC-RID chromatograms (Figs. 3–5) of rAHF suggest that three classes of monosaccharides are present in rAHF. They are neutral (Fuc, Gal, and Man), amino (Gal.Am and Glu.Am), and anionic (sialic acid) sugars.

The results from Table 1 show the precision of the assay. The percent coefficient of variation (% CV) for all six sugars of a single lot of rAHF was less than 10% (Table 1). To estimate the run variation, each assay was carried out on a different day. Each assay was also performed in triplicate. The intra-assay variations of three assays are shown in Table 2. These results show that this analytical method has an acceptable degree of precision.

The robustness of the assay was established by two analysts. Three independent assays were carried out by each analyst, and each assay was analyzed in triplicate (Table 3). The linearity of the assay was confirmed from the calibration curves of standard sugars at six different concentrations as shown in Figs. 6–8. The least mean square (r^2) values of all curves were > 0.99, suggesting good linearity over a wide range of sample amount. The linearity was established by two different means. In one case, the injection volume was constant throughout the study, and in the other, the injection volumes were altered as required by using the same stock solution.

The sensitivity (t-statistic values) of the assay for each sugar was estimated from the difference between the instrument response at the lowest concentration and the blank. It was also estimated from the response at two different concentrations of each sugar. The lowest detectable amounts of sialic acid, Fuc, Gal, Man, Gal.Am, and Glu.Am were 0.77, 0.4, 1.35, 1.35, 0.54, and 2.16 μg, respectively. In addition, this method was capable of detecting concentration changes at 0.16, 0.2, 0.67, 0.67, 0.27, and 1.08 μg for sialic acid, Fuc, Gal, Man, Gal.Am, and Glu.Am, respectively.

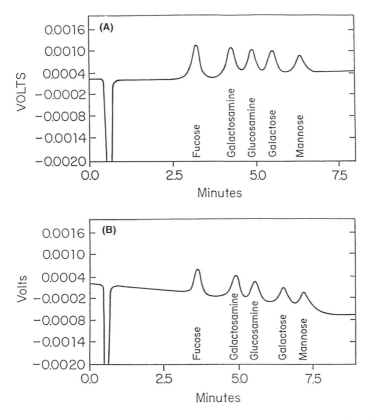

Figure 2 (A) Chromatographic traces of standard neutral and amino sugars. (B) Drift of chromatographic baseline due to change in column and/or detector temperature and flow rate.

The specificity of the assay was established on the basis of the retention time of each sugar. The column separation efficiency was established by comparing the retention time of each standard sugar individually chromatographed with the retention time of the same sugar when it was co-chromatographed in a mixture of standard sugars. In addition, the retention times of the standard sugars as a mixture were also compared with the retention times of the sugars in rAHF (Figs. 3–5). There was no statistically significant difference between the retention time of an individual standard sugar and the retention time of the same sugar in a mixture of standard sugars ($p > 0.05$). No statistically significant difference was observed between the retention times of the standard sugars in the mixture and the same sugars in rAHF ($p > 0.05$). A maximum of 4% difference in reten-

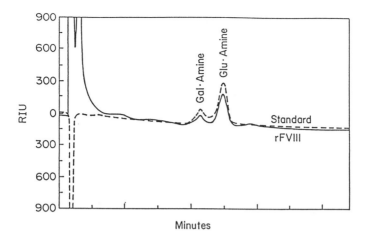

Figure 3 Chromatograhic traces of standard sugar mixture and rAHF digest of amino sugars.

tion times of standard sugars and sugars of rAHF was observed, and this difference varied from run to run (Table 4). Therefore, the equivalence in retention times of standard sugars and sugars of the glycoprotein was established by chromatographing the standard sugars before every analysis of the glycoprotein.

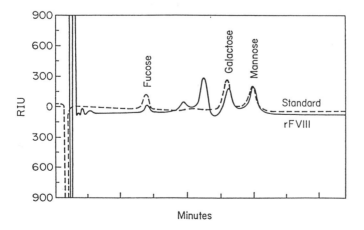

Figure 4 Chromatographic traces of standard neutral sugar mixture and rAHF digest of neutral sugars.

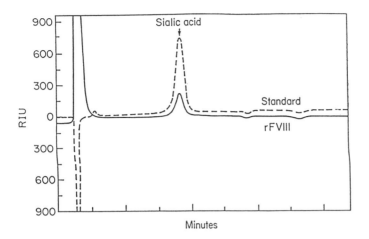

Figure 5 Chromatographic traces of standard anionic sugar and rAHF digest of anionic sugar.

In the case of sialic acid, it was the only analyte in the chromatogram. The differences between the retention times of standard sialic acid and protein sialic acid were ≤0.05 min (Fig. 5). This difference accounted for a maximum of 0.65% variation in retention times. In addition, when standard sialic acid was spiked into rAHF, it co-eluted with the sialic acid of rAHF.

Kobata and coworkers [20] have suggested that rAHF produced by

Table 1 Interassay Variation of Monosaccharide Analysis of rAHF

Assay	Fuc	Gal	Man	Gal.Am	Glu.Am	Sialic acid
1	8.18	29.53	48.73	11.43	44.73	25.53
2	7.12	30.74	42.14	10.36	49.61	25.93
3	7.07	35.45	39.74	11.27	47.54	23.19
Mean	7.46	31.90	43.54	11.02	47.28	24.88
STD	0.51	2.55	3.80	0.47	1.99	1.20
%CV	6.80	7.80	8.70	4.30	4.20	4.80

Each assay was performed under identical conditions in triplicate and on a single lot of rAHF. The protein used for each assay was 0.6 mg. The concentration of each sugar was estimated from the standard sugar curve constructed just before the analysis of glycoprotein. These three assays were carried out on different days.

Table 2 Intraassay Variation of Monosaccharide Analysis of rAHF

	%CV					
Assay	Fuc	Gal	Man	Gal.Am	Glu.Am	Sialic acid
1	3.22	3.06	1.44	3.93	6.53	3.44
2	7.88	8.06	3.39	5.52	1.92	1.84
3	5.08	4.74	3.95	7.63	6.53	6.06

The intraassay variation was calculated in terms of percent coefficient of variation in concentration (nmol/mL of protein) of each sugar from three runs of each assay. The injection volume was kept constant throughout the study. The amino and neutral sugar hydrolysates were analyzed on one system, whereas the sialic acid hydrolysates were analyzed on a different system.

different cell lines have different sugar chains, and that there are differences between plasma-derived FVIII and recombinant FVIII expressed in BHK cells. The signature of glycans in rAHF can be used to determine lot-to-lot consistency in the manufacture of this therapeutic glycoprotein. The present study on monosaccharide analysis of rAHF expressed in CHO cells suggests that the concentrations of Gal and Man are relatively high. This is an indication of the presence of lactosamine (complex), bi-, tri-, and tetraantennary N-linked oligosaccharides in rAHF. These results are fur-

Table 3 Comparison of Monosaccharide Assay Results by Two Analysts

Sugar	Analyst no. 1	Analyst no. 2	p
Fuc	7.46	7.69	0.75
Gal	31.90	33.02	0.72
Man	43.54	42.78	0.87
Gal.Am	11.02	10.36	0.13
Glu.Am	47.28	49.42	0.48
Sialic acid	24.88	23.21	0.18

Two analysts carried out three independent assays, and each assay was analyzed in triplicate. The mean value of three assays of each analyst is presented. The concentration of each sugar is expressed in terms of nmol/mL of protein. The statistical significant differences between two analysts' data were estimated from student t-test. The difference between two analysts' data is significant if the p value is less than 0.05.

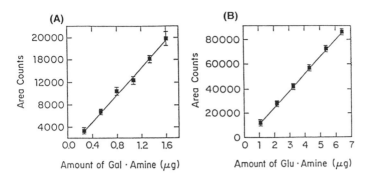

Figure 6 Standard curves of the amino sugars ((A) galactosamine and (B) glucosamine).

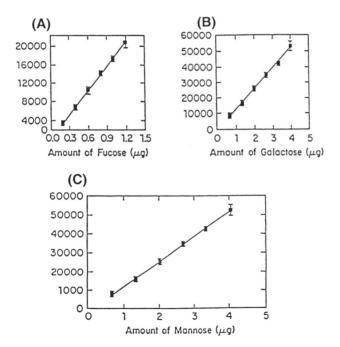

Figure 7 Standard curve of the neutral sugars ((A) fucose, (B) galactose, and (C) mannose).

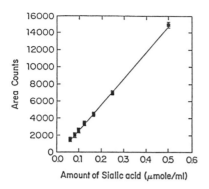

Amount of Sialic acid (μmole/ml)

Figure 8 Standard curve of anionic sugar (sialic acid).

ther supported by the presence of approximately equal amounts of Gal and Man. Galactosamine was also detected in rAHF, which indicates the presence of *O*-linked oligosaccharides. The ratio of sialic acid to Gal is an indication of the number of oligo-chains that terminate with sialic acid [21]. The present study suggests that most of the sugar chains in rAHF are

Table 4 Statistical Evaluation of Retention Times of Standard Sugars and Sugars of rAHF

Sugar	Percent difference in retention time	$P*$
Fuc	2.90	0.50
Gal	3.70	0.70
Man	3.50	0.70
Gal.Am	1.50	0.80
Glu.Am	0.50	0.40
Sialic acid	0.50	0.05

Standard sugars and glycoprotein were analyzed under identical conditions. The elution of sugars was monitored using RID. Sialic acid was the only analyte in the chromatogram, and retention time was also confirmed on the basis of spiking experiments. The sialic acid of the protein co-eluted with standard sialic acid. The p (paired t-test) values were calculated at the 95% confidence level. The difference in retention times is significant if the p value is less than 0.05.

sialylated. The complete sequence of *N*- and *O*-linked oligosaccharides of different lots of rAHF will provide another measure of the consistency of glycosylation for this recombinant glycoprotein.

IV. CONCLUSION

The main goal of this work was to develop an assay to monitor the consistency of gycosylation in different preparations of a recombinant glycoprotein. We have developed monosaccharide analysis of rAHF using HPAEC with RI detection. The results described here demonstrate that rAHF contains six sugars, Fuc, Gal, Man, Glu.Am, Gal.Am, and sialic acid. They also indicate the presence of *N*- and *O*-linked oligosaccharides in rAHF and that most of the termini are sialylated.

ACKNOWLEDGMENTS

The authors wish to thank Joan Brandt, Director, Quality Control, for her encouragement and support during the development of this assay and thanks to the entire staff of Test Technology for invaluable contributions. The authors also wish to express their gratitude to Cynthia H. Sarnoski, Jim Finn, and the QC staff of Genetics Institute for reviewing the entire chapter and giving constructive suggestions. We also want to thank Nancy Banks for helping in the preparation of this manuscript.

ABBREVIATIONS

AHF	Anti-Hemophilic Factor
BHK	Baby hamster kidney
CHO	Chinese hamster ovary
FVIII	factor VIII
Fuc	fucose
Gal	galactose
Gal.Am	galactosamine
GE	gel electrophoresis
Glu.Am	glucosamine
HPAEC	high pH anion-exchange chromatography
HPCE	high performance capillary electrophoresis
HPLC	high performance liquid chromatography
Man	mannose
MS	mass spectrometry
MW	molecular weight
PAD	pulsed amperometric detection

PBS	phosphate-buffered saline
rAHF	recombinant AHF
R Cell	reference cell
RI	refractive index
RID	refractive index detector

REFERENCES

1. Stanley, P. (1992). *Glycobiology* 2, 99–107.
2. Rudd, M. P., Woods, J. R., Wormald, R. M., Opdenakker, G., Downing, K. A., Campbellm, D. I., and Dwek, A. R. (1995). *Biochim. Biophys. Acta* 1248, 1–10.
3. Rademacher, T. W., Parekh, R. B., and Dwek, A. R. (1988). *Ann. Rev. Biochem.* 57, 785–838.
4. Birch, J. R., and Froud, S. J. (1994). *Biologicals* 22, 127–133.
5. Cumming, D. A. (1991). *Glycobiology* 1, 115–130.
6. Andersen, C. D., and Goochee, F. C. (1994). *Curr. Opin. Biotechnol.* 5, 546–549.
7. Watson, E., Shah, B., Leiderman, L., Hsu, Y.-R., Karkare, S., Lu, H. S., and Lin, F. K. (1994). *Biotechnol. Prog.* 10, 39–44.
8. Takeuchi, M., and Kobata, A. (1991). *Glycobiology* 1, 337–346.
9. Toole, J. J., Knopf, J. L., Wozney, J. M., Sultzman, L. A., Buecker, J. L., Pittman, D. D., Kaufman, R. J., Brown, E., Shoemaker, C., Orr, E. C., Amphlett, G. W., Foster, W. B., Coe, L. M., Knutson, G. J., Fass, D. N., and Hewick, R. M. (1984). *Nature* 312, 342–347.
10. Hardy, M. R. (1989). *Meth. Enzymol.* 179, 76–82.
11. Hardy, M. R., and Townsend, R. R. (1994). *Meth. Enzymol.* 230, 208–225.
12. Townsend, R. R. (1995). In El Rassi, Z., ed., *Carbohydrate Analysis: High Performance Liquid Chromatography and Capillary Electrophoresis*. Elsevier, New York, pp. 181–209.
13. Manzi, E. A., Diaz, S., and Varki, A. (1990). *Anal. Biochem.* 188, 20–32.
14. Honda, S., Iwase, S., Makino, A., and Fujiwara, S. (1989). *Anal. Biochem.* 176, 72–79.
15. El Rassi, Z., and Nashabeh, W. (1995). In El Rassi, Z., ed., *Carbohydrate Analysis: High Performance Liquid Chromatography and Capillary Electrophoresis*. Elsevier, New York, pp. 276–360.
16. Evangelista, A. R., Ming-Sun, Liu, and Chen A., Fu-Tai. (1995). *Anal. Chem.* 67, 2239–2245.
17. Settineri, C. A., and Burlingame, A. L. (1995). El Rassi, Z., ed., In *Carbohydrate Analysis: High Performance Liquid Chromatography and Capillary Electrophoresis*. Elsevier, New York, pp. 447–514.
18. Jackson, P. (1991). *Anal. Chem.* 196, 238–244.
19. Irene, R. M., Chuck, H., Robert, S., Ho, S., McAlister, S., Vincent, P., and Starr, C. M. (1995). *Trends Glycosci. Glycotechnol.* 7, 133–147.

20. Hironaka, T., Furukawa, K., Esmon, P. C., Yokota, T., Brown, J. E., Sawada, S., Fournel, A., Kato, M., Minaga, T., and Kobata, A. (1993). *Arch. Biochem. Biophys.* 307, 316–330.
21. Weitzhandler, M., Kadlecek, D., Avdalovic, N., Forte, G. J., Chow, D., and Townsend, R. R. (1993). *J. Biol. Chem.* 268, 5121–5130.

30
2-Aminobenzamide Labeling of Desialylated Oligosaccharides
A Sensitive Method for Monitoring Lot-to-Lot Consistency of Recombinant Glycoproteins

Kristina Kopp, Michael Schlüter, and Rolf G. Werner
Dr. Karl Thomae GmbH, a Company of Boehringer Ingelheim Pharma Germany, Biberach, Germany

I. INTRODUCTION

Glycosylation of proteins is one of the most common and important post-translational modifications found in eukaryotic secretory proteins. Carbohydrate moieties of glycoconjugates long were believed to carry no information and be devoid of biological function. Methods for analyzing carbohydrates were not available or at least labor-intensive. It has become clear only in recent years that basic glycosylation processes are both complicated and conserved during evolution, and that the type and extent of N-glycosylation often contributes to the physicochemical and recognition properties of glycoproteins [1,2]. Furthermore, it is now realized that the biological activity of glycoprotein hormones frequently depends on the attached N-linked oligosaccharides [3,4]. In recent years, the role of N-glycosylation in Chinese hamster ovary (CHO)–expressed human proteins has been studied extensively [5,6]. Some functional aspects of oligosaccharides in glycoproteins now appear to be universal (e.g., solubility [7] and circulatory lifetime [8,9]), but most other possible roles such as facilitating secretion, affecting biological activity, or increasing stability have to be investigated for each protein. CHO cells are the most commonly used eukaryotic cells for production of human proteins in biotechnology. They are adaptable for growth in suspension — which is inevitable for large scale production processes — and are robust to culture. Combined with the selection marker dihydrofolate reductase (DHFR), they represent a reliable expression system. Furthermore, they glycosylate proteins similarly to human cells. Although N-glycosylation is mainly governed by the type of host cell and the primary structure of the expressed glycoprotein, environmental factors also influence glycosylation [10–14]. Alterations in oligosaccharide structures occur either by affecting synthesis or by changing glycosidase activity after secretion [15,16]. Because changing the glycosylation pattern may have important consequences in glycoprotein pharmaceuticals, carbohydrate analysis is of utmost importance in product characterization to ensure consistent product quality.

Substantial progress in analyzing the fidelity of recombinant protein glycosylation has been made since the introduction of high pH anion-exchange chromatography with pulsed amperometric detection (HPAEC-PAD) in 1988 by Townsend and Hardy [17,18]. This technology bypasses refractive index (RI) detection or radioactive labeling of carbohydrates and has emerged as a valuable method in many laboratories. In biotechnology, HPAEC-PAD was the most suitable method for analyzing oligosaccharides. Charge differences of oligosaccharide structures are mainly based on differences in the number of terminal sialic acid residues. In addition to high performance liquid chromatographic (HPLC) mapping procedures, the sialylation status of glycoproteins can be determined by other methods such as isoelectric focusing and capillary electrophoresis. However, sialylation can conceal differences in oligosaccharide structures, which become more apparent after desialylation. To investigate the biosynthesis of *N*-linked oligosaccharides under different fermentation conditions, we expressed human glycoproteins in CHO cells and compared their glycosylation patterns using HPAEC-PAD fingerprinting of the desialylated oligosaccharides. We found that HPAEC was not optimal for analysis of typical CHO-derived glycosylation patterns. We then investigated an alternative tool to analyze the typical CHO oligosaccharide structures in more detail. Here, we report the use of an improved HPLC mapping method for 2-aminobenzamide (2-AB)-labeled, desialylated oligosaccharides.

II. MATERIALS AND METHODS

The strategy for analyzing glycans from recombinant glycoproteins is summarized in Fig. 1.

A. Preparation of Desialylated Oligosaccharides

Glycoprotein solutions were dialyzed overnight against 8 M urea, 0.3 M Tris, pH 8.6, containing 0.032 M EDTA. After determination of the sample volume, reduction with di-*thio*-threitol (10 mM) and carboxymethylation with iodoacetic acid (25 mM) was performed. The alkylation process was stopped by adding di-*thio*-threitol (50 mM). Reduced and carboxymethy-

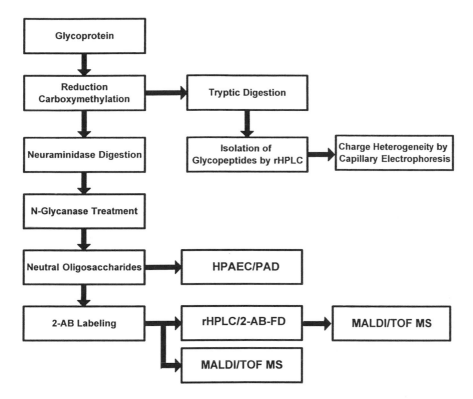

Figure 1 Flow chart to investigate N-linked oligosaccharides derived from recombinant glycoproteins.

lated protein samples were dialyzed against 0.05 M sodium acetate buffer, pH 5.5, containing 0.154 M NaCl and 9 mM CaCl$_2$ (6 h, room temperature). The glycoprotein was then incubated with neuraminidase (10 mU enzyme/mg protein, overnight at 37°C). After a second dialysis against a 0.05 M sodium acetate solution, pH 7.6, containing 0.154 M NaCl, (6 h, room temperature), the samples were incubated with N-glycosidase F (2 U enzyme/mg protein, 37°C, overnight). The protein was then precipitated by adding 3 volumes of ethanol per volume of sample and maintained at −20°C for 1 h. The glycan-containing supernatant was dried and dissolved in deionized water.

B. Labeling with 2-Aminobenzamide

The samples were labeled as described by the manufacturer of the labeling kit (Oxford GlycoSystems, Abingdon, United Kingdom). Glacial acetic acid

(150 µL) was added to 350 µL of dimethyl sulfoxide (DMSO) and mixed. This mixture (100 µL) was added to the lyophilized 2-aminobenzamide and mixed. The entire volume was then added to the lyophilized sodium cyanoborohydride and mixed vigorously. In the labeling step, 5 µL of this reagent mixture was added to 10 nmol of isolated glycans, mixed, incubated for 45 min at 75°C, and dissolved in 95 µL of deionized water. The labeled glycans were desalted by solid phase extraction with an aminopropyl cartridge containing 360 mg of packing material. After equilibration of the cartridge with 6.8 mL acetonitrile, the sample was loaded in 100 µL and washed with 20 mL of acetonitrile. The purified glycans were eluted with 4 mL of 50 mM ammonium acetate, pH 6.0. Samples were dried and reconstituted in deionized water. Based on an average occupancy of 2.5 mol of oligosaccharide/mol of tissue plasminogen activator (t-PA), approximately 1.5 nmol of glycans were analyzed by reversed phase (RP)-HPLC.

C. Chromatographic Separation

Aminobenzamide derivatives of oligosaccharides were separated on an octadecylsilyl (ODS) Hypersil HPLC column (3 µm, 4 × 250 mm) at a flow rate of 0.7 mL/min. The column was equilibrated for 30 min with 50 mM ammonium acetate, pH 6.0 (buffer A). Elution was performed using 8% acetonitrile in 40 mM ammonium acetate, pH 6.0 (buffer B), using the following gradient: 5 min, 15% B; 75 min, 75% B; 84 min, 100% B; and 1 min isocratic at the last condition. The column was then reequilibrated with 100% A for 15 min. Fluorescence detection was performed with excitation at 330 nm and emission at 420 nm. Isolated oligosaccharides were identified using standard oligosaccharides (Oxford Glycosystems) and data from matrix-assisted laser desorption ionization (MALDI) analyses (Table 1).

D. Mass Determination of Collected Peaks

In semipreparative runs, oligosaccharides (15 nmol) from each glycoprotein were separated by RP-HPLC and individual glycan peaks were collected. The fractions were dried and dissolved in 50 µL of deionized water. The oligosaccharides were analyzed using MALDI with a Kratos Kompact MALDI III (Shimadzu, Duisburg, Germany) instrument. The matrix was 2,5-dihydroxybenzoic acid (DHB) (Fig. 2).

III. RESULTS AND DISCUSSION

Previously, we have used HPAEC-PAD for carbohydrate mapping, in research projects, as well as for quality control of recombinant proteins. However, with respect to the separation and quantitation of typical asialo-

Table 1 MALDI-Mass Spectrometry of Oligosaccharides Isolated from CHO-Expressed Tissue Plasminogen Activator (t-PA) or Soluble Intercellular Adhesion Molecule (sICAM)[a]

Peak no.	Structure	Adduct	Expected mass	Observed mass
1	Man5	M + Na + 2-AB	1362	1351
2	Man6	M + K + 2-AB	1540	1540
3	Man7	M + K + 2-AB	1702	1699
4	Man8	M + K + 2-AB	1865	1865
5	Man9	M + K + 2-AB	2027	2028
6	Hybrid	M + K + 2-AB	1743	1741
7	NA2	M + K + 2-AB	1785	1784
8	NA2F	M + K + 2-AB	1931	1931
9	NA2F + 1 lact. repeat	M + K + 2-AB	2296	2293
10	NA3	M + K + 2-AB	2150	2150
11	NA3F	M + K + 2-AB	2296	2294
12	NA3F + 1 lact. repeat	M + K + 2-AB	2261	2261
13	NA3F + 2 lact. repeats	M + K + 2-AB	3027	3025
14	NA4	M + K + 2-AB	2515	2514
15	NA4F	M + K + 2-AB	2661	2658
16	NA4F + 1 lact. repeat	M + K + 2-AB	3027	3024
17	NA4F + 2 lact. repeats	M + K + 2-AB	3392	3389

[a]The peak fractions that were analyzed are shown in Figs. 4 and 5, respectively.

CHO-derived oligosaccharides (bi-, tri-, and tetraantennary, partially fucosylated complex-type oligosaccharides; hybrid and high-mannose-type oligosaccharides), the method has some disadvantages. Different classes of glycans (e.g., high-mannose and complex-type structures) co-eluted, hampering fractionation for subsequent analyses as well as a quantitation. On the other hand, 2-AB labeling and RP-HPLC with fluorescent detection (FD) resulted in improved resolution, facilitating rechromatography and mass determination of single peaks (Fig. 2). In general, the new mapping procedure clearly separated high-mannose and hybrid structures from complex-type oligosaccharides (Fig. 3A). A 10-fold amplification of the fluorescence signals (Fig. 3B) shows the large number of baseline separated structures. For the complex-type glycans of both t-PA and sICAM, nonfucosylated structures are well separated from their fucosylated counterparts, resulting in a more complex HPLC profile than observed using HPAEC-PAD (compare A and B in Fig. 4 and Fig. 5). All the major peaks found in HPAEC-PAD also appear in the RP-HPLC/2-AB-FD profile, but additional structures were observed. The direct comparison of the oligosaccha-

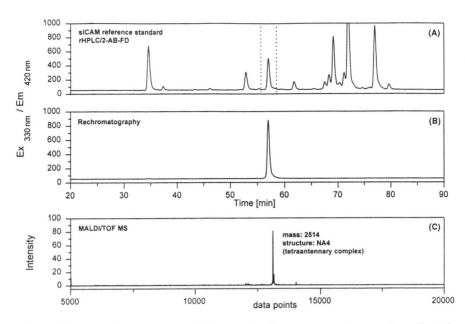

Figure 2 Mass determination of oligosaccharide structures. A manually collected peak (A) was characterized by rechromatography (B) and MALDI time-of-flight (TOF) mass spectrometry (C).

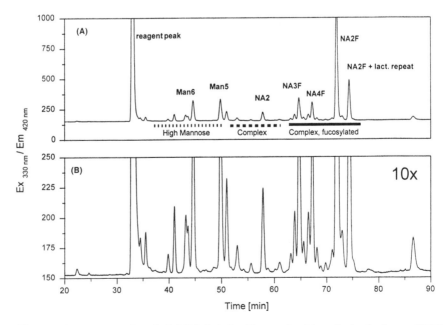

Figure 3 RP-HPLC of 2-AB-labeled oligosaccharides derived from CHO-expressed t-PA.

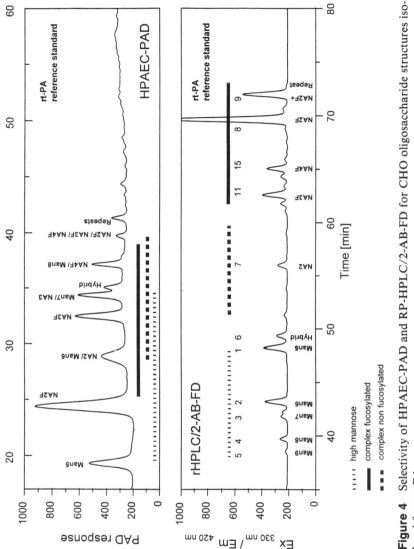

Figure 4 Selectivity of HPAEC-PAD and RP-HPLC/2-AB-FD for CHO oligosaccharide structures isolated from t-PA.

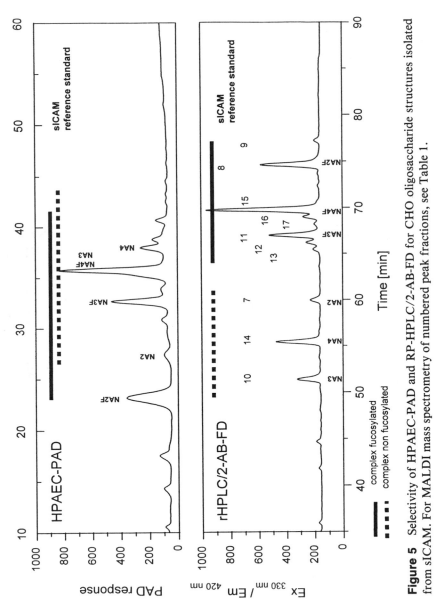

Figure 5 Selectivity of HPAEC-PAD and RP-HPLC/2-AB-FD for CHO oligosaccharide structures isolated from sICAM. For MALDI mass spectrometry of numbered peak fractions, see Table 1.

ride mapping methods revealed a higher resolution for the fluorescent method and a 3- to 5-fold higher sensitivity for both t-PA and sICAM. For HPAEC, 3–5 nmol of oligosaccharide (depending on sample preparation and detector sensitivity) was needed to obtain a profile without major baseline drifts. Using RP-HPLC/2-AB-FD, 1.5 nmol of labeled glycans was sufficient for the detection of even minor structures, with a very low baseline noise (Fig. 3). For t-PA, the structures identified agreed with those reported by Spellman et al. in 1989 [19]. The problem of differential electrochemical response factors with HPAEC [17] does not occur with fluorescence detection. Thus, the relative proportions of oligosaccharide structures can be deduced from peak areas.

CHO cells expressing t-PA and sICAM were cultivated in small scale experiments (2 L) under different fermentation conditions. Changes in oligosaccharide microheterogeneity resulting from different cell parameters (clone and age), fermentation parameters, and media conditions were investigated. All glycoprotein variants were isolated using the same purification procedures and compared with corresponding reference materials produced in large scale (2000–10,000 L). Milligram quantities of all glycoprotein vari-

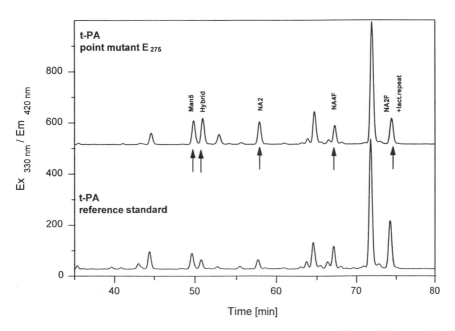

Figure 6 RP-HPLC of 2-AB-labeled oligosaccharides derived from different t-PA transfected cell clones.

ants allowed extensive protein and carbohydrate analysis, including oligosaccharide mapping. Among the cell parameters investigated, a clone expressing a t-PA point mutant influenced the synthesized oligosaccharides
only slightly (Fig. 6), and the age of the cell culture had little influence on
glycosylation (data not shown). Fermentation parameters such as scale and
cultivation mode (T-flask, spinner, or fermentor) did not change the oligosaccharide profiles of the two CHO-expressed proteins (Fig. 7). Two media
conditions affected glycosylation. The initial ammonia concentration in
the production medium (Fig. 8) as well as butyrate (Fig. 9) influenced
glycosylation patterns. These findings confirm data from earlier studies
with other glycoproteins [20,21]. The mechanism by which ammonia affects
glycosylation was not investigated in this study. Data from the literature
suggest, among other things, an accumulation in acidic organelles such as
lysosomes and Golgi, resulting in an increasing pH [15,22]. The extent
of apparent change in glycosylation, however, varied in our studies from
glycoprotein to glycoprotein, suggesting protein-dependent stability of glycosylation patterns. Another relevant parameter was the process time of

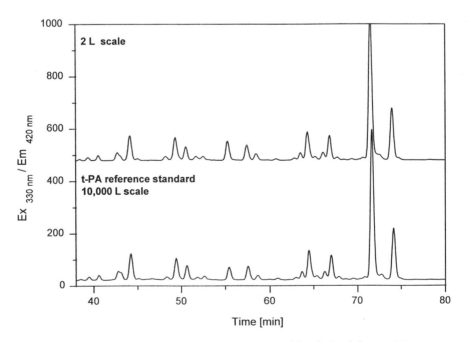

Figure 7   RP-HPLC of 2-AB-labeled oligosaccharides derived from t-PA manufactured on different scales.

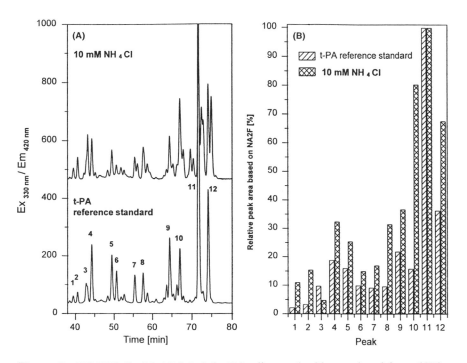

Figure 8 RP-HPLC of 2-AB-labeled t-PA oligosaccharides produced from CHO cells under an increased ammonia concentration. A 10 mM initial ammonia concentration influenced the glycosylation pattern of t-PA significantly (A). The peak areas were quantified in (B).

fermentation (Fig. 10). This factor has also been reported to influence IFN-γ product quality [23]. Considering the changing conditions during a batch culture, this finding is not surprising. With increasing process time, metabolites accumulate, viability of cells in batch cultures decreases, and protease or glycosidase activity may increase. Nevertheless, batch cultures are still preferable for large scale cell cultivation with regard to ease of handling (continuous processes require large reservoirs for both fresh media and harvest; moreover, the risk of contamination is higher). In this way, glycosylation patterns of recombinant products can be used as indicators of process control. The lot-to-lot consistency of different t-PA lots manufactured in large scale is shown in Fig. 11, indicating a reproducible process and consistent product quality.

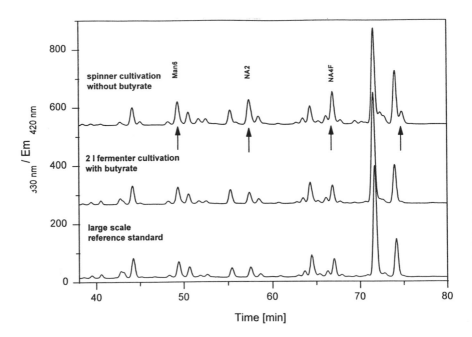

Figure 9 Influence of butyrate on the glycosylation pattern of t-PA.

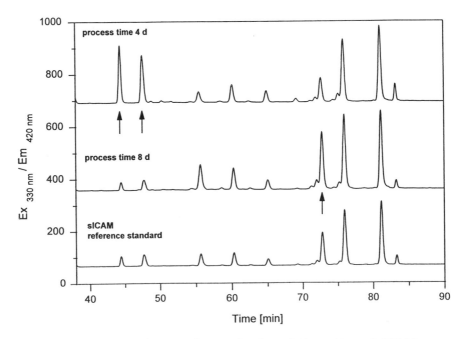

Figure 10 Influence of process time on the glycosylation pattern of sICAM.

486

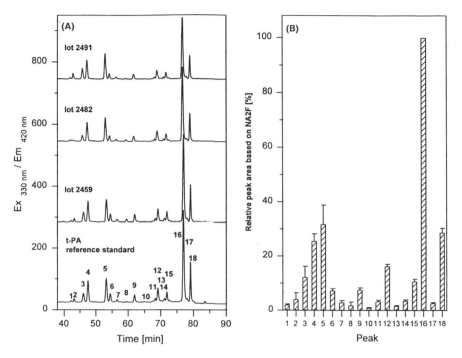

Figure 11   Lot-to-lot consistency of different t-PA production lots (A) and quantitative evaluation of peak areas (B).

IV.   SUMMARY AND CONCLUSIONS

The desialylated oligosaccharides from t-PA and sICAM expressed in CHO cells exhibited a protein-specific glycosylation pattern. Regardless of variations in fermentation conditions, glycoproteins tended to display the same oligosaccharide pattern. Most of the investigated fermentation parameters had little or no demonstrable influence on the synthesis of CHO cell asialo-oligosaccharides. From the resolution and reproducibility of the described mapping method, only small quantitative changes in oligosaccharide structures were observed. New oligosaccharide structures apparently did not appear due to varied fermentation conditions. In contrast, investigations with other proteins expressed in CHO cells have displayed different oligosaccharide profiles under such conditions. Therefore, the fidelity of glycosylation in response to process variations must be investigated on a case-by-case basis. The described HPLC mapping procedure for 2-AB-labeled asialo N-glycans can detect even small changes in protein glycosylation patterns. An important feature of the 2-AB-mapping method, in this respect, is the resolution of different structures that co-eluted using HPAEC.

ACKNOWLEDGMENTS

The authors are grateful to Iris Eisenmann for excellent technical assistance and chromatographic analyses, and to Otto Romer for MALDI-MS analyses.

ABBREVIATIONS

2-AB	2-aminobenzamide
CHO	Chinese hamster ovary
DHB	2,5-dihydroxybenzoic acid
DHFR	dihydrofolate reductase
DMSO	dimethyl sulfoxide
HPAEC-PAD	high pH anion-exchange chromatography with pulsed amperometric detection
HPLC	high performance liquid chromatography
Hybrid	$GalMan_4GlcNAc_3$
IFN-γ	interferon gamma
MALDI-MS	matrix-assisted laser desorption ionization mass spectrometry
Man5	$Man_5GlcNAc_2$
Man6	$Man_6GlcNAc_2$
Man7	$Man_7GlcNAc_2$
Man8	$Man_8GlcNAc_2$
Man 9	$Man_9GlcNAc_2$
NA2	$Gal_2Man_3GlcNAc_4$
NA2F	$Gal_2Man_3GlcNAc_4Fuc$
NA2F + 1 lact. repeat	$Gal_3Man_3GlcNAc_5Fuc$
NA3	$Gal_3Man_3GlcNAc_5$
NA3F	$Gal_3Man_3GlcNAc_5Fuc$
NA3F + 1 lact. repeat	$Gal_4Man_3GlcNAc_6Fuc$
NA3F + 2 lact. repeats	$Gal_5Man_3GlcNAc_7Fuc$
NA4	$Gal_4Man_3GlcNAc_6$
NA4F	$Gal_4Man_3GlcNac_6Fuc$
NA4F + 1 lact. repeat	$Gal_5Man_3GlcNAc_7Fuc$
NA4F + 2 lact. repeats	$Gal_6Man_3GlcNAc_8Fuc$
ODS	octadecylsilyl
RI	refractive index
RP-HPLC(2-AB-FD)	reversed phase high performance liquid chromatography (with 2-AB-fluorescence detection)
sICAM	soluble intercellular adhesion molecule
TOF	time of flight
t-PA	tissue plasminogen activator

REFERENCES

1. Montreuil, J. (1975). *Pure Appl. Chem.* 42, 431–477.
2. Montreuil, J. (1980). *Adv. Carbohydr. Chem. Biochem.* 37, 157–223.
3. Tsuda, E., Kawanishi, G., Ueda, M., Masuda, S., and Sasaki, R. (1990). *Eur. J. Biochem.* 188, 405–411.
4. Wölle, J., Jansen, H., Smith, L. C., and Chan, L. (1993). *J. Lip. Res.* 34, 2169–2176.
5. Parekh, R. B., Dwek R. A., Rudd, P. M., Thomas, J. R., Rademacher, T. W., Warren, T., Wun, T.-C., Hebert, B., Reitz, B., Palmier, M., Ramabhadran, T., and Tiemeier, D. C. (1989). *Biochemistry* 28, 7670–7679.
6. Elbein, A. D. (1991). *Trends Biotechnol.* 9, 346–352.
7. Berman, P. W., and Lasky, L. A. (1985). *Trends Biotechnol.* 3, 51–53.
8. Steer, C. F., and Ashwell, G. (1986). *Prog. Liver Dis.* 8, 99–123.
9. Takeuchi, M., Inoue, N., Strickland, T. W., Kubota, M., Wada, M., Shimizu, R., Hoshi, S., Kozutsumi, H., Takasaki, S., and Kobata, A. (1989). *Proc. Natl. Acad Sci. USA* 86, 7819–7822.
10. Goochee, C. F., and Monica, T. (1990). *Biotechnology* 8, 421–427.
11. Goochee, C. F., Gramer, M. J., Andersen, D. C., Bahr, J. B., and Rasmussen, J. R. (1991). *Biotechnology* 9, 1347–1355.
12. Gawlitzek, M., Conradt, H. S., and Wagner, R. (1995). *Biotechnol. Bioeng.* 467, 536–544.
13. Borys, M. C., Linzer, D. I. H., and Papoutsakis, E. T. (1994). *Biotechnol. Bioeng.* 43, 505–514.
14. Borys, M. C., Linzer, D. I. H., and Papoutsakis, E. T. (1993). *Biotechnology* 11, 720–724.
15. Gramer, M. J., and Goochee, C. F. (1993). *Biotechnol. Prog.* 9, 366–373.
16. Gramer, M. J., Schaffer, D. V., Sliwkowski, M. B., and Goochee, C. (1994). *Glycobiology* 4, 611–616.
17. Hardy, M. R., and Townsend, R. R. (1988). *Proc. Natl. Acad. Sci. USA* 85, 3289–3293.
18. Townsend, R. R., and Hardy, M. R. (1991). *Glycobiology* 1, 139–147.
19. Spellmann, M. W., Basa, L. J., Leonard, C. K., Chakel, J. A., O'Connor, J. V., Wilson, S., and van Halbeek, H. (1989). *J. Biol. Chem.* 264, 14100–14111.
20. Thorens, B. T., and Vassalli, P. (1986). *Nature* 321, 618–620.
21. Fishman, P. H., Bradley, R. M., and Henneberry, R. C. (1976). *Arch. Biochem. Biophys.* 172, 618–626.
22. Dean, R. T., Jessup, W., and Roberts, C. R. (1984). *Biochem. J.* 217, 27–40.
23. Curling, E. M., Hayter, P. M., Baines, A. J., Bull, A. T., Gull, K., Strange, P. G., and Jenkins, N. (1990). *Biochem. J.* 272, 333–337.

31

Characterization of Subnanomolar Amounts of *N*-Glycans by 2-Aminobenzamide Labeling, Matrix-Assisted Laser Desorption Ionization Time-of-Flight Mass Spectrometry, and Computer-Assisted Sequence Analysis

Rolf Nuck and Martin Gohlke
Free University of Berlin, Berlin-Dahlem, Germany

I. INTRODUCTION

Within the last decade interest in the structure of oligosaccharide moieties has increased along with a greater appreciation of their biological function (for review, see Ref. 1). In glycoproteins produced by biotechnology, glycosylation may be of importance for antigenicity [2], clearance [3], and biological activity [4,5]. Because the glycosylation is influenced by the type of cell line [6] and the culture conditions [7], structural analysis of glycans and control of glycosylation is important for economical and pharmaceutical reasons.

If sufficient material (50–500 nmol) is available, two-dimensional nuclear magnetic resonance analysis is the method of choice to determine composition, steric arrangement and sequence of oligosaccharides or glycopeptides [8]. However, because only a limited amount of material from biological sources is usually available, more sensitive techniques must be applied to the characterization of *N*-glycans. Usually, *N*-glycans are released from glycoproteins by chemical hydrazinolysis [9] or enzymatic hydrolysis [10] and are obtained as a complex mixture of heterogeneous oligosaccharides. Because most of the analytical methods require homogeneous glycans, in vivo labeling [11] or, alternatively, a sensitive radioactive [12] or fluorescent labeling [13] of the released carbohydrate chains is necessary to detect oligosaccharides during the different chromatographic purification steps.

We herein describe the application of some recently developed sensitive techniques that allow the characterization of subnanomolar amounts (50–200 pmol) of N-glycans derived from glycoproteins. The described strategy (Scheme 1) includes the nonselective fluorescent labeling of the enzymatically released N-glycans with 2-aminobenzamide, its purification by a three-dimensional separation technique, and characterization of the homogeneous glycans by size determination, matrix-assisted laser desorption ionization time-of-flight (MALDI-TOF) mass spectrometry and computer-assisted sequence analysis.

Glycoprotein

▼

Deglycosylation (Hydrazine, Trypsin/PNGase F)

▼

N-Linked Oligosaccharides

▼

Fluorescent Labelling (2-Aminobenzamide)

Anion Exchange HPLC (GlycoSep C)

▼

Neutral and Acidic 2-AB Glycans (N, A_1-A_X)

▼

Sialidase Digestion (*Arthrobacter ureafaciens*)

▼

Neutral 2-AB Glycans (N, A_1N-A_XN)

▼

Aminophase HPLC (APS2 Hypersil)

Reversed Phase HPLC (ODS Hypersil)

Size Exclusion Chromatography (Bio-Gel P-4)

▼

Separated 2-AB Glycans (A_1N_Y-A_XN_Z)

▼

Analysis

▼

Sizing (< 10 pmol)

MALDI TOF Mass Spectrometry (< 10 pmol)

RAAM 2000 Sequence Analysis (100-200 pmol)

Scheme 1   Preparation, labeling, separation, and analysis of N-glycans.

The described method is fast and allows a sufficient characterization of the N-glycosylation using a minimum of material. A further characterization by other techniques such as analysis of site-specific glycosylation [14], methylation analysis [15], or high pH anion-exchange chromatography with pulsed amperometric detection (HPAEC-PAD) [16] may also be performed, if the amount of glycans available is between 0.5 and 1.0 nmol.

II. MATERIALS AND METHODS

A. Release of N-Glycans

N-Glycans were released from proteins by hydrazinolysis [17] or enzymic digestion of tryptic peptides with PNGase F (5 mU/mL; Boehringer Mannheim, Mannheim, Germany) in 0.25 M sodium phosphate buffer, pH 8.6, for 20 h. Enzymatically hydrolyzed N-glycans were separated from enzyme by ultrafiltration using a Centricon-10 (Amicon, Witten, Germany) microconcentrator tube and purified from peptides and salt by reversed phase high performance liquid chromatography (HPLC) on a 4 mm × 25 cm UltraSep RP$_{18}$ column (6 μm, Bischoff, Lennberg, Germany) and gel filtration on a 1.6 × 80 cm Bio-Gel P$_2$ column (Bio-Rad, Munich, Germany) using water as eluent. Oligosaccharide-containing fractions, as proved by hydrolysis of an aliquot with 2 N trifluoroacetic acid for 4 h at 100°C and subsequent HPAEC-PAD [18], were pooled and evaporated to dryness in a vacuum centrifuge.

B. Fluorescent Labeling of N-Glycans with 2-Aminobezamide

Released N-glycans (5 to 10 nmol) were carefully concentrated from 10 μL of water to the bottom of a screw-capped microvial by evaporation at 30°C in a vacuum centrifuge. The oligosaccharides were fluorescently labeled by reductive amination with 2-aminobenzamide (2-AB) according to Bigge et al. [19] using a commercially available Signaling Kit (Oxford Glycosystems, Abingdon, England). Briefly, labeling of the oligosaccharides was performed with 10 μL of a reagent solution, containing 2-AB (250 μg) and sodium cyanoborohydride (500 μg), dissolved in dimethyl sulfoxide/acetic acid (7/4, by vol) by incubation for 2 h at 65°C in a heating oven in the dark. The reaction mixture was spotted (5 μL) onto paper strips (3 × 8 cm, Oxford Glycosystems) which were dried upright in a fume cupboard overnight. The labeled glycans were separated from unreacted 2-AB by paper chromatography in n-butanol/ethanol/water (4/1/1, by vol) for 1 h. The oligosaccharides were eluted from the origin in a Luer-locked syringe, equipped with a 0.45-μm HV-filter (Nihon Millipore, Tokyo, Japan), by incubation of the paper 3 times with 1 mL of water for 5 min followed by centrifugation for 5 min at 750 × g. The eluted oligosaccharides were evaporated to dryness, redissolved in 1 mL of water, and kept at −20°C in the dark until further use. Note: Working under illumination by fluorescent lights may cause a decrease of fluorescence in 2-AB-labeled glycans, even at 4°C. Therefore, if possible, all steps during the purification procedure should be performed to minimize exposure to fluorescent light, and the labeled oligosaccharides should be stored at −20°C in the dark.

C. Separation of Acidic N-Glycans by Anion-Exchange HPLC

Fluorescently labeled N-glycans were separated by charge on a 7.5 × 50 mm GlycoSep C column (5 μm, Oxford Glycosystems) equilibrated with water/acetonitrile (8/2, by vol) at a flow rate of 0.5 mL/min. After injection the column was eluted for 5 min with equilibration solvent followed by a linear gradient to 250 mM ammonium acetate buffer, pH 4.2, in acetonitrile (8/2, by vol) within 60 min. Fluorimetric detection of the oligosaccharides was employed at 330 nm excitation and 420 nm emission. Fractions containing the separated acidic glycans were pooled, desalted by gel filtration at 0.5 mL/min on a 0.5 × 20 cm Sephadex G_{25} Superfine column (Pharmacia, Freiburg, Germany), and treated with sialidase.

D. Sialidase Digestion

Sialylated N-glycans were incubated in 50 μL of 0.15 M ammonium acetate buffer, pH 5.0, with *Arthrobacter ureafaciens* sialidase (100 mU) for 18 h at 37°C. The reaction was terminated by heating to 95°C for 3 min. Oligosaccharides were separated from enzyme by ultrafiltration as described previously and from volatile buffer by repeated evaporation of the ultrafiltrate in vacuo.

E. Three-Dimensional Separation of Neutral N-Glycans

1. Aminophase HPLC

Desalted 2-AB-labeled neutral N-glycans were separated at 1.5 mL/min using a 4 mm × 25 cm APS 2-Hypersil column (3 μm, Life Sciences, Frankfurt/Main, Germany) equilibrated with acetonitrile. Acetonitrile (eluent 1) and 15 mM ammonium acetate buffer, pH 5.2 (eluent 2), were used as eluents. Oligosaccharides (1.5 nmol), dissolved in 30 μL of water, were introduced onto the column by an injection valve, carrying a 500-μL loop filled with 100% acetonitrile, and were eluted by a linear increase of eluent 2 to 20% over 10 min and then to 60% over 80 min.

2. Reversed Phase HPLC

2-AB labeled *N*-glycans, separated by aminophase HPLC, were desalted and purified from contaminating oligosaccharides on an 4 mm × 25 cm ODS-Hypersil column (3 μm, Life Sciences), equilibrated with water at a flow rate of 1 mL/min. After injection of a sample, the column was eluted with 5 mL of water, followed by a linear gradient to 15% (by vol) acetonitrile in water within 30 min. Purified *N*-glycans were collected, then immediately evaporated to dryness and kept at $-20°C$ until their further characterization by MALDI-TOF mass spectrometry or sizing on Bio-Gel P_4.

3. Gel Filtration and Sizing

Completely desalted 2-AB-labeled neutral *N*-glycans were applied to a 1 × 48 cm Bio-Gel P_4 column using a RAAM 2000 GlycoSequencer (Oxford Glycosystems) working at 55°C. (RAAM stands for reagent array analysis method.) Depending on the size of the glycan and resolution required, the oligosaccharides were eluted from the column at 100 μL/min ("RAAM" profile mode) or at 30 μL/min ("High Resolution" profile mode) using iron-free water (Oxford GlycoSystems, GlycoPure) as eluant. The size of *N*-glycans (measured in glucose units, GU) was determined by measurement of the hydrodynamic volume relative to a glucose oligomer mixture (corresponding to 150 μg of hydrolyzed dextran) used as an external or internal standard. Detection of fluorescently labeled oligosaccharides and of the unlabeled glucose oligomers was performed by dual measurement of the fluorescence and refraction index, respectively. Oligosaccharide-containing fractions were pooled, evaporated, and analyzed by MALDI-TOF mass spectrometry or sequencing as will be described.

F. RAAM Sequence Analysis of Purified 2-AB-Labeled *N*-Glycans

2-AB-labeled homogeneous (>95%) neutral *N*-glycans, as obtained by three-dimensional purification, were characterized by RAAM 2000 sequence analysis according to the sequencing method of radiolabeled oligosaccharides described by Edge et al. [20]. Briefly, 100 to 200 pmol of homogeneous *N*-glycans were divided into nine equal aliquots, which were separately digested in screw-capped microvials for 18 h at 37°C with 10 μL of nine different mixtures of exoglycosidases in a defined composition using the RAAM 2000 Neutral Sequencing Kit (Oxford Glycosystems). The reaction was terminated by the addition of 10 μL of pyridine, and the glycan fragments obtained were pooled and applied to a four-bed column containing three ion-exchange resins and one affinity resin according to the protocol of the manufacturer to remove enzymes and salts. The combined glycan

fragments were then dissolved in 80 μL of water, mixed with 20 μL of a
dextran hydrolysate (150 μg) and applied for automatic sequence analysis
on Bio-Gel P$_4$ using a RAAM 2000 GlycoSequencer. Computer-assisted
comparison of the coincidence of the elution profiles (signature) of the
digested N-glycan with the theoretical signatures of more than 200,000
N-glycans gave the most probable structure for each glycan.

G. MALDI-TOF Mass Spectrometry

Mass spectra were measured using a BIFLEX-MALDI-TOF mass spectrometer (Bruker, Bremen, Germany) in the reflectron and positive ionization mode. Five picomoles of 2-AB-labeled glycans were crystallized with 5
μg of 2,5-dihydroxybenzoic acid as a matrix in 1 μL acetonitrile/water (45/
55 by vol) on the target. The molecular masses of unknown glycans were
determined by calibrating the mass spectrometer with a mixture of reducing
glucose oligomers from dextran on the same target.

III. RESULTS AND DISCUSSION

Human α_1-acid glycoprotein (AGP), a well-characterized protein, was chosen as an example to investigate the suitability of three new tools: (1) fluorescent labeling with 2-AB; (2) matrix-assisted laser desorption ionization
mass spectrometry (MALDI-TOF MS) of 2-AB-labeled N-glycans; and (3)
computer-assisted sequence analysis by the reagent array analysis method
(RAAM) for the sensitive analysis of N-glycans.

N-Linked oligosaccharides were released from protein by hydrazinolysis or a two-step procedure consisting of proteolytic digestion of the protein with trypsin and enzymatic cleavage of the N-glycans with PNGase F.
In contrast, to chemical hydrazinolysis, which completely destroys the protein core, or enzymatic deglycosylation of the intact glycoprotein, which
often requires large amounts of enzyme (100 mU/mL) and the presence of
detergents [10], interfering with a subsequent sugar labeling, the release of
N-glycans from glycopeptides requires no detergent and low amounts of
enzyme ($<$5 mU/mL). The approach allows determination of site-specific
glycosylation by reversed phase HPLC in combination with enzymatic deglycosylation and amino acid sequence analysis [21].

The N-glycans corresponding to 20 μg of AGP were immediately
fluorescently labeled with 2-AB and separated into four acidic fractions,
A1 to A4, by anion-exchange HPLC on GlycoSep C (Fig. 1a). By sialidase
digestion and rechromatography of the digested pools A1N to A4N, it was
confirmed that the negative charge of the N-glycans in AGP was caused
completely by their content of sialic acid residues (Fig. 1b).

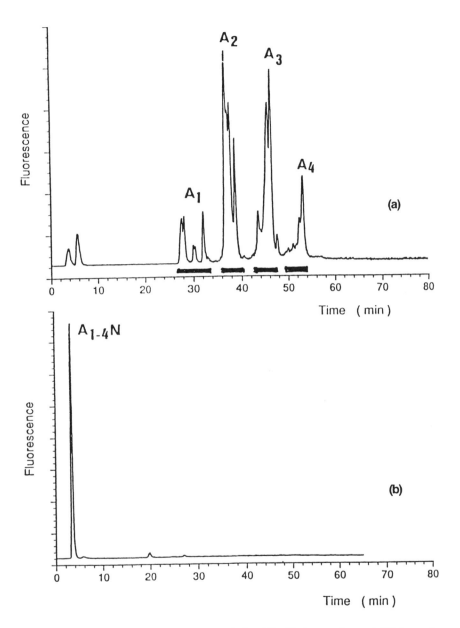

Figure 1 (a) Separation of 2-AB-labeled acidic *N*-glycans from AGP by anion-exchange HPLC on GlycoSep C (5 μm, 7.5 $\times$ 50 mm, Oxford Glycosystems) using an increasing linear gradient to 250 mM ammonium acetate, pH 4.2, in acetonitrile (8/2 by vol). (b) Acidic oligosaccharides in fractions A1 to A4 were digested with *Arthrobacter ureafaciens* sialidase and were applied to rechromatography on Glyco-Sep C. Fluorescence detection was employed at $\lambda_{exc} = 330$ nm and $\lambda_{em} = 420$ nm.

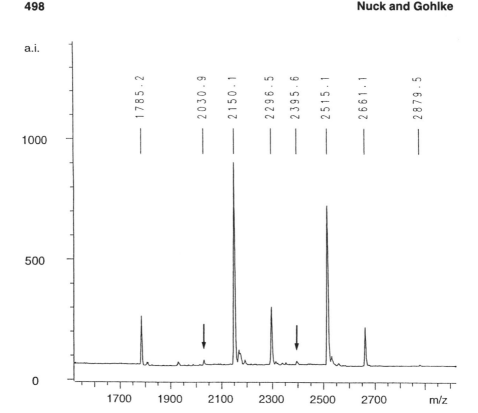

Figure 2 Improvement of the fluorescent labeling procedure of N-glycans by MALDI-TOF mass spectrometry. Desialylated N-glycans from AGP were labeled with 2-aminobenzamide for 2 h at 65°C, separated from excess label by paper chromatography and analyzed by MALDI TOF mass spectrometry. An incomplete labeling was recognized by increased signals at m/z 2030.9 and m/z 2395.6 (indicated by arrows), which corresponded to the unlabeled species of the major triantennary (m/z 2150.1) and tetraantennary (m/z 2515.1) 2-AB-labeled N-glycans in AGP. For mass spectrometry conditions, see the legend of Fig. 6.

For structure determination, oligosaccharides corresponding to 180 µg of AGP were first digested with *Arthrobacter ureafaciens* sialidase prior to fluorescent labeling with 2-AB, to avoid a partial loss of fluorescence during the desialylation procedure. The labeled neutral N-glycans were successfully separated from excess 2-AB by paper chromatography or reversed phase HPLC, and the completeness of the labeling procedure was confirmed by MALDI-TOF mass spectrometry (Fig. 2). Incomplete labeling, which is characterized by the occurrence of additional signals for the major components in the MALDI-TOF mass spectrum at a 120 Da lower mass

(Fig. 2), was avoided, by carefully concentrating oligosaccharides to the bottom of a microvial by reevaporation from 10 μL and not reducing reaction volume to less than 10 μL.

Because separation of the neutral *N*-glycans from AGP by direct size exclusion chromatography on Bio-Gel P₄ was found to be insufficient for obtaining homogeneous oligosaccharides (data not shown), the *N*-glycans were purified to homogeneity by a three-dimensional separation procedure prior to a further characterization. The neutral fluorescently labeled *N*-glycans were first separated into nine major peaks by amino phase HPLC (Fig. 3). Each of the oligosaccharide fractions was then desalted and separated from contaminating *N*-glycans by reversed phase HPLC on ODS-Hypersil (Fig. 4). Completely desalted *N*-glycans were further purified by size exclusion chromatography on Bio-Gel P₄ and characterized for their hydrodynamic volume by using a glucose oligomer mixture as an internal

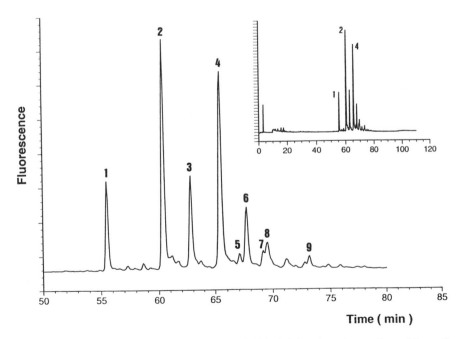

Figure 3 Separation of fluorescently labeled desialylated *N*-glycans from 25 μg of AGP by aminophase HPLC on APS 2-Hypersil (3 μm, 4 × 250 mm, Bischoff). Oligosaccharides were eluted by using an increasing gradient to 15 mM sodium phosphate, pH 5.2, in acetonitrile (6/4, by vol) at a flow rate of 1.5 mL/min. Fluorescence detection was employed at λ_{exc} = 330 nm and λ_{em} = 420 nm.

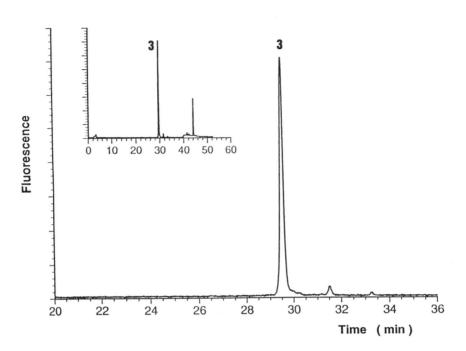

Figure 4 Reversed phase HPLC of fluorescently labeled neutral N-glycans. Glycan 3, as obtained from amino phase HPLC, was separated from salt and contaminating oligosaccharides by HPLC on ODS Hypersil (3 μm, 4 × 250 mm). N-Glycans were eluted with an increasing gradient of acetonitrile in water at a flow rate of 1 mL/min as described in Section II.

or external standard (Fig. 5 and Table 1). Complete desalting was necessary prior to sizing on Bio-Gel P_4, even though volatile buffers were used, because traces of salt caused the formation of spikes in the carbohydrate elution profile and may result in a size shifting of the investigated glycan.

The separated N-glycans were then characterized for their molecular mass by MALDI-TOF mass spectrometry in the reflection mode using 2, 5-dihydroxybenzoic acid as a matrix. In the positive ionization mode the oligosaccharides were observed as their adducts with sodium and potassium and a 120 Da increased mass, if compared with the corresponding unlabeled glycans (Fig. 6b). By calibration of the mass spectrometer with a glucose oligomer mixture from dextran (Fig. 6a), the deviation of the molecular mass measured was less than 1 Da compared with that calculated (Table 1). Under the conditions used, neither a loss of the fluorescent label nor fragmentation was observed. Therefore, MALDI-TOF mass spectrometry is not only a suitable tool to assess the completeness of the labeling reaction

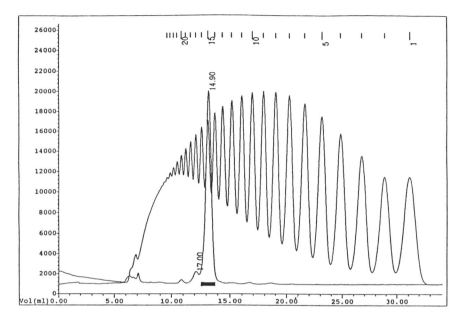

Figure 5 Fractionation and sizing of fluorescently labeled *N*-glycan 3 by size exclusion chromatography on Bio-Gel P$_4$ (10 × 480 mm). Prepurified and completely desalted *N*-glycans, as obtained by aminophase HPLC (Fig. 3) and reversed phase HPLC (Fig. 4), were applied to the RAAM 2000 Glycosequencer with a column temperature of 55°C at a flow rate of 100 μL/min. The size of *N*-glycans (measured in glucose units, GU) was determined by measurement of the hydrodynamic volume using a glucose oligomer mixture as an internal or external standard (see also Table 1).

but also allows a qualitative assay of the composition of the major components of complex glycan mixtures.

For a further characterization, purified *N*-glycans (100–200 pmol) were subjected to RAAM sequence analysis. Glycans were divided into nine aliquots and digested with a set of defined mixtures of exoglycosidases (RAAM Reagent Array, Oxford Glycosystems), containing nonspecific β-galactosidase, β-hexosaminidase, α-fucosidase, β(1-2)-specific hexosaminidase, and α1-3/4-specific fucosidase (Fig. 7a). By this method, according to the mixture of enzymes applied, a number of fragments, differing in size, was obtained for each *N*-glycan. As shown for glycan 3 (Fig. 7b), chromatography of combined fragments, performed on Bio-Gel P$_4$ using a RAAM 2000 Glycosequencer, yielded a unique profile (signature), characterized by the elution position and relative signal intensity of each fragment.

Table 1   Size, Molecular Mass, and Sialylation of 2-AB-Labeled *N*-glycans from AGP

Peak	Size (GU)[a]	M_r (exp.)[b]	M_r (theor.)[b]	Composition	Glycan type[c]	Sialylation[d]			
						A1	A2	A3	A4
1	11.3	1784.7	1784.3	Hex$_5$ HexNAc$_4$	Bi	+	+	–	–
2	14.2	2149.4	2150.0	Hex$_6$ HexNAc$_5$	Tri	+	+	+	–
3	14.9	2295.7	2296.1	Hex$_6$ HexNAc$_5$ dHex	Tri + 1 Fuc	+	+	+	–
4	16.4	2514.9	2515.3	Hex$_7$ HexNAc$_6$	Tetra	+	+	+	+
5	17.1	2661.5	2661.5	Hex$_7$ HexNAc$_6$ dHex	Tetra + 1 Fuc	+	+	+	+
6	17.2	2661.5	2661.5	Hex$_7$ HexNAc$_6$ dHex	Tetra + 1 Fuc	+	+	+	+
7	17.8	2807.5	2807.6	Hex$_7$ HexNAc$_6$ dHex$_2$	Tetra + 2 Fuc	+	+	+	+
8	19.6	2881.4	2880.7	Hex$_8$ HexNAc$_7$	Tetra + 1 Repeat	+	+	+	+
9	22.0	3246.0	3246.0	Hex$_9$ HexNAc$_8$	Tetra + 2 Repeat	–	+	+	+

[a]Size of *N*-glycans, given in glucose oligomer units (GU), as determined by gel filtration on Bio-Gel P$_4$.
[b]Molecular masses, as determined by MALDI-TOF mass spectrometry (M_r exp.) in comparison with the theoretical average masses (M_r theor.).
[c]Type of *N*-glycan as determined by RAAM sequence analysis.
[d]Sialylation of *N*-glycans as determined by anion-exchange HPLC, desialylation, and subsequent aminophase separation. Abbreviations: Hex = hexose, HexNAc = *N*-acetylhexosamine, dHex = deoxyhexose, Bi = biantennary *N*-glycan, Tri = triantennary *N*-glycan, Tetra = tetraantennary glycan, Fuc = fucose residue, Repeat = repeating unit (Galβ(1–4)GlcNAc).

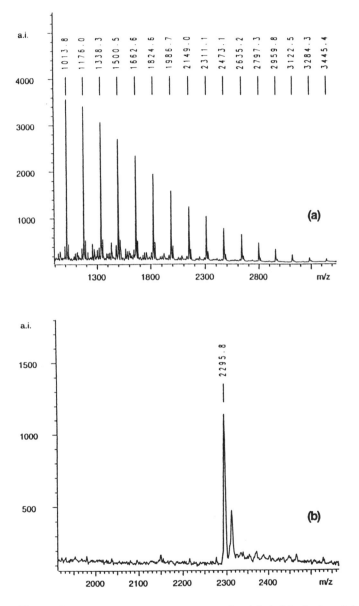

Figure 6 MALDI-TOF mass spectra of unlabeled hydrolyzed dextran (a) and the 2-AB-labeled fucosylated triantennary glycan 3 (b). Molecular masses of homogeneous 2-AB-labeled N-glycans (see Table 1) were determined in the reflectron and positive ionization mode, by using 2,5-dihydroxybenzoic acid as matrix and calibrating the mass spectrometer with a mixture of reducing glucose oligomers from dextran.

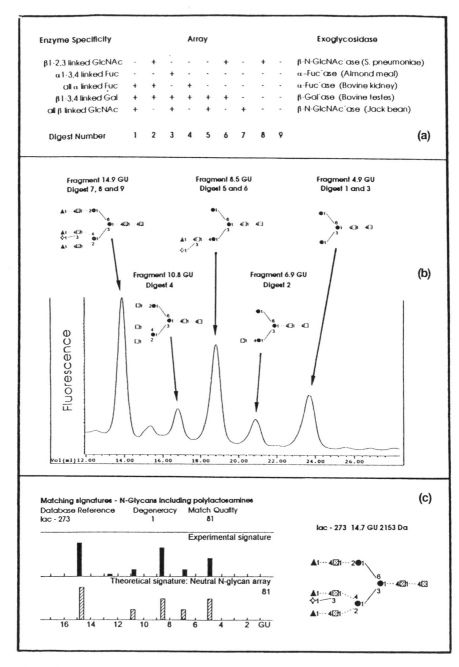

Figure 7 RAAM sequence analysis of glycan 3. Two hundred picomoles of 2-AB-labeled homogeneous glycan 3 were divided into nine equal aliquots, which were digested with nine different mixtures of exoglycosidases (a). The combined glycan fragments were characterized by automatic sequence analysis on Bio-Gel P₄ using a RAAM 2000 GlycoSequencer (b). Computer-assisted comparison of the elution profile (signature) of sequenced glycan 3 with the signatures of about 200,000 theoretical N-glycan structures yielded the structure of N-glycan 3 (c).

Computer-assisted comparative analysis of the signature (Fig. 7c) with more than 200,000 theoretical *N*-glycan structures revealed the most probable structures for *N*-glycans 1 to 9 as given in Fig. 8. The structures elucidated were in agreement with previous studies [22,23] or unpublished data [24] obtained by nuclear magnetic resonance (NMR) analysis. Our studies revealed that, compared with conventional stepwise exoglycosidase sequencing, RAAM sequence analysis was a rapid, sensitive, and convenient method for the structure elucidation of *N*-glycans. However, the results of RAAM sequencing must be carefully interpreted. Good matching signatures (>80%) require homogeneous *N*-glycans and almost complete digestion to a single glycan fragment in each of the nine digests. Results yielding lowered matching qualities (<75%) may be caused by minor contaminant

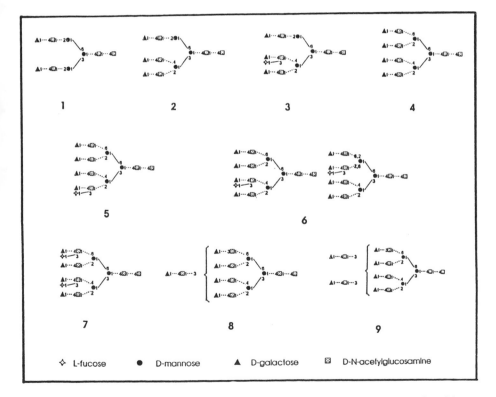

Figure 8 Proposed structures for the *N*-glycans 1 to 9 from AGP as analyzed by RAAM 2000 sequence analysis, size determination on Bio-GelP$_4$, and MALDI-TOF mass spectrometry.

glycans or incomplete digestion and may be improved by applying each digest separately to a sequencing run.

It should be noted that, by using the commercially available enzyme kit in the present formulation, the method is limited to N-glycans that are converted to the core structure (Man$_3$GlcNAcGlcNAc$_{2\text{-AB}}$) by enzymic digestion with a mixture of nonspecific β-galactosidase, β-hexosaminidase, and α-fucosidase. N-Glycans containing xylose residues, fucose $\alpha(1\text{-}2)$-linked or galactose $\alpha(1\text{-}3)$-linked to terminal galactose, must be converted by suitable exoglycosidases prior to RAAM sequence analysis. Moreover N-glycans containing type 1 chains (Gal$\beta(1\text{-}3)$GlcNAc) may not be distinguished from those containing type 2 chains (Gal$\beta(1\text{-}4)$GlcNAc). The quality of RAAM sequence analysis depends not only on the homogeneity ($>95\%$) of the investigated N-glycan, but on the correct enzyme concentration in the exoglycosidase mixtures. Therefore, the suitability of the enzymes to be used should be assessed by control sequencing of a suitable N-glycan of known structure, if possible a terminal fucosylated tetraantennary N-glycan such as glycan 6.

To investigate the degree of sialylation in the analyzed N-glycans, a study comprising anion-exchange HPLC (Fig. 1a), enzymatic desialylation of the four different sialylated glycan fractions A1 to A4, and a separate aminophase HPLC of the corresponding neutral oligosaccharide fractions A1N to A4N was performed. As summarized in Table 1, the data revealed that all oligosaccharides, including the four different N-glycans containing the Lewis[x] motif, were present as completely and as partially sialylated species.

IV. CONCLUSION

The methodology comprising fluorescent labeling, three-dimensional separation, MALDI-TOF mass spectrometry, and RAAM sequencing is a useful approach, which allows a fast and sensitive characterization of subnanomole amounts (100 pmol) of N-glycans. Current studies in our lab indicate that the described strategy may also be successfully applied to membrane glycoproteins, recombinant glycoproteins [25], or comparative studies on transformed cells and their normal counterparts.

ABBREVIATIONS

2-AB	2-aminobenzamide
AGP	α_1-acid glycoprotein
GU	glucose units

HPAEC-PAD	high pH anion-exchange chromatography with pulsed amperometric detection
HPLC	high performance liquid chromatography
MALDI-TOF	matrix-assisted laser desorption ionization time-of-flight
MS	mass spectrometry
NMR	nuclear magnetic resonance
RAAM	reagent array analysis method

REFERENCES

1. Lis, H., and Sharon, N. (1993). *Eur. J. Biochem.* 218, 1–27.
2. Hu, A., Cattaneo, R., Schwartz, S., and Norrby, E. (1994). *J. Gen. Virol.* 75 (Pt 5), 1043–1052.
3. Baenziger, J. U., Kumar, S., Brodbeck, R. M., Smith, P. L., and Beranek, M. C. (1992). *Proc. Natl. Acad. Sci. USA* 89, 334–338.
4. Dube, S., Fisher, J. W., and Powell, J. S. (1988). *J. Biol. Chem.* 263, 17516–17522.
5. Takeuchi, M., Inoue, N., Strickland, T. W., Kubota, M., Wada, M., Shimizu, R., Hoshi, S., Kozutsumi, H., Takasaki, S., and Kobata, A. (1989). *Proc. Natl. Acad. Sci. USA* 86, 7819–7822.
6. Parekh, R. B., Dwek, R. A., Thomas, J. R., Opdenakker, G., Rademacher, T. W., Wittwer, A. J., Howard, S. C., Nelson, R., Siegel, N. R., Jennings, M. G., Harakas, N. K., and Feder, J. (1989). *Biochemistry* 28, 7644–7662.
7. Gawlitzek, M., Valley, U., Nimtz, M., Wagner, R., and Conradt, H. S. (1995). *J. Biotechnol.* 42, 117–131.
8. Van Halbeek, H. (1994). *Meth. Enzymol.* 230, 132–168.
9. Patel, T., Bruce, J., Merry, A., Bigge, C., Wormald, M., Jaques, A., and Parekh, R. (1993). *Biochemistry* 32, 679–693.
10. Nuck, R., Zimmermann, M., Sauvageot, D., Josic, D., and Reutter, W. (1990). *Glycoconj. J.* 7, 279–286.
11. Nuck, R., Orthen, B., and Reutter,W. (1992). *Eur. J. Biochem.* 208, 669–676.
12. Takasaki, S., and Kobata, A. (1974). *J. Biochem. (Tokyo)* 76, 783–789.
13. Kuraya, N., and Hase, S. (1992). *J. Biochem. (Tokyo)* 112, 122–126.
14. Parekh, R. B., Tse, A. G., Dwek, R. A., Williams, A. F., and Rademacher T. W. (1987). *EMBO J.* 6, 1233–1244.
15. Geyer, R., Geyer, H., Kühnhardt, S., Mink, W., and Stirm, S. (1983). *Anal. Biochem.* 133, 197–207.
16. Hermentin, P., Witzel, R., Doenges, R., Bauer, R., Haupt, H., Patel, T., Parekh, R. B., and Brazel, D. (1992). *Anal. Biochem.* 206, 419–429.
17. Takasaki, S., Mizuochi, T., and Kobata, A. (1980). *Meth. Enzymol.* 83, 263–268.
18. Hardy, M. R., Townsend, R. R., and Lee, Y. C. (1988). *Anal. Biochem.* 170, 54–62.

19. Bigge, J. C., Patel, T. P., and Parekh, R. B. (1995). *Anal. Biochem.* 230, 229–238.

20. Edge, C. J., Rademacher, T. W., Wormald, M. R., Parekh, R. B., Butters, T. D., Wing, D. R., and Dwek, R. A. (1992). *Proc. Natl. Acad. Sci. USA* 89, 6338–6342.

21. Gohlke, M., Baude, G., Nuck, R., Grunow, D., Kannicht, C., Bringmann, P., Donner, P., and Reutter, W. (1996). *J. Biol. Chem.* 271, 7381–7386.

22. Fournet, B., Montreuil, J., Strecker, G., Dorland, L., Haverkamp, J., Vliegenthart, J. F. G., Binette, J. B., and Schmid, K. (1978). *Biochemistry* 17, 5206–5214.

23. Yoshima, H., Matsumoto, A., Mizuochi, T., Kawasaki, T., and Kobata, A. (1981). *J. Biol. Chem.* 256, 8476–8484.

24. Vliegenthart, J. F. G., personal communication.

25. Gohlke, M., Nuck, R., Kannicht, C., Grunow, D., Baude, G., Donner P., and Reutter, W. (1995). *Glycoconj. J.* 12, 415–416.

32

Isolation and Characterization of *N*-Linked Oligosaccharides from Yeast Glycoproteins

Robert B. Trimble
Wadsworth Center, New York State Department of Health, and State University of New York at Albany School of Public Health, Albany, New York

I. INTRODUCTION

Nearly all proteins trafficked through the secretory pathway, whether secreted from the cell or destined to become membrane- or extracellular matrix–associated resident components, are glycoproteins. The three major classes of protein-associated carbohydrates are found *N*-linked to specific asparagine residues in–Asn–Xaa–Thr/Ser–"sequons," *O*-linked to certain serine or threonine residues, or linked to the C terminus of some membrane proteins in the form of a glycosylphosphatidyl inositol, or "GPI," anchors. Although there are clearly differences in the structures and complexity in each of these three classes of glycans between the lower and more developed eukaryotes, there has been a remarkable conservation throughout the evolution of the fundamental cellular components involved in the biosynthesis of each type of glycosylation.

Numerous excellent reviews on the biosynthesis, processing, and biological roles of glycoprotein glycans have appeared. These have focused on the topology of protein glycosylation [1,2], the biosynthesis, structural analysis, and function of glycans [3–6], the association of certain glycan structures with normal and diseased states [7], and the occurrence and synthesis of GPI anchors [8]. Whereas these have summarized largely what has been learned in higher eukaryotic systems, reviews on glycoprotein glycosylation and processing in yeast have appeared recently by Lehle and Tanner [9], from Herscovics and Orlean [10], and from our laboratory [11], which summarize our current understanding of yeast glycoprotein metabolism.

Our ongoing glycoprotein studies have focused on two yeasts, *Saccharomyces cerevisiae*, the "budding" yeast, and *Schizosaccharomyces pombe*, the "fission" yeast. We have also investigated *Pichia pastoris*, a small methylotrophic budding yeast frequently exploited for the heterologous expression of secreted glycoproteins [12,13]. Although the former two yeasts are nearly 300 million years evolved from each other and equally distant from animal cells [14,15], the fundamental way in which the Glc_3-$Man_9GlcNAc_2$–PP–dolichol precursors are synthesized and nascent proteins are glycosylated with $Glc_3Man_9GlcNAc_2$ at–Asn–X–Thr/Ser–sequons has been highly conserved among them. Once the tetradecasaccharide is transferred to protein, the glucoses are rapidly removed in the endoplasmic reticulum (ER) of both yeasts, and in addition, a single specific mannose is trimmed in *S. cerevisiae*'s ER [16,17]. The $Man_{8,9}GlcNAc_2$ precursors are trafficked to the Golgi, where in *Saccharomyces* the Man_8 is elongated to a family of $Man_{10-14}GlcNAc_2$ completed "core" structures or to large "mannan" chains [18,19], whereas in *Schizosaccharomyces* the Man_9 is elongated to a novel family of galactomannans [17,20,21]. In the cases of *Pichia*, processing to $Man_8GlcNAc_2$ in the ER occurs as in *S. cerevisiae*, but subsequent elongation in the Golgi is much less extensive, leading to a family of $Man_{8-11}GlcNAc_2$ structures on heterologously expressed *S. cerevisiae SUC2* invertase [12].

Three families of yeast mutants have been particularly important in revealing the complexities in glycoprotein synthesis and trafficking: *alg*⁻ (asparagine-linked glycosylation) mutants isolated in Robbins's laboratory using a ³H-Man suicide protocol [22]; *sec*⁻ (secretion defective) mutants initially isolated in Schekman's laboratory as temperature-sensitive accumulators of secretory proteins after ethyl methanesulfonate mutagenesis [23]; and *mnn*⁻ mutants (mannoprotein) isolated serologically in Ballou's laboratory [19] as naturally occurring cells with nonconditional, but stable, alterations in surface carbohydrate epitopes. In addition to *S. cerevisiae, S. pombe* has received a great deal of attention, because, unlike the budding yeast, many of the signal transduction pathways involved in cell-cycle control and cytokinesis have been more highly conserved between *S. pombe* and mammalian cells compared with those of *S. cerevisiae* and mammalian cells. In addition, *S. pombe* has a well-developed Golgi apparatus [24], in which galactose is added to its glycans [17,20,21,25], providing another useful system to study glycoprotein processing and trafficking.

Our increasing awareness of the roles that carbohydrates play in cellular recognition phenomena and of the alterations seen in carbohydrate structures in numerous known disease states, including cancer, makes understanding the central mechanisms that determine glycan structures integral to being able to identify and intervene in additional carbohydate-based disorders. Several hundred million years of evolutionary conservation in the fundamental way that cells *N*-glycosylate and traffic their glycoproteins has allowed yeast to become an important system for studying the structural biochemistry and enzymology of glycoprotein processing, because of the ability to combine molecular, genetic, and biochemical approaches in this organism.

In the example provided here, Gal- and Man-containing N-linked $Hex_{10}GlcNAc$ oligosaccharide isomers are isolated from a crude $S.$ $pombe$ invertase/acid phosphatase preparation. Regardless of whether the individual researcher's focus is to study glycoprotein trafficking in novel yeast mutants, to perform sugar transferase assays using defined oligosaccharide acceptors, or to characterize an industrially important, heterologously expressed gene product, the strategy outlined here provides a facile approach to isolate and study N-linked glycans from diverse sources.

II. MATERIALS AND METHODS

A. Yeast Growth

$S.$ $pombe$ wild-type Strain 972^{H-} (ATCC 24843) was grown in 1-L batches of autoclaved (121 °C, 20 min) yeast extract medium (YE), consisting of 5 g yeast extract (Difco, Detroit, MI) and 30 g glucose, in 2-L flasks at 32 °C on a rotary platform shaker. Overnight growth from 12 L of medium, harvested on ice with 0.01% NaN_3 and washed once by centrifugation from 20 mM sodium acetate buffer, pH 4.0, yielded 200–220 g of packed cells. For extensive additional information on the growth and manipulation of $S.$ $pombe$ and $S.$ $cerevisiae$, the reader is referred to $Experiments$ $with$ $Fission$ $Yeast$ [26] and $Methods$ in $Yeast$ $Genetics$ [27], respectively.

B. Invertase/Acid Phosphatase Preparation

The packed $S.$ $pombe$ cells were resuspended by adding two volumes of ice cold 20 mM sodium acetate buffer, pH 4.0, and disrupted in ~200-mL batches (~70 g of cells) in a Bead Beater (Biospec Industries, Bartlesville, OK), in a ~300-mL vessel half full of ice cold acid-washed 0.5-mm glass beads. Four or five cycles of disruption, each for 1 min with 5 min chilling in ice between cycles, is usually sufficient for >90% disruption. Breakage may be confirmed by viewing a small drop of diluted suspension on a slide at 440× in the light microscope. The disrupted cells are nonrefractile.

The suspension was decanted from the beads and the beads retained for the next batch disruption. After the final batch, the beads were washed on a 600-mL filtration funnel (coarse frit). The combined suspensions were centrifuged at 27,000 × g in 250-mL polycarbonate bottles (DuPont, Newtown, CT Sorvall GSA rotor, or equivalent) for 15 min at 4 °C. Supernatants were decanted, concentrated with a 10-kDA cutoff spiral ultrafiltration cartridge (Amicon Corp., Beverly, MA) to about 200 mL, and brought to 85% saturation by gradual addition of solid ammonium sulfate (60.8 g/100 mL) with stirring at 4°C. After stirring an additional 60 min, the suspension was centrifuged at 48,000 × g in 50-mL polycarbonate tubes (Dupont Sorvall SS-34 rotor, or equivalent) for 15 min at 4°C. The supernatant was decanted and dialyzed against distilled water using the spiral ultrafiltration cartridge as previously described, equilibrated with 20 mM sodium citrate buffer, pH 5.5, and reduced in volume to a protein concentration range of 2.5–5 mg/mL (Bearden assay, [28]). This preparation contained about 50 mg protein from 200 g of cells, most of which is extracellular *S. pombe* invertase and acid phosphatase [29,30].

C. *N*-Linked Oligosaccharide Release and Isolation

1. Endoglycosidase Treatment

Trimming of *N*-linked glycans on wild-type yeast glycoproteins removes at most one mannose residue from the original Man$_9$ core [16]. Thus, these glycans remain sensitive to hydrolysis of the di-*N*-acetylchitobiose core by endo-β-*N*-acetyl glucosaminidase H (endo H, E.C.3.2.1.96) throughout their subsequent elongation in elements of the secretory pathway. The products of endo H hydrolysis are oligosaccharides with a single GlcNAc on their reducing ends, while a GlcNAc stub is left on the glycoprotein at the glycosylated asparagines. Endo H is available commercially, cost-effective, and very stable to denaturing conditions. Nevertheless, it may be desirable in certain instances, for example, to remove endo H–resistant oligosaccharides from certain mutant yeast glycoproteins [31] or to generate glycans with an intact di-*N*-acetylchitobiose on their reducing end, to digest glycoproteins with peptide-N^4-(*N*-acetyl-β-glucosaminyl) Asn amidase F (PNGase F, E.C.3.5.1.52). This enzyme is also commercially available, but it is considerably more expensive than endo H and much less stable to harsh treatment. The optimization, characterization, and application of endoglycosidases and asparagine amidases to the hydrolysis of glycoprotein glycans has been reviewed in depth [32–36].

For deglycosylation with endo H, the soluble *S. pombe* glycoprotein fraction (~50 mg protein) in 10 to 20 mL of 50 mM Na citrate buffer, pH 5.5, is denatured by heating at 100°C for 5 min in the presence of 0.1 M 2-mercaptoethanol and 1.2 mg sodium dodecyl sulfate (SDS)/mg protein. After cooling to room temperature, endo H is added at 50–100 mU/mL, and the reaction mixture is incubated overnight at either 30 or 37°C. We confirmed the retention of endo H activity at the end of the incubation by hydrolysis of a dansylated ovalbumin glycopeptide [18], but this could also be accomplished by incubating a small amount of the bulk reaction mixture with denatured molecular-weight markers containing ovalbumin and looking by SDS–polyacrylamide gel electrophoresis (PAGE) for an increase in its mobility relative to an undigested control.

2. Solvent Precipitation

Though it is possible to deglycosylate large amounts of glycoprotein with endoglycosidases, as a practical matter scaleup necessitates numerous chromatography steps, which become time consuming and cumbersone. To circumvent this problem, we have developed a solvent extraction procedure, which allows rapid sequential treatment of bulk glycoproteins with endo H to remove high-mannose structures, followed by PNGase F to remove endo H–resistant structures, and finally alkaline borohydride treatment to release O-linked glycans. For the study presented here, the proteins and liberated oligosaccharides in the endo H digestion mixture were precipitated by adding four volumes of ice cold acetone to samples in Corex centrifuge bottles (Corning Glass Works, Cambridge, MA) and placed at $-20°C$ overnight. Precipitates were centrifuged at $27,000 \times g$ for 15 min at 4°C, and the SDS-containing supernatant discarded. The pellet was suspended in 60% ice cold methanol, which solubilizes oligosaccharides containing ~25 or fewer hexoses. The suspension is centrifuged as previously described, and all remaining larger oligosaccharides are soluble in a 50% methanol wash of the residual pellet from the 60% methanol extraction [17,31]. The 60% methanol supernatant is reduced, by rotary evaporation, to 1–2 mL and passed over a small column (5 mL) of mixed bed ion-exchange resin (Baker (Phillipsburg, NJ) IONAC NM_{60}, or equivalent) to remove any residual salts, SDS, or peptides. This procedure also removed charged oligosaccharides, such as those with phosphate or acidic sugar residues. As a practical matter, this is not a problem in *S. pombe*, where sugar–phosphate residues are minimal (Gemmill and Trimble, unpublished data). The sample is concentrated to 0.7 mL by rotary evaporation for gel exclusion chromatography.

3. Bio-Gel P-4 Chromatography

Bio-Gel P-4 (extra fine) beads (Bio-Rad Laboratories, Hercules, CA) are hydrated in column buffer (0.1 M acetic acid/1% *n*-butanol), and on examination under the light microscopic at $100\times$ reveal, depending on the manufacturing lot number, a varying proportion of very small beads. These have a tendency to "plug" the space between larger beads, resulting in increased back pressure and, if severe enough, bed collapse and failure. These "fines" can be removed at room temperature by vigorously mixing a slurry of 1 part hydrated gel beads in 4 parts column buffer in a 2-L separatory funnel, and allowing the larger beads to settle for about 1 h. The settled beads are slowly drained from the bottom of the funnel, and the fines discarded. The procedure is repeated two or three additional times until the bulk of the suspended gel beads settles cleanly in 1 to 2 h. While this is a time-consuming procedure, it is well worth the effort, as a properly prepared and maintained sizing column will last for many years and provide superior resolution.

We used glass columns 1.6×95 cm (Pharmacia Biotech, Piscataway, NJ) fitted with 10-μm mesh low dead-volume bottoms and similarly equipped flow adaptors on top (Pharmacia Biotech). Columns are fitted with 1/16-in tubing and polyetheretherketone (PEEK) finger-tight connectors (Upchurch Scientific, Oak Harbor, WA). Columns are poured using a Beckman "Accu-Flo" piston pump set to deliver 8 mL/h at 7–9 lb/in^2. Although this setup provides a 24-h turnaround and very reproducible profiles, any precision peristaltic or piston pump should work as well. Column buffer was prepared from American Chemical Society (ACS) reagent-grade acetic acid, n-butanol, and Millipore (Bedford, MA) 19-megohm water, degassed under vacuum, and dispensed from a 6-L resevoir by gravity to the pump head inlets through Teflon tubing (1/18 in. i.d.) fitted with 10-μm submersible filters. Columns were calibrated using high-mannose oligosaccharides that had been separated by size and their structures confirmed by integration of anomeric protons by one-dimensional (1D) ^{1}H nuclear magnetic resonance (NMR) and, more recently, by matrix-assisted laser desorption ionization time-of-flight (MALDI-TOF) mass spectrometry (MS). These were reduced with alkaline NaB[^{3}H]$_4$ [18,37] for use as internal size markers. One must note that the alditols of oligosaccharides on Bio-Gel P-4 run about 0.5 hexose units larger (~ 3 fractions in our system) than their unreduced counterparts [38].

Although one may use a commercial sample-injection valve, 400–600 μL was loaded into the inlet tube at the column top through a silicone tubing sleeve with a 1-mL disposable pipette tip and gentle pressure from a 3-mL disposable syringe, fitted with an appropriate diameter of Tygon tubing as a seal. Care was taken to avoid introducing air into the column. Some resolution enhancement occurs at 55°C [38], but for most purposes collecting fractions of ~ 0.75 mL at room temperature was adequate. Under these conditions, the void volume was found at about tube 68 and monosaccharides elute at about tube 200.

Fractions were scanned for neutral hexose by a scaled-down version of the phenol–sulfuric acid assay [39] using sample made up to 200 μL with water plus 200 μL of 5% phenol (without preservatives) in 10×75 mm acid-washed glass tubes. Using a solvent-resistant Repipete, 1.2 mL of concentrated sulfuric acid was added and the tubes carefully "vortexed." The assay gave ~ 10 A$_{489}$/μmol mannose, and so 5 to 10 nmol of hexose was easily detectable. Fractions were pooled by size and rechromatographed exactly as previously described to isolate homogeneously sized fractions.

4. High pH Anion-Exchange Chromatography

Oligosaccharide isomers within a size class may be effectively resolved using the Dionex (Sunnyvale, CA) CarboPak PA-100 chromatography system [40]. Both analytical 4 × 250 mm and preparative 9 × 250 mm columns were utilized with a PA-100 guard column. For routine scanning of Bio-Gel P-4 column profiles, 1–3 nmol of oligosaccharide in 30 μL of water was introduced into a 25-μL sample loop of the Rheodyne autoinject valve and resolved on the analytical column at a flow rate of 1 mL/min. A gradient of 35 to 50 mM sodium acetate in 0.1N NaOH over 60 min was used. For preparative separation, nearly 500 nmol of $Hex_{10}GlcNAc$ was accommodated on the larger CarboPak PA-100 column. Sample (100 μL) was injected into a 200-μL sample loop of the Rheodyne autoinject valve and loaded onto the column at a flow rate of 1 mL/min. In this case, the gradient was 35 to 75 mM sodium acetate in 0.1 N NaOH over 120 min. Solvents were kept under helium in the Dionex Solvent Degas Module.

The effluent of preparative runs was desalted using a Dionex anion self-regenerating suppressor (ASRS-I, 4 mm), and fractions of 0.5 mL were collected and scanned for neutral hexose by the phenol–sulfuric acid assay previously described. Oligosaccharides were detected using a Dionex pulsed ampcromctric dctcction (PAD) cell with the following settings: $E_1 = 0.05$, $E_2 = 0.6$, $E_3 = 0.6$, $T_1 = 480$ msec, $T_2 = 120$ msec, $T_3 = 360$ msec. The output range was set at 1000 nA for analytical runs, whereas 3000 was used for preparative runs. Like fractions from multiple preparative runs were combined and chromatographed a final time on the Bio-Gel-P-4 column. Recoveries from the Bio-Gel P-4 and the preparative CarboPak PA-100 columns exceeded 95%.

D. Oligosaccharide Characterization

1. Monosaccharide Analysis

Compositional analysis provides useful information about separated oligosaccharide isomers. Methods applied in our laboratory are essentially the same as described by Hardy and Townsend [40], except that we find slightly more reproducible results using hydrolysis with HCl rather than trifluoroacetic acid (TFA). A monosaccharide standard mixture is available from Dionex Corp., and monosaccharides are resolved on either the Dionex CarboPak PA1 or the newer MA1 column fitted with the appropriate guard column and using the manufacturer's elution conditions.

Duplicate samples of oligosaccharide (0.1 to 1 nmol) were dried into the tip of 1.5-mL screw cap polymerase chain reaction (PCR) tubes (Bio-Rad Corp.), or equivalent, and hydrolyzed in 2 N HCl (0.3 mL) at 100°C for 2.5 h after bubbling the sample with prepurified N_2 for about 1 min. GlcNAc was analyzed as GlcN after hydrolysis of duplicate samples (~1 nmol) in 2 N HCl (0.5 mL) at 100°C for 16 h after flushing with N_2. Acid was removed either in a SpeedVac (Savant Industries, Holbrook, NY) without heat or transferred to a 15-mL pear-shaped flask for rotary evaporation at 30°C. Dried samples were reconstituted with 100 μL of water and 25 μL was injected onto the 0.4 × 25 cm Dionex PA1 column using a 100-μL sample loop. Although one may add an external standard, such as rhamnose, for overall recovery quantitation, the ratios of sugars present were determined quite accurately by normalizing values to GlcN set equal to 1.0, because endo H–released *S. pombe* and *S. cerevisiae* N-linked oligosaccharides have a single GlcNAc on their reducing ends prior to hydrolysis [17,18].

2. MALDI-TOF Mass Spectrometry

Although an oligosaccharide pool may consist of numerous $Hex_{10}GlcNAc$ isomers (mass ~1865), the branching and sugar composition of each can result in a unique hydrodynamic volume on Bio-Gel P-4, and a mixture of isomers can exhibit a fairly broad peak compared with a single isomer [38]. The ability to confirm mass values for oligosaccharides in Bio-Gel P-4 fractions greatly simplifies pooling peak fractions from complex profiles. Because the oligosaccharides are related to each other biosynthetically, forming a homologous series of structures, they gave relatively consistent intensities per amount present on laser desorption. Thus, on scanning across a profile, from $Hex_9GlcNAc$ to $Hex_{11}GlcNAc$, the ratio of component signals was a resonable measure of the proportions of the oligosaccharide sizes. The best general matrix for free oligosaccharide analysis by MALDI-TOF MS is 2,5-dihydroxybenzoic acid (DHB). Sample, at a concentration of 2–20 pmol/μL, was mixed with ~50 mM solution of DHB in water/acetonitrile (70%/30%). A volume of 0.4 to 1 μL of this mixture was applied to the sample stage and allowed to air dry. The Brüker (Billerica, MA) Reflex was operated in the linear mode at 28.5 kV. N-linked oligosaccharides were observed predominantly as the Na^+ adduct ion. Approximately 20 laser slots was sufficient to provide a good signal when ~1 pmol of glycan was analyzed.

3. ^1H NMR Spectroscopy

NMR is the only nondestructive method available to completely define a novel oligosaccharide's primary structure [41,42]. Methodology has advanced to the point where the structure of 10 to 20 nmol of a pure oligosaccharide can be determined by 1D ^{1}H NMR if similar compounds have already been defined [42]. Because a large library of ^{1}H chemical shifts exists for high-mannose type oligosaccharides [18,19] and because the Man$_{8,9}$GlcNAc core is common to larger yeast biosynthetic intermediates, assigning structures of isomers in a homologous series by 1D ^{1}H NMR is straightforward.

Samples from the final Bio-Gel P-4 chromatographic step (75–250 nmol) were exchanged three times with ^{2}H$_2$O by evaporation to dryness from 1 mL of 99.8% ^{2}H$_2$O (low in paramagnetics, Sigma Chemical Co., St. Louis, MO, or Cambridge Isotopes Lab., Andover, MA). The sample was dissolved in 1 mL of 99.96% ^{2}H$_2$O (Cambridge Isotopes Lab.), transferred to a 5-mL conical, acid-washed glass screw cap tube, and covered with a piece of laboratory film (e.g., Parafilm, American National Can Co., Greenwich, CT). Several holes were made with a 28-ga. needle, and the sample was shell-frozen in an alcohol–dry ice bath and lyophilized. Once dry, samples were stored over P$_2$O$_5$ in vacuo in a glass desiccator until time for NMR spectroscopy. The desiccator vacuum was replaced with prepurified dry N$_2$, the laboratory film removed, and the tubes quickly capped with Teflon-lined screw cap closures. Each tube was briefly opened, and 0.5 mL of 99.996% ^{2}H$_2$O from individual 0.7-mL ampules (Cambridge Isotopes Lab.) was introduced using a plastic pipette tip, immediately pre-washed with 99.996% ^{2}H$_2$O; followed by 5 μL of freshly prepared 1% acetone in 99.96% ^{2}H$_2$O (stored in glass) with immediate recapping as an internal signal. This provided ~1.35 mM acetone, the protons of which are found at 2.225 ppm [42].

After allowing time for the sample to dissolve completely, insolubles were removed by centrifugation at 1500 × g for 10 min in the swinging-bucket rotor of a clinical centrifuge. Caps were removed and a 480-μL aliquot of each sample, the amount required to cover the radiofrequency (rf) coil in the Brüker probes, was transferred to a 5-mm NMR tube (Wilmad Glass Works, 535 pp, Buena, NJ), prewashed in 99.8% ^{2}H$_2$O, dried, and stored over P$_2$O$_5$ as previously described, and flame sealed. As discussed earlier [42], moisture in the air is the NMR spectroscopist's enemy, but the methods described here provided samples with a relatively low residual HO^{2}H$_2$ signal without the use of a glove box or additional equipment. Spectra were taken on a pulse–Fourier transform (FT) NMR spectrometer, preferably at 500 or 600 MHz for ^{1}H, because of the superior resolution obtained for dilute samples. Unless the researcher is familiar with the unique aspects of carbohydrate NMR, it is best to consult with someone familiar with obtaining and interpreting NMR spectra. In the example provided in this chapter, four *S. pumbe* Hex$_{10}$GlcNAc isomers were given preliminary structural assignments from 1D ^{1}H NMR spectra, taken on samples ranging from 0.2 to 0.7 mM. Confirmation of these structures will employ two-dimensional double quantum filtered-correlation spectroscopy (DQF-COSY) experiments. Further strategies for the complete assignment of the primary structure of novel oligosaccharides using additional multidimensional NMR techniques have been described in detail by Van Halbeek [41,42].

E. Other Methods

1. Purification of Invertases

S. cerevisiae SUC2 extracellular invertases were purified to homogeniety from wild-type [43] and pmr1 [44] Saccharomyces strains as well as the heterologously expressed S. cerevisiae enzyme from Pichia pastoris [12].

2. Western and Carbohydrate Blotting of Invertases

Rabbit polyclonal antiserum was raised against highly purified nonglycosylated internal invertase [43], and the immunoglobin G (IgG) fraction was purified on Protein-A-Sepharose according to the manufacturer's protocol (Pharmacia Biotech.). An anti–rabbit IgG alkaline phosphatase kit (Promega Biotech, Madison WI) was used as described in the manufacturer's literature. Digoxigenin–Concanavaliln A (ConA) lectin and antidigoxigenin alkaline-phosphatase were from Boehringer Mannheim, Indianapolis, IN, and Biotin-*Bandeiraea simplicifolia*-Isolectin B$_4$ (BSL–IB$_4$) isolectin and avidin alkaline-phosphatase were from Vector Laboratories, Burlingame, CA. Invertases were run on 8% polyacrylamide gel slabs (0.5 mm thickness [45]), and then electroblotted onto polyvinylidene difluoride (PVDF) membranes (Immobilon, Millipore Corp.) according to the apparatus manufacturer's instructions (Bio-Rad Industries). Whereas 0.2–0.5 μg of invertase was required for Coomassie blue staining on SDS gels, only 25–50 ng is required on membrane blots for Western or lectin analysis. A review by Haselbeck and Hösel discusses applications and modifications of carbohydrate blotting in detail [46].

III. RESULTS AND DISCUSSION

A. Overview of S. pombe N-Linked Glycans

Unlike other yeast and higher eukaryotes, S. pombe does not express a detectable ER Man$_9$-α-mannosidase [17] and, therefore, does not trim the archetypal Man$_9$GlcNAc$_2$ to Man$_8$GlcNAc$_2$ prior to elongation with mannose and galactose residues in the Golgi apparatus [17,21]. The recent description of a number of new mutants in glycoprotein glycan synthesis in S.

pombe, including two galactosyl transferases Gma12p [25] and Gth1p (T. Chappell, personal communication), a probable mannosyltransferase, and UDP-Glc-4′-epimerase [47], provides a useful starting point for determining the role of these enzymes in glycan modeling in vivo. To this end, we have developed the strategy described here to isolate sufficient amounts of pure *N*-linked oligosaccharide isomers for structural assignment by NMR techniques, which should have general applicability to a wide variety of glycoprotein sources.

B. Enzymatic Deglycosylation Considerations

The single largest concern regarding deglycosylation of glycoproteins using either endo H or PNGase F is whether the reaction has gone to completion. Perhaps the simplest method to answer this question is to check for residual carbohydrate in the methanol-washed protein pellet by the phenol–sulfuric acid method should the total phenol–sulfuric absorbance recovered in the methanol washes not approximate the starting value. The protein pellet was solubilized in a suitable volume of 50 mM NaOH and brought to pH 8.5 by dropwise addition of 2 N H_3PO_4. This is an appropriate buffer for PNGase F. The acetone-precipitation, methanol-wash procedure (Section II.C.2) and the solubilization in NaOH fully denatures the protein(s), so that no detergents were required. For homogeneous glycoproteins, a small portion of the reaction was treated with PNGase F and run on an SDS–polyacrylamide gel to see whether there was an increase in mobility, in comparison with starting and endo H–digested material.

Another sensitive method to evaluate this question was carbohydrate blotting using lectins known to bind to the glycans of the test glycoproteins. It is a characteristic of yeast glycoproteins associated with both *N*- and *O*-linked sugars that they strongly interact with ConA. Figure 1 shows Western and lectin blots of invertases from three *S. cerevisiae* and one *S. pombe* strain. As shown in Fig. 1A, the invertase antibody, raised against the nonglycosylated wild-type *S. cerevisiae* enzyme, strongly reacted with control and endo H–treated wild-type invertases (lanes 1 and 2) and with underglycosylated invertases from the *S. cerevisiae pmr1* mutant (lane 3), which has a defect in a Ca^{2+}-ATPase [48], and the invertase heterologously expressed in *Pichia pastoris* (lane 4 [12]). Interestingly, invertase from *S. pombe* reacted weakly with the antibody, which might be due to the evolutionary divergence of these two yeasts [14].

Figure 1B shows that ConA binds to all of the invertases except the deglycosylated wild-type form (lane 2), signifying that only *N*-linked oligosaccharides were initially present and that the enzymatic hydrolysis reaction had gone to completion. The BSL–IB$_4$ isolectin, which is specific for termi-

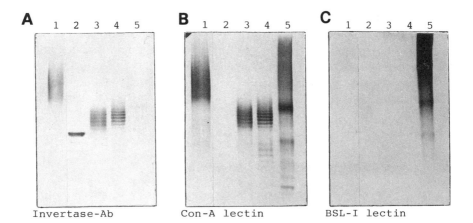

Figure 1 Western and lectin blotting of different invertase preparations. Three identical SDS polyacrylamide gels were run and electroblotted onto PVDF membranes: (A) rabbit anti-invertase Western blot; (B) digoxigenin–ConA blot; (C) biotin–BSL–IB$_4$ isolectin blot. Each lane contained 50 ng of invertase protein from *S. cerevisiae* wild-type before (lane 1) and after (lane 2) endo H treatment; *S. cerevisiae pmr1* mutant (lane 3); *S. cerevisiae SUC2* gene expressed in *Pichia pastoris* (lane 4); and *S. pombe* wild-type (lane 5). Color development was with alkaline phosphatase conjugated to goat anti-rabbit IgG (A), to rabbit antidigoxigenin (B), and to avidin (C) according to the supplier's protocols. (See Section II.)

nal α-linked galactose [49], reacted only with the *S. pombe* enzyme preparation (Fig. 1C). Figures 1B and C also reveal that *S. pombe*'s invertase has a much greater size heterogeneity than even the wild-type *S. cerevisiae* invertase. Some of the galactomannans on *S. pombe* invertase have been estimated to have molecular weights of > 100,000 [20].

It is important to note that residual carbohydrate found by using either the phenol–sulfuric acid assay or by lectin blotting could mean the presence of *O*-linked sugars rather than endo H–resistant *N*-linked glycans or simply a failed digestion. Again, the PNGase F digestion as described previously can serve as a useful test. Applying the solvent precipitation method and phenol–sulfuric acid assay or lectin blotting of a transfer membrane for residual glycan may provide an immediate answer. Nevertheless, if there is still residual carbohydrate, it may be prudent to perform an alkaline β-elimination reaction on the bulk protein sample and spectrophotometrically check for increased absorbance at 240 nm due to formation of anhydro-serine and/or anhydro-threonine [50], or to carry out this reaction in the presence of NaB[^{3}H]$_4$ to label any released *O*-linked glycans at their reducing ends [37]. These can be separated on the Bio-Gel P-4 column, which will sharply resolve all *O*-linked forms including the monomeric aldi-tol [50].

C. Chromatographic Isolation of a $Hex_{10}GlcNAc$ Pool from an *S. pombe* Invertase/Acid Phosphatase Preparation

The endo H digestion of the enriched *S. pombe* invertase/acid phosphatase preparation was carried through the solvent precipitation scheme (Section II.C.2.); the 60% methanol wash of the acetone pellet enriched in the *N*-linked oligosaccharides, with less than about 25 hexose residues, was rotary evaporated to dryness and dissolved in 2 mL of water. The 50% methanol wash containing larger galactomannans was treated similarly, and the total phenol–sulfuric acid absorbance in both fractions was compared with the starting value to ensure that the endo H hydrolysis was complete. Each fraction was then passed through a Sep-Pak C_{18} RP filter cartridge (Waters Corp., Milford, MA) to remove peptides and/or any residual de-tergents so as not to contaminate the Bio-Gel P-4 sizing column. The fil-tered 60% methanol fraction was taken to dryness in a 15-mL pear-shaped flash by rotary evaporation, dissolved in 0.6 mL of 0.1 N acetic acid/ 1% *n*-butanol column buffer, supplemented with $Man_5GlcNAc[^3H]$-ol and $Man_{13}GlcNAc[^3H]$-ol as internal markers, and loaded onto the Bio-Gel P-4 column as described in Section II.C.3.

Fractions were scanned for neutral hexose by the phenol–sulfuric acid method and for radioactivity in the markers by liquid scintillation counting. The data are shown in Fig. 2; the majority of the oligosaccharides eluted as a broad saw-toothed peak of sizes from $Hex_9GlcNAc$ to about $Hex_{17}GlcNAc$. A smaller peak of oligosaccharides at the void volume of ~ $Hex_{25}GlcNAc$ was present. The smallest *N*-linked oligosaccharide in *S. pombe* was $Man_9GlcNAc$, found in tubes 134–138 (Fig. 2), which was char-acterized in our previous work [17]. A quick way to initially resolve individ-ual-sized glycans from such a profile is to pool alternate four- to five-fraction segments representing each peak into individual "odd" and "even" pools, that is, $Man_{9,11,13}GlcNAc$ and $Man_{10,12,14}GlcNAc$. These are then re-chromatographed separately on the Bio-Gel P-4 column, which provided near baseline resolution between the even- and odd-numbered species (data not shown).

For the example reported here, tubes 130–133 were pooled from Fig. 2 and run on the Bio-Gel P-4 column again to remove the $Man_9GlcNAc$ and $Hex_{11}GlcNAc$ "tails" that overlapped with the $Hex_{10}GlcNAc$ pool (not shown). The $Hex_{10}GlcNAc$ was again pooled, supplemented with the $Man_5.GlcN[^3H]$-ol as an internal reference marker, and then chromatographed on the Bio-Gel P-4 column.The phenol–sulfuric acid and radioactivity scans

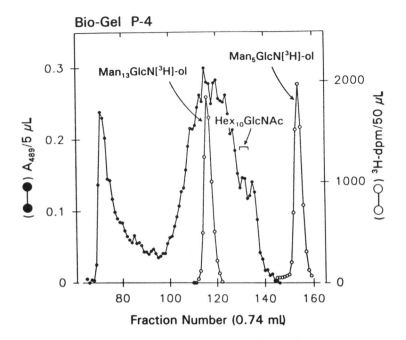

Figure 2 Bio-Gel P-4 chromatography of the *S. pombe* invertase/acid phosphatase oligosaccharides soluble in the 60% methanol extract of an acetone-precipitated endo H digestion. The thyroglobulin Man$_5$GlcN[^{3}H]-ol and yeast Man$_{13}$GlcN[^{3}H]-ol markers were prepared according to previous reports [16,18]. The Hex$_{10}$GlcNAc region in fractions 130–133 (bracket) was pooled for further study.

are plotted in Fig. 3. Hex$_{10}$GlcNAc fractions 129–133 were pooled, which represented the central 95% of the symmetrical peak. The center of the Man$_5$GlcN[^{3}H]-ol marker peak eluted in tube 155 in Fig. 2 and is a split peak centered at tube 154.5 in Fig. 3. The center of the Hex$_{10}$GlcNAc peak eluted 23.5 tubes earlier in both profiles, attesting to the reproducibility of this chromatographic system (see also Ref. 38).

D. Characterization of the *S. pombe* Hex$_{10}$GlcNAc Pool

The recovery of Hex$_{10}$GlcNAc from the Bio-Gel P-4 column (Fig. 3), isolated from the invertase/acid phosphatase preparation from ∼200 g cells in the initial work-up, was 750 nmol. To assure that this pool was homogeneous by size, 1 to 2 pmol were placed on the stage of a Brüker Reflex MALDI-TOF Spectrometer in DHB matrix as described in Section II.D.2, and the signal from 20 laser shots directed at amorphous crystalline areas

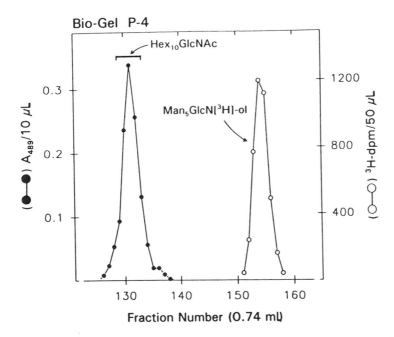

Figure 3 Final Bio-Gel P-4 sizing of the *S. pombe* Hex$_{10}$GlcNAc pool. Fractions 129–133 (bracket), representing >95% of the Hex$_{10}$GlcNAc oligosaccharides present, were pooled.

seen with the Brüker Scout sample visualization system was collected. The calculated mass of [Hex$_{10}$GlcNAc + Na]$^+$ is 1865.7, and as shown in Fig. 4, the Hex$_{10}$GlcNAc pool gave a single signal at m/z 1865. Note the absence of any apparent contamination with [Hex$_9$GlcNAc + Na]$^+$ at m/z 1704 or [Hex$_{11}$GlcNAc + Na]$^+$ at m/z 2028.

Because *S. pombe* elongates the precursor Man$_9$GlcNAc with both galactose and mannose, the Hex$_{10}$GlcNAc pool could be composed of a number of isomers. To determine how many species might be present, about 2 nmol of the Hex$_{10}$GlcNAc pool was chromatographed exactly as described in Section II.C.4 using the analytical Dionex PA-100 column. Figure 5 shows the profile, in which there are four main peaks, a, b, c, and g. The remainder of the Hex$_{10}$GlcNAc pool was divided in half (~375 nmol oligosaccharide each) and separated into isomers on the preparative Dionex PA-100 column as described in Section II.C.4. The fractions were scanned for neutral hexose, and the like peaks from the two preparative runs were pooled. Table 1 summarizes the recovery of isomers a–c and g and their monosaccharide composition as determined using the Dionex PA1 column

MALDITOF MS

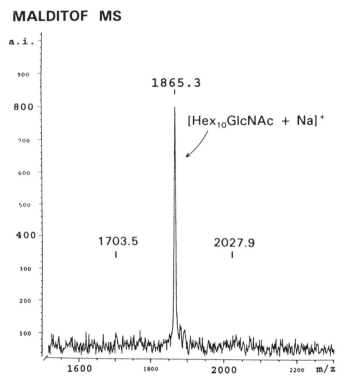

Figure 4 MALDI-TOF mass spectrometry of 1–2 pmol of the *S. pombe* Hex$_{10}$Glc-NAc pool from Fig. 3.

after mild acid hydrolysis (see Section II.D.1). The total recovery of Hex$_{10}$-GlcNAc in peaks a–c and g was 730 nmol, or >95% of the initial pool of ~750 nmol from the Bio-Gel P-4 chromatography (Fig. 3). Recoveries of peaks d–f were insufficient for further structural characterization.

E. ^{1}H NMR Spectroscopy of *S. pombe* Hex$_{10}$GlcNAc Isomers

Oligosaccharide peaks a–c and g were exchanged with ^{2}H$_2$O as described (Section II.D.3) and 500-MHz 1D ^{1}H NMR spectra taken at 305 K. Figure 6 is a montage of the spectra for the individual Hex$_{10}$GlcNAc peaks from high pH anion-exchange chromatography (HPAEC) in comparison with the precursor Man$_9$GlcNAc. The structures deduced for these glycans on the basis of 1D ^{1}H spectroscopy [17,19,21], and their presumed biosynthetic relationship to the Man$_9$GlcNAc precursor are shown in Fig. 7. Isomer a is

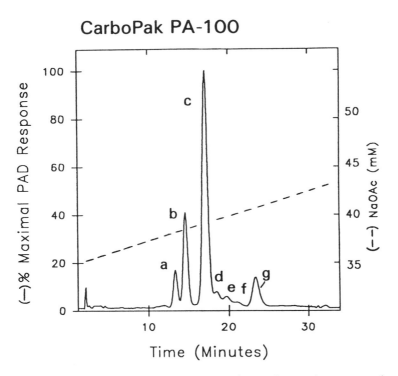

Figure 5 Analytical Dionex high pH anion-exchange chromatography on a PA-100 column (4.5 mm × 25 cm) of 2 nmol of the *S. pombe* $Hex_{10}GlcNAc$ pool from Fig. 3. Peaks a–c and g were isolated using the larger Dionex PA-100 (9 mm × 25 cm) as decribed in Section II. A pulsed amperometric detector (PAD) response of 100% equaled 429 nA.

Table 1 Characterization of the Major $Hex_{10}GlcNAc$ Isomers from *S. pombe* Invertase/Acid Phosphatase

Peak	nmol	Composition[a]			
		GlcN	Man	Gal	Glc
a	100	1.0	9.8	—	—
b	130	1.0	8.6	1.1	—
c	375	1.0	9.6	—	—
d	87	1.0	8.6	0.6	0.3

[a]Compositions are expressed relative to GlcN, the product of GlcNAc on acid hydrolysis, set to 1.0.

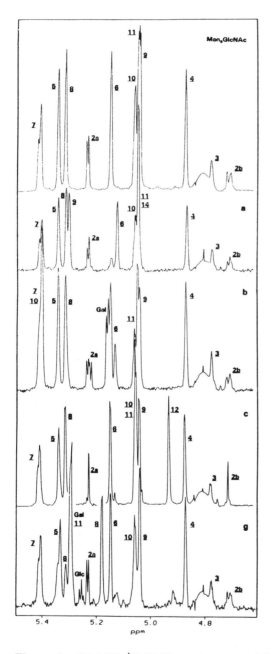

Figure 6 500-MHz ^{1}H NMR spectroscopy of *S. pombe* Hex$_{10}$GlcNAc isomers a–c, g, and the precursor Man$_9$GlcNAc. Samples exchanged in 99.996% ^{2}H$_2$O were in concentrations from 0.2 to 0.7 mM.

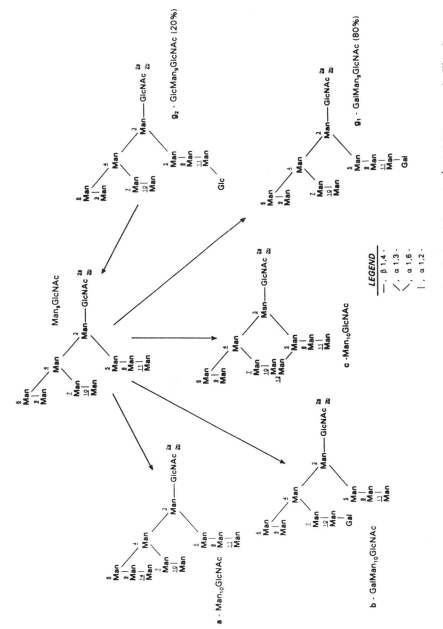

Figure 7 Structures of oligosaccharides present in peaks a–c and g, deduced from the ¹H NMR spectra in Fig. 6.

identical to an oligosaccharide described in *Pichia* with a new $\alpha(1,2)$-linked upper-arm mannose (residue 14), except for the retention of the middle-arm α (1,2)-linked terminal mannose residue 10 [12]. Isomer b is a novel species, in which an $\alpha(1,2)$-linked terminal galactose, anomeric proton signal observed at 5.157 ppm, (Fig. 6) has been added to residue 10, resulting in the shift of this residue's anomeric proton from 5.048 to 5.403 ppm. Isomer c has the new $\alpha(1,6)$-linked mannose residue 12 found at 4.932 ppm, which is linked to the core $\alpha(1,3)$-linked mannose 5. This is the $\alpha(1,6)$-linked backbone branch-initiating residue seen in all yeast studied to date [11].

Finally, isomer g is a mixture of two species estimated to be in an 80: 20 ratio by integrating the ^1H anomeric resonance peaks. The major species is another galactose-containing isomer with the $\alpha(1,2)$-linked galactose on the lower-arm mannose residue, 11. Its anomeric proton resonance is found at 5.290 ppm and has shifted 11's anomeric proton from 5.054 to 5.290 ppm. A through-space shielding effect of galactose's ring protons on residue 8 shifted 0.8 mol of the anomeric proton of 8 upfield from 5.304 to 5.178 ppm [12]. The second isomer in peak g is the $Man_9GlcNAc$ core with an $\alpha(1,3)$-linked glucose on lower-arm mannose residue 11, which is found at 5.257 ppm [51,52] and integrated to 0.2 mol. This species appears to be a small amount of residual glucosylated precursor (Fig. 7) that escaped trimming of the last glucose before leaving the endoplasmic reticulum. These assignments agreed closely with the monosaccharide composition (Table 1) determined for each of the peaks from Fig. 5.

In summary, the isolation and characterization of N-linked glycans from yeast glycoproteins is a relatively straightforward matter. A particularly nice characteristic of yeast is that even minor isomers, such as d–f in Fig. 5, can be isolated in sufficient quantities for structural characterization simply by increasing the amount of starting material. Future studies in our laboratory will focus on verifying these preliminary $Hex_{10}GlcNAc$ structural assignments using two-dimensional NMR techniques [17,41,42,51,52], performing similar studies on the larger size classes, and determining the biosynthetic relationships among the isomers.

ACKNOWLEDGMENTS

This work was supported in part USPHS NIGMS grant GM23900 to RBT. The author is grateful to Frederick Ziegler, Mary Fran Verostek, and Catherine Lubowski, who have been involved in the experimental aspects of the studies. Also gratefully acknowledged is the use of the Wadsworth Center Biological Mass Spectrometry Facility under the direction of Charles Hauer and the Structural NMR Facility under the direction of John Cavanagh.

Lynda Jury and Tracy Godfrey are thanked for processing the original manuscript.

ABBREVIATIONS

alg⁻	asparagine-linked glycosylation
ASRS-I	anion self-regeneration suppressor
BSL	*Bandeireae simplicifolia*
ConA	concanavalin A
DFQ-COSY	double quantum filtered-correlation spectroscopy
DHB	2,5-dihydroxybenzoic acid
endo H	endo-β-N-acetyl glucosaminidase H
ER	endoplasmic reticulum
FT	Fourier transform
GPI	glycosyl phosphatidyl inositol
HPAEC	high pH anion-exchange chromatography
IB$_4$	Isolectin-B$_4$
IgG	Immunoglobulin G
MALDI-TOF	matrix-assisted laser desorption ionization time-of-flight
mnn⁻	mannoprotein mutants
MS	mass spectrometry
NMR	nuclear magnetic resonance
1D	one-dimensional
PAD	pulsed amperometric detection
PAGE	polyacrylamide gel electrophoresis
PCR	polymerase chain reaction
PEEK	polyetheretherketone
PNGase F	peptide-N^4-(N-acetyl-β-glucosaminyl) Asn amidase F
PVDF	polyvinylidene difluoride
rf	radiofrequency
SDS	sodium dodecyl sulfate
sec⁻	secretion defective
TFA	trifluoroacetic acid
YE	yeast extract medium

REFERENCES

1. Hirschberg, C. B., and Snider, M. D. (1987). *Ann. Rev. Biochem.* 56, 63–87.
2. Abeijon, C., and Hirshberg, C. B. (1992). *TIBS* 17, 32–36.
3. Kornfeld, R., and Kornfeld, S. (1985). *Ann. Rev. Biochem.* 54, 631–664.
4. Schachter, H. (1991). *Glycobiology* 1, 453–461.

5.   Dwek, R. A., Edge, C. J., Harvey, D. J., Wormald, M. R., and Parekh, R. B. (1993). *Ann. Rev. Biochem.* 62, 65–100.
6.   Varki, A. (1993). *Glycobiology* 3, 97–130.
7.   Radermacher, T. W., Parekh, R. B., and Dwek, R. A. (1988). *Ann. Rev. Biochem.* 57, 785–838.
8.   Englund, P. T. (1993). *Ann. Rev. Biochem.* 62, 121–138.
9.   Lehle, L., and Tanner, W. (1995). In Montreuil, J., Vliegenthart, J. F. G., and Schachter, H., eds., *Glycoproteins*. Elsevier Science, New York, pp. 475–509.
10. Herscovics, A., and Orlean, P. (1993). *FASEB J.* 7, 540–550.
11. Trimble, R. B., and Verostek, M. F. (1995). *TIGG* 7, 1–30.
12. Trimble, R. B., Atkinson, P. H., Tschopp, J. F., Townsend, R. R., and Maley, F. (1991). *J. Biol. Chem.* 266, 22807–22817.
13. Wegner, G. H. (1990). *FEMS Microbiol. Rev.* 87, 279–283.
14. Russel, P., and Nurse, P. (1986). *Cell* 45, 781–782.
15. Pidoux, A. L., and Armstrong, J. (1992). *EMBO J.* 11, 1583–1591.
16. Byrd, J. C., Tarentino, A. L., Maley, F., Atkinson, P. H., and Trimble, R. B. (1982). *J. Biol. Chem.* 257, 14657–14666.
17. Ziegler, F. D., Gemmill, T., and Trimble, R. B. (1994). *J. Biol. Chem.* 269, 12527–12535.
18. Trimble, R. B., and Atkinson, P. H. (1986). *J. Biol. Chem.* 261, 9815–9824.
19. Ballou, C. E. (1990). *Meth. Enzymol.* 185, 440–470.
20. Moreno, S., Ruiz, T., Sanchez, Y., Villaneuva, J. R., and Rodriguez, L. (1985). *Arch. Mikrobiol.* 142, 370–374.
21. Ballou, C. E., Ballou, L., and Ball, G. (1994). *Proc. Natl. Acad. Sci. USA* 91, 9327–9331.
22. Huffaker, T. C., and Robbins, P. W. (1983). *Proc. Natl. Acad. Sci. USA* 80, 7466–7470.
23. Novick, P., Field, C., and Schekman, R. (1980). *Cell* 21, 205–215.
24. Ayscough, K., Hajibagheri, N. M. A., Watson, R., and Warren, G. (1993). *J. Cell. Sci.* 106, 1227–1237.
25. Chappell, T. G., Hajibagheri, M. A. N., Ayscough, K., Pierch, M., and Warren, G. (1994). *Mol. Biol. Cell* 5, 519–528.
26. Alfa, C., Fantes, P., Hyams, J., McLeod, M., and Warbrick, E. (1993). *Experiments with Fission Yeast*. Cold Spring Harbor Laboratory Press, New York.
27. Rose, M. D., Winston, F., and Hieter, P. (1990). *Methods in Yeast Genetics*. Cold Spring Harbor Laboratory Press, New York.
28. Bearden, J. C., Jr. (1978). *Biochim. Biophys. Acta* 533, 525–529.
29. Moreno, S., Sanchez, Y., and Rodriquez, L. (1990). *Biochem. J.* 267, 697–702.
30. Dibenedetto, G. (1972). *Biochim. Biophys. Acta* 286, 363–374.
31. Verostek, M. F., Atkinson, P. H., and Trimble, R. B. (1993). *J. Biol. Chem.* 268, 12095–12103.
32. Trimble, R. B., and Maley, F. (1984). *Anal. Biochem.* 141, 515–522.

33. Tarentino, A. L., Trimble, R. B., and Plummer, T. H., Jr. (1989). *Meth. Cell Biol.* 32, 111–139.
34. Maley, F., Trimble, R. B., Tarentino, A. L., and Plummer, T. H., Jr. (1989). *Anal. Biochem.* 180, 195–204.
35. Tarentino, A. L., and Plummer, T. H., Jr. (1994). *Meth. Enzymol.* 230, 44–57.
36. Trumbly, R. J., Robbins, P. W., Ziegler, F. D., Maley, F., and Trimble, R. B. (1985). *J. Biol. Chem.* 260, 5683–5690.
37. McLean, C., Werner, D. A., and Aminoff, D. (1973). *Anal. Biochem.* 55, 72–84.
38. Kobata, A. (1994). *Meth. Enzymol.* 230, 200–208.
39. Dubois, M., Gilles, K. A., Hamilton, J. K., Rebers, T. A., and Smith, F. (1956). *Anal. Chem.* 28, 350–356.
40. Hardy, M. D., and Townsend, R. R. (1994). *Meth. Enzymol.* 230, 208–225.
41. Van Halbeek, H. (1993). *Meth. Molec. Biol.* 17, 115–148.
42. Van Halbeek, H. (1994). *Meth. Enzymol.* 230, 132–168.
43. Williams, R. S., Trumbly, R. J., MacColl, R., Trimble, R. B., and Maley, F. (1985). *J. Biol. Chem.* 260, 1334–1341.
44. Verostek, M. F., and Trimble, R. B. (1995). *Glycobiology* 5, 671–681.
45. Laemmli, U. K. (1970). *Nature* 227, 680–685.
46. Haselbeck, A., and Hösel, W. (1993). *Meth. Molecular Biol.* 14, 161–173.
47. Huang, K. M., and Snider, M. (1995). *Mol. Biol. Cell* 6, 485–496.
48. Rudolph, H. K., Antebi, A., Fink, G. R., Buckley, C. M., Dorman, T. E., LeVitre, S., Davidow, L. S., Mao, L. S., and Moir, D. T. (1989). *Cell* 58, 133–145.
49. Murphy, L. A., and Goldstein, I. J. (1977). *J. Biol. Chem.* 252, 4739–4742.
50. Trimble, R. B., Maley, F., and Watorek, W. (1981). *J. Biol. Chem.* 256, 10037–10043.
51. Verostek, M. F., Atkinson, P. H., and Trimble, R. B. (1993). *J. Biol. Chem.* 268, 12104–12115.
52. Trimble, R. B., and Atkinson, P. H. (1992). *Glycobiology* 2, 57–75.

33
High Resolution Electrophoretic Analysis of Carbohydrates Using a DNA Sequencer

Michael G. O'Shea and Matthew K. Morell
Co-operative Research Centre for Plant Science, Canberra, A.C.T., Australia

I. INTRODUCTION

This chapter describes the application of a high resolution slab gel electrophoretic technique for the analysis of polysaccharide structures. This technique was developed primarily for the analysis of plant-derived polysaccharides (particularly the amylopectin component of starch), but it is clear that many other polysaccharide structures can be investigated using this method. Suitable examples were chosen to demonstrate the utility of the method.

Most polysaccharides possess heterogeneity in branching, sequence of component monosaccharides, and molecular weight distribution. Many polysaccharides can be thought of as containing a repeating unit (some more so than others), and in a number of cases, structural information has been determined by the analysis of oligomeric fragments obtained either by chemical or enzymatic degradation of the polysaccharide. This is because much of the heterogeneity of the original structure remains in oligosaccharides obtained after limited degradation. The analysis of polysaccharide structure by enzymatic degradation has been the subject of a comprehensive review by McCleary and Matheson [1].

Starches isolated from a variety of plants are known to possess both different physical characteristics and chemical structures. Starch is a heterogeneous population of molecules that is divided into two distinct subpopulations, amylose and amylopectin [2] (Fig. 1). Both of these possess linear chains consisting of α-(1,4)-linked glucose units and α-(1,6) linkages, which appear as branch points. Amylose consists of lightly branched molecules with a degree of polymerization (DP) of 4000–6000, and an average chain length (ACL) of 200–500 [3]. Amylopectin consists of much larger molecules (at least DP 200,000) that are heavily branched, thereby possessing a considerably shorter ACL (usually 15–25) [3], with the widely accepted

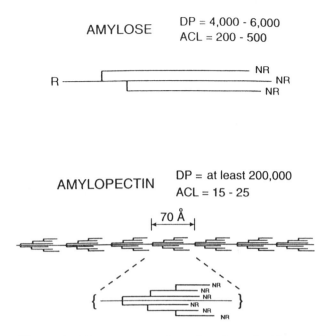

Figure 1 The structures of starch polysaccharides, amylose and amylopectin (R = reducing end, NR = nonreducing end).

"cluster model" illustrated in Fig. 1 [2]. With respect to their chemical structures, the most common measurements used to differentiate starches are the amylose–amylopectin ratio and amylopectin ACL. ACL remains one of the most widely measured characteristics used to relate starch, amylopectin structure, and the functional properties of starch. As examples, the effects of amylopectin chain length have been studied in association with the paste properties of starch [4] and the crystalline structure of starch granules [5,6]. ACL is usually measured after debranching by colorimetric means as the ratio of total carbohydrate as glucose [7] to the number of reducing ends [8]. Despite its widespread use, ACL does not permit comparison of the relative distributions of individual oligomers within amylose and amylopectin. However, the properties of starch are profoundly influenced by the chain length distribution and the frequency and spacing of the branch points that link the linear chains into complex macromolecules.

The predominant reasons for the lack of more detailed information on amylopectin chain lengths are the problematic separation of individual glucose oligomers with a DP in the range of 6–100, and the inherently low sensitivity of the commonly used mass-based detection systems. Typically, these two problems have not been solved simultaneously. For previously reported separations of high resolution involving high pH anion-exchange chromatography and pulsed amperometric detection (HPAEC-PAD) [9, 10], there have been quantification problems for maltooligosaccharides because response factors increase with increasing DP [9,11]. Furthermore, it has not been possible to isolate maltooligosaccharides standards above DP 17 due to the limited solubility of these oligosaccharides in aqueous solutions [11]. Unfortunately, most of the separations amenable to quantification provide unsatisfactory resolution across the required range of chain lengths (examples here include thin-layer chromatography, high performance liquid chromatography (HPLC), polyacrylamide gel electrophoresis (PAGE), and size exclusion chromatography), with the exception of capillary electrophoresis where the rapid separation and resolution of maltooligosaccharides up to DP 70 has been achieved [12]. High resolution separations are desirable because measurement of the ACL can mask subtle differences between oligosaccharide distributions, which is precisely where much of the information being sought is contained. In this chapter, we present a method that can satisfactorily deal with these problems in the analysis of maltooligosaccharides obtained by the debranching of amylopectins. To illustrate the broader applicability of the technique to polysaccharide structural analysis, distributions of oligosaccharides obtained from other polysaccharides (galactomannan, arabinan, glucuronoxylan, and pullulan) have also been included.

The procedure utilizes laser-induced fluorescence (LIF) detection to quantify the bands produced by the high resolution slab gel electrophoresis of 8-amino-1,3,6-pyrenetrisulfonic acid (APTS) reducing end–labeled oligosaccharides using a DNA sequencer. In this way, substantially higher resolution and sensitivity are obtained over traditional slab gel PAGE techniques. The principal advantage of this method is that it provides an opportunity for high resolution analysis of carbohydrates without the need for expensive dedicated instrumentation, so long as one has access to a molecular biology laboratory with a DNA sequencer. As we shall demonstrate, this method provides resolution and sensitivity equivalent to those of HPAEC-PAD [9-11,13] and capillary electrophoresis [12,14,15]. The identification of distinct differences in the amylopectin chain length distributions of starches from a variety of sources has been previously used to illustrate the efficiency of this method [16].

II. MATERIALS AND METHODS

A. Materials

Glycogen (bovine liver), glucose, maltotriose, and maltopentaose were obtained from Sigma (St. Louis, MO). Wheat starch and potato starch were generous gifts from Ian Batey (CSIRO Division of Plant Industry, Sydney,

Australia). Maltohexaose and maltoheptaose were obtained from Boeh-ringer Mannheim GmbH (Germany). Isoamylase was obtained as a suspension in 3.2 M ammonium sulfate (200 U/mL) from Megazyme (Sydney), whereas distributions of oligosaccharides obtained from the enzymatic digestion of carob galactomannan, sugar-beet arabinan, pullulan, and birch-wood glucuronoxylan as well as the standard oligosaccharides necessary to characterize them (1,4-β-D-mannotriose, 6^1-α-D-galactosylmannotriose, 1,5-α-L-arabinopentaose, 6^3-α-D-glucosylmaltotriose, and 1,4-β-D-xylotriose) were generous gifts from Barry McCleary (Megazyme). APTS was purchased from Lambda Fluoreszenztechnologie GmbH (Graz, Austria). Sodium cyanoborohydride was obtained from Aldrich (Milwaukee, WI). Biotechnology grade 20–50 mesh AG 501-X8(D) mixed bed ion-exchange resin was purchased from Bio-Rad (Hercules, CA).

B. Preparation of Labeled Oligosaccharides

Wheat starch, potato starch, and glycogen (bovine liver) were debranched with isoamylase according to the following method. The polysaccharides (50 mg) were suspended in 0.25 N NaOH (1 mL) and heated in a vigorously boiling water bath for 5 min before cooling to room temperature. Glacial acetic acid (32 μL) was added, followed by 50 mM sodium acetate buffer (4 mL, pH 4.0) and isoamylase (25 μL, 5 U), and the mixture was incubated at 37°C for 2 h before heating at 100°C for 20 min to denature the enzyme. The mixture was centrifuged at 14,000 × g (2 min, room temperature) and the supernatant desalted by the addition of 0.2 g/mL of a mixed bed ion-exchange resin (AG 501-X8(D)). This solution was stirred for 5 min, decanted, and diluted with distilled water to a total volume of 10 mL. Appropriate aliquots for fluorescent labeling (1–50 nmol) were determined from a reducing end assay [8], and evaporated to dryness in a centrifugal vacuum evaporator. The reaction scheme for the labeling of oligosaccharide reducing ends is outlined in Fig. 2.

These aliquots or oligosaccharide standards (5 nmol) were fluorescently labeled by the sequential addition of 5 μL of a solution of 0.2 M APTS in aqueous glacial acetic acid (15%) and 5 μL of freshly prepared 1 M aqueous sodium cyanoborohydride solution. The reaction mixture was incubated at 37°C for 15 h and diluted 100- to 1000-fold with an electrophoresis sample buffer consisting of 6 M urea in 40 mM boric acid and 40 mM tris–(hydroxy-methyl)aminomethane base buffer (pH 8.6) before loading.

C. Electrophoresis Using a DNA Sequencer

Sample volumes of 1–5 μL were loaded into the wells of uniform polyacryl-amide gels within the concentration range of 5–10% (37.5 : 1 ratio of acryl-amide to N,N'-methylenebisacrylamide as cross-linker) containing 8.3 M

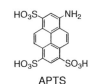

APTS

8-amino-1,3,6-pyrenetrisulfonic acid

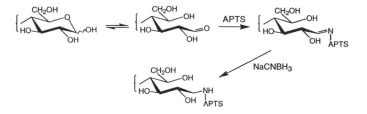

Figure 2 Reaction scheme depicting reductive amination of reducing end saccharides using 8-amino-1,3,6-pyrenetrisulfonic acid (APTS).

urea. They were electrophoresed using a buffer containing 89 mM Tris base, 89 mM boric acid, and 2 mM EDTA for 15 h at 40°C at a constant current of 30 mA using an Applied Biosystems (Perkin-Elmer Corporation, Foster City, California) 373A DNA sequencer. Fluorescence data was collected by the 373A DNA sequencer at the rate of 10 scans per min and analyzed using 672 Genescan software. The detection system of the DNA sequencer was calibrated by electrophoresing known amounts of APTS-derivatized maltoheptaose. The optimum polyacrylamide concentration required to maximize resolution of glucose oligosaccharides in the range DP 6–80 was determined to be 10%.

D. Efficiency of Labeling

To test the efficiency of labeling across a range of chain lengths, a solution containing equimolar amounts of glucose, maltotriose, maltopentaose, and maltoheptaose was prepared, and three 5-nmol aliquots (total reducing ends) removed for labeling as previously described. The relative labeling efficiencies for these compounds are shown in Table 1; they were calculated by comparing the relative peak areas with the maltopentaose peak, which was arbitrarily assigned as 100%.

III. RESULTS AND DISCUSSION

A. Derivatization by Reductive Amination

Due to the inherent lack of functionality of most carbohydrates, a prederivatization step is often required to create a means for either separation or

Table 1 Labeling Efficiencies of Maltooligosaccharides Using APTS

Chain length	Labeling efficiency (%)^{a,b}	Standard deviation
1	104.6	0.3
3	100.0	2.2
5	100.0	0.0
7	103.3	1.0

^{a}Values were calculated by comparing relative peak areas obtained from the average of three identical APTS labeling reactions.
^{b}Labeling efficiencies are relative to maltopentaose, which has had the value of 100% arbitrarily assigned.

increased detection sensitivity or, in some cases, both. By far the most commonly used procedure for labeling carbohydrates possessing a reducing end group is reductive amination, allowing the introduction of a single suitable chromophoric or fluorophoric label (Fig. 2). From this reaction, only a single fluorophore can be introduced per reducing end and, therefore, per oligomer. For a separatory procedure such as that described here, the label is required to confer both a charge for electrophoretic movement and increased detectability to the carbohydrate molecule. Polycyclic aromatic hydrocarbons bearing multiple sulfonic acid substituents have been commonly utilized for this purpose [17–19].

APTS was chosen as the preferred label because it possesses suitable excitation and emission wavelengths for the scanning laser detection system of the DNA sequencer. The literature contains a number of examples where similar labels have been carefully chosen to match as closely as possible the operating parameters of LIF excitation systems [17, 20]. In addition, APTS is readily available, simply introduced in high yield via reductive amination, and exhibits an intense fluorescence upon excitation at 488 nm [21].
excitation at 488 nm [21].

Fluorescent labeling of the oligosaccharide reducing ends with APTS was accomplished after adapting the originally reported conditions [22–24]. Guttman et al. [22] optimized the labeling reaction and observed a labeling efficiency of 97% for a number of different reducing end saccharides. However, by adding the reducing agent as a freshly prepared aqueous solution instead of as a tetrahydrofuran (THF) solution, the crude products of the labeling reaction produced reproducibly cleaner oligosaccharide profiles upon electrophoresis.

B. Efficiency of Labeling

For the oligosaccharide profiles shown in Fig. 3, if one assumes that the efficiency of APTS labeling at the reducing end is not influenced by chain length, the peak areas therefore allow direct characterization on a molar basis. Additionally, because the molecular weight of each oligomer is known, the results can also be calculated on a mass basis. To test the

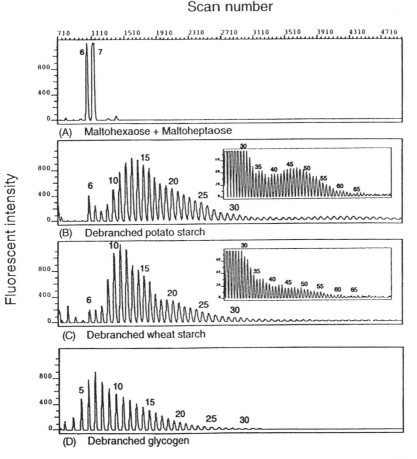

Figure 3 Electropherograms of APTS-labeled debranched polysaccharides obtained from a DNA sequencer. The numbers on each distribution indicate the chain length of the appropriate oligosaccharide as deduced by comparison with labeled maltohexaose and maltoheptaose. The insets are expansions of the appropriate regions of the electropherograms depicted in the main panels.

labeling efficiency with respect to increasing chain length, a freshly prepared solution containing equimolar amounts of oligomers with 1, 3, 5, and 7 glucose units was labeled with APTS under the standard conditions, and the peak areas compared. As can be seen in Table 1, a constant peak area was obtained for each oligomer, unlike the results of Klockow et al. [25] for the capillary electrophoretic separation of 8-amino-1,3,6-naphthalenetrisulfonic acid (ANTS) derivatives of maltooligosaccharides. Our results also exhibit a high level of reproducibility between three replicate samples, thereby providing the basis for quantification of these results. Despite these results, further work is required to prove that constant labeling exists to a chain length of 80 glucose units; however, relative trends between different samples can still be examined as we will outline.

C. Amylopectin Chain Length Profiles

As can be seen in Fig. 3, there is excellent resolution of APTS-labeled glucose oligomers obtained from debranched starches. Broadly speaking, the distributions are similar for all starches, with subtle but significant differences revealed upon treatment of the data obtained (to be discussed). Baseline resolution is observed over the range of chain lengths DP 6–70. In fact, increased resolution up to DP 80 is possible when larger loadings of the labeled samples are used (results not shown). Within this range of chain lengths, the incremental elution time between oligomers remains constant (Fig. 4), indicating that further detection of longer chains than those described here is hindered only by the constitution of the samples used, not by the available resolution of the technique. Close examination of the profiles reveals the peak shapes to be somewhat asymmetric; however, this is not apparent in the primary data and is evident only upon printing of the distributions when the data is compressed by the Genescan 672 software package. Importantly, the profiles depicted are reproducibly observed, with less than 5% variation in the absolute peak areas observed per oligomer among replicate samples, and much of this variation can be explained by the technical difficulty in attempting to reproducibly load 1–2 μL aliquots of labeled carbohydrates into the wells of the slab gel.

The samples illustrated were carefully chosen, as previous work has shown that the ACL varies greatly among all three [6,26–28], and the distribution profiles obtained for glycogen and wheat and potato amylopectins clearly highlight these differences. Glycogen contains the shortest chain length at the maximum of the distribution (DP 7), and it also contains shorter chains overall. Wheat and potato amylopectins show a similar distribution length to one another but vastly different profiles, with the latter displaying longer chain lengths at the maximum (DP 13) together with an

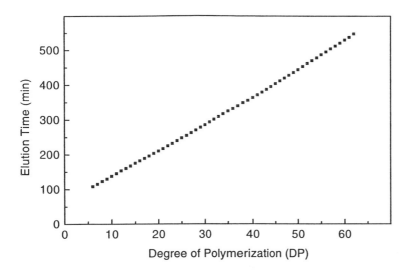

Figure 4 Plot of elution time of APTS-labeled oligomers from debranched potato starch versus degree of polymerization, highlighting the constant separation between oligomers. On this basis, baseline resolution is anticipated to continue for the separation of longer oligomers.

increased proportion of longer chains (particularly those from DP 40 to 60), which is in agreement with results generated using HPAEC-PAD [11].

As discussed in Section I, this information is far more detailed than the traditionally used average chain length determinations. Also, the colorimetric assays involved contain some degree of variation based on who performs them; for example, the average chain length of wheat amylopectin has been reported over a relatively wide range (ACL 18–23) [5].

As shown in the scheme for reductive amination using APTS (Fig. 2), there is only one fluorescent label introduced per reducing end, thus enabling quantification of the results using fluorescent detection. Importantly, the fact that baseline resolution between oligosaccharides was observed allows for differences between samples to be ascribed to differences in the populations of individual oligosaccharides. Before such calculations can be used, the issue of labeling efficiency with respect to chain length must be addressed. The results shown in Table 1 for the analysis of APTS-labeled glucose, maltotriose, maltopentaose, and maltoheptaose indicate that over those chain lengths, labeling via reductive amination is sufficiently constant to allow comparisons among quantities of oligomers present. Naturally, this does not preclude the possibility of a decrease in labeling efficiency for considerably longer chains, such as those in the typical de-

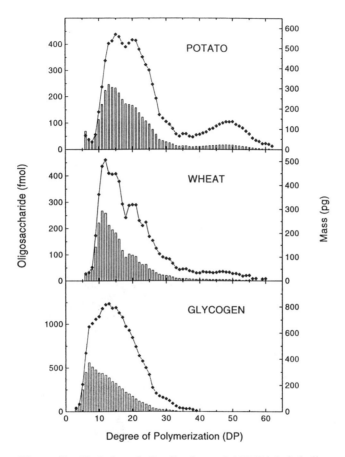

Figure 5   Chain length distributions of APTS-labeled oligomers from debranched polysaccharides in terms of both mass (--) and molar (histogram) responses.

branched amylopectin distribution (i.e., DP 6–80), although it does enable relative trends to be established among samples, if not an absolute description of minor differences.

Assuming equal efficiency of labeling across all chain lengths, the data obtained from the oligosaccharide profiles can be plotted on both a mass and molar basis (Fig. 5). Here, the molar data has been plotted as a histogram indicating a distribution with respect to each oligomer, in a similar fashion to the results of Koizumi et al. [11], who quantitated oligomer chain lengths up to DP 17 for a variety of amylopectins. A similar approach has been reported by Hanashiro et al. [13], who compared entire HPAEC-

PAD distributions by assuming equivalent detector response across all chain lengths. The plots on the basis of mass response are observed to approximate those reported throughout the literature where mass-based detection systems have been employed [6,28].

Additionally, direct comparisons can be made between samples over the full range of the oligomeric distributions. For these calculations, the percentage of the total mass present for each oligomer was determined for each distribution. Taking wheat starch as the yardstick, the previously determined oligomer percentages for glycogen and potato starch were subtracted from the corresponding values of wheat starch; the results are shown in Fig. 6. In this way, a positive value on the Y axis infers that a greater proportion of mass is present in that chain length than for wheat amylopectin, whereas a corresponding negative value indicates those chain lengths with a lower mass contribution. Here the observed differences in the distributions can be readily seen. With respect to wheat amylopectin, potato amylopectin contains less mass in chains in the range DP 8–16, but greater mass is present in longer chains (DP 17–28 and DP > 40). These results are very similar to the three size classes of potato maltooligosaccharides identified by Hanashiro et al. relative to an arrowhead chain length profile [13]. In the case of glycogen, a greater proportion of mass is found in lower chain lengths (below DP 10), with a lower contribution from the chains of longer than 20 glucose units.

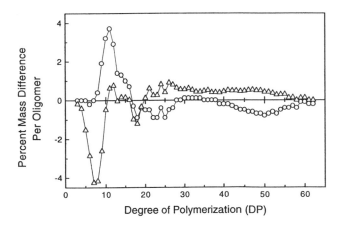

Figure 6 Comparison of the normalized mass distributions of APTS-labeled oligosaccharides from both debranched potato starch (-○-) and glycogen(-△-), with that of wheat starch. The calculations were made by subtracting the percentages of total mass present per oligomer of debranched potato starch and glycogen from the respective values for debranched wheat starch.

D. Oligosaccharides Produced from Other Polysaccharides

Although the principal reason behind the development of our technique was as a tool to aid starch structural analysis, other polysaccharides lend themselves to a similar examination, especially as the reductive amination labeling reaction is amenable for any reducing saccharide. To illustrate this, oligosaccharide distributions have been obtained from enzymatic digestions of nonstarch polysaccharides as listed in the following:

1. Limit digestion of carob galactomannan with guar β-mannanase followed by removal of galactose residues provided a series of β-D-mannosaccharides.
2. The endo-arabinase-mediated partial hydrolysis of previously debranched sugar-beet arabinan gave a series of α-L-arabinooligosaccharides.
3. The partial hydrolysis of birchwood glucuronoxylan with endo-xylanase provided a series of β-D-xylooligosaccharides.
4. The partial hydrolysis of pullulan with pullulanase and β-amylase gave a series of oligomers containing a repeating maltotriose unit.

Originally, Jackson reported the utility of reductive amination labeling with ANTS for a variety of mono- and oligosaccharides [17]. In a similar way, the foregoing examples have been chosen to illustrate the effectiveness of the method for polysaccharide analysis. The first three of these oligomeric distributions possess mannose, arabinose, and xylose as their respective reducing end carbohydrates, whereas pullulan has been included because it contains an interesting maltotriose repeat unit and will exhibit a longer oligosaccharide distribution than the others.

The oligosaccharide profiles obtained from the DNA sequencer under the standard conditions are shown in Fig. 7. As can be readily observed, the standard ladders of oligosaccharides obtained from samples 1–3 exhibit baseline resolution and a regular separation between oligomers. It should be noted that these distributions contain far shorter oligomeric chains than those we discussed for starch amylopectins.

The detailed characterization of oligosaccharides produced upon enzymatic treatment of carob galactomannan with β-D-mannanase has demonstrated that both a series of β-D-mannooligosaccharides and a variety of galactose-containing mannosaccharides are produced [29]. To simplify the observed profile for this study, the galactosyl residues were enzymatically removed from the oligomers before fluorescent labeling. The resultant distribution of β-D-mannooligosaccharides was observed within the range DP 2–9 (Fig. 7A), although very small quantities of longer chains up to DP 17

Figure 7 Electropherograms of APTS-labeled oligosaccharides derived from non-starch polysaccharides obtained from the DNA sequencer. The numbers on the distributions indicate the degree of polymerization of that particular oligosaccharide, which was deduced by comparison with appropriate standards. In the case of the pullulan digestion, the oligosaccharides have been identified in terms of the number of maltotriose units contained; thus, GM5 contains the glucosyl stub followed by five distinct maltotriose units all linked in a 1,6-fashion to one another (16 glucose units), and GM is 6^3-α-D glucosyl-maltotriose. The insets are expansions of the appropriate regions of the electropherograms depicted in the main panels.

were observed when the detection threshold was drastically lowered (not shown). The monosaccharide observed eluting before mannobiose may be mannose itself, or possibly galactose, which was cleaved but not separated from the oligomers prior to labeling.

The series of α-L-arabinooligosaccharides obtained from sugar-beet arabinan after debranching and partial hydrolysis with endo-arabinase shows a much longer profile, with oligomers detectable up to DP 35 (Fig. 7B). Most of the distribution consists of short chain material of DP < 10, and the profile contains only very small quantities of longer oligomers. The ladder of xylooligosaccharides produced from birchwood glucuronoxylan after partial hydrolysis with endo-xylanase is particularly short, with very little material present with a chain length longer than six units (Fig. 7C). In this series, a number of smaller contaminating bands can be observed, which may be explained by the presence of glucuronic acids.

From the pullulan digestion, the products obtained from treatment with pullulanase and β-amylase are no longer those of a straight chain oligomeric series of compounds. Pullulan is a neutral polysaccharide produced by the black yeast *Aureobasidium pullulans*, and it can be envisaged as a polymer consisting of maltotriose units joined to one another by terminal α-(1,6) linkages, thereby creating a macromolecule with a highly kinked structure as proposed to the branched structures of amylopectin or glycogen. Digestion with pullulanase cleaves only the α-(1,6) linkages, thereby exposing a maltotriose unit that can be cleaved by β-amylase, thus creating the structures shown in Fig. 8. Partial digestion with pullulanase ensures that a reasonable size range of these oligomers was obtained as shown in Fig. 7D. Naturally, the most abundant product was 6^3-α-D-glucosylmaltotriose, with ever decreasing amounts of longer oligomers. By decreasing the detection threshold, oligomers up to GM16 (DP 49) can be observed, highlighting the high resolution and sensitivity of the technique. The detection of longer chain oligomers than GM16 was not possible due to the

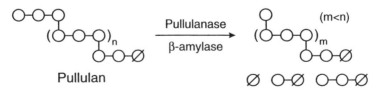

Figure 8 The structure of pullulan and the anticipated products obtained upon simultaneous treatment with pullulanase and β-amylase (–○– represents a D-glucose unit, whereas ⊘ represents a reducing terminal D-glucose unit).

nature of the enzymatic digest; however, their detection is possible when a shorter pullulanase digestion period was used (results not shown). Size exclusion chromatographic separation of these same fragments using parallel columns of Bio-Gel P-6 and P-2 has been reported; however, only resolution up to GM8 was obtained [30]. Stefansson and Novotny [12] used capillary electrophoresis to separate similarly labeled products obtained following pullulan isoamylolysis.

E. Resolution and Sensitivity

Our technique compares very favorably with those currently employed to perform oligosaccharide separations. Improved results are obtained with respect to previous HPAEC-PAD results, which had the limitation of being unable to quantitate maltooligosaccharides above DP 17 [9–11]. Through use of post-column enzyme reactor, quantitative HPAEC-PAD analysis across all chain lengths of debranched amylopectin is now reported in this volume (Chapter 34). The results obtained using a DNA sequencer are also comparable with those obtained for the resolution of maltooligosaccharides separated by capillary electrophoresis [12,14,15,31].

Resolution of ANTS-labeled oligosaccharides on high concentration polyacrylamide slab gel systems has been achieved by Jackson [17]; however, the method has limited range in the separation of glucose oligomers (up to DP 26). This is due to the nature of the detection system, which requires that all oligomers to be detected be present on the gel at the same time for photography, thus providing decreasing resolution as chain length increases. The DNA sequencer circumvents this by having a fixed scanning LIF detection system past which all oligomers are required to be electrophoresed, ensuring that no decrease in resolution is observed for longer chain lengths. This eliminates detection systems requiring either photography or digital imaging of the completed gel. A brief summary of the advantages and disadvantages of the aforementioned separatory techniques is given in Table 2.

As well as providing high resolution, our technique provides extremely high sensitivity, due to the detection system of the DNA sequencer. Dilution analysis of prelabeled samples allows the detection of individual oligomers down to a level of 1 fmol of labeled carbohydrate per peak in the electropherogram (data not shown). To put this in perspective, the amount of labeled carbohydrate electrophoresed to generate the distributions shown previously (15 ng per sample) constitutes less starch than is present in a single 30 μg diameter starch granule. Capillary electrophoresis and a LIF detector are capable of single attomole carbohydrate detection [31].

Table 2 Comparison of Methods for Oligosaccharide Separation

Characteristic	DNA sequencer	HPAEC-PAD [13]	Capillary electrophoresis [12]	Glyko FACE[a] system [17]
Prederivatization required	Yes	No	Yes	Yes
Maximum resolution (glucose oligomers)	DP = 80	DP = 81	DP = 70	DP = 26
Quantification possible	Yes	Yes	Yes	Yes
Resolution versus chain length	Constant	Constant	Constant	Decreasing
Potential for resolution of longer chains	Yes	Yes	Yes	No

[a]FACE = fluorophore-assisted carbohydrate electrophoresis.

F. Analysis Time

With regard to sample throughput of the DNA sequencer, the slab gels are electrophoresed for 15 h to obtain the full oligomeric distribution for amylopectin analysis, but a number of samples are run in parallel (up to 24 samples per gel), allowing an excellent rate of sample throughput with respect to traditional HPLC techniques. Naturally, the run times are considerably shorter for samples not requiring optimum resolution up to dp 80. Run times of 15 min for resolution of maltooligosaccharides up to dp 70 were possible with capillary electrophoresis [12].

G. Future Possibilities

Other types of carbohydrates that could be analyzed with the DNA sequencer include N-linked glycans. There are commercially available slab gel electrophoretic systems that are largely marketed as analytical tools in the structural identification of unknown glycans (most commonly used for the analysis of N-linked glycans from glycoproteins). In these systems, the unknown N-linked glycan is fluorescently labeled and various aliquots treated independently with one of a number of exoglycosidase mixtures, and these are applied separately to the lanes of a polyacrylamide slab gel. Structural information can then be deduced from the relative mobilities of the enzymatic digests. This type of analytical separation would be ideally suited to the DNA sequencer and would possess at least the same resolution as the commercially available instruments (with respect to both resolving power and detection sensitivity). There is already a report comparing such a system with capillary electrophoretic separations of similarly generated substrates [32].

Some technical aspects of the DNA sequencer present further possibilities with respect to this work. The data capture system of the sequencer obviously has the capability to distinguish between fluorescent labels with different emission characteristics; this may permit an internal standard to be electrophoresed in the same lane, thereby enabling sizing of oligomers using the existing Genescan software package. It may also prove possible to label samples at a number of stages during analysis with spectrally discrete fluorescent labels, which will provide more information on the molecular structure, from only a single electrophoretic run.

IV. CONCLUSIONS

This chapter has described a novel technique possessing general utility in the analysis of oligosaccharide chain lengths. In short, the technique provides significant advantages over commonly used methods for analyzing chain lengths because of the high resolution obtained and a highly sensitive laser-induced fluorescence detection system. The method currently permits baseline resolution of individual glucose oligosaccharides from DP 5 to 80, with a sensitivity down to the femtomole level. There is the additional possibility of analyzing chain lengths in excess of DP 80, although at this stage optimization has been performed with the analysis of debranched starches in mind.

Most of the work described herein involved the investigation of oligosaccharide chain length distributions, particularly the analysis of starch structures. Examples have been included to show the utility of our method for the analysis of other plant-derived oligosaccharides. Other oligosaccharides that possess a reducing end that can be efficiently labeled via a reductive amination protocol are also candidates for analysis with this method.

Perhaps one of the greatest advantages possessed by our method of electrophoresis is that no hardware alterations to a standard DNA sequencer are necessary in order to electrophorese labeled oligosaccharides. Such instruments are now commonplace in modern research institutions because of their importance as one of a suite of tools available to molecular biologists. In this way, carbohydrate chemists can have access to extremely high resolving electrophoretic techniques for the separation of oligosaccharides without the requirement of an expensive dedicated piece of equipment.

ACKNOWLEDGMENTS

The authors thank Ian Batey and Rudi Appels for helpful discussions and Lynette Preston and Michael Samuel for providing technical assistance when running the DNA sequencer and the associated software package. Special thanks to Barry McCleary at Megazyme, who kindly provided the non-starch-derived oligosaccharide distributions, the standard oligosaccharides necessary to characterize them, and helpful discussions. This research was supported by the Australian Grains Research and Development Corporation Grant CPS1.

ABBREVIATIONS

ACL	average chain length
ANTS	8-amino-1,3,6-napthalenetrisulfonic acid
APTS	8-amino-1,3,6-pyrenetrisulfonic acid
DP	degree of polymerization
FACE	fluorophore-assisted carbohydrate electrophoresis
HPAEC-PAD	high pH anion-exchange chromatography with pulsed amperometric detection
HPLC	high performance liquid chromatography
LIF	laser-induced fluorescence
PAGE	polyacrylamide gel electrophoresis
THF	tetrahydrofuran

REFERENCES

1. McCleary, B. V., and Matheson, N. K. (1986). *Adv. Carbohydr. Chem. Biochem.* 44, 147–276.
2. French, D. (1984). In Whistler, R. L., BeMiller, J. N., and Paschall, E. F., eds., *Starch: Chemistry and Technology*. Academic Press, Orlando, Florida, pp. 183–247.
3. Bruinenberg, P. M., Jackson, E., and Visser, R. G. F. (1995). *Chem. and Ind.* no. 21, 881–884.
4. Jane, J.-L., and Chen, J.-F. (1992). *Cereal Chem.* 69, 60–65.
5. Hizukuri, S., Kaneko, T., and Takeda, Y. (1983). *Biochim. Biophys. Acta* 760, 188–191.
6. Hizukuri, S. (1985). *Carbohydr. Res.* 141, 295–306.
7. Koehler, L. H. (1952). *Anal. Chem.* 24, 1576–1579.
8. Hizukuri, S., Takeda, Y., Yasuda, M., and Suzuki, A. (1981). *Carbohydr. Res.* 94, 205–213.
9. Koizumi, K., Kubota, Y., Tanimoto, T., and Okada, Y. (1989). *J. Chromatogr.* 464, 365–373.

10. Wong, K. S., and Jane, J. (1995). *J. Liq. Chromatogr.* 18, 63–80.
11. Koizumi, K., Fukuda, M., and Hizukuri, S. (1991). *J. Chromatogr.* 585, 233–238.
12. Stefansson, M., and Novotny, M. (1994). *Carbohydr. Res.* 258, 1–9.
13. Hanashiro, I., Abe, J.-I., and Hizukuri, S. (1996). *Carbohydr. Res.* 283, 151–159.
14. Stefansson, M., and Novotny, M. (1994). *Anal. Chem.* 66, 1134–1140.
15. Chiesa, C., and Horvath, C. (1993). *J. Chromatogr.* 645, 337–352.
16. O'Shea, M. G., and Morell, M. K. (1996). *Electrophoresis*, 17, 681–686.
17. Jackson, P. (1990). *Biochem. J.* 270, 705–713.
18. Lee, K. B., Al-Hakim, A., Loganathan, D., and Linhardt, R. J. (1991). *Carbohydr. Res.* 214, 155–168.
19. Chiesa, C. N., and O'Neill, R. A. (1994). *Electrophoresis* 15, 1132–1140.
20. Stack, R. J., and Sullivan, M. T. (1992). *Glycobiology* 2, 85–92.
21. Evangelista, R. A., Liu, M.-S., and Chen, F.-T. A. (1995). *Anal. Chem.* 67, 2239–2245.
22. Guttman, A., Chen, F.-T. A., Evangelista, R. A., and Cooke, N. (1995). *Glycobiology* 5, 717.
23. Chen, F.-T. A., and Evangelista, R. A. (1995). *Anal. Biochem.* 230, 273–280.
24. Evangelista, R. A., Liu, M.-S., and Chen, F.-T. A. (1995). *Anal. Chem.* 67, 2239–2245.
25. Klockow, A., Widmer, H. M., Amado, R., and Paulus, A. (1994). *Fresenius J. Anal. Chem.* 350, 415–425.
26. Rani, M. R. S., Shibanuma, K., and Hizukuri, S. (1992). *Carbohydr. Res.* 227, 183–194.
27. Gunja-Smith, Z., Marshall, J. J., and Whelan, W. J. (1970). *FEBS Lett.* 12, 96–100.
28. Hizukuri, S. (1986). *Carbohydr. Res.* 147, 342–347.
29. McCleary, B. V., and Nurthen, E. (1983). *Carbohydr. Res.* 118, 91–109.
30. Kainuma, K., Nogami, A., and Mercier, C. (1976). *J. Chromatogr.* 121, 361–369.
31. Liu, J., Shirota, O., Wiesler, D., and Novotny, M. (1991). *Proc. Natl. Acad. Sci. USA* 88, 2302–2306.
32. Guttman, A., and Starr, C. (1995). *Electrophoresis* 16, 993–997.

34

Recent Developments in Starch Structure Analysis Using High-Performance Anion-Exchange Chromatography with Pulsed Amperometric Detection

Kit-Sum Wong and Jay-lin Jane
Iowa State University, Ames, Iowa

I. INTRODUCTION

Analysis of starch and amylodextrin, which are homopolymers of glucose with no unique chromophore or fluorophore for direct detection, has always been a challenge. Many analytical techniques have been applied to starch analysis. High performance size exclusion chromatography (HPSEC) with laser light scattering (LLS) detection and refractive index (RI) detection [1–5] allows direct detection of carbohydrates and provides molecular weight distribution information. However, the nonspecific response of the RI detector prohibits the use of gradient eluents and hampers the separation resolution of amylodextrins. High performance capillary electrophoresis (HPCE) with laser-induced fluorescence (LIF) detection provides attractive separation resolution of amylodextrins [6,7], but a precolumn derivatization of amylodextrins is required to produce spectroscopically active compounds. The invention of the pulsed amperometric detector (PAD) [8] has allowed direct detection of carbohydrate at alkaline pH and the use of gradient eluents for chromatographic separation. HPCE and high performance anion-exchange chromatography (HPAEC) have been coupled with PAD for carbohydrate analysis [9–16]. The HPCE-PAD system has been applied to separate sugar alcohol and oligosaccharide mixtures [9,10], and its application to separation of polysaccharide mixtures remains to be studied. The HPAEC-PAD system, however, has already proven to be a powerful analytical technique for amylodextrins. Separation

Journal Paper No. J-16771 of the Iowa Agriculture and Home Economics Experiment Station, Ames, Iowa. Project No. 3258.

and detection of amylodextrins with degree of polymerization (dp) > 50 can be easily obtained [11,13,15,16].

Most HPAEC-PAD studies reported in the literature have been conducted with sodium acetate as the pushing agent. Few studies have evaluated the relative efficiency of various pushing agents. In addition, a major problem of using PAD to detect amylodextrins is that the mass sensitivity of PAD decreases with the increase of dp [17]. Therefore, PAD has failed to provide quantitative results for amylodextrin analysis.

To obtain quantitative results, we have improved the separation and detection of amylodextrin mixtures by HPAEC-PAD. Studies of different pushing agents to improve the separation of the amylodextrins and the application of a post-column enzyme reactor to achieve quantitative detection of amylodextrins are discussed in this chapter.

II. SELECTION OF PUSHING AGENT

For HPAEC analysis, a pushing agent is needed in the mobile phase to control the capacity factor, k', of ionic compounds so that a good separation within a reasonable period of time can be achieved. Two studies have been reported on the evaluation of pushing agents for HPAEC-PAD. Rocklin and Pohl [11] recommended sodium acetate as the pushing agent for the system, because it has a stationary phase affinity similar to that of hydroxide, and nitrate, sulfate, and carbonate reduced the load capacity of the column. Lu et al. [18] have suggested that sodium nitrate is a better pushing agent, because it is more effective in decreasing the retention time of the compounds to be analyzed while maintaining their baseline separation.

Either acetate or nitrate may be used as a pushing agent for the analysis of maltooligosaccharide mixtures with low dp. The selection of a pushing agent, however, is critical for the analysis of amylodextrin mixtures of high dp. Sodium acetate and sodium nitrate were compared as pushing agents in the analysis of branch chain-length distribution of normal maize amylopectin [19]. The isoamylase-debranched normal maize amylopectin (amylodextrin) was analyzed with a HPAEC-PAD system that included a DX300 gradient pump equipped with a PAD, and CarboPac PA$_1$ guard and analytical columns (Dionex, Sunnyvale, CA). For a complete separation of amylodextrin within 100 min, a moderate gradient of the concentration of nitrate (from 30 to 125 mM) was sufficient when nitrate was used as a pushing agent, but a gradient with higher elevation of the concentration of acetate (125 to 400 mM) was needed when acetate was used as a pushing agent (Fig. 1) [19]. The chromatograms of chain-length distribution of debranched normal maize amylopectin obtained by using the acetate and

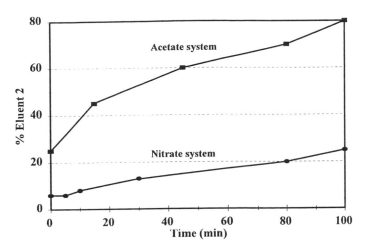

Figure 1 Gradient programs used in HPAEC-PAD for the separation of amylo-dextrins. Flow rate is 1.0 mL/min. Acetate gradient system: eluent 1 is 150 mM sodium hydroxide; eluent 2 is 150 mM sodium hydroxide and 500 mM sodium acetate. Nitrate gradient system: eluent 1 is 150 mM sodium hydroxide; eluent 2 is 150 mM sodium hydroxide and 500 mM sodium nitrate. (Adapted from Ref. 19, p. 66, courtesy of Marcel Dekker, Inc.)

the nitrate systems are shown in Fig. 2 [19]. The chromatogram using nitrate as the pushing agent (Fig. 2b) shows that minor peaks/shoulder (at dp 10 to 18), which may be isomers (carrying short stubs) of the adjacent major peaks [14], were separated from the linear amylodextrin peaks. With a signal-to-noise ratio of 2 as the criterion for the lowest measurable peak, amylodextrin (1.5 mg/mL) resolution, using the acetate and the nitrate systems, reached dp 58 and 66, respectively [19]. These results indicate that the nitrate system provides better resolution and high dp detection for amylodextrin analysis. In a study of PAD signal ratios for various amylode-xtrin sample concentrations, the use of the nitrate system also results in better accuracy (Fig. 3). With a column regeneration time of 10 min be-tween sample analyses, the retention time of each amylodextrin peak ob-tained by using the nitrate system had better consistency (lower standard deviation) than that obtained by the acetate system (higher standard devia-tion) (Fig. 4) [19]. Because of the small difference in the initial and final concentrations of nitrate in the gradient, the time required for column regeneration between runs was shorter for the nitrate system. Another ad-vantage of using nitrate as the pushing agent was that nitrate has a lower relative viscosity than acetate [20]; thus, less column back pressure was generated [18].

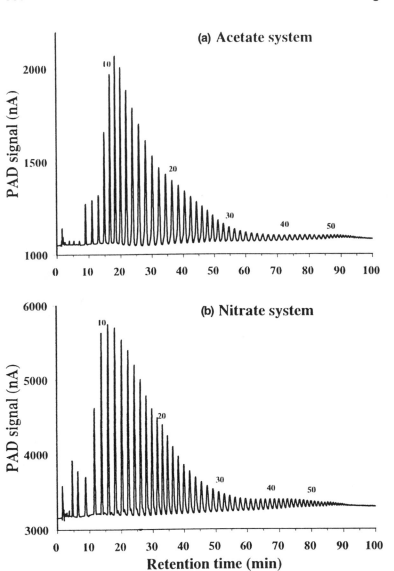

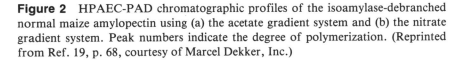

Figure 2 HPAEC-PAD chromatographic profiles of the isoamylase-debranched normal maize amylopectin using (a) the acetate gradient system and (b) the nitrate gradient system. Peak numbers indicate the degree of polymerization. (Reprinted from Ref. 19, p. 68, courtesy of Marcel Dekker, Inc.)

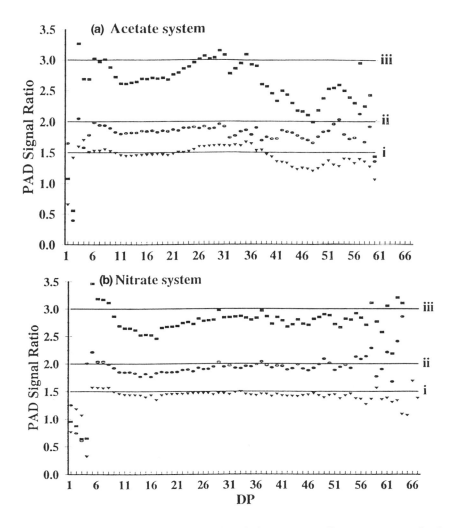

Figure 3 The signal ratios versus dp of the corresponding components in the debranched normal maize amylopectin with sample concentrations of (▼) 1.5 and 1.0 mg/mL, (●) 1.0 and 0.5 mg/mL, and (■) 1.5 and 0.5 mg/mL under (a) the acetate system and (b) the nitrate system. The solid lines represent the true values of the corresponding ratios. (Reprinted from Ref. 19, p. 77, courtesy of Marcel Dekker, Inc.)

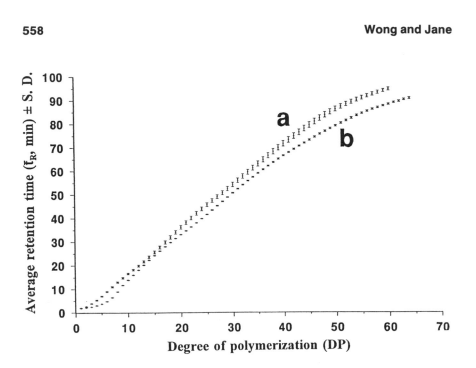

Figure 4 The plot of average retention time (from nine analyses) and its standard deviation versus dp of the corresponding component in the chromatograms of the debranched normal maize amylopectin with various concentrations under (a) the acetate system and (b) the nitrate system. (Reprinted from Ref. 19, p. 72, courtesy of Marcel Dekker, Inc.)

These data indicate that nitrate is a better pushing agent for the analysis of amylodextrin mixtures. Branch chain-length distributions of wheat, normal maize, waxy maize, amaranth, and tapioca debranched amylopectins were studied by using HPAEC-PAD with sodium nitrate as the pushing agent. The results are shown in Fig. 5. In comparison with amylopectins of other starches, amaranth and wheat amylopectins were enriched in chain lengths of dp 6–11 and dp 8–11, respectively. Waxy maize and normal maize amylopectins had similar amounts of short chains (A and B_1), but normal maize amylopectin had substantially more long chains (dp > 54) than waxy maize amylopectin (Fig. 5). Tapioca amylopectin had the most long chains (dp 45–67). The abundant short A and B_1 chains of wheat and amaranth amylopectins correlate with the extremely low viscosities of the two starch pastes and the low gelatinization temperature of wheat starch. More studies are needed to reveal whether the very short A and B_1 branches are responsible for these unique functional properties.

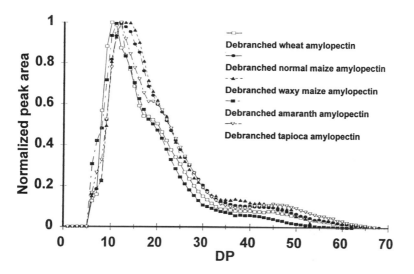

Figure 5 Branch chain-length distribution of amylopectin of various starches by using HPAEC-PAD with nitrate as the pushing agent.

III. QUANTITATIVE ANALYSIS BY HIGH PERFORMANCE ANION-EXCHANGE CHROMATOGRAPHY WITH POST-COLUMN ENZYME REACTION AND PULSED AMPEROMETRIC DETECTION (HPAEC-ENZ-PAD)

The sensitivity of PAD to a substance depends on the net charge and the diffusion coefficient of the substance to be measured [17]. Consequently, the detection responses of PAD to amylodextrins of different dp vary. To obtain a consistent detection response for a homologous series of amylodextrins, one may convert each HPAEC-separated amylodextrin into the same electrochemically active species, such as glucose [17]. Compared with acid hydrolysis [21,22], enzyme hydrolysis of starch and amylodextrin is selective and effective [23]. Among the starch-hydrolyzing enzymes, amyloglucosidase (AMG) produces exclusively glucose and, hence, is the best choice for the post-column enzyme hydrolysis of amylodextrin.

When an immobilized AMG reactor (2 mm i.d. × 23 mm) was installed in-line with the HPAEC-PAD, a quantitative analysis of branch chain-length distribution of amylopectin was achieved [24]. Although the enzyme activity of AMG dropped 40% after immobilization, the immobilized AMG reactor proved to be effective in converting debranched normal maize amylopectin into glucose.

The experimental setup of the HPAEC-ENZ-PAD for the analysis of debranched amylopectin is shown in Fig. 6 [24]. In this setup, nitrate was used as the pushing agent to separate amylodextrins and an AMG reactor operated at pH 4.5 was used to hydrolyze each HPAEC-separated amylodextrin into glucose. The effectiveness of the system was examined by using maltooligosaccharide standards (G1 to G7), and quantitative conversion of each HPAEC-separated maltooligosaccharide to glucose was obtained. When the HPAEC-ENZ-PAD system was applied to analyze the debranched tapioca amylopectin (0.5 mg/mL), the result shown in Fig. 7 was obtained [24]. The amylodextrin detector response was greatly enhanced without affecting the resolution of the homologous amylodextrin peaks. As a result, debranched chains with dp up to 77 were detected by this system.

Studies on the effects of reactor length, reactor temperature, and an additional α-amylase reactor have shown that the chromatographic profile of the debranched tapioca amylopectin sample was basically unaffected by increasing either the reactor length from 23 to 46 mm or the reactor temperature from 25°C to 50°C (the optimum reaction temperature for AMG) [24]. The increase in reactor length, however, decreased the detection sensitivity of amylodextrin, which may be attributed to diffusion. When an in-line α-amylase reactor was placed before the AMG reactor, a negative

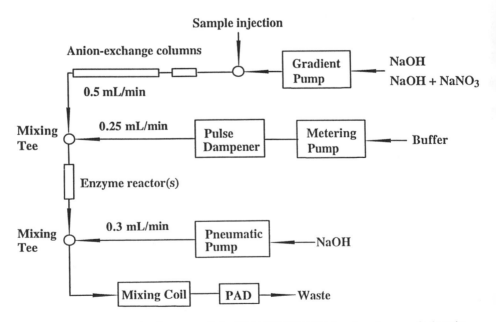

Figure 6 Schematic diagram of the HPAEC-ENZ-PAD setup for amylodextrin analysis. (Adapted from Ref. 24, p. 301, courtesy of Marcel Dekker, Inc.)

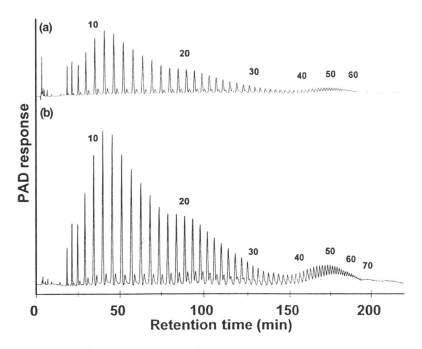

Figure 7 Chromatograms of 0.5 mg/mL debranched tapioca amylopectin obtained by using HPAEC-PAD without (a) and with (b) an AMG reactor. (Reprinted from Ref. 24, p. 304, courtesy of Marcel Dekker, Inc.)

effect on the AMG hydrolysis of amylodextrin resulted. In this study, to minimize the dilution and diffusion effects introduced by an additional pH-controlling buffer, both reactors were controlled at pH 5. Because AMG is more active on oligosaccharides with longer chains than on maltose and maltotriose [17,21], the decrease of the AMG reactor efficiency may be attributed to the decrease of AMG activity toward the maltose and maltotriose, produced by α-amylase at pH 5.

To investigate the quantitative analysis of the debranched amylopectin sample, we have compared the branch chain-length distribution of tapioca amylopectin obtained by using HPAEC-ENZ-PAD with that obtained using HPSEC-RI and gel permeation chromatography (GPC)–anthrone-sulfuric acid (Fig. 8) [24]. HPSEC-RI utilized TSK-GEL PW_{XL} guard column, and G_{4000} and G_{3000} columns (Tosohaas, Montgomeryville, Pennsylvania) for the separation. GPC utilized a 1.5 cm i.d. × 80 cm Bio-Gel P_6 column (Bio-Rad Laboratories, Richmond, CA) for the separation. The percentage peak areas of F1 and F2 obtained by HPAEC-ENZ-PAD

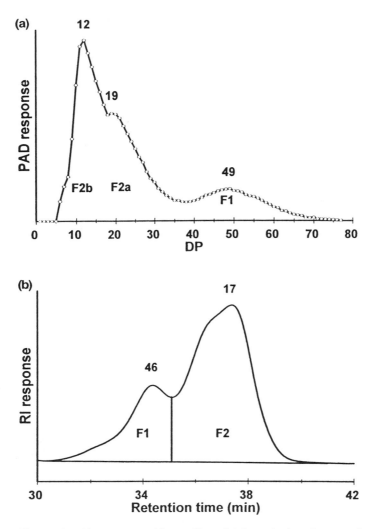

Figure 8 Chromatographic profiles of debranched tapioca amylopectin obtained by using (a) HPAEC-ENZ-PAD with one AMG reactor, (b) HPSEC-RI, and (c) GPC–anthrone–sulfuric acid. Peak numbers indicate the degree of polymerization. (Reprinted from Ref. 24, p. 306, courtesy of Marcel Dekker, Inc.)

and HPSEC-RI were in agreement (Table 1) [24]. These data indicate that quantitative analysis of branch chain-length distribution of amylopectin is achieved by using HPAEC-PAD with a post-column AMG reactor. It is not clear why the GPC–anthrone–sulfuric acid result was different from the results of the other two techniques. However, the difference in the molecu-

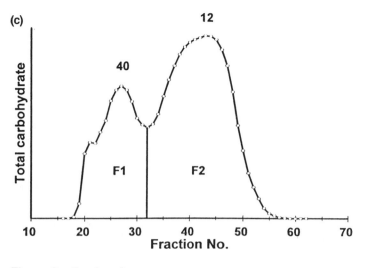

Figure 8 Continued

Table 1 Characterization of Branch Chain-Length Distribution of Tapioca Amylopectin Using Different Chromatographic Techniques

| Analytical methods | Peak dp[a] | | |
| | F1 | F2 | |
		F2a	F2b
HPAEC-ENZ-PAD[b]	48 ± 1[c] (21)	19[c] (32)	12[c] (47)
HPSEC-RI	46 ± 1[d] (17)	16.0 ± 0.5[d] (83)	
GPC	41 ± 3[e] (32)	12.5 ± 0.5[e] (68)	

[a]Average ±S.D.; numbers in parentheses represent the percentage peak area of the corresponding fraction.
[b]F1: DP 36 to 77; F2a: DP 19 to 35; F2b: DP 6 to 18.
[c]Three repetitions.
[d]Two repetitions.
[e]Three repetitions.
(Reprinted from Ref. 24, p. 307, courtesy of Marcel Dekker, Inc.)

lar weight cutoff between the gel permeation column and the high performance size exclusion columns may contribute to the differences observed.

IV. CONCLUSIONS

HPAEC-PAD is one of the analytical techniques used in starch structure analysis. When sodium nitrate, instead of sodium acetate, was used as the pushing agent in the HPAEC-PAD system, the separation resolution of amylodextrin peaks improved. As a result, higher dp detection, better accuracy, and better consistency in amylodextrin analysis were obtained. When a post-column AMG reactor was connected in-line with the HPAEC-PAD system (HPAEC-ENZ-PAD), the system separated individual homologous amylodextrins from de-branched amylopectin and gave quantitative detection results. The knowledge of the fine structure of amylopectin obtained by this powerful technique will lead to a better understanding of the correlation between the structure of a starch and its functional properties.

ACKNOWLEDGMENTS

The authors thank W. R. LaCourse, T. J. Lu, and D. Johnson for their suggestions and discussions, and Grain Processing Corporation for financial support of the study.

ABBREVIATIONS

AMG	amyloglucosidase
dp	degree of polymerization
ENZ	enzyme reaction
GPC	gel permeation chromatography
HPAEC	high performance anion-exchange chromatography
HPCE	high performance capillary electrophoresis
HPSEC	high performance size exclusion chromatography
LIF	laser-induced fluorescence
LLS	laser light scattering
PAD	pulsed amperometric detection
RI	refractive index

REFERENCES

1. Takagi, T., and Hizukuri, S. (1984). *J. Biochem. (Tokyo)* 95, 1459–1467.
2. Hizukuri, S. (1985). *Carbohydr. Res.* 141, 295–306.

3. Hizukuri, S., and Maehara, Y. (1990). *Carbohydr. Res.* 206, 145–159.
4. Takeda, C., Takeda, Y., and Hizukuri, S. (1993). *Carbohydr. Res.* 246, 273–281.
5. Ong, M. H., Jumel, K., Tokarczuk, P. F., Blanshard, J. M. V., and Harding, S. E. (1994). *Carbohydr. Res.* 260, 99–117.
6. Liu, J., Shirota, O., and Novotny, M. (1991). *Anal. Chem.* 63, 413–417.
7. Stefansson, M., and Novotny, M. (1994). *Anal. Chem.* 66, 1134–1140.
8. Hughes, S., Meschi, P. L., and Johnson, D. C. (1981). *Anal. Chim. Acta* 132, 1–10.
9. O'Shea, T. J., Lunte, S. M., and LaCourse, W. R. (1993). *Anal. Chem.* 65, 948–951.
10. Lu, W., and Cassidy, R. M. (1993). *Anal. Chem.* 65, 2878–2881.
11. Rocklin, R. D., and Pohl, C. A. (1983). *J. Liq. Chromatogr.* 6, 1577–1590.
12. Koizumi, K., Kubota, Y., Tanimoto, T., and Okada, Y. (1989). *J. Chromatogr.* 464, 365–373.
13. Koizumi, K., Fukuda, M., and Hizukuri, S. (1991). *J. Chromatogr.* 585, 233–238.
14. Ammeraal, R. N., Delgado, G. A., Tenbarge, F. L., and Friedman, R. B. (1991). *Carbohydr. Res.* 215, 179–192.
15. Murugesan, G., Hizukuri, S., Fukuda, M., and Juliano, B. O. (1992). *Carbohydr. Res.* 223, 235–242.
16. Kasemsuwan, T., Jane, J., Schnable, P., Stinard, P., and Robertson, D. (1995). *Cereal Chem.* 72, 457–464.
17. Larew, L. A., and Johnson, D. C. (1988). *Anal. Chem.* 60, 1867–1872.
18. Lu, T. J., Jane, J., and LaCoruse, W. R., unpublished data.
19. Wong, K.-S., and Jane, J. (1995). *J. Liq. Chromatogr.* 18, 63–80. Marcel Dekker, Inc., New York.
20. Weast, R. C., ed. (1989). *CRC Handbook of Chemistry and Physics*, 70th ed. CRC Press, Boca Raton, Florida.
21. Van Beynum, G. M. A., and Roels, J. A., eds. (1985). *Starch Conversion Technology*. Marcel Dekker, Inc., New York, pp. 101–142.
22. Östergärd, K., Björck, I., and Gunnarsson, A. (1988). *Starch/Stärke* 40, 58–66.
23. Banks, W., Greenwood, C. T., and Muir, D. D. (1973). *Starch/Stärke* 25, 405–408.
24. Wong, K.-S., and Jane, J. (1997). *J. Liq. Chromatogr.* 20, 297–310. Marcel Dekker, Inc., New York.

35

Red Seaweed Galactans
Methodology for the Structural Determination of Corallinan, a Different Agaroid

Carlos A. Stortz, Marcelo R. Cases, and Alberto S. Cerezo
University of Buenos Aires, Buenos Aires, Argentina

I. INTRODUCTION

A. Red Seaweed Galactan Structural Diversity

Galactans are the main matrix polysaccharides in most of the Rhodophyceae. Essentially, they consist of linear chains of alternating 3-linked β-galactose residues (A units) and 4-linked α-galactose residues (B units) [1]. The A units always belong to the D-series, whereas the B units may include residues of the D- or the L-series. These structures are usually masked by substitution with sulfate esters, pyruvic acid ketals, methoxyl groups, a 3,6-anhydro ring on the B unit, and variable length side chains [1]. Although these substitutions may produce several combinations, yielding radically different polysaccharides, the simplicity of a common backbone allows a clear classification to be made on the basis of the configuration of the B units: carrageenans if they belong to the D-series, agars for those with L-configuration, and the so-called "intermediate", hybrids with D- and L-galactose B units interspersed on the same molecules [1]. This simple arrangement allows for the placement of "corallinans" and other unusual polysaccharides within the galactan scheme, in spite of the complicated pattern arising from substitution.

The differences in substitution dramatically modify the properties of the extracted products. With so many structural variations possible, a wide range of properties are exhibited, ranging from stiff gels to products that do not gel at all. For example, the presence of the 3,6-anhydro ring induces gelling; major replacement of this unit by galactose 6-sulfate yields products without gelling properties. This emphasizes the practical need of determining the fine structure of these galactans.

B. Carrageenans

Carrageenans have commercial importance, mostly for food industry applications. Carrageenans appear in most of the red seaweeds from the order Gigartinales, with *Gigartina, Chondrus, Iridaea, Hypnea,* and *Eucheuma* the most commercially important genera. Separation of two carrageenan constituents (κ- and λ-carrageenan) was reported based on their differential solubility in potassium chloride solutions [2]. Later it was determined that carrageenan is not a mixture of two or more polysaccharides, but a family with continuously varying structures between idealized extremes [3]. However, from the biosynthetic point of view, the κ- and λ-groupings exist, because each is present in a different stage of the life cycle of families Gigartinaceae and Phyllophoraceae in the order Gigartinales [4,5].

The κ-family, produced by the gametophytes or by representatives of taxonomic families [6–9] that do not yield different products in the two life cycle stages, includes polysaccharides sulfated on C-4 of the A unit (Table 1). The B unit may include a 3,6-anhydro ring, thus rendering the polysac-

Table 1 Classification of Carrageenans According to Idealized Repeating Units

	Unit A 3-linked β-D-galactose	Unit B 4-linked α-D-galactose
κ-family		
κ (kappa)	4-sulfate	3,6-anhydro
ι (iota)	4-sulfate	3,6-anhydro 2-sulfate
μ (mu)	4-sulfate	6-sulfate
ν (nu)	4-sulfate	2,6-disulfate
λ-family		
λ (lambda)	2-sulfate	2,6-disulfate
ξ (xi)	2-sulfate	2-sulfate
π (pi)	2-sulfate, 4,6-(1-carboxy- ethyliden)	2-sulfate
θ (theta)	2-sulfate	3,6-anhydro 2-sulfate
β-family		
β (beta)	—	3,6-anhydro
α (alpha)	—	3,6-anhydro 2-sulfate
γ (gamma)	—	6-sulfate
δ (delta)	—	2,6-disulfate
ω-family		
ω (omega)	6-sulfate	3,6-anhydro
ψ (psi)	6-sulfate	6-sulfate

charide insoluble in KCl solutions, or a 6-sulfated galactose residue (Table 1), the biological precursor of the 3,6-anhydro ring, and a soluble product. Most of the products extracted from natural sources are actually hybrids that contain both 3,6-anhydrogalactose and its precursor.

The λ-family (Table 1), produced only by the sporophytes of some species of the Gigartinaceae and Phyllophoraceae, comprises polysaccharides devoid of 3,6-anhydrogalactose and galactose 4-sulfate [10] but sulfated on C-2 of the A unit (Table 1). Although originally accepted as soluble in KCl solutions, it was later shown that λ-carrageenans precipitate at high concentrations of this salt [11–13].

Two other families of carrageenans were defined from only isolated species. The β-family [14,15] (Table 1) is characterized by the lack of sulfate on the A unit, whereas the ω-family [16,17] contains a sulfate on C-6 of that unit. Other authors [18] classify these minor families differently.

It should be pointed out that most of the carrageenans are devoid of methoxyl or pyruvate groups, and side groups were observed in minute amounts in only a few representatives [19].

C. Agars and Agaroids

Classification of agars is not so well defined as that of carrageenans. However, some systematization has been described. At one extreme is agarose, the strongly gelling, uncharged polysaccharide produced by agarophytes with an idealized repeating structure of 3-linked β-D-galactose and 4-linked 3,6-anhydro-α-L-galactose [20]. However, partial replacement of the agarose B unit by L-galactose 6-sulfate is possible [21], with the other idealized extreme (the so-called "Yaphe's third extreme") recognized by the lack of 3,6-anhydrogalactose [22]. These structures are altered by the presence of sulfate, methoxyl, and pyruvyl groups. Agar constituents also include 4,6-O-(1-carboxyethylidene)-D-galactose, 6-O-Methyl-D-galactose, and 4-O-methyl-L-galactose [22,23]. For instance, the products from some species of the genus *Porphyra* have significant proportions of the A units methoxylated on C-6 (B units are L-galactose 6-sulfate). The nomenclature of agars is still confusing in the literature. Agars were defined as the gelling polysaccharides the main structure of which is that of agarose and its methoxylated, pyruvylated, sulfated, or glycosylated derivatives, whereas all the others (nongelling polysaccharides), including agar precursors (porphyran, etc.), were grouped as "agaroids" [18].

Most of the commercial agars are produced by representatives of the orders Gelidiales and Gigartinales (family Gracilariaceae). However, it has been shown that at least some carrageenophytes carry minor amounts of agar-like polysaccharides [13,24,25].

D. Variants

Many variants of these main structures of agars and carrageenans are recognized. It is noteworthy that the variants from the carrageenan group derive mainly from the order Cryptonemiales [26], whereas the agar variants come mainly from the order Ceramiales [18].

Within the Ceramiales, many agaroid products have also been reported. A common structural pattern is found in the family Rhodomelaceae. Significant levels of A units are methoxylated on C-6 and some B units are methoxylated on C-2. In *Laurencia pinnatifida*, most of the D-galactose units are sulfated on C-2, whereas most of the L-galactose units, and their 2-O-methylethers, are sulfated on C-6 (precursors of agarose polysaccharides) [27,28]. In *Polysiphonia lanosa*, the same pattern was found [29], although unusual A units were encountered (D-galactose 6-sulfate, and 6-O-methyl-D-galactose 4-sulfate). A polysaccharide (odonthalan) obtained from another seaweed of the Rhodomelaceae also contained 6-O-methyl-D-galactose 4-sulfate [30]. In *Laurencia undulata*, β-D-xylosyl side chains are found on C-4 of A units [31]. These side chains were also present in *Chondria macrocarpa* [32] and in *Laurencia nipponica* [33,34] on C-3 of 6-sulfated B units, which are then not cyclizable to 3,6-anhydro rings. Another ceramialean species (*Laingia* or *Okinamura pacifica*, now reclassified as *Delesseria crassifolia* [18]) from the Delesseriaceae produces a complex polymer (also containing glucuronic acid, 3-O-methylgalactose, 4-O-methylgalactose (both enantiomeric forms), and unidentified monosaccharides) in which xylosyl side chains were reported on C-2 of 3,6-anhydrogalactose units and C-6 of A units [35–38]. In the Ceramiaceae, a polysaccharide from *Ceramium rubrum* shows characteristics common to agar-like polysaccharides that are heavily methoxylated on C-6 of A units and on C-2 of 3,6-anhydro-L-galactose units [39]. However, a new characteristic of these polysaccharides is that many B units (more than one-third) are unsubstituted L-galactose units and thus are unable to generate 3,6-anhydrogalactose.

Within the order Cryptonemiales, many species from six different families were studied [26]. Mostly carrageenan-type structures were produced by these algae, but some common peculiarities included minor amounts of galactose with the L-configuration, methoxylation (uncommon in Gigartinales carrageenans), and sulfation only in the A units. For example, the aeodan polysaccharides produced by members of the Cryptonemiaceae [40–44] are characterized by A unit sulfation mostly on C-2 (less on C-4) and methoxylation on C-6, with the B units (never 3,6-anhydrogalactose) partially methoxylated on C-2, substituted with 4-O-methyl-L-galactose side chains on C-6, and devoid of sulfate. Minor amounts of agar-like disaccharides are interspersed in aeodans, whereas methoxyl groups are concentrated in blocks. Other species from the Cryptonemiaceae, like *Phyllimenia cornea* [45,46] and *Pachymenia carnosa* [47–49], produce similar polysaccharides, whereas those produced by *Phyllimenia hieroglyphica* differ by being devoid of methoxylated units, containing much less sulfate, and being xylosylated on C-6 of the B units [50]. On the other hand, *Pachymenia hymantophora* produces a polysaccharide in which the A units are mostly 2,6-disulfated, according to one study [50]. However, another study disagreed with that finding [51]. Both studies reported that no methoxyl groups were present and some B units included 3,6-anhydrogalactose and/or its precursor [50,51]. It is noteworthy that, for some of these polymers, structures with larger amounts of 1 → 3 linked galactose relative to 1 → 4 linked galactose were postulated, although oligosaccharides without the strictly alternating structure were not isolated [49]. The presence of 1 → 3 linked galactose blocks also was postulated for the polysaccharide of another member of the Cryptonemiales, namely *Dilsea edulis* (family Dumontiaceae) [52,53]. The polysaccharide produced by *Aeodes ulvoidea* was reported to contain 1 → 6 galactose linkages [43]. However, because this linkage has been reported in a green alga [54], the isolated disaccharide may come from a contaminant endophytic member of the Chlorophyceae. As mentioned, most of the aeodans have some repeating units of the agar family (B units of the L series). Other polysaccharides form the genus *Grateloupia*, particularly those from *G. divaricata* and *G. turuturu*, are really carrageenan–agar hybrids (not mixtures), in which at least 60% of the B units belong to the D-series [55–57], and also contain D-xylose and 3,6-anhydrogalactose substituents. Another interesting hybrid was found in *Pachymenia lusoria* [58] in which many of the A units are methoxylated on C-6 and/or sulfated on C-2. Other substitution patterns arise from pyruvyl groups on 4,6-positions, disulfation on 2,6-positions, and xylosyl groups on C-6. Unusual B unit structures: those cyclized to 3,6-anhydrogalactose or its 2-*O*-methyl ether belong to the L-series, whereas the uncycled 2-*O*-methylgalactose belongs to the D-series and the unsubstituted galactose belongs to both series, with the D-series predominating [58].

The polysaccharides from some other species of the Cryptonemiales have not received sufficient attention to draw structural conclusions, and others are very similar to those already mentioned. It was observed that those produced by the genus *Gloiopeltis* (family Endocladiaceae) [51,59–61] belong to the agar group.

Besides the already mentioned polysaccharides from the Cryptonemiales, other hybrid agar–carrageenan structures have been reported. While the exact structure of the polysaccharide from *Anatheca dentata* (Solieraceae, Gigartinales) was not established [62–64], the presence of L-galactose units (sulfated on C-3 or C-6 but not alkali-labile), xylosyl side chains, and traces of 3-*O*-methylgalactose characterized this polymer. Additionally, the replacement of some β-D-galactose units by β-D-glucuronic acid was postulated. In a polysaccharide from a family (Lomentariaceae, Rhodymeniales) not reported previously to produce galactan sulfates, a novel agar-predominant agar–carrageenan hybrid structure was found in *Lomentaria catenata* [65]. This polysaccharide was not methoxylated, but α-D-glucose (possibly on C-3 of B units) and glucuronic acid (on C-4 of A units) side chains were present [65].

As explained earlier, xylose side chains are present at different positions in many different polysaccharides. In the agar from *Melanthalia abscissa* (Gracilariaceae), xylose was located on C-4 or C-6 of A units, or C-3 of B units [66]. However, the xylogalactan polysaccharides from *Corallina officinalis* [67,68] have the highest levels of xylose yet reported [1].

II. STRUCTURAL DETERMINATION OF THE POLYSACCHARIDE FORM *CORALLINA OFFICINALIS*

A. Polysaccharide Preparation

Corallina officinalis is a cosmopolitan calcareous red seaweed. It belongs to the family Corallinaceae, included originally in the order Cryptonemiales but now recognized as a separate order (Corallinales) [69]. The milled seaweed was treated with dilute HCl to destroy the calcium carbonate covering [68]. Further extraction with water (room and boiling temperatures), centrifugation, precipitation with alcohol, and purification rendered products obtained in 0.5 and 0.4% yields. The products are similar; they are composed of monomethylated galactoses on the four available positions, and they lack 3,6-anhydrogalactose and its biological precursor.

The product was fractionated by means of ion-exchange chromatography (Fig. 1) [68]. Eleven fractions were isolated, with the analytical characteristics shown in Table 2 [68]. Most of the product (83% of the recovered) was eluted at NaCl concentrations between 0.3 and 0.5 M. Some minor products with low molecular weight were obtained; they were also the richest in protein [68].

Initially, composition analysis was performed based on the flame ionization detector response of the aldononitrile acetates of the component sugars [68]. However, it was later determined by several methods that, using those derivatives, the proportion of mono-*O*-methylgalactoses was severely overestimated. Table 3 shows the values obtained with this method and those obtained by derivatization as the alditol acetates and the 1-deoxy-1-(2-hydroxypropylamino)alditol acetates [70]. The latter method (to be

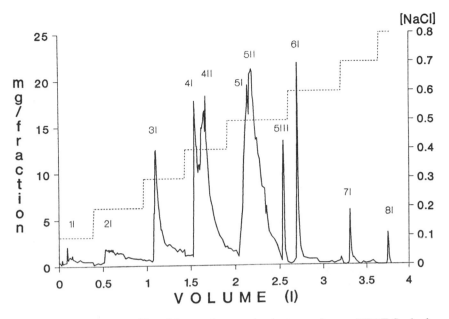

Figure 1 Elution profile of ion-exchange chromatography on DEAE-Sephadex A-50 of the polysaccharide extracted from *Corallina officinalis*. Fraction labels are listed above the peaks. (Reprinted from Ref. 68, with kind permission from Elsevier Science Ltd, The Boulevard, Langford Lane, Kidlington OX5 1GB, UK.)

discussed) also assesses the configuration of the sugars; the estimates for 2-*O*- and 3-*O*-methylgalactose are roughly one-half of those obtained using the aldononitrile acetates method. The values obtained with alditol acetates are in between those obtained from the other two derivatives. For other methylated sugars the same trend was observed, but in a lower proportion. The data obtained by ethylation analysis (to be discussed) shows the best coincidence with an average of the three methods.

The D/L ratio was estimated as close to 1 and the optical rotation was strongly negative, thus indicating that this polysaccharide has an alternating 3-linked β-D-galactose, 4-linked α-L-galactose agaroid structure.

Recently, the polysaccharides from other Corallinaceae seaweeds [71,72] were examined. Apparently, the xylogalactans have similar characteristics to those reported for *Corallina officinalis*, although methoxylated groups were not observed. Floridean starch and alginic acid also were isolated from these algae and characterized [71,72]. A similar xylogalactan [73] is produced by a species of a radically different order (Nemaliales) that coexists with the xylans and xylomannans typical of this alga. Additionally,

Table 2   Yields and Analyses of the Fractions Obtained by Ion-Exchange Chromatography of the Polysaccharide from *Corallina officinalis*

Fraction	Yield (%)	Carbohydrate (% anhydrous)	Sulfate (% SO$_3$Na)	Protein (%)	$[\alpha]_D$ (°)	M_n
1I	2.4	47.7	5.2	15.5	−17.6	1,500
2I	5.5	53.9	6.4	9.4	−46.8	2,700
3I	11.9	56.0	6.3	5.3	−52.7	5,500
4I	6.2	84.6	9.8	2.9	−70.0	12,500
4II	18.4	75.5	12.5	1.0	−90.7	21,000
5I	8.0	81.0	12.3	1.3	−85.9	33,000
5II	20.9	85.2	14.7	0.7	−99.5	49,000
5III	2.3	83.6	15.4	3.0	−84.7	12,500
6I	4.6	75.2	14.6	1.6	−82.9	29,000
7I	1.2	66.8	15.1	1.9	−67.8	15,000
8I	0.7	50.0	19.3	2.2	−48.1	15,000

Source: Reprinted from Ref. 68, with kind permission from Elsevier Science Ltd, The Boulevard, Langford Lane, Kidlington 0X5 1GB, UK.

a complex sulfated xylogalactomannan was reported from other species [74].

B.   Gas Chromatographic Separation of Enantiomeric Galactoses and Methyl Ethers

The traditional methods for determining monosaccharide configuration are optical rotation, circular dichroism, and enzymatic treatment [75]. However, these methods suffer from the disadvantage of requiring large amounts of hydrolysates in pure form or the availability of specific enzymes. With the advent of higher resolution capillary gas chromatography (GC) columns, a different approach was used based on the reaction of a sugar with an optically pure chiral secondary alcohol and the generation of mixed anomers of the alkyl glycosides formed. This mixture is diastereomeric to that obtained by the enantiomer of the sugar and, thus, can be separated by GC [76,77].

Each sugar gives rise to as many as four peaks. This may be regarded as an advantage, as it yields a monosaccharide-specific fingerprint for each enantiomer. However, for polysaccharides constituted by many monosaccharides, co-elution of peaks is difficult to avoid. This is especially true for polymers like those of *Corallina officinalis*, in which galactose coexists with its mono-*O*-methylethers, which usually produce closely eluting peaks [68]. Therefore, an approach using a chiral reagent yielding only one peak per

Table 3 Composition Analysis[a] of the Fractions of the *Corallina officinalis* Polysaccharide Derivatized as Acetylated Aldononitriles, Alditols, and 1-Amino-1-Deoxyalditols[a]

	1I[b]	2I	3I	4I	4II	5I	5II	5II	6I	7I	8I
Sulfate	30	26	27	26	33	32	36	37	39	43	72
Aldononitriles											
Xylose	24	19	23	25	28	32	34	25	21	4	8
2-O-Methylgalactose	25	30	34	25	24	23	24	26	29	40	42
3-O-Methylgalactose	10	5	4	7	7	9	7	5	4	3	2
6-O-Methylgalactose	7	5	tr	tr	tr	—	—	4	6	10	12
4-O-Methylgalactose	10	12	19	13	2	1	1	1	3	7	2
Galactose	58	60	62	68	69	68	69	65	61	47	44
Alditols											
Xylose	22	21	15	20	24	33	32	23	18	nd	nd
2-O-Methylgalactose	17	22	20	16	15	15	14	16	18	nd	nd
3-O-Methylgalactose	4	3	2	3	4	5	2	3	2	nd	nd
6-O-Methylgalactose	4	2	1	1	1	1	1	2	4	nd	nd
4-O-Methylgalactose	5	7	10	5	1	1	—	1	2	nd	nd
Galactose	75	73	77	80	80	79	83	79	76	nd	nd

Table 3 Continued

	1I[b]	2I	3I	4I	4II	5I	5II	5II	6I	7I	8I
Sulfate	30	26	27	26	33	32	36	37	39	43	72
Aminodeoxyalditols											
D-Xylose	25	19	17	20	28	34	34	29	21	8	12
2-O-Methyl-D-galactose	–	–	tr	tr	–	tr	–	–	–	–	–
2-O-Methyl-L-galactose	13	15	16	12	11	10	10	12	14	22	24
3-O-Methyl-L-galactose	3	3	2	3	4	5	4	3	2	1	tr
6-O-Methyl-D-galactose	5	2	1	2	1	1	1	2	4	9	11
4-O-Methyl-D-galactose	5	7	11	5	3	2	1	2	3	5	2
4-O-Methyl-L-galactose	2	2	1	1	1	tr	tr	tr	tr	tr	–
D-Galactose	46	42	38	46	54	51	53	52	48	41	43
L-Galactose	33	38	43	37	30	33	32	31	32	27	22

[a] Expressed as mol/100 mol galactose + 2-O-methylgalactose + 3-O-methylgalactose + 6-O-methylgalactose.
[b] Fraction 1I also showed significant amounts of D-Man, D-Glc, L-Rha, L-Ara, and L-Fuc. Some of these sugars also appeared in other fractions, usually in trace amounts.
Source: Adapted from Refs. 70 (with kind permission of Elsevier Science -NL, Sara Burgerhartstraat 25, 1055 KV, Amsterdam, Netherlands) and 86.

sugar was required. The procedure utilized involved the reaction of the sugar aldehyde group with a chiral amine to yield an imine [78], which is then reductively aminated to give an open chain 1-amino-1-deoxyalditol (Fig. 2), and finally acetylated [70]. Reaction of D-galactose with racemic 1-amino-2-propanol produced the same amounts of both diastereomers (by ^{13}C nuclear magnetic resonance (NMR) spectroscopy), which following acetylation were resolved by capillary GC as two equal-size peaks the structures of which were confirmed by GC-mass spectrometry. After a careful study to improve the yield of the aminoalditols, it was concluded that the reaction should be carried out using a molar ratio of amine : sugar : NaBH$_3$CN : AcOH of 5 : 1 : 1.1 : 10 in methanol, for 1 hr at 65°C.

The method was applied condensing the (S)-1-amino-2-propanol (and the (R)-, (S)-racemic mixture) to different monosaccharides. Comparing the results obtained with the chiral amino alcohol and the racemic mixture, the GC retention times of the derivatives of both enantiomers were assigned (those produced by the R amino alcohol being equivalent to those produced by the enantiomer of the sugar with the S amino alcohol). The derivatives of most of the enantiomeric sugars gave baseline resolution (Table 4) [70], although those of 2-O-methylgalactose and glucose were not separated. Some co-elution between 4-O-methyl-L-galactose and 3-O-methyl-D-galactose, L-rhamnose and D-fucose, and D-mannose with D-glucose occurred (Table 4). No enantioselectivity was observed. Because the 2-O-methylgalactose pair was not separated and because 2-O-methylgalactose is an important constituent of many seaweed polysaccharides, the same technique was applied using (S)-α-methylbenzylamine [78] as the chiral amine. With this amine, a pronounced enantioselectivity was observed, which was especially marked for galactose (the D sugar reacts with a yield 2.3 times larger than the L one). Otherwise, sugars with the *manno* configuration were resolved much better, and the 2-O-methylgalactose enantiomers were well resolved (Table 5) compared with the previous amine [70]. No enantio-

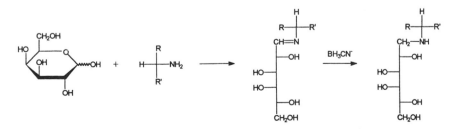

Figure 2 Reaction between a sugar (D-galactose) and a chiral amine to yield a 1-amino-1-deoxyalditol.

Table 4 Retention Times[a] and Separation Factors (r) of Acetylated Aminoalditols of Enantiomeric Sugars Derivatized with (S)-1-Amino-2-Propanol with Different Chromatographic Programs[b]

	Program A			Program B		
	D	L	r	D	L	r
Rhamnose	1.644	1.653	1.005	1.738	1.750	1.007
Fucose	1.656	1.667	1.006	1.757	1.770	1.007
Arabinose	1.694	1.686	1.005	1.803	1.792	1.006
Xylose	1.720	1.729	1.006	1.840	1.854	1.008
Glucose	2.289	2.294	1.002	2.605	2.609	1.002
Mannose	2.291	2.304	1.006	2.598	2.612	1.005
Galactose	2.341	2.356	1.007	2.642	2.659	1.006
2-O-Methylgalactose	2.103		1			
3-O-Methylgalactose	2.234	2.250	1.007			
4-O-Methylgalactose	2.214	2.225	1.005	2.503	2.519	1.006
6-O-Methylgalactose	1.995	2.006	1.006			

[a]Relative to peracetylated *myo*-inositol = 1 (20.95 min in Program A, 19.40 min in Program B).
[b]The column used was Ultra-2, 50 m × 0.2 mm. An HP-5 column gives similar results.
Source: Reprinted from Ref. 70, with kind permission of Elsevier Science -NL, Sara Burgerhartstraat 25, 1055 KV, Amsterdam, Netherlands.

selectivity was observed with these latter examples using (S)-α-methylbenzylamine.

Preparation of the acetylated 1-deoxy-1-(2-hydroxypropylamino) alditols is as easy as that of the acetylated aldononitriles or alditols, the usual choices for composition analysis of polysaccharides. Furthermore, this new derivatization procedure permits working with as little as 1 mg of polysaccharide (roughly the same as traditional methods). It should be noted that enantiomeric sugars placed in a diastereomeric environment are as different as other isomers (e.g., the 2-epimers, D-glucose, and D-mannose) [79]. Therefore, it is advisable to use this technique routinely, even if there are no suspicions of rare enantiomers [70].

C. Methylation Analysis

Methylation analysis of the 11 fractions [80] was carried out using the Hakomori method on the triethylammonium salts [81], obtained by passage through a cationic resin. Attempts to obtain them by dialysis against triethylamine hydrochloride were unsuccessful. The results are shown in Table 6 [80]. Larger amounts of 2,3,4,6-tetra-O-methylgalactose were present in

Table 5   Retention Times[a], Separation Factors (*r*), and Enantioselectivity Ratio $r_{D/L}$ of Acetylated Aminoalditols Originated in Enantiomeric Sugars Derivatized with (*S*)-α-Methylbenzylamine Using the Column Ultra-2 (Program C)

	D-	L-	*r*	$r_{D/L}$[b]
Rhamnose	2.377	2.316	1.026	0.9
Fucose	2.343	2.365	1.009	1.6
Arabinose	2.389	2.375	1.006	0.7
Xylose	2.440	2.448	1.003	1.5
Glucose		3.109	1	
Mannose	3.107	3.048	1.019	0.7
Galactose	3.094	3.114	1.006	2.3
2-*O*-Methylgalactose	2.833	2.867	1.012	1.0
3-*O*-Methylgalactose		3.009	1	
4-*O*-Methylgalactose	2.953	2.968	1.005	3.1
6-*O*-Methylgalactose	2.728	2.748	1.007	1.8

[a]Relative to peracetylated *myo*-inositol = 1 (20.60 min).
[b]Area ratio of the peaks corresponding to both enantiomers.
Source: Reprinted from Ref. 70, with kind permission of Elsevier Science -NL, Sara Burgerhartstraat 25, 1055 KV, Amsterdam, Netherlands.

the fractions that were also rich in 4-*O*-methylgalactose, supporting the conclusion that this sugar is present as a single sugar side chain [80].

D. Desulfation–Methylation Analysis

Attempts at desulfating by the solvolytic method were unsuccessful. Some fractions were submitted to methanolic-HCl desulfation, which gave 90% yield of products that were about 50% desulfated [80]. Methylation analysis of these fractions (Table 7) indicated that sulfate was located on positions 2 and 3 of 4-linked units and on position 6 of 3-linked units [80]. Because desulfation was not complete, it was hard to determine if 4-linked units only contained sulfate and methoxyl groups as substituents, although other results strongly support this conclusion [80].

E. Ethylation Analysis

Given the significant proportions of natural monomethylated galactoses in the *Corallina officinalis* polymers, methylation analysis leaves uncertainties, which can be overcome using trideuteromethyl iodide or ethyl iodide as the alkylating agents. The former method gives the same GC separation patterns and, therefore, requires the use of GC-mass spectrometry for inter-

Table 6 Methylation Analysis of the Corallinans

Position of O-methyl	Fraction[a]										
	1I	2I	3I	4I	4II	5I	5II	5III	6I	7I	8I
Xylose											
2,3,4-	15	13	14	16	28	30	26	27	20	8	6
Galactose											
2,3,4,6-	22	19	23	12	3	2	1	2	5	8	4
3,4,6-	8	6	10	4							
2,3,4-	6	2									
2,4,6-	16	13	10	9	5	4	5	7	8	14	15
2,3,6-	21	19	20	25	26	25	24	28	28	34	33
2,6-	5	11	13	9	11	10	10	11	10	7	6
4,6-	2	1		2	1						
3,6-	2	3	3	5	7	9	9	7	5	3	3
2,4-	18	20	18	33	44	48	51	45	40	32	36
2,3-					1						
6-		2	2								
2-	2	2	1	1	2	2			4	2	3

[a]Expressed as mol/100 mol galactoses. Percentages smaller than 1% are not shown.
Source: Reprinted from Ref. 80, with kind permission of Elsevier Science -NL, Sara Burgerhartstraat 25, 1055 KV, Amsterdam, Netherlands.

Table 7 Methylation Analysis of Desulfated Corallinans

Position of O-methyl	Fraction[a]					
	3I	4I	4II	5I	5II	5III
Xylose						
2,3,4-	19 (+5)	25 (+9)	71 (+43)	44 (+14)	36 (+10)	38 (+8)
Galactose						
2,3,4,6-	26 (+3)	10 (−2)	4 (+1)	4 (+2)		3 (+1)
3,4,6-	9 (−1)	4 (0)				
2,4,6-	14 (+4)	12 (+3)	14 (+9)	12 (+8)	11 (+6)	20 (+13)
2,3,6-	25 (+5)	33 (+8)	28 (+2)	43 (+18)	39 (+15)	38 (+10)
2,6-	9 (−4)	7 (−2)	4 (−7)	2 (−8)	3 (−7)	2 (−9)
3,6-	<1 (−3)	1 (−4)	(−6)	1 (−8)	2 (−7)	<1 (−7)
2,4-	16 (−2)	33 (0)	48 (+4)	39 (−9)	44 (−7)	36 (−9)

[a]Expressed as mol/100 mol galactoses. In parentheses are the differences from original values.

Source: Reprinted from Ref. 80, with kind permission of Elsevier Science -NL, Sara Burgerhartstraat 25, 1055 KV, Amsterdam, Netherlands.

pretation and quantitation of the chromatograms. The latter approach was already used by Albersheim and coworkers [82,83] as an alternative to methylation to achieve different chromatographic separations. Besides using a cheaper reagent, the ethyl iodide procedure produced a significant change in the chromatographic behavior, although it yielded similar separation order of the derivatives formed. Ethylation of polysaccharides was accomplished in the same way that they were methylated, but a second addition of carbanion and iodoalkane was needed to complete the alkylation [84]. Tables 8 and 9 show the relative retention times of ethylated and methylated galactoses derivatized as alditol acetates and aldononitrile acetates, respectively [84]. Column age contributed to differences in resolution of the sugar derivatives. The values on the tables are those for an aged column (~3 years). The column age effect was, however, not excessive [84]. Despite the molecular weight increment, ethylated derivatives appeared earlier than the corresponding methylated ones, due to a decrease in their polarity. The ethylated alditol acetates produced more co-eluting

Table 8 Relative Retention Times[a] of Methylated and Ethylated Galactoses, Derivatized as Alditol Acetates, on an Aged SP-2330 Capillary Column[b] Using Two Different Programs

Position of O-alkyl	Isothermal		Programmed	
	Methylated	Ethylated	Methylated	Ethylated
2,3,4,6-	358	235	602	420
2,4,6-	514	367	763	625
3,4,6-	553	371	799	633
2,3,6-	568	380	810	646
2,3,4-	692	415	892	687
2,6-	735	562	908	811
4,6-	738	571	908	815
3,6-	847	634	955	861
2,3-	1053	724	1016	908
2,4-	1083	773	1024	929
6-	933	844	980	951
2-	1319	1098	1078	981
3- (=4-)	1694	1362	1156	1084
—	1728	1741	1164	1167

[a]Relative to per-O-acetylated xylitol = 1000.
[b]Approximately three years old.
Adapted from Ref. 84, with kind permission of Elsevier Science -NL, Sara Burgerhartstraat 25, 1055 KV, Amsterdam, Netherlands.

Table 9 Relative Retention Times[a] of Methylated and Ethylated Galactoses, Derivatized as Aldononitrile Acetates, on an Aged SP-2330 Capillary Column[b], Using Two Different Programs

Position of O-alkyl	Isothermal		Programmed	
	Methylated	Ethylated	Methylated	Ethylated
2,3,4,6-	328	238	535	399
2,4,6-	498	390	729	633
3,4,6-	541	377	773	620
2,3,6-	596	440	815	691
2,3,4-	633	425	846	674
2,6-	797	664	929	868
4,6-	863	680	955	875
3,6-	902	695	968	882
2,3-	1163	898	1045	968
2,4-	1081	852	1020	953
6-	1135	1021	1043	1002
2-	1514	1385	1123	1094
3-	1880	1548	1200	1128
4-	2050	1697	1230	1157
—	2279	~2260	1279	1292

[a]Relative to per-O-acetylated xylitol = 1000.
[b]Approximately three years old.
Adapted from Ref. 84, with kind permission of Elsevier Science -NL, Sara Burgerhartstraat 25, 1055 KV, Amsterdam, Netherlands.

peaks than the aldononitrile derivatives, though the same trend was observed for methylated products. The less volatile aldononitrile derivatives separated better, requiring longer run times. The mass spectrometry fragmentation patterns of ethylated derivatives were easily distinguishable based on those of the methylated derivatives [84].

Ethylation analysis (Table 10) confirmed many structural features by matching the D/L configuration determination [80]. Assuming an alternating galactan structure, 4-O-methylgalactose appears to be a side chain, though minor amounts of 2-linked double side chains cannot be excluded. The 2-O-methylgalactose was detected mostly in unsubstituted 4-linked L-galactose B units, though in some fractions it was present as nonreducing termini and/or in 6-substituted A units (corresponding to the small amounts of 2-O-methyl-D-galactose observed by GC). The 3-O-methyl-galactose was also detected in unsubstituted 4-linked B units, although in fraction 1I (low M_r) it also appears in other locations. The 6-O-methyl-

Table 10 Ethylation Analysis of the Corallinans

Position of O-alkyl	Position of O-methyl	Fraction[a]								
		1I	2I	3I	4I	4II	5I	5II	5III	6I
Xylose										
2,3,4-	—	17	20	31	25	51	47	56	46	25
Galactose										
2,3,4,6-	—	7	10	10	4					1
2,3,4,6-	2-	7	6	6	1					
2,3,4,6-	3-	3								
2,3,4,6-	4-	5	7	12	7	3			2	4
3,4,6-	—	6	4	8	1					
3,4,6-	4-	4	3	2	1					
2,3,4-	—	3	1		2					
2,3,4-	3-	2								
2,4,6-	—	7	9	10	7	5	4	5	4	5
2,4,6-	6-	3	3	1	2	1			2	4
2,3,6-	—	12	9	8	12	13	12	11	13	14
2,3,6-	2-	11	10	9	11	12	12	12	14	14
2,3,6-	3-	2	2		3	4	5	4	4	3
2,6-	—	7	10	10	13	11	11	11	13	12
4,6-	—	1								
3,6-	—	3	4	1	6	9	12	9	8	5
2,4-	—	12	14	21	22	42	42	48	35	36
2,4-	2-	1	4		5		2		4	2
2,3-	—	2	1		1					
6-	—	2	3	2	2				1	

[a]Expressed as mol/100 mol galactoses. Percentages smaller than 1% are not shown.
Source: Ref. 80, with kind permission of Elsevier Science -NL, Sara Burgerhartstraat 25, 1055 KV, Amsterdam, Netherlands.

galactose only appeared in 3-linked unsubstituted units. It was observed that only low levels of methyl groups were present in monomeric units containing another substituent [80].

F. NMR Spectroscopy

Unfortunately, NMR spectra of these complex xylogalactans are not as simple as those from seaweed galactans that have a regular repeating disaccharide unit. The ^{13}C NMR spectra of three fractions were obtained with a 400-MHz instrument at high temperature [85]. The most striking feature was the appearance of five sharp signals corresponding to the β-D-xylosyl side chains (106.0, 75.6, 78.3, 71.9, and 67.7 ppm), suggesting that they are attached to just one position (i.e., C-6 of D-galactose units) [80]. Possibly, the mobility of the 1 → 6 linkage was responsible for the sharpness of the signals. The signals of methoxyl groups appeared around 59–60 ppm, and signals corresponding to unsubstituted C-6 were present at 63 ppm. Between 69 and 80 ppm, the signals corresponding to carbons carrying secondary hydroxyl groups and substituted C-6 appeared, whereas the glycosylated carbons were observed at 80–85 ppm. Fractions 5I and 4II clearly illustrated the signals corresponding to the anomeric carbons (106.0 and 105.8 ppm) assigned to β-D-galactosyl units substituted on C-6 by xylose and sulfate. At 103.2 and 103.0 ppm two separate signals appeared, which should correspond to the α-L-galactose units, possibly split due to the presence or absence of C-3 substitution. The signals at 101.2 and 100.8 ppm were produced by sulfate and methoxyl groups 2-substituted on α-L-galactose residues. In the spectrum of fraction 3I, a signal around 98.9 ppm appeared and that at 101.2 disappeared, suggesting that part of this fraction had a sharply different substitution pattern.

The ^{1}H NMR spectrum of fraction 5I contained signals corresponding to the methoxyl groups (3.45–3.55 ppm). In the anomeric region, three signals at δ 4.44–4.48 ppm represented the β-D-xylosyl and β-D-galactosyl units, whereas four separate signals were observed for the α-L-galactosyl units (also observed in the ^{13}C NMR spectra). Substitution on C-2 and/or C-3 with sulfate and methoxyl groups produced the 5.28, 5.35, 5.48, and 5.54 ppm signals. An almost negligible signal due to pyruvic acid ketals was also found.

III. THE STRUCTURE OF THE CORALLINANS

Conclusions on the structure of the corallinans were drawn from the analytical, composition, methylation, desulfation–methylation, ethylation, and

NMR data [80]. The alternating [→3)-β-D-Gal$p$-(1 → 4)-α-L-Gal$p$-(1 →] structure was modulated by substitution with sulfate and β-D-xylosyl side chains (plus minor amounts of methoxyl and 4-O-methylgalactosyl side chains) at position 6 of A units, and at positions 2 and 3 of B units by methoxyl and sulfate groups. The proportion of the substituents on the backbone could have been calculated for each fraction. However, the fractions were organized into several groups [86]. Table 11 shows the substitution patterns suggested for these groups. The main group, formed by fractions 4II, 5I, and 5II, represented almost 60% of the corallinans. The fundamental structure of the main group included C-6 of A units almost completely substituted, and about the same proportions of substitution on C-2 and C-3 of B units with sulfate groups, methoxyl groups, or not at all (but never a sulfate group and a methoxyl group together on the same galactose unit). The other fractions were actually deviations from the fundamental structure of the main group. The variant group I, formed by the late-eluting fractions, was arbitrarily divided in two subgroups, with fraction 5III and 6I (group I-A) being structurally in between the main group

Table 11   Structure of the Corallinans[a]

Group	Main (4II, 5I, 5II)	I-A (5III & 6I)	I-B (7I & 8I)	II-A (4I)	II-B (2I & 3I)	II-C (1I)
Unit A						
β-D-Gal						
6-O-(β-D-Xyl)[b]	62	45	12	47	38	30
6-O-(4-O-MeGal)[b]	4	5	6	21	26	8
6-Sulfate[b]	24	35	52	11	–	12
6-O-Methyl	1	6	21	3	4	9
Unsubstituted	9	9	9	18	32	40
Unit B						
α-L-Gal						
2-Sulfate	20	15	7	14	15	12
3-Sulfate[b]	23	25	15	27	37	18
2-O-Methyl	28	33	63	30	36	31
3-O-Methyl	10	6	3	8	5	8
Unsubstituted	19	21	12	21	7	31

[a]Expressed per 100 disaccharide units.
[b]Part of the glycosyl substituents assigned on C-6 of A units may actually be on C-3 of B units; the equivalent amount of sulfate would be then substituting C-6 of A units instead of that ascribed to C-3 on B units.

and group I-B. The quantity of methoxyl groups on the B units increased at the expense of the sulfate groups, whereas in the A units the proportions of methoxyl groups and sulfate groups increased at the expense of xylosyl side chains. Somehow, the overall polarity of the substituents was compensated for in the whole polysaccharides, although the presence of domains of more hydrophobic and hydrophilic nature may have been evident in the different fractions [80]. Continuous structural variation was observed from fraction 4II to fraction 8I.

The structure of the early-eluting fractions was not so obvious. Desulfation–methylation suggested the presence of glycosyl substituents on C-3 of B units. Additionally, other structurally obscure factors appeared (discussed later).

Fraction 4I (group II-A) acted as a hinge with the other group II fractions. The B unit substitution was identical to that of the main group, but in the A unit, sulfate and xylosyl side chains on C-6 were replaced by 4-O-methylgalactosyl side chains and hydrogen. In group II-B, this trend was accentuated, but the B unit also showed changes such as the increasing substitution with polar groups on C-3 at the expense of methoxyl groups and hydrogen, whereas a slight decrease in the proportion of sulfate and hydrogen on C-2 was compensated for by an increase in methoxyl group content. Fraction 1I (group II-C) represented only a small proportion of the total polysaccharide. However, the A unit had less total substitution, though the amount of methoxyl groups increased, and the B unit also was less substituted, especially by sulfate groups.

The variant group II was characterized by low molecular weights, the presence of sugars other than galactose [82] (Glc, Man, Rha, Ara, Fuc), and large amounts of 4-O-methylgalactose. Alkylation studies of these fractions (Table 6 and 10) revealed significant amounts of 2,3,4,6-tetra- 3,4,6-tri-, and 2,3,4-tri-O-alkylgalactoses [80], which represent the unsubstituted nonreducing termini, and substituted on C-2 and C-6, respectively. However, because their proportions were higher than those expected from molecular weight considerations, these fractions possibly contained 1 → 2 and/or 1 → 6 linkages [80]. After ethylation, the 4-O-methylgalactose was represented by both the tetra-O-alkylated and the 3,4,6-alkylated products, suggesting that if this sugar acts only as a backbone substituent, it may be partially arranged as double side chains linked through C-2, or single side chains sulfated on C-2.

The 2,3,4,6-tetra-O-alkylgalactose was not only produced from 4-O-methylgalactose but also from the 2- and 3-methyl ethers of galactose, suggesting that the ethylation studies revealed new diverging branching patterns.

IV. CONCLUSIONS

A. Biological Role of the Galactan Sulfate in *Corallina officinalis*

In spite of the unusual characteristics of the xylogalactans from *Corallina officinalis* and other members of the family, these galactans share many common characteristics with other red seaweed galactans. However, it should be noted that Corallinaceae consists of calcareous seaweeds, and calcium carbonate crystals represent ca. 90% of the dry weight of these algae [87]. Algal $CaCO_3$ deposits are found on or within the cell wall or mucilage layer [87]. An acidic polysaccharide produced by *Emiliania huxleyi* (Coccolithophoridae) was found to be the key substance in the Ca^{2+} binding process [88,89]. In the Corallinaceae, the cell wall was shown to influence the orientation of the $CaCO_3$ crystals [87]. Many compounds, including amino acids, peptides, and acidic polysaccharides, were shown to have an effect on $CaCO_3$ crystal nucleation. For *Corallina officinalis*, a fraction was found that inhibited calcite crystallization [90], whereas similar extracts from noncalcareous seaweeds were not inhibitors, or were promoters, of crystal growth [90]. This *Corallina officinalis* fraction was an undefined protein–polysaccharide complex that was not covalently linked, but is was rich in acidic amino acids and resistant to pronase [90]. The rather unexpected report of alginic acid [71] in a red seaweed may also be related to the calcification process. However, because brown seaweeds are not calcified, the role of alginic acid may be combined with those of the corallinans and acidic protein components of the cell wall [71].

B. The Galactan "Hexagon" Closes?

As explained earlier, two extreme structures can be postulated for agar: one is that of "ideal" agarose, and the other is a polymer in which all the 3,6-anhydro-L-galactose was replaced by its precursor, L-galactose 6-sulfate. Porphyrans and other polysaccharides from the order Ceramiales have structures intermediate (with partial replacements) between these two idealized extremes and contain methoxyl groups and other substituents.

Carrageenans show equivalent counterparts. β-Carrageenan is diastereomeric with agarose, and it represents one extreme of carrageenan structure. In the same position, more common carrageenans such as κ and ι appear by sulfate substitution. Another extreme of the carrageenan structure is represented by γ-carrageenan (or its more sulfated equivalents, μ-, ν-, or λ-carrageenan), in which the B units do not carry 3,6-anhydrogalactose, but its precursor, a 6-sulfated D-galactose unit. Most of the natural gametophytic carrageenans lie in between these extremes. Aeodans and ξ-carrageenan represent a third extreme that is devoid of 3,6-anhydrogalactose and its precursor, while other substituents (such as sulfate in both polysaccharides, methoxyl groups and 4-O-methylgalactose side chains in aeodan) are present.

The polysaccharide from *Ceramium rubrum* exhibited the same trend in the agar family, as 20% of the 4-linked residues were neither 3,6-anhydro-L-galactose nor is precursor [39]. However, because the corallinans are completely devoid of 3,6-anhydro rings and of substituents on C-6 of B units, they also represent this extreme. In this way, corallinans represent the sixth vertex of the hexagon (Fig. 3), a position for which the aeodans represent its diastereomeric counterpart in the carrageenan family. Furthermore, aeodans (polysaccharides with a D-backbone) are substituted with 4-O-methyl-L-galactose, whereas in corallinans (B unit with the L-configuration) the same side chain sugar has the D-configuration.

It should be noted that within this hexagon model, some polysaccha-

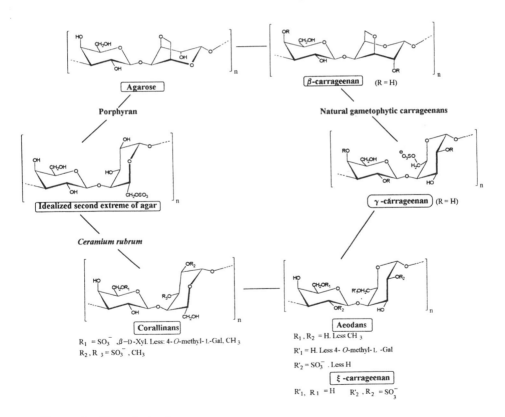

Figure 3 The galactan hexagon.

rides, such as those from *Grateloupia divaricata* or *Lomentaria catenata*, which possess B units with both the D- and L-configurations, should be located inside or at the horizontal sides of the hexagon.

REFERENCES

1. Painter, T. J. (1982). In Aspinall, G. O., ed., *The Polysaccharides*, vol. 2. Academic Press, Orlando, Florida, pp. 195–285.
2. Smith, A. B., and Cook, W. H. (1953). *Arch. Biochem. Biophys.* 45, 232–233.
3. Pernas, A. J., Smidsrød, O., Larsen, B., and Haug, A. (1967). *Acta Chem. Scand.* 21, 98–110.
4. McCandless, E. L., Craigie, J. S., and Walter, J. A. (1973). *Planta* 112, 201–212.
5. Pickmere, S. E., Parsons, M. J., and Bailey, R. W. (1973). *Phytochemistry* 12, 2441–2444.
6. Hosford, S. P. C., and McCandless, E. L. (1975). *Can. J. Bot.* 53, 2835–2841.
7. Doty, M. S., and Santos, G. A. (1978). *Aquat. Bot.* 4, 143–150.
8. Di Ninno, V. L., and McCandless, E. L. (1978). *Carbohydr. Res.* 66, 85–93.
9. Bert, M., Ben Said, R., Deslandes, E., and Cosson, J. (1989). *Phytochemistry* 28, 71–72.
10. Rees, D. A. (1969). *Adv. Carbohydr. Chem. Biochem.* 24, 267–332.
11. Stortz, C. A., and Cerezo, A. S. (1988). *Carbohydr. Res.* 172, 139–146.
12. Matulewicz, M. C., Ciancia, M., Noseda, M. D., and Cerezo, A. S. (1989). *Phytochemistry* 28, 2937–2941.
13. Stortz, C. A., and Cerezo, A. S. (1993). *Carbohydr. Res.* 242, 217–227.
14. Greer, C. W., and Yaphe, W. (1984). *Bot. Mar.* 28, 473–478.
15. Zablackis, E., and Santos, G. A. (1986). *Bot. Mar.* 29, 319–322.
16. Usov, A. I., Yarotskii, S. V., and Shashkov, A. S. (1980). *Biopolymers* 19, 977–990.
17. Mollion, J., Morvan, H., Bellanger, F., and Moreau, S. (1988). *Phytochemistry* 27, 2023–2026.
18. Craigie, J. S. (1990). In Cole, K. M., and Sheath, R. G., eds., *Biology of the Red Algae.* Cambridge University Press, New York, pp. 221–257.
19. Stortz, C. A., and Cerezo, A. S. (1986). *Carbohydr. Res.* 145, 219–235.
20. Araki, C., and Arai, K. (1957). *Bull. Chem. Soc. Japan* 30, 287–293.
21. Araki, C. (1966). *Proc. Int. Seaweed Symp.* 5, 3–17.
22. Duckworth, M., and Yaphe, W. (1971). *Carbohydr. Res.* 16, 189–197.
23. Araki, C., Arai, K., and Hirase, S. (1967). *Bull. Chem. Soc. Japan* 40, 959–962.
24. Craigie, J. S., and Rivero-Carro, H. (1992). *Abstr. XIVth. Int. Seaweed Symp.*, p. 71.
25. Ciancia, M., Matulewicz, M. C., and Cerezo, A. S. (1993). *Phytochemistry* 34, 1541–1543.
26. Chopin, T., Hanisak, M. D., and Craigie, J. S. (1994). *Bot. Mar.* 37, 433–444.

27. Bowker, D. M., and Turvey, J. R. (1968). *J. Chem. Soc.*, 983–988.
28. Bowker, D. M., and Turvey, J. R. (1968). *J. Chem. Soc.*, 989–992.
29. Batey, J. F., and Turvey, J. R. (1975). *Carbohydr. Res.* 43, 133–143.
30. Shashkov, A. S., Usov, A. I., and Yarotskii, S. V. (1978). *Bioorg. Khim.* 4, 74–81.
31. Hirase, S., Watanabe, K., Takano, R., and Tamura, J. (1982). *Int. Carbohydr. Symp. Abstr.* 11, 3–12.
32. Furneaux, R. H., and Stevenson, T. T. (1990). *Hydrobiologia* 204/205, 615–620.
33. Usov, A. I., Ivanova, E. G., and Elashvili, M. Ya. (1989). *Bioorg. Khim.* 15, 1259–1267.
34. Usov, A. I., and Elashvili, M. Ya. (1991). *Bot. Mar.* 34, 553–560.
35. Kochetkov, N. K., Usov, A. I., and Miroshnikova, L. I. (1967). *Zh. Obshch. Khim.* 37, 792–796.
36. Kochetkov, N. K., Usov, A. I., and Miroshnikova, L. I. (1970). *Zh. Obshch. Khim.* 40, 2469–2473.
37. Kochetkov, N. K., Usov, A. I., and Miroshnikova, L. I. (1970). *Zh. Obshch. Khim.* 40, 2473–2478.
38. Kochetkov, N. K., Usov, A. I., Miroshnikova, L. I., and Chizhov, O. S. (1973). *Zh. Obshch. Khim.* 43, 1832–1839.
39. Turvey, J. R., and Williams, E. L. (1976). *Carbohydr. Res.* 49, 419–425.
40. Nunn, J. R., and Parolis, H. (1968). *Carbohydr. Res.* 6, 1–11.
41. Nunn, J. R., and Parolis, H. (1968). *Carbohydr. Res.* 8, 361–362.
42. Allsobrook, A. J. R., Nunn, J. R., and Parolis, H. (1971). *Carbohydr. Res.* 16, 71–78.
43. Allsobrook, A. J. R., Nunn, J. R., and Parolis, H. (1974). *Carbohydr. Res.* 36, 139–145.
44. Allsobrook, A. J. R., Nunn, J. R., and Parolis, H. (1975). *Carbohydr. Res.* 40, 337–344.
45. Nunn, J. R., and Parolis, H. (1969). *Carbohydr. Res.* 9, 265–276.
46. Nunn, J. R., and Parolis, H. (1970). *Carbohydr. Res.* 14, 145–150.
47. Farrant, A. J., Nunn, J. R., and Parolis, H. (1971). *Carbohydr. Res.* 19, 161–168.
48. Farrant, A. J., Nunn, J. R., and Parolis, H. (1972). *Carbohydr. Res.* 25, 283–292.
49. Parolis, H. (1978). *Carbohydr. Res.* 62, 313–320.
50. Parolis, H. (1981). *Carbohydr. Res.* 93, 261–267.
51. Lawson, C. J., Rees, D. A., Stancioff, D. J., and Stanley, N. F. (1973). *J. Chem. Soc. Perkin Trans.* 1, 2177–2182.
52. Barry, V. C., and McCormick, J. E. (1957). *J. Chem. Soc.*, 2777–2783.
53. Rees, D. A. (1961). *J. Chem. Soc.*, 5168–5171.
54. Hirst, E. L., Mackie, W., and Percival, E. E. (1965). *J. Chem. Soc.*, 2958–2967.
55. Usov, A. I., Miroshnikova, L. I., and Barbakadze, V. V. (1975). *Zh. Obshch. Khim.* 45, 1618–1624.
56. Barbakadze, V. V., and Usov, A. I. (1978). *Bioorg. Khim.* 4, 1100–1106.

57. Usov, A. I., and Barbakadze, V. V. (1978). *Bioorg. Khim.* 4, 1107–1115.
58. Miller, I. J., Falshaw, R., and Furneaux, R. H. (1995). *Carbohydr. Res.* 268, 219–232.
59. Penman, A., and Rees, D. A. (1973). *J. Chem. Soc. Perkin Trans* 1, 2182–2187.
60. Hirase, S., and Watanabe, K. (1972). *Proc. Int. Seaweed Symp.* 7, 451–454.
61. Whyte, J., Hosford, S., and Englar, J. (1985). *Carbohydr. Res.* 140, 336–341.
62. Nunn, J. R., Parolis, H., and Rusell, I. (1971). *Carbohydr. Res.* 20, 205–215.
63. Nunn, J. R., Parolis, H., and Rusell, I. (1973). *Carbohydr. Res.* 29, 281–289.
64. Nunn, J. R., Parolis, H., and Rusell, I. (1981). *Carbohydr. Res.* 95, 219–226.
65. Takano, R., Nose, Y., Hayashi, K., Hara, S., and Hirase, S. (1994). *Phytochemistry* 37, 1615–1619.
66. Furneaux, R. H., Miller, I. J., and Stevenson, T. T. (1990). *Hydrobiologia* 204/205, 645–654.
67. Turvey, J. R., and Simpson, P. R. (1966). *Proc. Int. Seaweed Symp.* 5, 323–327.
68. Cases, M. R., Stortz, C. A., and Cerezo, A. S. (1992). *Phytochemistry* 31, 3897–3900.
69. Silva, P. C., and Johansen, H. W. (1986). *Br. Phycol. J.* 21, 245–254.
70. Cases, M. R., Cerezo, A. S., and Stortz, C. A. (1995). *Carbohydr. Res.* 269, 333–341.
71. Usov, A. I., Bilan, M. I., and Klochkova, N. G. (1995). *Bot. Mar.* 38, 43–51.
72. Usov, A. I., and Bilan, M. I. (1995). *Eur. Carbohydr. Symp. (EUROCARB)* 8, B59.
73. Matulewicz, M. C., Haines, H. H., and Cerezo, A. S. (1994). *Phytochemistry* 36, 97–103.
74. Usov, A. I., Yarotsky, S. V., and Estevez, M. L. (1981). *Bioorg. Khim.* 7, 1261–1270.
75. Aspinall, G. O. (1982). In Aspinall, G. O., ed., *The Polysaccharides*, vol. 1. Academic Press, Orlando, Florida, pp. 36–131.
76. Leontein, K., Lindberg, B., and Lönngren, J. (1978). *Carbohydr. Res.* 62, 359–362.
77. Gerwig, G. J., Kamerling, J. P., and Vliegenthart, J. F. G. (1979). *Carbohydr. Res.* 77, 1–7.
78. Oshima, R., Kumanotani, J., and Watanabe, C. (1983). *J. Chromatogr.* 259, 159–163.
79. Aspinall, G. O. (1982). In Aspinall, G. O., ed., *The Polysaccharides*, vol. 2. Academic Press, Orlando, Florida, pp. 1–9.
80. Cases, M. R., Stortz, C. A., and Cerezo, A. S. (1994). *Int. J. Biol. Macromol.* 16, 93–97.
81. Stevenson, T. .T, and Furneaux, R. H. (1991). *Carbohydr. Res.* 210, 277–298.
82. Sweet, D. P., Albersheim, P., and Shapiro, R. H. (1975). *Carbohydr. Res.* 40, 199–216.
83. Sweet, D. P., Shapiro, R. H., and Albersheim, P. (1974). *Biomed. Mass Spectrom.* 1, 263–268.

84. Cases, M. R., Stortz, C. A., and Cerezo, A. S. (1994). *J. Chromatogr.* 662, 293–299.
85. Cases, M. R. Stortz, C. A., Cherniak, R., Bacon, B. E., and Cerezo, A. S., unpublished results.
86. Cases, M. R. (1995). Ph.D. thesis, University of Buenos Aires, Argentina.
87. Borowitzka, M. A. (1987). *CRC Crit. Rev. Plant Sci.* 6, 1–45.
88. de Jong, E. W., Bosch, L., and Westbroek, P. (1976). *Eur. J. Biochem.* 70, 611–621.
89. Westbroek, P., de Jong, E. W., van der Wal, P., Borman, A. H., de Vrind, J. P. M., Kok, D., de Bruijn, W. C., and Parker, S. B. (1984). *Phil. Trans. R. Soc. Lond. B* 304, 435–444.
90. Somers, J. A., Tait, M. I., Long, W. F., and Williamson, F. B. (1990). *Hydrobiologia* 204/205, 491–497.

36

Some New Methods to Study Plant Polyuronic Acids and Their Esters

Nicholas C. Carpita
Purdue University, West Lafayette, Indiana

Maureen C. McCann
John Innes Centre, Norwich, England

I. INTRODUCTION

The plant's primary, or "growing," cell wall is thought to comprise a high tensile strength network of cellulose microfibrils tethered by cross-linking glycans and embedded in an independent but interactive pectin matrix of uronic acid–rich polysaccharides [1,2]. The pectin matrix has several structural and physiological roles in plants, and among the pectic polysaccharides are some of the more complex polymers found in the plant cell wall. Pectins establish domains of fixed anionic environments within the wall, creating a weakly acidic extracellular milieu that is the major cation-exchange matrix of the plant. The pectins also establish the "pore size" of the wall by generating anionic channels of about 4 nm in diameter [3]. There is evidence for the presence of a large proportion of galactose in some cell walls [4] and a recent nuclear magnetic resonance (NMR) study [5] suggests that this neutral galactan may be further involved in limiting porosity by forming short flexible rods anchored at one end that protrude into the pores of the network.

Chemically, there are two principal pectic polymer backbones, homogalacturonan (polygalacturonan, PGA), consisting of chains of α-D-(1 $\rightarrow$ 4)-galacturonic acid, and rhamnogalacturonan (RG I), a backbone polymer of (1 $\rightarrow$ 2)-α-L-rhamnosyl-(1 $\rightarrow$ 4)-α-D-galacturonic acid disaccharide repeating units [6]. The PGA and RG I domains may exist within a single backbone [7]. Several types of neutral sugar polymers such as 5-linked arabinan, 4-linked galactan, and type I and type II arabinogalactans are attached to the *O*-4 of a portion of the rhamnosyl (Rha) units of RG I [8]. The RG I macromolecules may also possess "smooth" domains, which have little or no side group substitution, and "hairy" regions, which have extensive substitution of the Rha units with oligomeric side chains [9,10]. Another type of rhamnogalacturonan, RG II, is a rhamnose-rich highly branched polysaccharide with a homogalacturonan backbone [11] and contains several rare sugars such as 2-O-methyl xylose, 2-O-methyl fucose, apiose, aceric acid [12], and even deoxy-D-mannooctulosonic acid (KDO) [13]. A newly characterized xylogalacturonan contains a homogalacturonan backbone with nonreducing terminal xylosyl units attached at the O-3 positions of about half of the galacturonosyl units [14]. The wide variation in the distribution of the various polymers attached to pectic backbones in different cell types has aroused much suspicion that these polymers, or epitopes within them, are responsible for many of the subtle biological roles of the pectin matrix. They provide a spectrum of surface markers that foretell developmental fate and release elicitors of plant defense responses against predatory insects and invading fungal, bacterial, and viral pathogens [1]. RG I has been shown to be a developmental marker for the non-hair-forming cells in epidermal cells of *Arabidopsis* roots [15]. In this chapter, we briefly review the current status of standard methodologies to analyze pectins and their esters and new techniques that make it possible to examine the specific pectin organization and dynamics within a single cell wall during development.

II. CHEMICAL ANALYSIS OF PECTINS

A. Fractionation of Pectic Substances from Plant Cell Walls

Pectins are classically defined as uronic acid–rich polysaccharides extracted by chelating agents such as EDTA, EGTA, CDTA, and ammonium oxalate [7,16]. Within much larger polymers are regions of homogalacturonan consisting of a minimum of about 10 to 14 α-D-galacturonic acid (GalA) units that are cross-linked by 5 to 7 Ca^{2+} ions to form "junction zones" linking two antiparallel chains [17]. If sufficient Ca^{2+} is present, some esterified GalA can be tolerated in the stable junction zone. With excess Ca^{2+} available, four-chain or higher-order stacking of PGA chains is possible, but such stacking is unlikely to occur in vivo [7,18]. The chelating agents have different calcium ion binding constants and thus differ in their efficacy in extracting pectins. Purified cell wall preparations are normally boiled in neutral 0.5% ammonium oxalate solutions, whereas the other three reagents can be used at ambient or chilling temperatures. Although CDTA has been shown to be a superior extractant of pectins [19], it has proven difficult to remove by dialysis, perhaps because of the formation of stable

$CDTA-Ca^{2+}$-pectin complexes [20]. Another effective solvent for pectins is imidazole, which will extract chelator-soluble pectins at ambient temperatures but is easily dialyzed against deionized water [20].

Chelating agents are presumed to extract pectins by disrupting the Ca^{2+} cross-bridges. However, a substantial portion of the pectins extracted by chelating agents include PGA, highly esterified with methyl groups, and RG I, which is not cross-linked by calcium ions [7]. The simple explanation is that RG I and PGA represent regions of very large molecules, and that regions representing a small percentage of the molecules are deesterified and cross-linked with calcium [7]. A significant proportion of the pectins that remain after exhaustive extraction with chelating agents are subsequently extracted with weak alkali, an indication that much of the pectin matrix is attached to the wall matrix covalently, possibly through other kinds of ester linkages. Because the dilute alkali used to extract the remaining pectins also hydrolyzes the methyl esters, the degree of esterification of the covalently bound pectin fraction is assayed in the insoluble matrix remaining after extraction of the chelator-soluble pectins before any other treatment [21]. After sequential extractions with chelators and weak alkali, the wall matrix remaining comprises cellulose microfibrils and their cross-linking glycans, but often additional pectic-like polymers are extracted with greater than 1 M alkali. Rather than a subclass of pectic substances held by covalent bonds, this observation may be a result of a swelling of the matrix to release "encaged" pectin molecules as the hydrogen bonds between cross-linking glycans and cellulose are broken.

B. Colorimetric Assays of Uronic Acids in Pectic Substances

After bulk extraction of pectic substances from plant tissues, the content of uronic acid is determined using one of two chromogenic reagents, carbazole or *m*-hydroxybiphenyl. The carbazole assay [22] produces an intense pink color in the presence of uronic acid. It is a reasonably sensitive assay, but one prone to interference by other substances. Boiling cell wall polysaccharides in concentrated sulfuric acid results in browning of the reaction mix by neutral sugars and increased background absorbance at 525 nm, the maximum absorbance of the chromagen. The hydrolysis of the galactosyluronic acid units of pectins requires concentrated sulfuric acid and is an essential step in any colorimetric assay. This leads to some browning by decomposition of the neutral sugars, but it is the strong browning reaction upon addition of the carbazole reagent and subsequent heating of the reaction mixture that is the real problem. This problem may be overcome to some extent by using *m*-hydroxybiphenyl, which produces a pink color without the second heating [23]. Galambos [24] also showed that a substantial reduction of the browning by neutral sugars during the primary and secondary boiling could be eliminated by inclusion of small amounts of sulfamate in the reaction mixtures. A survey of the literature will find that the *m*-hydroxybiphenyl method of Blumenkrantz and Asboe-Hansen is used by the vast majority of researchers, whereas the carbazole–sulfamate method of Galambos [24] produces much lower background color than the *m*-hydroxybiphenyl method [25]. A combination of use of *m*-hydroxybiphenyl and sulfamate is preferred in attempts to measure uronic acids in the presence of great excesses of neutral sugar [25].

C.   Determination of Methyl-Esterified Pectins

The degree of methyl esterification of uronic acids is also determined in bulk samples of purified pectin by colorimetric assay. The assay relies on quantitation and comparison of relative amounts of methanol, which is released by saponification [26], and total uronic acid, which is measured by another assay. The methanol is oxidized to formaldehyde by potassium permanganate, and formaldehyde is used to form a yellow adduct with a dione reagent [26]. We have modified the procedure slightly to increase sensitivity, about twofold, and reproducibility [21]. We found that keeping the mixtures of methylated PGA capped in small Eppendorf centrifuge tubes during the saponification, acidification, and centrifugations prior to addition of permanganate will prevent evaporative loss of the methanol.

D.   Preparation of Oligomer Standards

Oligomeric standards are made by heating 2% PGA solutions, prepared in water adjusted to pH 4.5, at 121°C for 1 h. The browned solution is clarified by addition of charcoal, and pectin oligomers of various size fractions are precipitated by sequential addition of ethanol to 20, 50, and 80% (v/v) and collecting the newly precipitated oligogalacturonides after each increment. The pellets are dissolved in water and freeze-dried, and portions of the pectins are fully methyl-esterified by addition of methanolic HCl, made by addition of 0.2 mL of acetyl chloride to 1 mL of chilled methanol, and stirring the mixture at ice cold temperatures overnight [27]. A range of methyl-esterified oligomers can then be made by brief treatments with dilute sodium carbonate.

E.   Analysis of Total Versus Methyl-Esterified Pectins

The determination of methyl-esterified pectin by the colorimetric assays assumes that galacturonic acid is the only uronic acid and methyl groups are the only substituent esters. These criteria may be established empirically using methods that chemically reduce the uronic acids to their respective neutral sugars [28], followed by separation of derivatives of neutral sugars by gas–liquid chromatography (GLC) [29]. Uronic acids are quantified by the increase in abundance of the neutral sugar or a decrease in uronic acid content. The carboxylic acid groups are activated by addition of water-soluble carbodiimides to form reaction products that are subsequently reduced to their primary alcohols [28]. The most effective carbodiimide activator is 1-cyclohexyl-3-(2-morpholinoethyl)carbodiimide metho-*p*-toluenesulfonate (CMC), and the fundamental technique has been modified so that additions of the carbodiimide to the PGA solutions are maintained near pH 4.75 with 100 mM acetic acid–sodium acetate. After formation of the carbodiimide reaction products, the solutions are brought to 1 M chilled imidazole, pH 7, just prior to addition of the $NaBH_4$. This maintains a slightly alkaline solution for optimal reduction and reduces alkaline decomposition of the pectic backbones [29]. Sensitivity of detection of uronic acids is increased substantially by reduction of the carbodiimide-activated carboxyl groups with $NaBD_4$ to generate the 6,6-dideuterio neutral sugars, which can be differentiated from their respective sugars by GLC–electron impact mass spectrometry (EIMS) [30].

Maness et al. [29] discovered that $NaBH_4$ not only will reduce the carbodiimide reaction products but also any nascently methyl-esterified uronic acid. The method was revised to selectively reduce the esterified uronic acids to their 6,6-dideuterio derivatives with $NaBD_4$ (Fig. 1). Using this method, the glucuronic acids of cereal glucuronoarabinoxylans were found to never be esterified, but the galacturonic acid units were highly esterified during rapid growth of maize coleoptiles [21].

Because all esters of uronic acids are reduced, total ester is determined rather than only the methyl esters. Comparison of the total ester and methyl ester determinations showed that, in maize, the chelator-soluble esterified pectins were largely methyl esters, whereas the fraction of pectin covalently attached to the wall matrix contained a high proportion of nonmethyl esters (Table 1) [21]. Similar dynamics in the state of esterification accompany growth in tobacco cells in liquid culture [31]. Brown and Fry [32] discovered an unusual nonmethyl hydrophobic substituent ester of GalA, but they did not identify the alcohol moiety. The methyl groups may be "activated" carboxyl groups that are substrates in the wall for transesterases that form other esters. Lamport [33] suggested that some of the novel esters could be sugar substituents on other cell wall polymers.

III. PROBING THE FINE STRUCTURE OF ISOLATED PECTINS

Quantitation of homogalacturonan and its esters is fairly straightforward, but understanding the more complex RGs and their associated polysaccha-

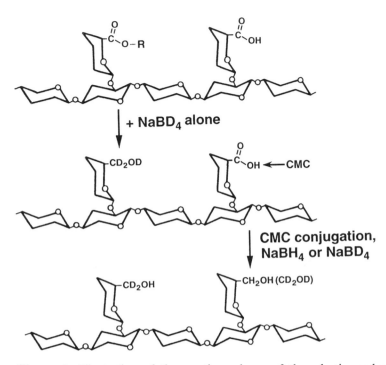

Figure 1 Illustration of the reaction scheme of the selective reduction of the esterified uronic acids in a model polysaccharide. Only the esterified uronic acids can be reduced with NaBD$_4$. The free acids are conjugated with a water-soluble diimide, and one fraction is reduced with NaBD$_4$ (to calculate total uronic acid) and a paired fraction is reduced with NaBH$_4$ (to calculate the percentage of total esterified uronic acid) [21]. Alditol acetates are prepared and separated by GLC. The uronic acid and neutral sugars are calculated from the [M]$^+$ and [M$_{D2}$]$^+$ of diagnostic m/z pairs revealed by electron impact mass spectrometry [21].

rides is a much greater challenge. Linkage structure is determined by methylation analysis after reduction of the carbodiimide-activated uronic acids with NaBD$_4$ [30]. This analysis will provide a rough estimate of the PGA and RG I backbone distribution, the proportion of rhamnosyl units with side chains, and the type and size of the associated polysaccharides such as 5-linked arabinans, 4-linked galactans, type I arabinogalactans, and type II 3,6-linked arabinogalactans. Linkage information can be obtained noninvasively by ^{1}H and ^{13}C nuclear magnetic resonance (NMR) spectroscopy [34], and solution and solid state cross-polarization-magic angle spinning (CP-MAS) NMR has been used to examine the three-dimensional configu-

Table 1 Comparison of Percentage (mol %) of Total- and Methyl-Esterified Galactosyluronate Esters in the Cell Wall and Pectin Fractions from Maize

Fraction	Total GalA	Total GlcA	GalA	
			Methyl ester	Total ester
Total cell wall:	9	3	40	68
Chelator-soluble	41	1	54	65
Remaining cell wall	7	2	29	73

Cell walls were purified from maize coleoptiles at the most rapid rate of cell elongation. The total cell wall pectins, the pectins extracted with ammonium oxalate (chelator-soluble), and the pectins in the cell wall remaining after extraction with chelator were each subjected to the double-reduction technique as described in Fig. 1 [21]. The methyl ester was determined by paired colorimetric assays of total uronic acid [25] and methanol after saponification [26], with proportions of galacturonic acid (GalA) and glucuronic acid (GlcA) from GLC/EIMS of alditol acetates [21]. The total ester was determined from ratios of the proportions of 6,6-dideuteriogalactose in galactose after carbodiimide-activated reduction with $NaBH_4$ and $NaBD_4$, respectively (Fig. 1) [21]. The nonmethyl esters of the galacturonic acids are enriched in pectins unable to be extracted with chelator.

rations, sugar–sugar interactions, and mobilities of molecules in the wall matrix before and after pectic extractions [35–37].

Determination of microheterogeneity of the RG I structures has also been explored using hydrolysis of selective linkages in anhydrous HF at subzero temperatures [38], and by use of sequence-dependent enzymes that cleave specific regions of RG I [9,10,39]. In general, about half of the rhamnosyl units of RG I have side chains, but this ratio can vary with cell type and physiological state. Further, as many as a third of the GalA units of RG I are acetylated [40]. Excellent separation of oligosaccharides by high performance liquid chromatography (HPLC) is now made possible by high pH anion-exchange chromatography with selective detection of sugars and related substances by pulsed amperometry [41]. The polystyrene-based anion-exchange Dionex columns, now used routinely for separation of oligosaccharides at high pH, are also quite suitable for the separation of a broad size range of oligogalaturonides at neutral or slightly acidic pH [42]. Resolution of RG I oligosaccharides has also been accomplished at high pH [9,10,43]. Thermospray mass spectrometry, electrospray mass spectrometry, and matrix-assisted laser desorption ionization time-of-flight (MALDI-TOF) mass spectrometry are major advances in mass spectrometry technology that make it possible to analyze sugar compositions and even sequence structures of oligosaccharides. With a limit of over m/z 2000 and the introduction of multiple charges on a single molecule, molecular masses of the order of 100 kDa can be determined. Primary and secondary fragmentations of negative ions yield linkage information in addition to molecular mass. The disaccharides (1 → 2)-linked sophorose, (1 → 3)-linked laminaribiose, (1 → 4)-linked cellobiose, and (1 → 6)-linked gentiobiose each give distinctive fragmentations [44]. Schols et al. [45] separated RG I oligosaccharides up to nine units by anion-exchange HPLC and detected them by in-line thermospray mass spectrometry.

One need not rely strictly on chemical analyses to determine molecular size. Isolated wall polymers can also be visualized by spraying the polymers in glycerol onto a freshly cleaved surface of mica, drying them down in vacuo, and then rotary-shadowing the preparation with platinum/palladium and carbon [46]. The measured length of the molecule provides a minimum estimate of degree of polymerization, as each sugar residue will contribute roughly 0.5 nm to the imaged length. It is also possible to adsorb polymers to plastic-filmed gold grids, to immunolabel with a primary antibody followed by a colloidal-gold-conjugated secondary antibody, and then to negatively stain to see if the distribution of particular epitopes is uniform along the polymer [46]. Elucidation of the sequence structure of the neutral side chains of the smooth and hairy regions of RG I may give insight into function.

IV. PROBING PECTIN STRUCTURES OF INDIVIDUAL CELLS

Within a single wall are zones of different architecture: the middle lamella, plasmodesmata, Casparian strips, thickenings, channels, pit fields, and the cell corners. There are also domains across a wall where the degree of pectin esterification is modified [46]. The size of microdomains in the wall compared with the sizes of the polymers that must be accommodated in these regions implies that mechanisms must exist for packaging and positioning of large molecules. For example, unesterified pectins can range up to 700 nm in length and yet, in some cell types, are accommodated in a middle lamella of 10 to 20 nm width, and so they must at least be constrained to lie parallel to the plasma membrane. Pectic gels made from extracted pectins have 100fold greater volumes than the cell walls from which they are extracted. In some species, onion [2], tomato, and sugar beet [47], the interface regions between cells, that is, the middle lamella and the cell corners, are rich in relatively unesterified pectins; these may function in cell–cell adhesion and play an important structural role in tissue integrity. Pathogens also tend to attack cell corners, and the complex mixture of pectins and arabinogalactan proteins found there may have a role in enmeshing the invading organism as well as signaling defense mechanisms to operate through the release of small pectic fragments. The complexity and diversity in pectin structure and localization in cell walls of different species and tissues, and in different wall domains around a single cell [2], may reflect a functional diversity in the fine control of gel rheology. In specialized cells, such as those of many fruits and vegetables, the content of pectic substances in the cell wall is greatly increased, and the structure of the individual components differs among the cells of the tissue. During ripening, additional changes occur as the pectins in certain cells are modified chemically via esterase and polysaccharide hydrolase activities. Because of the broad variation in structure and the dynamics of the domain structure during cell development, we need many new probes to identify the microheterogeneous domains of pectins in situ.

Given the variety in cell wall composition among tissues, and even between walls bordering a single cell [2], it is important to demonstrate that data obtained from bulk chemical analysis is applicable to the walls of all cells in the population sampled. Fourier transform infrared (FTIR) microspectroscopy and electron microscopy are two methodologies that permit analysis at the level of a single cell wall and that are suitable for validating the conclusions from bulk analysis [31].

A. Specific Probes for Cell Wall Polymers Demonstrate the Domain Nature of Cell Walls

Probes specific for cell wall epitopes, such as monoclonal antibodies coupled to fluorescent markers, or to enzyme conjugates, the presence of which is detected by addition of substrate, are used to label sections of plant tissue [47,48] or tissue-prints of soluble components on nitrocellulose [49]. Antibody probes, coupled to colloidal gold, are used at the electron microscope level to show the distribution of components within a single wall [50].

Varner and Taylor [51] found that Ni^{2+} or Co^{2+} could replace the Ca^{2+} interlocking the unesterified PGA. A black precipitate that forms upon incubation with Na_2S provides a simple means to localize junction zones directly in the cells or in tissue blots. Vreeland and colleagues [27] found that a fluorescein-labeled oligogalacturonate is a convenient probe that recognizes unesterified PGA in a Ca^{2+}-dependent manner. These methods are quite useful to determine the relative enrichment of unesterified PGA at the tissue level, but neither provides the resolution needed to locate unesterified PGA within a single cell wall domain.

Monoclonal antibodies against specific polysaccharide epitopes, methyl-esterified and unesterified PGA [47], and RG I [15] have been used to determine the locations of pectins in the wall by immuno-electron microscopy. The antibody approach provides the needed resolution and has an additional advantage in that the epitopes recognized are often oligomeric sequences rather than single units or linkages. Gold-labeled antibodies specific to either unesterified or methyl-esterified PGAs have been used to localize each of these pectins within the cell walls of several cell types, and they reveal a stark variation in the distribution of both forms [47,52,53]. These antibodies have also been used to reveal the different sites of synthesis of PGA and subsequent methyl esterification in the membranes of the Golgi apparatus, the organelle that synthesizes all noncellulosic polysaccharides of the plant [54]. Although unesterified pectins are not necessarily cross-linked with Ca^{2+} [55], Liners et al. [56] prepared monoclonal antibodies that recognize Ca^{2+}-bound PGA dimers (with each monomer at least a nonagalacturonide [57] but not the unbound polymers. These antibodies are potentially useful to resolve where PGA junction zones are located in the walls of single cells.

B. Spectroscopic Methods for Analysis of Pectins in Single Cell Walls

FTIR microspectroscopy is an extremely rapid, noninvasive vibrational spectroscopic method that detects quantitatively carboxylic esters, phenolic esters, protein amides, and carboxylic acids, providing a complex "fingerprint" of carbohydrate constituents and their organization (Fig. 2) [31,58–60]. Infrared spectroscopy is a well-established technique, which has only recently been applied to biological samples following the advent of rapid Fourier transform data acquisition technology that permits subtraction of water absorbance. In the Fourier transform instrument, all frequencies are scanned simultaneously, making data collection extremely rapid and resulting in an improved signal-to-noise ratio. In the spectrometer, an infrared source emits radiation with a range of frequencies that is then passed through the sample. Particular chemical bonds in the sample absorb infrared radiation of specific frequencies from the beam, and this is plotted as an absorbance spectrum against wavenumber [61]. The real power of this technique is in the attachment of a microscope accessory whereby the beam is diverted through a sample resting on the microscope stage. This permits us to define precisely the area of a single cell wall as small as $10 \times 10~\mu m$, from which infrared data can be obtained [58]. Polarizers can be inserted in the path of the infrared beam to determine whether band frequencies of specific functional groups are oriented transversely or longitudinally with respect to the long axis of the cell [62]. Polarized FTIR microspectroscopy shows that both cellulose and free acid stretches attributable to pectin are

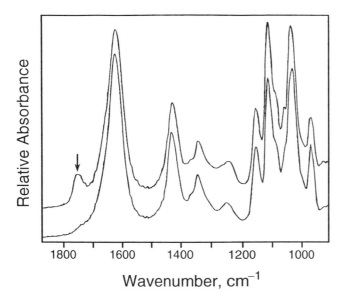

Figure 2 FTIR spectrum of a PGA and methyl-esterified PGA. Esterification of pectins is easily detected by FTIR because of the characteristic carboxylic ester absorbance around 1745 cm^{-1} (arrow), which is present in methylated PGA (upper trace) and absent from PGA (lower trace). Methyl esterfication also affects the relative heights of absorbances in the fingerprint region of the spectrum (below 1500 cm^{-1}). With a series of appropriate standards, it is possible to estimate the degree of esterification for particular pectins by calculating the ratio of the area under the ester peak to the area under the spectral region 1160 to 970 cm^{-1} [60].

oriented transversely to the direction of cell elongation during growth of both carrot and tobacco suspension cells [31,62].

In the infrared region 1200–900 cm^{-1}, pectins have a similar profile to that of polygalacturonic acid with peaks in common at 1140, 1095, 1070, 1015, and 950 cm^{-1} in the carbohydrate region of the spectrum, and these peaks can easily be distinguished from nonpectic polysaccharides (Fig. 2). When the pectins are treated with alkali, the ester linkages are hydrolyzed and two absorptions may be seen at about 1600 and 1420 cm^{-1} (antisymmetric and symmetric COO^{-}/stretches); these two peaks are then diagnostic for pectin in salt form. The most diagnostic peak in the infrared spectra of cell walls is the peak centered at about 1745 cm^{-1}, arising from the ester carbonyl stretching associated with pectin (Fig. 2). Its exact frequency and bandshape reveal the type (saturated-alkyl or aryl esters) and/or environment of the ester groups. Different kinds of esters may absorb at slightly different frequencies (Fig. 3).

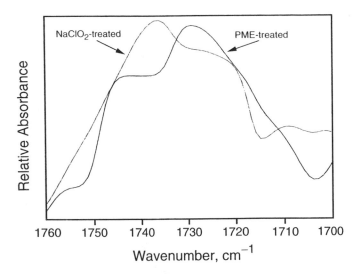

Figure 3 Different esters are lost when the walls are treated with pectin methyl esterase (PME) versus NaClO$_2$. In samples of plant cell wall material, several kinds of carboxylate esters may be present, and they can be resolved to a limited extent by FTIR. Growing maize seedlings have two principal esters, the galacturonate esters from the pectins and the ferulate esters attached the arabinosyl units of glucuronoarabinoxylans [1]. Treatment of maize cell wall with a pectin methyl esterase selectivity removes a portion of the galacturonate esters, whereas destruction of aromatic rings with sodium chlorite selectively removes the ferulate esters.

The infrared spectra are sensitive to the local molecular environment of the bonds, and we have used this to detect conformational changes upon drying in a pectin extracted from onion parenchyma cell walls [58]. Unfortunately, many polysaccharides have absorbances in the fingerprint region of the spectrum, where many complex vibrational modes overlap and peaks cannot be assigned uniquely. In this region, the spectra constitute species-specific and tissue-specific fingerprints of cell walls, reflecting even subtle differences in composition [60,62,63].

Because FTIR microspectroscopic analysis is so rapid and can be done on dried but otherwise intact plant tissues, we have been using the technique to screen for pectin mutants in flax and Arabidopsis. Pairs of intact, hot-ethanol-extracted leaves are placed on KBr windows and mounted in a spectrometer interfaced to a IR-Plan microscope. The infrared transmission is recorded between 2000 and 900 cm^{-1} at a resolution of 8 cm^{-1}, and 64-interferograms are co-added. A principal component and cluster analysis is applied to a database of acquired spectra from control plants [63].

This type of analysis easily discriminates between potential mutants and natural variants in the wild-type flax population (Fig. 4). Seven were identified to have unusually high amounts of methyl ester, and one had low amounts of ester. The spectra reveal even subtle differences in composition and differences among cell wall mutants. Ester and free acid peaks are well resolved in the spectrum, making FTIR spectroscopy an approach complementary to sugar analysis in screening for pectin mutants. Future

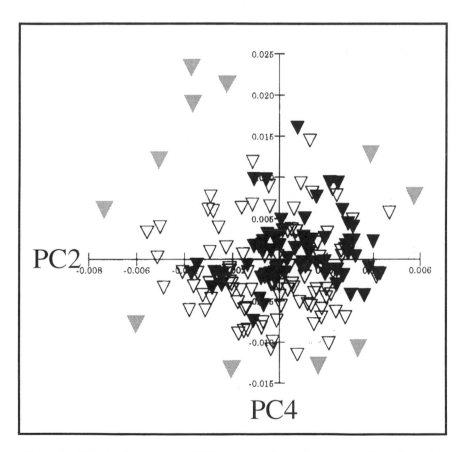

Figure 4 Principal component (PC) analysis of putative mutants from flax using one spectrum that accentuates differences in the pectic components (PC2) versus one that accentuates the carbohydrate fingerprint region (PC4). A flax wild-type cluster (▼) is compared with a portion of the M_2 population (▽). Eleven M2 plants (▼) were selected, which, by cluster analysis, were clearly different from wild type. Seven of these were shown by other analyses to have altered pectin composition.

studies will include screening by FTIR of mutations of cell morphology that result from alterations in the microfibril orientation mechanisms.

V. CONCLUSIONS

Technological advances in recent years have made it possible to deduce the unit structures of many complex pectic polysaccharides, but for true sequence determinations we need methods similar to those available for DNA and proteins. Because pectin oligosaccharides are anionic, they are amenable to electrophoretic separation. Coupling this technique with treatment with specific glycanases to sequentially remove subtending neutral sugars would allow development of molecular sequencing methods of pectins. As we understand more about the fine structure of pectins, parallel studies are beginning to unravel the biological functions of many of the specific pectin side groups. The novel ester linkages revealed by selective reduction have not yet been identified. The correlation of the appearance of these non-methyl esters during rapid plant growth demands further studies to characterize this linkage, to help elucidate the biological function. Biological functions of pectins can be best investigated through identification of mutations in the fine structure of the complex polysaccharides. These mutants are now being selected by use of FTIR and other rapid screening techniques sensitive to cell wall carbohydrate structure.

ABBREVIATIONS

CMC	1-cyclohexyl-3-(2-morpholinoethyl)carbodiimide metho-p-toluenesulfonate
CP-MAS	cross-polarization-magic angle spinning
EIMS	electron impact mass spectrometry
FTIR	Fourier transform infrared
GalA	galacturonic acid
GLC	gas–liquid chromatography
GlcA	glucuronic acid
HPLC	high performance liquid chromatography
KDO	deoxy-D-mannooctulosonic acid
MALDI-TOF	matrix-assisted laser desorption ionization time-of-flight
NMR	nuclear magnetic resonance
PC	principal component
PGA	polygalacturonan (homogalacturonan)
PME	pectin methyl esterase
RG I, II	rhamnogalacturonan I, II
Rha	rhamnose

REFERENCES

1. Carpita, N. C., and Gibeaut, D. M. (1993). *Plant J.* 3, 1–30.
2. McCann, M. C., and Roberts, K. (1992). In C. W. Lloyd, ed., *The Cytoskeletal Basis of Plant Growth and Form*. Academic Press, London, pp. 109–129.
3. Baron-Epel, O., Gharyl, P. K., and Schindler, M. (1988). *Planta* 175, 389–395.
4. Redgwell, R. J., and Selvendran, R. R. (1986). *Carbohydr. Res.* 157, 183–199.
5. Foster, T. J., Ablett, S., McCann, M. C., and Gidley, M. J. (1996). *Biopolymers*, 39, 51–66.
6. Lau, J. M., McNeil, M., Darvill, A. G., and Albersheim, P. (1985). *Carbohydr. Res.* 137, 111–125.
7. Jarvis, M. C. (1984). *Plant Cell Environ.* 7, 153–164.
8. Bacic, A., Harris, P. J., and Stone, B. A. (1988). In Priess, J., ed., *The Biochemistry of Plants*, vol. 14, Academic Press, New York, pp. 297–371.
9. Mutter, M., Beldman, G., Schols, H. A., and Voragen, A. G. J. (1994). *Plant Physiol.* 106, 241–250.
10. Schols, H. A., Voragen, A. G. J., and Colquhoun, I. J. (1994). *Carbohydr. Res.* 256, 97–111.
11. Stevenson, T. T., Darvill, A. G., and Albersheim, P. (1988). *Carbohydr. Res.* 182, 207–226.
12. Melton, L. D., McNeil, M., Darvill, A. G., Albersheim, P., and Dell, A. (1986). *Carbohydr. Res.* 146, 279–305.
13. York, W. S., Darvill, A. G., McNeil, M., and Albersheim, P. (1985). *Carbohydr. Res.* 138, 109–126.
14. Schols, H. A., Bakx, E. J., Schipper, D., and Voragen, A. G. J. (1995). *Carbohydr. Res.* 279, 265–279.
15. Freshour, G., Clay, R. P., Fuller, M. S., Albersheim, P., Darvill, A. G., and Hahn, M. G. (1996). *Plant Physiol.*, 110, 1413–1429.
16. O'Neill, M., Albersheim, P., and Darvill, A. (1990). *Meth. Plant Biochem.* 2, 415–441.
17. Powell, D. A., Morris, E. R., Gidley, M. G., and Rees, D. A. (1982). *J. Mol. Biol.* 155, 517–531.
18. Walkinshaw, M. D., and Arnott, S. (1981). *J. Mol. Biol.* 153, 1075–1085.
19. Jarvis, M. C., Hall, M. A., Threlfall, D. R., and Friend, J. (1981). *Planta* 152, 93–100.
20. Mort, A. J., Moerschbacher, B. M., Pierce, M. L., and Maness, N. O. (1991). *Carbohydr. Res.* 215, 219–227.
21. Kim, J.-B., and Carpita, N. C. (1992). *Plant Physiol.* 98, 646–653.
22. Dische, Z. (1950). *J. Biol. Chem.* 183, 489–494.
23. Blumenkrantz, N., and Asboe-Hansen, G. (1973). *Anal. Biochem.* 54, 484–489.
24. Galambos, J. T. (1967). *Anal. Biochem.* 19, 119–132.
25. Filisetti-Cozzi, T. M. C. C., and Carpita, N. C. (1991). *Anal. Biochem.* 197, 157–162.
26. Wood, P. J., and Siddiqui, I. R. (1971). *Anal. Biochem.* 39, 418–428.

27. Vreeland, V., Morse, S. R., Robichaux, R. H., Miller, K. L., Hua, S.-ST., and Laetsch, W. M. (1989). *Planta* 177, 435–446.

28. Taylor, R. L., and Conrad, H. E. (1972). *Biochemistry* 11, 1383–1388.

29. Maness, N. O., Ryan, J. D., and Mort, A. J. (1990). *Anal. Biochem.* 185, 346–352.

30. Carpita, N. C., and Shea, E. M. (1989). In C. J. Biermann and G. D. McGinnis, eds., *Analysis of Carbohydrates by GLC and MS*. CRC Press, Boca Raton, Florida, pp. 157–216.

31. McCann, M. C., Shi, J., Roberts, K., and Carpita, N. C. (1994). *Plant J.* 5, 773–785.

32. Brown, J. A., and Fry, S. C. (1993). *Plant Physiol.* 103, 993–999.

33. Lamport, D. T. A. (1970). *Ann. Rev. Plant Physiol.* 21, 235–270.

34. Grasdalen, H., Bakøy, O. E., and Larsen, B. (1988). *Carbohydr. Res.* 184, 183–191.

35. Bush, C. A. (1988). *Bull. Magn. Reson.* 10, 73–95.

36. Gidley, M. J. (1992). *Trends Food Sci. Technol.* 3, 231–236.

37. Mackay, A. L., Wallace, J. C., Sasaki, K., and Taylor, I. E. P. (1988). *Biochemistry* 27, 1467–1473.

38. Mort, A. J., Komalavilas, P., Rorrer, G. L., and Lamport, D. T. A. (1989). In Linskens, H. F., and Jackson, J. F., eds., *Modern Methods of Plant Analysis, Vol. 10, Plant Fibers*, Springer-Verlag, Berlin, pp. 37–69.

39. Lerouge, P., O'Neill, M. A., Darvill, A. G., and Albersheim, P. (1993). *Carbohydr. Res.* 243, 359–371.

40. Komalavilas, P., and Mort, A. J. (1989). *Carbohydr. Res.* 189, 261–272.

41. Lee, Y. C. (1990). *Anal. Biochem.* 189, 151–162.

42. Hotchkiss, A. T., and Hicks, K. B. (1990). *Anal. Biochem.* 184, 200–206.

43. An, J., O'Neill, M. A., Albersheim, P., and Darvill, A. G. (1994). *Carbohydr. Res.* 252, 235–243.

44. Garozzo, D., Impallomeni, G., Spina, E., Green, B., and Hutton, T. (1992). *Carbohydr. Res.* 221, 253–257.

45. Schols, H. A., Mutter, M., Voragen, A. G. J., Niessen, W. M. A., Van der Hoeven, R. A. M., Van der Greef, J., and Bruggink, C. (1994). *Carbohydr. Res.* 261, 335–342.

46. McCann, M. C., Wells, B., and Roberts, K. (1992). *J. Microscopy* 166, 123–136.

47. Knox, J. P., Linstead, P. J., King, J., Cooper, C., and Roberts, K. (1990). *Planta* 181, 512–521.

48. Knox, J. P. (1990). *J. Cell. Sci.* 96, 557–561.

49. Ye, Z.-H., Song, Y.-R., Marcus, A., and Varner, J. E. (1991). *Plant J.* 1, 175–183.

50. McCann, M., Roberts, K., Wilson, R. H., Gidley, M. J., Gibeaut, D. M., Kim, J.-B., and Carpita, N. C. (1995). *Can. J. Bot.* 73(suppl.), S103–S113.

51. Varner, J. E., and Taylor, R. (1989). *Plant Physiol.* 91, 31–33.

52. Lynch, M. A., and Staehelin, L. A. (1992). *J. Cell Biol.* 118, 467–479.

53. Moore, P. J., Darvill, A. G., Albersheim, P., and Staehelin, L. A. (1986). *Plant Physiol.* 82, 787–794.

54. Zhang, G. F., and Staehelin, L. A. (1992). *Plant Physiol.* 99, 1070–1083.
55. Lin, L.-S., Yuen, H. K., and Varner, J. E. (1991). *Proc. Natl. Acad. Sci. USA* 88, 2241–2243.
56. Liners, F., Letesson, J.-J., Didembourg, C., and Van Cutsem, P. (1989). *Plant Physiol.* 91, 1419–1424.
57. Liners, F., Thibault, J.-F., and Van Cutsem, P. (1992). *Plant Physiol.* 99, 1099–1104.
58. McCann, M. C., Hammouri, M., Wilson, R., Belton, P., and Roberts, K. (1992). *Plant Physiol.* 100, 1940–1947.
59. Morikawa, H., and Senda, M. (1978). *Plant Cell Physiol.* 19, 1151–1159.
60. Séné, C. F. B., McCann, M. C., Wilson, R. H., and Grinter, R. (1994). *Plant Physiol.* 106, 1623–1631.
61. Williams, D. H., and Fleming, I. (1980). *Spectroscopic Methods in Organic Chemistry*, 3rd ed., McGraw-Hill, New York, pp. 35–73.
62. McCann, M. C., Stacey, N. J., Wilson, R., and Roberts, K. (1993). *J. Cell Sci.* 106, 1347–1356.
63. Kemsley, E. K., Belton, P. S., McCann, M. C., Ttofis, S., Wilson, R. H., and Delgadillo, I. (1994). *Food Control*, 5, 241–243.

37

Analysis of Red Algal Extracellular Matrix Polysaccharides
Cellulose and Carrageenan

Michael R. Gretz and Jean-Claude Mollet
Michigan Technological University, Houghton, Michigan

Ruth Falshaw
Industrial Research Ltd., Lower Hutt, New Zealand

I. INTRODUCTION

Cell walls of red algae are primarily polysaccharide with a variety of polymers represented including galactans, cellulose, mannans, xylans, and complex mucilages containing glucuronsyl, galactosyl, xylosyl, a variety of *O*-methylated glycosyls, and many other residues [1,2]. As red algae inhabit aquatic environs quite different from those typical of land plants, it is, perhaps, not unexpected to discover that the composition and organization of their extracellular matrices are distinct from those commonly encountered in the latter. The majority of red algae examined share with higher plants the general cell wall form, including cellulosic microfibrils embedded in matrix polysaccharides and other material; however, cellulose is not a major component of the red algal cell wall. In three instances, cellulose has been found to be not present. Cellulose is replaced with xylan microfibrils in one life cycle phase of *Bangia* and *Porphyra* [3-5], and a crystalline microfibrillar component appears to be absent from one of the few unicellular red algal genera, *Porphyridium* [6]. In the economically important marine red algae, the "matrix" polysaccharide is almost entirely made up of galactan (carrageenans or agars), which may be highly substituted with sulfate ester groups [1]. Much discussion has centered on the physiological significance of this extracellular matrix in mechanical, hydration, and osmotic regulation in marine environments [2].

In addition to polysaccharides, relatively minor amounts of proteinaceous and lipoidal components have been demonstrated from highly purified cell walls of selected red algae [5,7,8] and several proteoglycans have

613

been partially characterized [1], but relatively little is known of the structure or function of proteins in red algal cell walls. A unique proteinaceous multilamellar cuticle is present as the outermost layer of the thallus of several marine red algae [9,10] and is thought to lend these algae their iridescent sheen.

This chapter provides a summary of selected techniques utilized to determine the nature of and to localize cellulose and carrageenans in the red algal extracellular matrix.

II. CELLULOSE

Cellulosic cell walls have been reported or inferred from a number of genera of Rhodophyta, based primarily on x-ray analysis, cytochemical staining, and solubility considerations. Applied individually, these methods may not be definitive for the presence of cellulose, which we define as crystalline arrays of 4-linked β-glucan. Based on this criterium, red algal genera in which cellulose has been documented include *Agardhiella* [11], *Bangia* [4], *Boldia* [11], *Compsopogon* [11], *Corallina* [12], *Erythrocladia* [13], *Gracilaria* [14,15], *Griffithsia* [3], *Lemanea* [16], *Porphyra* [5,17], and *Palmaria* [3,18]. Cellulose analysis in red algae has had significant taxonomic import, as at one time the presence or absence of cellulose was a major distinguishing character separating the two classes of the Rhodophyta [4].

Determination of 4-linked β-glucan has been commonly accomplished by monosaccharide and methylation analysis. Yields of glucose from cellulosic samples during acid hydrolysis increased dramatically following pretreatment with cold concentrated acids (fuming HCl for 90 min at 4°C) prior to hydrolysis with 4 N HCl (14 h, 100°C) or other acids. To enhance detection of noncellulosic polymers, hydrolysis using 2 N trifluoroacetic acid (121°C, 3 h) was also conducted. Per-*O*-methylation was performed according to standard methods [19] including premethylation steps to increase solubility in dimethyl sulfoxide (DMSO) and minimize undermethylation. Efficient purification of methylated polymers was accomplished using C_{18} Sep-Pak columns (Millipore, Corp., Milford) and was followed by hydrolysis, reduction, and acetylation to partially methylated alditol acetates [19,20]. We have adopted a method of per-*O*-methylation utilizing butyllithium [21,22] that significantly reduced preparation time and provided yields from cellulose comparable with the foregoing method without the need for premethylation. Gas chromatography/mass spectrometry (GC/MS) on a high polarity SP-2330 capillary column (150–245°C at 4°C/min, with 245°C hold) provided relative retention time data that assisted in the identification of electron impact mass spectra of partially methylated alditol acetates. Glycosyl linkages/substitution sites were assigned according to characteristic mass ion fragments [20,23].

Electron microscopy has been utilized to visualize the microfibrillar components of the extracellular matrix and their organization relative to each other and to other wall components. Freeze fracture methods involving quick freezing and transmission electron microscope (TEM) observation of replicas containing fracture planes through frozen red algal cell walls revealed multilayered microfibrillar regions with the microfibrils commonly arranged in random patterns within each layer [24]. Chemical or physical removal of matrix polymers yielded microfibrillar preparations that also showed random patterns when imaged with TEM [3,5,25], although some disruption of in situ patterns may have occurred upon extraction. Scanning electron microscopy (SEM) of isolated microfibrillar fractions of the freshwater red algal genera *Compsopogon* and *Boldia* revealed random arrangement of microfibrils [26]. The lack of highly ordered patterns of microfibril orientation in red algae contrasts sharply with those typically found in higher plants and green algae.

The morphology of the microfibrils of red algae has been described for 29 species as either "cylindric" or "flat ribbon-like" based on freeze fracture analysis [24]. Microfibrils isolated from *Compsopogon* and *Boldia* show the flat ribbon structure when observed with TEM following negative staining with uranyl acetate (Fig. 1). These microfibrils are rectangular ribbons approximately 25–30 nm × 4–8 nm in cross section and appear similar to those shown with freeze fracture for *Erythrocladia subintegra* [13]. In contrast, negative staining of the freshwater red alga *Lemanea* showed more squarely shaped microfibrils approximately 8–12 nm in cross section, with microfibril substructure less organized and microfibrils significantly shorter than those of *Boldia* and *Compsopogon* [26]. Linear substructure exhibited by microfibrils from *Boldia* and *Compsopogon* (Fig. 1) and other red algae [24] suggests a subfibril construction. Models for cellulose biogenesis in higher plants and algae include synthesis via plasma membrane–associated integral membrane particles (IMPs) arranged in a rectangular array (linear) or a rosette and referred to as the terminal complex (TC) [27,28]. Linear TCs have been described for two red algae, *Porphyra* [29] and *Erythrocladia* [13,30], and smaller six-IMP arrays have been described for *Laurencia* and *Radicilingua* [24]. If a direct correlation between microfibril dimensions and TC organization does indeed exist [27,31–33], then we would expect linear TCs to be also present in the plasma membranes of *Boldia* and *Compsopogon*.

The commonly occurring polymorph or allomorph (crystalline form) of 4-linked β-glucan occurring in red algae (and in nature) is cellulose I. Almost all organisms that produce a cellulosic extracellular matrix synthesize cellulose I microfibrils. The microfibrillar fractions from *Compsopogon* and *Boldia* are cellulose I, as demonstrated by x-ray diffraction (Fig.

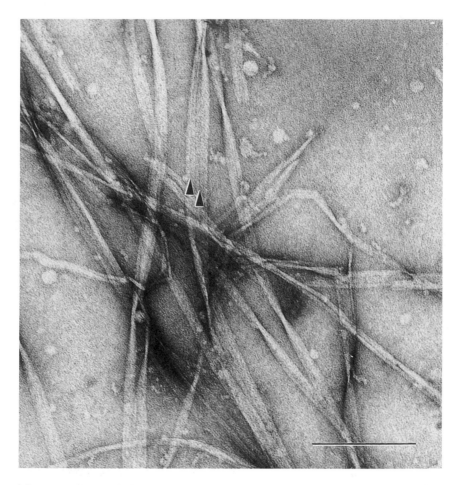

Figure 1 Transmission electron microscopy of negatively stained cellulose micro-fibrils from the freshwater red alga *Compsopogon*. Arrows indicate fibril substructure. Scale bar = 0.2 μm.

2), and the presence of 4-linked β-glucan was determined by GC/MS. Another allomorph, cellulose II, rarely occurs, but it has been reported as being the native wall 4-linked β-glucan produced by green algae including *Halicystis* and *Enteromorpha*, by the fungi *Oomyces* and *Dictyostelium*, and by the bacteria *Sarcina* and a mutant of *Acetobacter* [34]. We have determined that cellulose I is not the native crystalline form of 4-linked β-glucan in *Lemanea* [26]. To avoid possible mercerization by alkali, we used acetic–nitric acid reagent [35] to extract cellulose from mechanically

Figure 2 X-ray diffraction Debye–Scherrer powder pattern of cellulose I from *Compsopogon*.

isolated cell walls, packed a small amount into glass capillary tube, and observed x-ray diffraction patterns corresponding to cellulose II using a Debye–Scherrer powder camera with Cu K_α radiation. The use of a Debye–Scherrer camera for x-ray diffraction was inexpensive and required very small amounts of material for each analysis. We routinely performed x-ray diffraction followed by GC/MS analysis on the same 0.5-mg sample. Of course, as in any microanalysis, exclusion of contaminants became critical when working with trace amounts of material. Standard cellulose I samples included *Acetobacter* and *Valonia* cellulose. Microcrystalline rayon and cellulose precipitated from cuprammonium served as cellulose II standards. Solid state ^{13}C nuclear magnetic resonance (NMR) investigations of red algal walls [36] allowed determination of the relative amounts of the α- and β-allomorphs of cellulose I in situ. High I_α content correlated with organisms that synthesize flat ribbon-type microfibrils [37]. Therefore, we may expect that the cellulose from *Boldia* and *Compsopogon* will be predominately I_α. Analysis of Fig. 2 indicates that *Compsopogon* cellulose has the requisite high crystallinity correlated with linear TC–derived celluloses [28].

III. CARRAGEENANS

The most intensively studied polysaccharides from red algae are galactans isolated from marine genera. These commercially important polysaccharides are based on a backbone built of alternating 3-linked β-D-galactopyranosyl and either 4-linked α-galactopyranosyl or 4-linked 3,6-anhydro-α-galactopyranosyl residues, and they are separated into two major groups, carrageenans and agars (or agarocolloids) based on the absolute configuration of the 4-linked residues (D- or L-, respectively) [1,38]. The backbone structures are most often found in nature masked with various substituents including ester-linked sulfate groups in various patterns, methyl esters, or/and pyruvic acid as the 4,6-O-carboxyethylidene group. Agarose and related polymers from the agar group are critically important biotechnological reagents [39], and methods of analysis of this group will be discussed in another chapter in this volume (see Chapter 35). The commercial importance of carrageenans derives primarily from their use as food additives, with particular application as thickening, stabilizing, and emulsifying agents in dairy and other products [40].

A range of carrageenans exists in nature, defined by the positions of various substituents on the D-galactopyranosyl backbone. Some idealized structures are shown in Fig. 3. The positions of these substituents on the galactan backbone determines the different properties of the various carrageenans. In reality, different structural patterns may occur within one polysaccharide extract, due possibly to mixtures of or, alternatively, hybrid carrageenan structure. Determining the structures of carrageenans is also important for taxonomic purposes. To define the structures observed, Greek letters have traditionally been assigned to idealized disaccharide units (Fig. 3), but a new system of nomenclature has recently been developed that defines each unit independently, namely, κ-carrageenan as [G4S-DA] and λ-carrageenan as [G2S-D2S,6S] [38].

A major problem with chemical analysis of carrageenans is the instability of 3,6-anhydrogalactosyl units under acidic conditions, which leads to significant losses of material and structural information. The development of techniques that preserve these units has revolutionized the chemical analysis of carrageenans (and also related agars). With reductive hydrolysis, glycosyl units (including 3,6-anhydrogalactosyl units), formed during hydrolysis with 2 M trifluoroacetic acid (TFA), are stabilized by in situ reduction with 4-methylmorpholine borane (MMB) (Fig. 4). Following peracetylation, the resulting volatile alditol peracetates were analyzed by GC and GC/MS to determine the constituent sugars present. Using this method, Sigma κ-carrageenan (Sigma Chemical Co., St. Louis, Missouri) had a 3,6-anhydrogalactose : galactose mole ratio of 1 : 1, whereas for Sigma λ-carrageenan it was 1 : 19 [41].

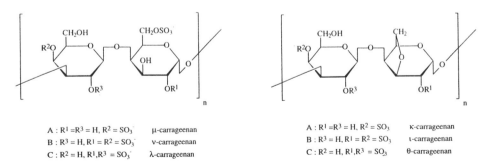

A : $R^1 = R^3 = H$, $R^2 = SO_3^-$ μ-carrageenan   A : $R^1 = R^3 = H$, $R^2 = SO_3$ κ-carrageenan
B : $R^3 = H$, $R^1 = R^2 = SO_3^-$ ν-carrageenan   B : $R^3 = H$, $R^1 = R^2 = SO_3$ ι-carrageenan
C : $R^2 = H$, $R^1, R^3 = SO_3^-$ λ-carrageenan   C : $R^2 = H$, $R^1, R^3 = SO_3$ θ-carrageenan

Figure 3 Idealized structures of carrageenan repeating disaccharide units.

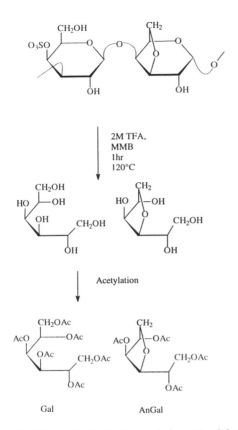

Figure 4 Reductive hydrolysis method for analysis of κ-carrageenan.

If the polysaccharide is methylated prior to reductive hydrolysis, analysis of the resulting partially methylated, partially acetylated alditol acetates provides information regarding the positions of linkages/substituents, but no differentiation is made between the two. The glycosyl linkage positions can be clarified by removal of the sulfate ester groups. Methanolic HCl is an efficient desulfating agent and has been used on certain carrageenans [42], but its acidic nature again causes problems with the loss of 3,6-anhydrogalactosyl units. Solvolytic desulfation of red algal polysaccharide sulfates in the pyridinium salt form using hot DMSO with a little pyridine and water or methanol causes less degradation and is quite effective for a range of structures: for example, λ-carrageenans [43,44], κ-carrageenan [45], and a sulfated xylogalactan [46]. Subsequent methylation identifies the remaining glycosyl linkages and any pyruvate acetal substitution. The positions of sulfate groups are then determined by the difference from methylation analysis of the native polysaccharide. In particularly complex cases where ambiguity still exists, a combination of methylation, desulfation, and then trideuteriomethylation may resolve the structure. This protocol was used successfully to determine the existence of unusual 4-linked galactopyranosyl-2,3-disulfate units in *Champia novae-zealandiae* [47]. An alternative recently developed technique, reductive cleavage, has been successfully applied to agarose [G-LA] after permethylation. In this method, 1,5-anhydroalditols were prepared using triethylsilane and a mixture of trimethylsilylmethanesulfonate and boron trifluoride etherate. Following acetylation, the partially methylated partially acetylated 1,5-anhydroalditols were analyzed by GC [48]. It may be possible to use this technique on carrageenans and other sulfated polysaccharides.

Of particular interest is the presence of sulfate groups on the 6 position of 4-linked galactopyranosyl units, as in many cases the presence of these units can adversely affect the overall gelling properties of the polysaccharide. Treatment with hot alkali causes the intramolecular displacement of sulfate from C-6 by O-3 with corresponding ring closure to produce the 3,6-anhydride. Thus, μ-carrageenan [G4S-D6S] is converted to κ-carrageenan [G4S-DA], and η-carrageenan [G4S-D2S-6S] to ι-carrageenan [G4S-DA2S]. In fact, this conversion may occur unintentionally during methylation (which is undertaken in basic conditions), resulting in underestimation of 4-linked galactosyl-6-sulfate units. These can be determined indirectly from the difference in constituent 3,6-anhydrogalactose between the native and alkali-treated samples. The presence of a 2-sulfate on the adjacent 3-linked galactosyl unit surprisingly but significantly hinders 3,6-anhydrogalactose formation in λ-type carrageenans (e.g., [G2S-D2S,6S]), and it does not occur during methylation [43,49,50].

The configurations of galactosyl units (including naturally methylated ones) are determined by preparation and GC analysis of (S)-(+)-2-butyl glycosides as the trimethylsilyl ether derivatives. Again, this technique cannot be applied to 3,6-anhydrogalactose due to its instability during the acidic butanolysis step. However, by using mild hydrolysis conditions (0.5 M TFA) with in situ reduction, the 3,6-anhydrogalactosidic, but not the galactosidic, bonds are cleaved. This reductive partial hydrolysis results in biitols if the anhydrogalactosyl units alternate with galactosyl ones. Carrabiitol (3,6-anhydro-4-O-β-D-galactosyl-D-galactitol) and agarobiitol (3,6-anhydro-4-O-β-D-galactosyl-L-galactitol) are diastereomers and their peracetates are separable by GC [49], so that the configuration of the 3,6-anhydrogalactosyl units can be determined. By permethylating the polysaccharides prior to reductive partial hydrolysis, information can be obtained about the substitution patterns of adjacent units. This approach has been used to prove the existence of α-carrageenan [G-DA2S] in the polysaccharide from *Catenella nipae* [51] and [G2S-LA(2Me)] units in the complex carrageenan/agar polysaccharide from *Pachymenia lusoria* [52].

The presence of pyruvate acetals can be inferred by the presence of 1,(2),3,4,6-linked/substituted units from methylation analysis of a native polysaccharide and a corresponding amount from the desulfated material. Direct methods involve hydrolytic cleavage of the acetals followed by detection as a colored derivative [53] or through enzymatic reduction [54]. Also, a diagnostic signal is usually visible at ~ 25 ppm in the ^{13}C NMR spectrum of the polymer in hot aqueous solution.

Most methods for determining sulfate content involve removal of sulfate from the polysaccharide by hydrolysis or combustion followed by detection of the sulfate by turbidimetry [55], colorimetry [56], or ion-exchange chromatography [57]. In all cases, it is important to remove extraneous sources of sulfate.

IV. LOCALIZATION OF POLYMERS

Cytological localization of extracellular matrix polymers in situ provides valuable information complementary to chemical and physical characterization of extracted components. The interrelation of cellulose, carrageenans,

Table 1   Common Cytochemical Reagents Used for In Situ Polymer Localization

Target	Reagent	Color
Carboxylated and sulfated polysaccharides	Alcian blue 0.1% in 3% acetic acid, pH 2.5	Blue–green
	Toluidine blue 0.05% in 0.1 M acetate buffer, pH 4.4	Red–green turquoise
Sulfated polysaccharides	Alcian blue 0.1%, 0.9 M MgCl$_2$ in 0.1 M HCl, pH 1	Blue–green
	Toluidine blue 0.05% in 0.1 M HCl, pH 1	Red–green turquoise
Anionic polysaccharides	Ruthenium red 0.2% in distilled water	Red
β-Glycan	Calcofluor white 0.04% in distilled water	Fluorescent blue
Cellulose	Chlor–zinc–iodine^a	Blue to violet
	I$_2$KI and 65% H$_2$SO$_4$	Pale blue
Protein	Coomassie blue G-250 0.25% in 7% acetic acid	Blue

^{a}ZnCl$_2$ 5 g, KI 1.6 g, and I to saturation in 1.7 mL of distilled water.

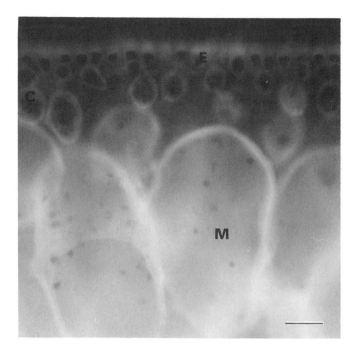

Figure 5 Transverse section of *Kappaphycus alvarezii* stained with Tinopal (calcofluor white) for β-glycan detection. Note the strong staining of the cell wall of the
large medullary cells (M). E = epidermis; C = cortex. Scale bar = 40 μm.

other polysaccharides and proteins in the cell wall can be observed using a
variety of cytochemical reagents (Table 1), or more specific molecules such
as lectins, enzymes, antibodies, and hybridization probes. Considering the
demonstrated limits of specificity of many cytochemical reagents, correlation between the chemistry of extracted polymers and localization patterns
is an important aspect in critical work.

Distribution of sulfated polysaccharides within the thallus of several
carrageenophytes such as *Chondrus crispus* [58] and agarophytes such as
Pterocladia capillacea [59] seems to follow a gradient with a greater amount
localized in external regions and less in the internal parts of the thallus [2].
However, it seems that these results are not applicable to every species, as
sulfated polysaccharides were found equally distributed in all regions of the
Eucheuma nudum thallus [60]. Chemical analysis of *Eucheuma cottonii*
indicated a typical two-polysaccharide cell wall (cellulose and carrageenan)
common among carrageenophytes of commercial import. Our cytochemical
staining of *Eucheuma cottonii* transverse sections revealed a strong reaction

in the external part of the thallus for anionic polysaccharides, whereas inner parts showed less sulfated carbohydrates. Chlor–zinc–iodine, Tinopal (calcofluor white) (Fig. 5), and I_2Kl/H_2SO_4 reagents localized cellulose in internal regions. These areas were also birefringent as indicated by polarization microscopy, suggesting highly ordered structure in medullary and axial core walls (Fig. 6).

In an innovative approach, the cationic specificities of different carrageenans were utilized to hybridize fluorescein isothiocyanate (FITC)–carrageenan oligosaccharides to wall components of *Kappaphycus alvarezii* (previously known as *Eucheuma cottonii*) [61]. These results confirmed the localization of highly sulfated polymers in the external part of the thallus, previously observed with cytochemical staining. ι-Carrageenan units were found in the epidermal cell walls, and κ-carrageenan in all other parts of the thallus. Polysaccharide-specific monoclonal antibodies conjugated to FITC have been used to localize κ-, ι-, and a putative λ-carrageenan in the thallus of *Kappaphycus alvarezii* [62]. These specific probes were also used to

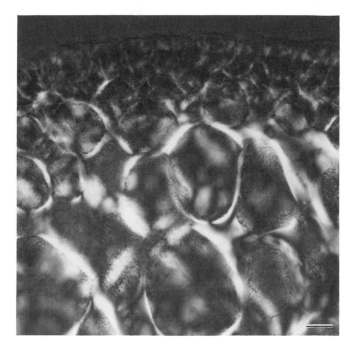

Figure 6 Polarization optics of transverse section of *Kappaphycus alvarezii*. Note the strong birefringence of the medulla cell wall material. Scale bar = 40 μm.

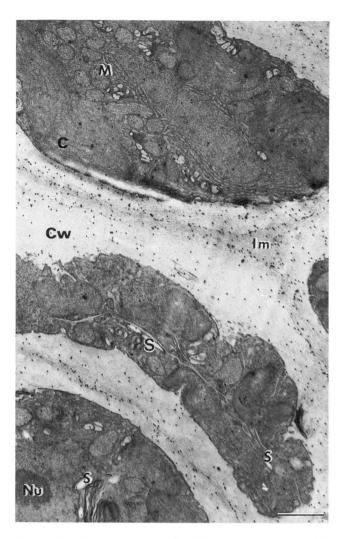

Figure 7 Electron micrograph of *Kappaphycus alvarezii* labeled with cellobiohy-drolase–gold probe. Note the absence of specific intracellular binding. Cw = cell wall, Im = intercellular matrix, M = mitochondria, C = chloroplast, Nu = nucleus, S = floridean starch. Scale bar = 1 μm.

follow the cell wall regeneration of protoplasts from the same species [63]. Cytochemical stains, such as calcofluor white and toluidine blue, have also been used to indicate cellulose and sulfated galactan deposition during cell wall regeneration of protoplasts isolated from agar- [64,65] and carrageenan-producing algae [66].

Electron microscopy requires electron dense conjugates, the most popular being colloidal gold. Enzyme-, antibody-, and lectin–gold conjugates have the potential to provide details about polymer distributions in the wall and at the intracellular level. Cellobiohydrolase (CBH)–gold was used to localize 4-linked β-glucan in the extracellular matrix of red algae such as the ι-carrageenan-producing *Agardhiella subulata* [67] and the κ-carrageenophytes *Chondrus crispus* gametophyte [11] and *Kappaphycus alvarezii* (Fig. 7). In all cases, specific intracellular labeling was not found, which supports models of plasma membrane–mediated cellulose polymerization in red algae. Electron microscopical localization of sulfated galactans has been accomplished using a range of enzyme- and antibody-probes conjugated with colloidal gold [11,67,68]. An agarase produced from *Psuedomonas* coupled to colloidal gold bound intensely to the cell wall of *Gracilaria verrucosa*, but poorly to the more highly sulfated cell wall of *Porphyra leucosticta* [67]. Monoclonal anti-ι-carrageenan–gold showed a strong affinity for *Agardhiella subulata* cell wall and intercellular matrix, demonstrating that cellulose and ι-carrageenan are intermixed in the cortical cell wall [67,68]. Localization of the monoclonal antibody in the *trans* Golgi cisternae indicated that biosynthesis of ι-carrageenan is completed intracellularly. Our recent results using κ-carrageenase–gold with *Kappaphycus alvarezii* showed labeling present in the intercellular matrix but not in the cell wall, indicating the presence of another sulfated polymer in the epidermal cell wall, possibly the methylated ι-carrageenan identified as a minor constituent in extracts from this seaweed [69]. Specific gold probes make it possible to identify qualitative spatial variations in thallus composition, to determine cell wall polymer interrelations and organization, and to better understand extracellular matrix biogenesis.

ACKNOWLEDGMENTS

Supported in part by the National Science Foundation (Grants BSR-8618847 and BSR-8919009 to MRG and Milton Sommerfeld) and New Zealand FRST Contract 8302.

ABBREVIATIONS

CBH	cellobiohydrolase
DMSO	dimethyl sulfoxide

FITC   fluorescein isothiocyanate
GC   gas chromatography
IMP   integral membrane particle
MMB   4-methylmorpholine borane
MS   mass spectrometry
NMR   nuclear magnetic resonance
SEM   scanning electron microscopy
TC   terminal complex
TEM   transmission electron microscopy
TFA   trifluoroacetic acid

REFERENCES

1. Craigie, J. S. (1990). In Cole, K. M., and Sheath, R. G., eds., *Biology of the Red Algae*. Cambridge University Press, New York, pp. 221–257.
2. Kloareg, B., and Quatrano, R. S. (1988). *Oceanoqr. Mar. Biol. Ann. Rev.* 26, 259–315.
3. Preston, D. (1974). *The Physical Biology of Plant Cell Walls*. Chapman and Hall, London.
4. Gretz, M. R., Aronson, J. M., and Sommerfeld, M. R. (1980). *Science* 207, 779–781.
5. Mukai, L. S., Craigie, J. S., and Brown, R. G. (1981). *J. Phycol.* 17, 192–198.
6. Arad, S. M., Kolani, R., Simonberkovitch, B., and Sivan, A. (1994). *Phycologia*. 33, 158–162.
7. Gretz, M. R., Aronson, J. M., and Sommerfeld, M. R. (1986). *Bot. Mar.* 29, 91–96.
8. Gretz, M. R., Sommerfeld, M. R., and Aronson, J. M. (1982). *Bot. Mar.* 25, 529–535.
9. Hanic, L. A., and Craigie, J. S. (1969). *J. Phycol.* 5, 89–102.
10. Craigie, J. S., Correa, J. A., and Gordon, M. E. (1992). *J. Phycol.* 28, 777–786.
11. Gretz, M. R., and Vollmer, C. M. (1989). In Schuerch, C., ed., *Cellulose and Wood: Chemistry and Technology*. John Wiley and Sons, New York, pp. 623–637.
12. Turvey, J. R., and Simpson, P. R. (1966). *Proc. Int. Seaweed Symp.* 5, 323–327.
13. Tsekos, I., and Reiss, H. D. (1992). *Protoplasma* 169, 57–67.
14. Lahaye, M., Revol, J. F., Rochas, C., McLachlan, J., and Yaphe, W. (1988). *Bot. Mar.* 31, 491–501.
15. Bellanger, R., Verdus, M. C., Henocq, V., and Christiaen, D. (1990). *Hydrobiologia* 204/205, 527–531.
16. Gretz, M. R., Sommerfeld, M. R., Athey, P. V., and Aronson, J. M. (1991). *J. Phycol.* 27, 232–240.
17. Gretz, M. R., Aronson, J. M., and Sommerfeld, M. R. (1984). *Phytochemistry* 23, 2513–2514.

18. Lahaye, M., Michel, C., and Barry, J. L. (1993). *Food Chem.* 47, 29–36.
19. Harris, P. J., Henry, R. J., Blakeney, A. B., and Stone, B. A. (1984). *Carbohydr. Res.* 127, 59–73.
20. Waeghe, T. J., Darvill, A. G., McNeil, M., and Albersheim, P. (1983). *Carbohydr. Res.* 123, 281–304.
21. Kvernheim, A. L. (1987). *Acta Chem. Scand.* B41, 150–152.
22. Pas Parente, J., Cardon, P., Leroy, Y., Montreuil, J., and Fournet, B. (1985). *Carbohydr. Res.* 141, 41.
23. Jansson, P., Kenne, L., Leidgren, H., Lindberg, B., and Lönngren, J. (1976). *Dept. Organic Chem. Stockholm Univ. Chem. Commun.*, No. 8, 1–75.
24. Tsekos, I., Reiss, H. D., and Schnepf, E. (1993). *Acta Bot. Neer.* 42, 119–132.
25. Myers, A., and Preston, R. D. (1959). *Proc. Roy. Soc. Lond. B* 150, 456–459.
26. Gretz, M. R., Roberts, E. M., and Sommerfeld, M. R. (1993). *J. Phycol* 29(Suppl.), 47(abstr.).
27. Hotchkiss, A. T., Jr. (1989). In Lewis, N. G., and Paice, M. G., eds., *Plant Cell Wall Polymers: Biogenesis and Biodegradation.* American Chemical Society, Washington, D.C., pp. 232–247.
28. Delmer, D. P., and Amor, Y. (1995). *Plant Cell* 7, 987–1000.
29. Tsekos, I., and Reiss, H. D. (1994). *J. Phycol.* 30, 300–310.
30. Okuda, K., Tsekos, I., and Brown, R. M., Jr. (1994). *Protoplasma* 180, 49–58.
31. Kuga, S., and Brown, R. M., Jr. (1988). In Schuerch, C., ed., *Cellulose and Wood: Chemistry and Technology.* John Wiley and Sons, New York, pp. 677–688.
32. Giddings, T. H., Brower, D. L. J., and Staehelin, L. A. (1980). *J. Cell Biol.* 84, 327–339.
33. Itoh, T. (1990). *J. Cell Sci.* 95, 309–319.
34. Roberts, E. M., Saxena, I. M., and Brown, R. M., Jr. (1988). In Schuerch, C., ed., *Cellulose and Wood: Chemistry and Technology.* John Wiley and Sons, New York, pp. 689–704.
35. Updegraff, D. M. (1969). *Anal. Biochem.* 32, 420–424.
36. Toffanin, R., Knutsen, S. H., Bertocchi, C., Rizzo, R., and Murano, E. (1994). *Carbohydr. Res.* 262, 167–171.
37. Atalla, R. H., and Vanderhart, D. L. (1989). In Schuerch, C., ed., *Cellulose and Wood: Chemistry and Technology.* John Wiley and Sons, New York, pp. 169–188.
38. Knutsen, S. H., Myslabodski, D. E., Larsen, B., and Usov, A. I. (1994). *Bot. Mar.* 37, 163–169.
39. Renn, D. W. (1990). *Hydrobiol.* 204, 7–13.
40. Indergaard, M., and Østgaard, K. (1991). In Guiry, M. D., and Blunden, G., eds., *Seaweed Resources in Europe: Uses and Potential.* John Wiley and Sons, New York, pp. 169–183.
41. Stevenson, T. T., and Furneaux, R. H. (1991). *Carbohydr. Res.* 210, 277–298.
42. Usov, A. I., Rekhter, M. A., and Kochetkov, N. K. (1969). *Zh. Obshch. Khim.* 39, 905–911.

43. Falshaw, R., and Furneaux, R. H. (1994). *Carbohydr. Res.* 252, 171–182.
44. Falshaw, R., and Furneaux, R. H. (1995). *Carbohydr. Res.* 276, 155–165.
45. Bhattacharjee, S. S., and Yaphe, W. (1978). *Carbohydr. Res.* 60, C1–C3.
46. Furneaux, R. H., and Stevenson, T. T. (1990). *Hydrobiol.* 204, 615–620.
47. Miller, I. J., Falshaw, R., and Furneaux, R. H. (1996). *Hydrobiologia*, in press.
48. Kiwitthaschemie, K., Heims, H., Steinhart, H., and Mischnick, P. (1993). *Carbohydr. Res.* 248, 267–275.
49. Falshaw, R., and Furneaux, R. H. (1995). *Carbohydr. Res.* 269, 183–189.
50. Noseda, M. D., and Cerezo, A. S. (1995). *Carbohydr. Polym.* 26, 1–3.
51. Falshaw, R., Furneaux, R. H., Wong, H., Liao, M., Bacic, A., and Chandrkrachang, S. (1996). *Carbohydr. Res.*, 285, 81–98.
52. Miller, I. J., Falshaw, R., and Furneaux, R. H. (1995). *Carbohydr. Res.* 268, 219–232.
53. Furneaux, R. H., and Miller, I. J. (1982). *New Zealand J. Sci.* 25, 15–18.
54. Duckworth, M., and Yaphe, W. (1970). *Chem. Ind.* 23, 747–748.
55. Dodgson, K. S., and Price, R. G. (1962). *Biochem. J.* 84, 106–110.
56. Silvestri, L. J., Hurst, R. E., Simpson, L., and Settine, J. M. (1985). *Anal. Biochem.* 123, 303–309.
57. Grojian, Jr., Padmos-Hicks, P. A., and Keel, B. A. (1986). *Carbohydr. Res.* 367, 367–375.
58. Gordon, E. M., and McCandless, E. L. (1973). *Proc. Nova Scot. Inst. Sci.* 27, 111–133.
59. Rascio, N., Mariani, P., Vecchia, F. D., and Trevisan, R. (1991). *Bot. Mar.* 34, 177–185.
60. La Claire, J. W., and Dawes, C. J. (1976). *J. Phycol.* 12, 368–375.
61. Zablackis, E., Vreeland, V., Doboszewski, B., and Laetsch, W. M. (1991). *J. Phycol.* 27, 241–248.
62. Vreeland, V., Zablackis, E., and Laetsch, W. M. (1992). *J. Phycol.* 28, 328–342.
63. Zablackis, E., Vreeland, V., and Kloareg, B. (1993). *J. Exp. Bot.* 44, 1515–1522.
64. Mollet, J. C., Verdus, M. C., Kling, R., and Morvan, H. (1995). *J. Exp. Bot.* 46, 239–247.
65. Cheney, D. P., Mar, E., Saga, N., and van der Meer, J. (1986). *J. Phycol.* 22, 238–243.
66. Le Gall, Y., Braud, J. P., and Kloareg, B. (1996). *Plant Cell Report* 8, 582–585.
67. Wu, Y. (1996). Ph.D. dissertation, Michigan Technological University, Houghton, Michigan.
68. Gretz, M. R., Wu, Y., Vreeland, V., and Scott, J. (1990). *J. Phycol.* 26(suppl.), 14a(abstr.).
69. Bellion, C., Brigand, G., Prome, J., and Bociek, D. W. S. (1983). *Carbohydr. Res.* 119, 31–48.

Index